Collective Classical and Quantum Fields

in Plasmas, Superconductors, Superfluid ^{3}He, and Liquid Crystals

Collective Classical and Quantum Fields

in Plasmas, Superconductors, Superfluid ³He, and Liquid Crystals

Hagen Kleinert

Freie Universität Berlin, Germany

 **World Scientific**

NEW JERSEY · LONDON · SINGAPORE · BEIJING · SHANGHAI · HONG KONG · TAIPEI · CHENNAI · TOKYO

Published by

World Scientific Publishing Co. Pte. Ltd.

5 Toh Tuck Link, Singapore 596224

USA office: 27 Warren Street, Suite 401-402, Hackensack, NJ 07601

UK office: 57 Shelton Street, Covent Garden, London WC2H 9HE

British Library Cataloguing-in-Publication Data
A catalogue record for this book is available from the British Library.

COLLECTIVE CLASSICAL AND QUANTUM FIELDS
in Plasmas, Superconductors, Superfluid ^{3}He, and Liquid Crystals

ISBN 978-981-3223-93-6
ISBN 978-981-3223-94-3 (pbk)

For any available supplementary material, please visit
http://www.worldscientific.com/worldscibooks/10.1142/10545#t=suppl

Printed in Singapore

To Annemarie and Hagen II

Preface

Strongly interacting many-body systems behave often like a system of weakly interacting collective excitations. When this happens, it is theoretically advantageous to replace the original action involving the fundamental fields (electrons, nucleons, ^{3}He, ^{4}He atoms, quarks etc.) by another action in which only certain collective excitations appear as independent quantum fields. Mathematically, such replacements can be performed in many different ways without changing the physical content of the initial theory. Experimental understanding of the important processes involved can help theorists to identify the *dominant* collective excitations. If they possess only weak residual interactions, these can be treated perturbatively. The associated collective field theory greatly simplifies the approximate description of the physical system.

It is the purpose of this book to discuss some basic techniques for deriving such collective field theories. They are based on Feynman's functional integral formulation of quantum field theory. In this formulation, the transformation to collective fields amounts to mere changes of integration variables in functional integrals.

Systems of charged particles may show excitations of a type whose quanta are called *plasmons*. For their description, a real field depending on one space and one time variable is most convenient. If the particles form bound states, a complex field depending on two spacetime coordinates renders the most economic description. Such fields are *bilocal*, and are referred to as *pair fields*. If the attractive potential is of short range, the bilocal field simplifies to a local field. This has led to the field theory of superconductivity by Ginzburg and Landau. A bilocal theory of this type has been used in elementary-particle physics to explain the observable properties of strongly interacting mesons.

The change of integration variables in path integrals will be shown to correspond to an exact resummation of the perturbation series, thereby accounting for phenomena which cannot be described perturbatively in terms of fundamental particles. The path formulation has the great advantage of translating all quantum effects among the fundamental particles completely into the field language of collective excitations. All fluctuation corrections may be computed using only propagators and interaction vertices of the collective fields.

The method becomes unreliable if several collective effects compete with each other. An example is a gas of electrons and protons at low density where the attractive forces can produce hydrogen atoms. They are absent in a description involving a plasmon field. A mixture of plasmon and pair effects is needed to describe these.

Another example is superfluid ^{3}He, where pairing forces are necessary to produce the superfluid phase transition. Here plasma-like magnetic excitations called *paramagnons* provide strong corrections. In particular, they are necessary to obtain the pairing in the first place. If we want to tackle such mixed phenomena, another technique must be used called *variational perturbation theory*.

In Chapter 1, I explain the mathematical method of changing from one field description to another by going over to *collective fields* representing the dominant collective excitations. In Chapters 2 and 3, I illustrate this method by discussing simple systems such as an electron gas or a superconductor. At the end of Chapter 3, I had good help from my collaborator S.-S. Xue, with whom I wrote the basic strong-coupling paper (arxiv:cond-mat/1708.04023), that is cited as Ref. [89] on page 143. In Chapter 4, I apply the technique to superfluid ^{3}He. In Chapter 5, I use the field theoretic methods to study physically observable phenomena in liquid crystals. In Chapter 6, finally, I illustrate the working of the theory by treating some simple solvable models.

I want to thank my wife Dr. Annemarie Kleinert for her great patience with me while writing this book. Although her field of interest is French Literature and History (her homepage https://a.klnrt.de), and thus completely different from mine, her careful reading detected many errors. Without her permanent reminding me of the still missing explanations of certain questions I could never have completed this work. My son Michael, who just received his PhD in experimental physics, deserves the credit of asking many relevant questions and making me improve my sometimes too formal manuscript.

Berlin, December 2017
H. Kleinert

Contents

Preface   **vii**

1 Functional Integral Techniques   **1**
  1.1 Nonrelativistic Fields 2
    1.1.1 Quantization of Free Fields 2
    1.1.2 Fluctuating Free Fields 4
    1.1.3 Interactions 8
    1.1.4 Normal Products 10
    1.1.5 Functional Formulation 14
    1.1.6 Equivalence of Functional and Operator Methods 15
    1.1.7 Grand-Canonical Ensembles at Zero Temperature 16
  1.2 Relativistic Fields 22
    1.2.1 Lorentz and Poincaré Invariance 22
    1.2.2 Relativistic Free Scalar Fields 27
    1.2.3 Electromagnetic Fields 31
    1.2.4 Relativistic Free Fermi Fields 34
    1.2.5 Perturbation Theory of Relativistic Fields 37
  Notes and References 39

2 Plasma Oscillations   **41**
  2.1 General Formalism 41
  2.2 Physical Consequences 45
    2.2.1 Zero Temperature 46
    2.2.2 Short-Range Potential 47
  Appendix 2A Fluctuations around the Plasmon 48
  Notes and References 49

3 Superconductors   **50**
  3.1 General Formulation 52
  3.2 Local Interaction and Ginzburg-Landau Equations 59
    3.2.1 Inclusion of Electromagnetic Fields into the Pair Field Theory 69
  3.3 Far below the Critical Temperature 72
    3.3.1 The Gap 73
    3.3.2 The Free Pair Field 77
  3.4 From BCS to Strong-Coupling Superconductivity 91
  3.5 Strong-Coupling Calculation of the Pair Field 92

3.6 From BCS Superconductivity near T_c to the onset of pseudogap behavior 100
3.7 Phase Fluctuations in Two Dimensions and Kosterlitz-Thouless Transition 105
3.8 Phase Fluctuations in Three Dimensions 111
3.9 Collective Classical Fields 112
    3.9.1 Superconducting Electrons 115
3.10 Strong-Coupling Limit of Pair Formation 117
3.11 Composite Bosons 122
3.12 Composite Fermions 127
3.13 Conclusion and Remarks 129
Appendix 3A Auxiliary Strong-Coupling Calculations 131
Appendix 3B Propagator of the Bilocal Pair Field 133
Appendix 3C Fluctuations Around the Composite Field 135
Notes and References 138

4 Superfluid ^{3}He   **145**
4.1 Interatomic Potential 145
4.2 Phase Diagram 147
4.3 Preparation of Functional Integral 149
    4.3.1 Action of the System 149
    4.3.2 Dipole Interaction 149
    4.3.3 Euclidean Action 150
    4.3.4 From Particles to Quasiparticles 151
    4.3.5 Approximate Quasiparticle Action 152
    4.3.6 Effective Interaction 155
    4.3.7 Pairing Interaction 158
4.4 Transformation from Fundamental to Collective Fields 159
4.5 General Properties of a Collective Action 164
4.6 Comparison with O(3)-Symmetric Linear σ-Model 169
4.7 Hydrodynamic Properties Close to T_c 170
4.8 Bending the Superfluid ^{3}He-A 178
    4.8.1 Monopoles 179
    4.8.2 Line Singularities 182
    4.8.3 Solitons 184
    4.8.4 Localized Lumps 187
    4.8.5 Use of Topology in the A-Phase 188
    4.8.6 Topology in the B-Phase 190
4.9 Hydrodynamic Properties at All Temperatures $T \leq T_c$ 193
    4.9.1 Derivation of Gap Equation 194
    4.9.2 Ground State Properties 199
    4.9.3 Bending Energies 208
    4.9.4 Fermi-Liquid Corrections 218
4.10 Large Currents and Magnetic Fields in the Ginzburg-Landau Regime 227

4.10.1   B-Phase 228
4.10.2   A-Phase 239
4.10.3   Critical Current in Other Phases for $T \sim T_c$ 240
4.11   Is ^{3}He-A a Superfluid? 248
4.11.1   Magnetic Field and Transition between A- and B-Phases . . 272
4.12   Large Currents at Any Temperature $T \leq T_c$ 274
4.12.1   Energy at Nonzero Velocities 274
4.12.2   Gap Equations 275
4.12.3   Superfluid Densities and Currents 283
4.12.4   Critical Currents 285
4.12.5   Ground State Energy at Large Velocities 289
4.12.6   Fermi Liquid Corrections 289
4.13   Collective Modes in the Presence of Current at all Temperatures
$T \leq T_c$ 292
4.13.1   Quadratic Fluctuations 292
4.13.2   Time-Dependent Fluctuations at Infinite Wavelength 295
4.13.3   Normal Modes 298
4.13.4   Simple Limiting Results at Zero Gap Deformation 301
4.13.5   Static Stability 303
4.14   Fluctuation Coefficients 304
4.15   Stability of Superflow in the B-Phase under Small Fluctuations for
$T \sim T_c$ 307
Appendix 4A Hydrodynamic Coefficients for $T \approx T_c$ 312
Appendix 4B Hydrodynamic Coefficients for All $T \leq T_c$ 315
Appendix 4C Generalized Ginzburg-Landau Energy 319
Notes and References 319

5   Liquid Crystals   323
5.1   Maier-Saupe Model and Generalizations 324
5.1.1   General Properties 324
5.1.2   Landau Expansion 326
5.1.3   Tensor Form of Landau-de Gennes Expansion 327
5.2   Landau-de Gennes Description of Nematic Phase 328
5.3   Bending Energy 336
5.4   Light Scattering 338
5.5   Interfacial Tension between Nematic and Isotropic Phases 347
5.6   Cholesteric Liquid Crystals 351
5.6.1   Small Fluctuations above T_1 354
5.6.2   Some Experimental Facts 355
5.6.3   Mean-Field Description of Cholesteric Phase 357
5.7   Other Phases 362
Appendix 5A Biaxial Maier-Saupe Model 365
Notes and References 368

6 Exactly Solvable Field-Theoretic Models   **371**
  6.1   Pet Model in Zero Plus One Time Dimensions 371
       6.1.1   The Generalized BCS Model in a Degenerate Shell 379
       6.1.2   The Hilbert Space of the Generalized BCS Model 390
  6.2   Thirring Model in 1+1 Dimensions 393
  6.3   Supersymmetry in Nuclear Physics 397
  Notes and References 397

Index   **399**

List of Figures

1.1   Contour C in the complex z-plane 19

2.1   The pure current piece of the collective action 43
2.2   The non-polynomial self-interaction terms of plasmons 44
2.3   Free plasmon propagator 44

3.1   Time evolution of critical temperatures of superconductivity 51
3.2   Fundamental particles entering any diagram 54
3.3   Free pair field following the Bethe-Salpeter equation 56
3.4   Free pair propagator 58
3.5   Self-interaction terms of the non-polynomial pair Lagrangian 59
3.6   Free part of pair field Δ Lagrangian 62
3.7   Energy gap of a superconductor as a function of temperature 76
3.8   Temperature behavior of the superfluid density ρ_s/ρ (Yoshida func-
       tion) and the gap function $\bar{\rho}_s/\rho$ 87
3.9   Temperature behavior of the inverse square coherence length $\xi^{-2}(T)$ 88
3.10   Gap function Δ and chemical potential μ at zero temperature as
       functions of the crossover parameter $\hat{\mu}$ 96
3.11   Temperature dependence of the gap function in three (a) and two
       (b) dimensions 97
3.12   Dependence of T^* on the crossover parameter in three (a) and two
       (b) dimensions 98
3.13   Dependence of the pair-formation temperature T^* on the chemical
       potential 109
3.14   Qualitative phase diagram of the BCS-BEC crossover as a function
       of temperature T/ϵ_F and coupling $1/k_F a$ 118
3.15   Qualitative phase diagram in the unitarity limit 126

4.1   Interatomic potential between ^{3}He atoms as a function of the dis-
       tance r 145
4.2   Imaginary part of the susceptibility caused by repeated exchange of
       spin fluctuations 146
4.3   Phase diagram of ^{3}He plotted against temperature, pressure, and
       magnetic field 148
4.4   Three fundamental planar textures, splay, bend, and twist of the
       director field in liquid crystals 172

4.5   Sphere with one, two, or no handles and their Euler characteristics   176

4.6   Local tangential coordinate system $\mathbf{n}, \mathbf{t}, \mathbf{i}$ for an arbitrary curve on the surface of a sphere 176

4.7   The $\mathbf{l} \| \mathbf{d}$-field lines in a spherical container 179

4.8   Two possible parametrizations of a sphere 180

4.9   Spectra of Goldstone bosons versus gauge bosons 181

4.10   Cylindrical container with the $\mathbf{l} \| \mathbf{d}$-field lines spreading outwards when moving upwards 183

4.11   Field vectors in a composite soliton 186

4.12   Nuclear magnetic resonance frequencies of a superfluid ^{3}He-A sample in an external magnetic field 187

4.13   Vectors of orbital and spin orientation in the A-phase of superfluid ^{3}He 188

4.14   Parameter space of ^{3}He-B containing the parameter space of the rotation group 192

4.15   Possible path followed by the order parameter in a planar texture (soliton) when going from $z = -\infty$ to $z = +\infty$ 193

4.16   Another possible class of solitons 193

4.17   Fundamental hydrodynamic quantities of superfluid ^{3}He-B and -A, shown as a function of temperature 198

4.18   Condensation energies of A- and B-phases as functions of the temperature 203

4.19   The temperature behavior of the condensation entropies in B- and A-phases 205

4.20   Specific heat of A- and B-phases as a function of temperature 207

4.21   Temperature behavior of the reduced superfluid densities in the B- and in the A-phase of superfluid ^{4}He 211

4.22   Superfluid stiffness functions K_t, K_b, K_s of the A-phase as functions of the temperature 218

4.23   Superfluid densities of B- and A-phase after applying Fermi liquid corrections 224

4.24   Coefficients $c = c^{\|}$ and their Fermi liquid corrected values in the A-phase as a function of temperature 225

4.25   Coefficient K_s for splay deformations of the fields, and its Fermi liquid corrected values in the A-phase as a function of temperature . 226

4.26   Remaining hydrodynamic parameters for twist and bend deformations of superfluid ^{3}He-A 227

4.27   Hydrodynamic parameters of superfluid ^{3}He-B together with their Fermi liquid corrected values, as functions of the temperature . . . 228

4.28   Shape of potential determining stability of superflow 231

4.29   Superflow in a torus 248

4.30   In the presence of a superflow in ^{3}He-A, the l-vector is attracted to the direction of flow 254

4.31 Doubly connected parameter space of the rotation group corresponding to integer and half-integer spin representations 254

4.32 Helical texture in the presence of a supercurrent 255

4.33 Three different regions of equilibrium configurations of the texture at $H = 0$ (schematically) 257

4.34 Pitch values for stationary helical solutions as a function of the angle of inclination β_0 260

4.35 Regions of stable helical texture, II- and II+ 262

4.36 Regions of stable helical texture (shaded areas) 263

4.37 Shrinking of the regions of stability when dipole locking is relaxed . 265

4.38 As a stable helix forms in the presence of superflow in ^{3}He-A 266

4.39 Angle of inclination as a function of the magnetic field at different temperatures 267

4.40 Sound attenuation parametrized in terms of three constants 269

4.41 Velocity dependence of the gap in the A- and B-phases 281

4.42 Current as a function of velocity 292

4.43 Collective frequencies of B-phase in the presence of superflow of velocity v 303

5.1 Molecular structure of PAA 323

5.2 Graphical solution of the gap equation 326

5.3 Phase diagram of general Landau expansion in the (a_3, a_2)-plane . . 330

5.4 Biaxial regime in the phase diagram of general Landau expansion of free energy in the (a_3, a_2)-plane 331

5.5 Jump of the order parameter φ from zero to a nonzero value $\varphi_>$ in a first-order phase transition at $T = T_1$ 333

5.6 Different configurations of textures in liquid crystals 337

5.7 Experimental setup of the light-scattering experiment 343

5.8 Inverse light intensities as a function of temperature 345

5.9 Behavior of coherence length as a function of temperature 346

5.10 Relevant vectors of the director fluctuation 347

5.11 Contour plots of constant reduced free energy $\tilde{f}_{\mathrm{ext}}$ 354

5.12 Momentum dependence of the gradient coefficients 355

5.13 Momenta and polarization vectors of a body-centered cubic phase of a cholesteric liquid crystal 363

5.14 Regimes in the plane of α, τ, where the phases, cholesteric, hexatic, or bcc are lowest 363

5.15 Momenta and polarization vectors for an icosahedral phase of a cholesteric liquid crystal 364

5.16 Density profile with five-fold symmetry 364

5.17 Density profile with seven-fold symmetry 365

5.18 Blue phases in a cholesteric liquid crystal 365

6.1   Level scheme of the BCS model in a single degenerate shell of multiplicity $\Omega = 8$ 381

List of Tables

4.1 A factor of roughly 1000 separates the characteristic length scales of
superconductors and ^{3}He 148

4.2 Pressure dependence of Landau parameters F_1, F_0, and F_0^S of ^{3}He
together with the molar volume v and the effective mass ratio m^*/m 153

4.3 Parameters of the critical currents in all theoretically known phases . 243

1

Functional Integral Techniques

An important goal of many-body physics is the study of collective phenomena in systems of many bosons or fermions. The interactions are typically caused by two-body forces. In their field-theoretic description, such forces emerge in a perturbation theory from the exchange of virtual particles, such as photons or phonons. More complicated forces can also be generated by the exchange of virtual particles carried by higher tensor fields. So far, all forces in nature between particles can be reduced to such exchange processes. Depending on the detailed properties of forces and thermodynamic parameters such as density, pressure, and temperature, bosons or fermions may exhibit different collective behaviors. In an electron gas, for example, one may observe density fluctuations or pair condensation. The first type is found if the exchanged particles couple strongly to other particles or holes. Examples are plasma oscillations in a degenerate electron gas. The second type of behavior is found if the forces favor the formation of bound states between pairs of particles. This is usually observed below a certain *critical* temperature T_c. Examples are excitons in a semiconductor or Cooper pairs in a superconductor.

For systems showing plasma type of excitations, real fields depending on space and time are most convenient to describe the physical phenomena. To describe pair condensation, complex fields render the most economic description of such phenomena. They contain the two spatial arguments of the constituents and their common time coordinates. Such fields will be called *bilocal*. In relativistic systems, also the time coordinates of the constituents may be different. If the potential has a sufficiently short range, the bilocal field degenerates into a local field. The most important example for the latter case is the collective pair-field theory of superconducting electrons which is known as the Ginzburg-Landau theory.

A bilocal field theory is useful in elementary particle physics where it allows to study the transition from inobservable quark fields to observable meson fields (see [1] or Chapter 26 in the textbook [2]). The new basic field quanta of the converted theory are no longer the fundamental particles but the set of all quark-antiquark meson bound states which are obtained by solving a Bethe-Salpeter bound-state equation in the so-called ladder approximation. They are called *bare mesons*. Such a formulation can also be given to quantum electrodynamics of electrons and positrons, where the bare mesons are positronium atoms [1].

1.1   Nonrelativistic Fields

Let us begin with the description of functional methods that can be used for the study of many-body physics of relativistic particles. We shall follow the historic development.

1.1.1   Quantization of Free Fields

Consider free nonrelativistic particles, whose energy ε depends on the momentum $\mathbf{p}$ by some function $\varepsilon(\mathbf{p})$. In free space, this has the form $\varepsilon(\mathbf{p}) = \mathbf{p}^2/2m$. For a particle moving in a periodic solid, the momentum dependence is usually more complicated. However, for many purposes it can be approximated by the same quadratic behavior, provided that we exchange the mass m by an effective mass parameter $m^* \neq m$ called the effective mass. The action of a free nonrelativistic field describing an ensemble of these particles reads

$$\mathcal{A}_0 = \int d^3x dt \, \psi^*(\mathbf{x}, t) \left[i\hbar\partial_t - \epsilon(-i\hbar\boldsymbol{\nabla})\right] \psi(\mathbf{x}, t). \tag{1.1}$$

By extremizing this, we find the equation of motion

$$\frac{\delta\mathcal{A}_0}{\delta\psi^*(\mathbf{x}, t)} = \left[i\hbar\partial_t - \epsilon(-i\hbar\boldsymbol{\nabla})\right]\psi(\mathbf{x}, t) = 0, \tag{1.2}$$

which coincides with the Schrödinger equation for a single free particle.

In the Lagrange formalism of classical mechanics, each dynamical variable possesses a canonically conjugate variable called *momentum* variable. For the action (1.14), this is the field momentum

$$\pi(\mathbf{x}, t) \equiv \hbar\frac{\delta\mathcal{A}}{\delta\partial_t\psi(\mathbf{x}, t)} = \psi^\dagger(\mathbf{x}, t). \tag{1.3}$$

According to the rules of quantum mechanics, the fields and their conjugate momenta are turned into operators $\hat{\psi}(\mathbf{x}, t)$ and $\hat{\pi}(\mathbf{x}, t)$, which satisfy the equal-time commutation rules:

$$[\hat{\psi}(\mathbf{x}, t), \hat{\psi}(\mathbf{x}')] = 0, \tag{1.4}$$

$$[\hat{\pi}(\mathbf{x}, t), \hat{\pi}(\mathbf{x}')] = 0, \tag{1.5}$$

$$[\hat{\pi}(\mathbf{x}, t), \hat{\psi}(\mathbf{x}')] = -i\hbar\delta^{(3)}(\mathbf{x} - \mathbf{x}'). \tag{1.6}$$

Inserting (1.3), these become commutation rules of independent creation and annihilation operators $\hat{a}_\mathbf{x}^\dagger(t) \equiv \hat{\psi}^\dagger(\mathbf{x}, t)$ and $\hat{a}_\mathbf{x}(t) \equiv \hat{\psi}(\mathbf{x}, t)$ of harmonic oscillators situated at each space point $\mathbf{x}$:

$$[\hat{a}_\mathbf{x}(t), \hat{a}_{\mathbf{x}'}(t)] = [\hat{\psi}(\mathbf{x}, t), \hat{\psi}(\mathbf{x}', t)] = 0, \tag{1.7}$$

$$[\hat{a}_\mathbf{x}^\dagger(t), \hat{a}_{\mathbf{x}'}^\dagger(t)] = [\hat{\psi}^\dagger(\mathbf{x}, t), \hat{\psi}^\dagger(\mathbf{x}', t)] = 0, \tag{1.8}$$

$$[\hat{a}_\mathbf{x}(t), \hat{a}_{\mathbf{x}'}^\dagger(t)] = [\hat{\psi}(\mathbf{x}, t), \hat{\psi}^\dagger(\mathbf{x}', t)] = \delta^{(3)}(\mathbf{x} - \mathbf{x}'). \tag{1.9}$$

For each oscillator, there exists a ground state $|0\rangle_{\mathbf{x}}$ defined by the condition $\hat{\psi}(\mathbf{x}, t)|0\rangle = 0$. The excited states are obtained by multiplying $|0\rangle_{\mathbf{x}}$ with $n_{\mathbf{x}}$ creation operators $\hat{\psi}^{\dagger}(\mathbf{x}, t)$. They are denoted by $(a^{\dagger})_{\mathbf{x}}^{n_{\mathbf{x}}}|0\rangle_{\mathbf{x}} = [\hat{\psi}^{\dagger}(\mathbf{x}, t)]^{n_{\mathbf{x}}}|0\rangle_{\mathbf{x}}$, where $n_{\mathbf{x}}$ are integer quantum numbers $n_{\mathbf{x}} = 0, 1, 2, 3, \ldots$. These are interpreted as the numbers of particles at point $\mathbf{x}$.

Thus, by quantizing the field and converting it to a field *operator*, the single-particle Schrödinger theory changes into a theory of arbitrarily many identical oscillators at all space point $\mathbf{x}$. The ground state of the system is the direct product of the ground states of all these oscillators: $|0\rangle \equiv \prod_{\mathbf{x}} |0\rangle_{\mathbf{x}}$. The resulting many-particle Hilbert space is called the *Fock space*, and the procedure of field quantization is called *second quantization*. The usual quantization is ensured by the correspondence rule $\mathbf{p} \rightarrow -i\hbar\boldsymbol{\nabla}$ in the single-particle Schrödinger equation (1.14) and the action (1.1).

The free quantum field $\hat{\psi}(\mathbf{x}, t)$ can be expanded into a Fourier series

$$\hat{\psi}(\mathbf{x}, t) = \sum_{\mathbf{p}} \frac{e^{i\mathbf{p}\mathbf{x} - i\epsilon(\mathbf{p})t}}{\sqrt{V}} a_{\mathbf{p}}, \tag{1.10}$$

where V is the volume of the system and $\mathbf{p}$ are the discrete momenta in it. The operators $a_{\mathbf{p}}$ and their hermitian conjugates $a_{\mathbf{p}}^{\dagger}$ are annihilation and creation operators of single particles in momentum space. From (1.7)–(1.9), we find that these satisfy the oscillator commutation rules

$$[\hat{a}_{\mathbf{p}}(t), \hat{a}_{\mathbf{p}'}(t)] = 0, \tag{1.11}$$

$$[\hat{a}_{\mathbf{p}}^{\dagger}(t), \hat{a}_{\mathbf{p}'}^{\dagger}(t)] = 0, \tag{1.12}$$

$$[\hat{a}_{\mathbf{p}}(t), \hat{a}_{\mathbf{p}'}^{\dagger}(t)] = \delta^{(3)}(\mathbf{x} - \mathbf{x}'). \tag{1.13}$$

An important quantity of free fields is the free Green function $G_0(\mathbf{x}, t; \mathbf{x}', t')$, which satisfies the inhomogeneous version of the field equation (1.14):

$$[i\hbar\partial_t - \epsilon(-i\hbar\boldsymbol{\nabla})] G_0(\mathbf{x}, t; \mathbf{x}', t') = i\delta^{(3)}(\mathbf{x} - \mathbf{x}', t - t'). \tag{1.14}$$

This can be solved by the spectral representation, that has, for translationally invariant systems at hand, a Fourier decomposition:

$$G_0(\mathbf{x}, t; \mathbf{x}', t') = \int \frac{dE}{2\pi} \int \frac{d^3p}{(2\pi)^4} e^{-i[E(t-t') - \mathbf{p}(\mathbf{x}-\mathbf{x}')]} \frac{i}{E - \epsilon(\mathbf{p}) + i\eta}. \tag{1.15}$$

The solution of Eq. (1.14) is not unique, since there are various ways to carry the contour of the energy integration past the pole at $E = \epsilon(\mathbf{p})$. Different ways produce Green functions with different boundary properties. The differences are solutions of the homogeneous field equation. In integral (1.15) we have chosen a contour of integration which passes *above* the pole, where the denominator in (1.15) diverges. This is indicated by adding a term $-i\eta$ to the energy $\epsilon(\mathbf{p})$, where η is a positive infinitesimal number, a procedure called the *iη-prescription*. With that choice, the

Green function (1.15) coincides with the vacuum expectation value of the time-ordered product of field operators (1.10):

$$G_0(\mathbf{x}, t; \mathbf{x}', t') = \langle 0|\hat{T}\hat{\psi}(\mathbf{x}, t)\hat{\psi}^\dagger(\mathbf{x}', t')|0\rangle. \tag{1.16}$$

The time-ordering operator $\hat{T}$ is defined to change the position of the operators behind it in such a way that fields with later time arguments stand to the left of those with earlier time arguments. The expectation value on the right-hand side is also called the *free propagator* of the quantum field $\hat{\psi}(\mathbf{x}, t)$. By inserting the expansion (1.10) into (1.16) it is easy to verify that the evaluation of the expectation value (1.16) gives exactly the expression (1.15).

1.1.2   Fluctuating Free Fields

There exists an equivalent approach to second quantization where the thermodynamic partition of the above system is expressed as a functional integral over all possible fluctuating fields [4, 5]. For free fields, we define a partition function

$$Z_0 = N \int \mathcal{D}\psi^*(\mathbf{x}, t)\mathcal{D}\psi(\mathbf{x}, t) \exp\left\{i\mathcal{A}_0[\psi^*, \psi]\right\}, \tag{1.17}$$

where N is some constant which will play no role in all subsequent discussions. From here on we shall work with natural units in which $\hbar = 1$.

The functional formulation was found by Richard Feynman. He observed that the amplitudes of diffraction phenomena of light are obtained by summing over the individual amplitudes for all paths which the light could possibly have taken. Each path is associated with a pure phase depending only on the action of the light particle along the path. For fields, this principle leads to Formula (1.17) for the partition function.

The functional integral may conveniently be defined by grating the spacetime into a finer and finer cubic lattice of spacing δ with corners at $(x_{i_1}, y_{i_2}, z_{i_3}, t_{i_4}) = (i_1, i_2, i_3, i_4)\,\delta$. The fields are characterized by their values at the nearest lattice point:

$$\psi_{i_1 i_2 i_3 i_4} \equiv \psi\left(x_{i_1}, y_{i_2}, z_{i_3}, t_{i_4}\right)\sqrt{\delta}^4. \tag{1.18}$$

The measure in the functional integral in (1.17) is then defined by the product of all integrals over the cubus around each lattice point:

$$\int \mathcal{D}\psi^*(\mathbf{x}, t)\mathcal{D}\psi(\mathbf{x}, t) \equiv \prod_{\substack{i_1 i_2 i_3 i_4 \\ i_1' i_2' i_3' i_4'}} \int\int \frac{d\psi^\dagger_{i_1 i_2 i_3 i_4} d\psi_{i_1' i_2' i_3' i_4'}}{\sqrt{2\pi i}\sqrt{2\pi i}}. \tag{1.19}$$

The double integral over complex variables $\int\int d\psi^* d\psi$ symbolizes the real integrals

$$\int_{-\infty}^{\infty}\int_{-\infty}^{\infty} d\left(\frac{\psi + \psi^*}{\sqrt{2}}\right) d\left(\frac{\psi - \psi^*}{\sqrt{2i}}\right). \tag{1.20}$$

This naive definition of path integration is straightforward for Bose fields. If we want to use the functional technique to describe also the statistical properties of fermions, some modifications are necessary. Then the fields must be taken to be anticommuting c-numbers. In mathematics, such objects form a *Grassmann algebra*. If ξ, ξ' are any two real elements in this algebra, they satisfy the anticommutation relation

$$\{\xi, \xi'\} \equiv \xi\xi' + \xi'\xi = 0. \tag{1.21}$$

A trivial consequence of this condition is that the square of each Grassmann element vanishes, i.e., $\xi^2 = 0$. If $\xi = \xi_1 + i\xi_2$ is a complex Grassmann variable, then $\xi^2 = -\xi^*\xi = -2i\xi_1\xi_2$ is nonzero, but $(\xi^*\xi)^2 = (\xi\xi)^2 = 0$.

All results to be derived later will make use of only one simple class of integrals over Grassmann variables. For boson fields, they are generalizations of the elementary Gaussian (or Fresnel) formula for $A > 0$ [6]:

$$\int_{-\infty}^{\infty} \frac{d\xi}{\sqrt{2\pi i}} \exp\left(\frac{i}{2}\xi A\xi\right) = A^{-1/2}. \tag{1.22}$$

The first generalization concerns the dimension. For a D-dimensional real space of vectors $\boldsymbol{\xi} \equiv (\xi_1, \xi_2, \ldots, \xi_D)$, and a diagonal matrix A with diagonal elements A_k, the integral (1.22) becomes

$$\int_{-\infty}^{\infty} \frac{d^D\xi}{\sqrt{2\pi i}^D} e^{i\boldsymbol{\xi}^T A \boldsymbol{\xi}/2} \equiv \prod_k \left[\int_{-\infty}^{\infty} \frac{d\xi_k}{\sqrt{2\pi i}}\right] \exp\left(\frac{i}{2}\sum_k \xi_k A_k \xi_k\right) = \left[\prod_k A_k\right]^{-1/2}. \tag{1.23}$$

Next we generalize the exponent to the matrix form $(i/2)\sum_{k,l}\xi_k A_{kl}\xi_l$, where A_{kt} is an arbitrary symmetric positive matrix. An orthogonal transformation of the ξ_k's can be used to bring A_{kl} to a diagonal form. The orthogonality ensures that the measure of integration remains invariant. Thus an equation like (1.23) is still valid with the right-hand side becoming the product of eigenvalues of the matrix A_{kl}. This can be written as a determinant, so that we obtain the formula

$$\int_{-\infty}^{\infty} \frac{d^D\xi}{\sqrt{2\pi i}^D} e^{i\boldsymbol{\xi}^T A \boldsymbol{\xi}/2} \equiv \prod_m \left[\int_{-\infty}^{\infty} \frac{d\xi_m}{\sqrt{2\pi i}}\right] \exp\left(\frac{i}{2}\sum_{k,l} \xi_k A_{kl} \xi_l\right) = [\det A]^{-1/2}. \tag{1.24}$$

Even more generally, we allow ξ to be a complex variable, and A to be a hermitian and positive matrix. Then the result (1.24) follows separately for the real and for the imaginary part, yielding

$$\int_{-\infty}^{\infty} \frac{d^D\xi^\dagger d^D\xi}{2\pi i^D} e^{i\boldsymbol{\xi}^\dagger A \boldsymbol{\xi}/2} \equiv \prod_m \left[\int \frac{d\xi_m^* d\xi_m}{\sqrt{2\pi i}\sqrt{2\pi i}}\right] \exp\left(i\sum_{k,l} \xi_k^* A_{kl} \xi_l\right) = [\det A]^{-1}. \tag{1.25}$$

For the study of fermion systems, the integrals are performed over anticommuting real or complex variables ξ or $\xi^*\xi$. In this case, the right-hand sides of Eqs. (1.24) and (1.25) are replaced by their inverses $[\det A]^{1/2}$, $[\det A]^{1}$.

Let us prove this for complex variables. After bringing the matrix A_{kl} to a diagonal form via a unitary transformation, the integral reads

$$\int \prod_m \left[\frac{d\xi_m^* d\xi_m}{\sqrt{2\pi i}\sqrt{2\pi i}} \right] \exp\left(i \sum_n \xi_n^* A_n \xi_n \right) = \prod_m \int \frac{d\xi_m^* d\xi_m}{\sqrt{2\pi i}\sqrt{2\pi i}} \exp\left(i\xi_m^* A_m \xi_m \right). \quad (1.26)$$

Expanding the exponentials into a power series leaves only the first two terms, since $(\xi_m^* \xi_m)^2 = 0$. Thus the integral becomes

$$\prod_m \int \frac{d\xi_m^* d\xi_m}{\sqrt{2\pi i}\sqrt{2\pi i}} (1 + i\xi_m^* A_m \xi_m). \quad (1.27)$$

Each of these integrals can be performed trivially by defining two basic integrals over the Grassmann variables, from which all the others follow using the linearity of integrals. For real Grassmann variables, these rules are

$$\int \frac{d\xi}{\sqrt{2\pi i}} = 0, \quad \int \frac{d\xi}{\sqrt{2\pi i}} \xi = 1. \quad (1.28)$$

The integrals over higher powers vanish trivially due to the anticommutation property (1.21):

$$\int \frac{d\xi}{\sqrt{2\pi i}} \xi^n = 0, \quad n > 1. \quad (1.29)$$

The two rules (1.28) and (1.29) determine the integrals over any function $F(\xi)$ of a real Grassmann variable ξ. They ensure that such a function is determined by only two parameters: the zeroth- and the first-order Taylor coefficients. Indeed, due to the property $\xi^2 = 0$, the Taylor series can only possess the first two terms $F(\xi) = F_0 + F'\xi$, where $F_0 = F(0)$ and

$$F' \equiv dF(\xi)/d\xi. \quad (1.30)$$

But according to (1.28), the integral yields also F':

$$\int \frac{d\xi}{\sqrt{2\pi i}} F(\xi) = F'. \quad (1.31)$$

Remarkably, this property makes the linear operation of integration over Grassmann variables in (1.28) identical to the linear operation of *differentiation*. As a consequence, a linear change of Grassmann integration variables multiplies the integral by the *inverse* of the Jacobian. For example, going from a real ξ to another Grassmann variable $\xi' = a\xi$, the integrals over ξ' have again the properties (1.28):

$$\int \frac{d\xi'}{\sqrt{2\pi i}} = 0, \quad \int \frac{d\xi'}{\sqrt{2\pi i}} \xi' = 1. \quad (1.32)$$

In order to be compatible with (1.28), the measure must change as follows:

$$\int \frac{d\xi'}{\sqrt{2\pi i}} = \frac{1}{a} \int \frac{d\xi}{\sqrt{2\pi i}}. \tag{1.33}$$

This is in contrast to ordinary integrals where the factor on the right-hand side would be a.

From the real rules (1.28), we derive the integrals involving complex Grassmann variables:

$$\int \frac{d\xi}{\sqrt{2\pi i}} = 0, \quad \int \frac{d\xi^*}{\sqrt{2\pi i}} \frac{d\xi}{\sqrt{2\pi i}} i\xi^*\xi = 1, \quad \int \frac{d\xi^*}{\sqrt{2\pi i}} \frac{d\xi}{\sqrt{2\pi i}} (\xi^*\xi)^n = 0, \quad n > 1. \tag{1.34}$$

The integration rules (1.28) imply that the right-hand side of (1.27) becomes the product of eigenvalues A_m:

$$\prod_m A_m = \det A, \tag{1.35}$$

which is exactly the inverse of the bosonic result (1.25), thus proving the statement after Eq. (1.25).

For real Fermi fields, the proof is slightly more involved, since now the hermitian matrix A_{kl} can no longer be diagonalized by a unitary transformation, so that the invariance of the measure of integration $\prod_m (d\xi_m/\sqrt{2\pi i})$ is no longer automatically guaranteed. However, the integral can be done after all by observing that A_{kl} may always be assumed to be antisymmetric. If there is any symmetric part, it cancels in the quadratic form $\sum_{kl} \xi_k A_{kl} \xi_l$ due to the anticommutativity of the Grassmann variables. Now, an antisymmetric hermitian matrix can always be written as $A = -iA_R$, where A_R is real and antisymmetric. Such a matrix is standard in symplectic spaces. It can be brought to a canonical form $\mathbf{C}$ which is zero everywhere except for 2×2 matrices,

$$c = i\sigma^2 = \begin{pmatrix} 0 & 1 \\ -1 & 0 \end{pmatrix}, \tag{1.36}$$

along the diagonal. Here σ^2 is the second Pauli matrix. Then A can be written as

$$A = -iT^T \mathbf{C} T \tag{1.37}$$

where the hermitian matrix $-i\mathbf{C}$ contains only σ^2-matrices along the diagonal. This matrix has a unit determinant so that $\det T = \det^{1/2}(A)$. Thus, under a linear transformation of Grassmann variables $\xi'_k \equiv T_k \xi_l$, the measure of integration changes according to

$$\prod_k d\xi_k = (\det T) \prod_k d\xi'_k, \tag{1.38}$$

as a direct consequence of the integration rule (1.33). With the help of the rules (1.28) and (1.29), the Grassmann version of the functional integral (1.24) can now be evaluated as follows:

$$\prod_m \left[\int \frac{d\xi_m}{\sqrt{2\pi i}} \right] \exp \left(i \sum_{k,l} \xi_k A_{kl} \xi_l \right)$$

$$= (\det T) \prod_m \left[\int \frac{d\xi'_m}{\sqrt{2\pi i}} \right] \exp\left(-\sum_{kl} \xi'_k C_{kl} \xi'_l\right)$$

$$= (\det A)^{1/2} \prod_n^\infty \left[\int \frac{d\xi'_{2n}}{\sqrt{2\pi i}} \right] \frac{d\xi'_{2n+1}}{\sqrt{2\pi i}} \left(1 + \xi'_{2n+1}\xi'_{2n}\right) = (\det A)^{1/2}. \quad (1.39)$$

Thus the right-hand side is the inverse of the bosonic result (1.24), as announced after Eq. (1.25).

In order to apply these formulas to fields $\psi(\mathbf{x}, t)$ defined on a continuous space-time, both formulas have to be written for the lattice field (1.18) in such a way that the limit of infinitely fine lattice spacing $\delta \to 0$ can be performed without problem. For this we recall the useful matrix identity

$$[\det A]^{\mp 1} = \exp[i(\pm i \mathrm{Tr}\log A)], \quad (1.40)$$

where $\log A$ may be expanded in the standard fashion as

$$\log A = \log\left[1 + (A - 1)\right] = -\sum_{n=1}^\infty \left[-(A-1)\right]^n \frac{1}{n}. \quad (1.41)$$

This formula reduces the calculation of the determinant to a series of matrix multiplications. In each of these, the limit $\delta \to 0$ can easily be taken. One simply replaces all sums over lattice indices by integrals over $d^3x dt$, for instance

$$\mathrm{tr}A^2 = \sum_{kl} A_{kl}A_{lk} \longrightarrow \mathrm{Tr}\,A^2 = \int d^3x dt d^3x' dt'\, A(\mathbf{x}, t; \mathbf{x}', t')A(\mathbf{x}', t', \mathbf{x}, t). \quad (1.42)$$

With this in mind, the field versions of (1.24) and (1.25) amount to the following functional formulas for boson and fermion fields:

$$\int \mathcal{D}\varphi(\mathbf{x}, t) \exp\left[\frac{i}{2} \int d^3x dt d^3x' dt'\, \varphi(\mathbf{x}, t) A(\mathbf{x}, t; \mathbf{x}', t')\varphi(\mathbf{x}', t')\right]$$

$$= \exp\left[i\left(\pm\frac{i}{2}\mathrm{Tr}\log\left\{\begin{matrix} 1 \\ i \end{matrix}\right\} A\right)\right], \quad (1.43)$$

$$\int \mathcal{D}\psi^*(\mathbf{x}, t)\mathcal{D}\psi(\mathbf{x}, t) \exp\left[i \int d^3x dt d^3x' dt'\, \psi^*(\mathbf{x}, t) A(\mathbf{x}, t; x', t')\psi(\mathbf{x}', t')\right]$$

$$= \exp\left[i(\pm i \mathrm{Tr}\log A)\right]. \quad (1.44)$$

Here φ, ψ are arbitrary real and complex fields, with the upper sign holding for bosons, the lower for fermions. The same result is of course true if φ and ψ have several components (describing, for example, spin) and A is a matrix in the corresponding space.

1.1.3 Interactions

Consider now a many-particle system described by an action of the form (in natural units with $\hbar = 1$):

$$\mathcal{A} \equiv \mathcal{A}_0 + \mathcal{A}_{\mathrm{int}} = \int d^3x dt \psi^*(\mathbf{x}, t) \left[i\partial_t - \epsilon(-i\boldsymbol{\nabla})\right]\psi(\mathbf{x}, t) \quad (1.45)$$

$$- \frac{1}{2} \int d^3x dt d^3x' dt'\, \psi^*(\mathbf{x}', t')\psi^*(\mathbf{x}, t)V(\mathbf{x}, t; \mathbf{x}', t')\psi(\mathbf{x}, t)\psi(\mathbf{x}', t').$$

The fundamental field $\psi(x)$ may describe bosons or fermions. The interaction potential is usually translationally invariant in space and time:

$$V(\mathbf{x}, t; \mathbf{x}', t') = V(\mathbf{x} - \mathbf{x}', t - t'). \tag{1.46}$$

In nonrelativistic many-body systems, the potential is often instantaneous in time:

$$V(\mathbf{x}, t; \mathbf{x}', t') = \delta(t - t') V(\mathbf{x} - \mathbf{x}'). \tag{1.47}$$

This property simplifies many calculations. It is in general fulfilled only approximately.

For instance, the attraction between electrons in a low-temperature superconductor is caused by phonon exchange which contains retardation effects due to the finite speed of sound.

The complete information on the physical properties of the system resides in its Green functions. In the field operator language, one uses the so-called *Heisenberg picture*, where the Green functions are given by the expectation values of the time-ordered products of the field operators

$$G\left(\mathbf{x}_1, t_1, \ldots, \mathbf{x}_n, t_n; \mathbf{x}_{n'}, t_{n'}, \ldots, \mathbf{x}_{1'}, t_{1'}\right) \tag{1.48}$$
$$= \langle 0 | \hat{T} \left(\hat{\psi}_H(\mathbf{x}_1, t_1) \cdots \hat{\psi}_H(\mathbf{x}_n, t_n) \hat{\psi}_H^\dagger(\mathbf{x}_{n'}, t_{n'}) \cdots \hat{\psi}_H^\dagger(\mathbf{x}_{1'}, t_{1'}) \right) | 0 \rangle,$$

where $\hat{\psi}_H(\mathbf{x}, t)$ are the Heisenberg operators of the interacting field. The time-ordering operator $\hat{T}$ changes the position of the operators behind it in such a way that fields with later time arguments stand to the left of those with earlier time arguments. To achieve this order, a number of field transmutations are necessary. If F denotes the number of transmutations of Fermi fields, the final product receives a sign factor $(-1)^F$.

It is convenient to view all Green functions (1.48) as derivatives of the *generating functional*

$$Z[\eta^*, \eta] = \langle 0 | \hat{T} \exp \left\{ i \int d^3x\, dt \left[\hat{\psi}_H^\dagger(\mathbf{x}, t) \eta(\mathbf{x}, t) + \eta^*(\mathbf{x}, t) \hat{\psi}_H(\mathbf{x}, t) \right] \right\} | 0 \rangle, \tag{1.49}$$

namely

$$G\left(\mathbf{x}_1, t_1, \ldots, \mathbf{x}_n, t_n; \mathbf{x}_{n'}, t_{n'}, \ldots, \mathbf{x}_{1'}, t_{1'}\right) \tag{1.50}$$
$$= (-i)^{n+n'} \left. \frac{\delta^{n+n'} Z[\eta^*, \eta]}{\delta \eta^*(\mathbf{x}_1, t_1) \cdots \delta \eta^*(\mathbf{x}_n, t_n) \delta \eta(\mathbf{x}_{n'}, t_{n'}) \cdots \delta \eta(\mathbf{x}_{1'}, t_{1'})} \right|_{\eta = \eta^* \equiv 0}.$$

Physically, the generating functional describes the amplitude that the vacuum remains a vacuum in spite of the presence of external perturbations.

The calculation of this Green functional is usually performed in the interaction picture which can be summarized by the operator formulation for $Z[\eta^*, \eta]$:

$$Z[\eta^*, \eta] = N \langle 0 | \hat{T} \exp \left\{ i \mathcal{A}_{\text{int}}[\hat{\psi}^\dagger, \hat{\psi}] + i \int d^3x\, dt \left[\hat{\psi}^\dagger(\mathbf{x}, t) \eta(\mathbf{x}, t) + \text{h.c.} \right] \right\} | 0 \rangle. \tag{1.51}$$

In the interaction picture, the field operators $\hat{\psi}(\mathbf{x}, t)$ move according to the free-field equation of motion (1.14). The time-ordered product of two of these field operators coincide with the free-field propagator calculated before in (1.16).

The normalization constant N is determined by the condition [which is trivially true for (1.49)]:

$$Z[0, 0] = 1. \tag{1.52}$$

The calculation may now proceed perturbatively. One expands the exponential $\exp\{i\mathcal{A}_{\text{int}}\}$ in (1.51) in a power series and obtains

$$Z[\eta^*, \eta] = N \sum_{n=0}^{\infty} \frac{1}{n!} Z_n[\eta^*, \eta], \tag{1.53}$$

where the contribution of order n is given by

$$Z_n[\eta^*, \eta] \equiv N \langle 0 | \hat{T} \left\{ \left(i\mathcal{A}_{\text{int}}[\hat{\psi}^\dagger, \hat{\psi}] \right)^n \exp \left(i \int d^3x dt \left[\hat{\psi}^\dagger(\mathbf{x}, t) \eta(\mathbf{x}, t) + \text{h.c.} \right] \right) \right\} | 0 \rangle. \tag{1.54}$$

This expression is further expanded in powers of η^* and η. The resulting vacuum expectation values of time-ordered products of field operators can be expanded in products of Green functions of the free field operators. The rules for doing this is provided by *Wick's theorem* [5, 6, 7]. This theorem states that any time ordered product of free field operators $\psi(\mathbf{x}, t)$ and its hermitian conjugate $\psi^\dagger(\mathbf{x}, t)$ can be expanded into a sum of *normal products* with all possible contractions taken via Feynman propagators.

1.1.4   Normal Products

Given an arbitrary set of n free field operators $\phi_1(x_1) \cdots \phi_n(x_n)$, each of them consists of a creation and an annihilation part:

$$\phi_i(x_i) = \phi_i^c(x_i) + \phi_i^a(x_i). \tag{1.55}$$

Some ϕ_i may be commuting Bose fields, some anticommuting Fermi fields. The normally ordered product or *normal product* of n of these field operators will be denoted by $\hat{N}(\phi_1(x)\phi(x_2) \cdots \phi(x_n))$. In the present context, a function symbol is more convenient than the earlier double-dot notation. The normal product is a function of a product of field operators which has the following two properties:

i) **Linearity:** The normal product is a linear function of all its n arguments, i.e., it satisfies

$$\hat{N}\left((\alpha\phi_1 + \beta\phi_1')\phi_2\phi_3 \cdots \phi_n\right) = \alpha\hat{N}(\phi_1\phi_2\phi_3 \cdots \phi_n) + \beta\hat{N}(\phi_1'\phi_2\phi_3 \cdots \phi_n). \tag{1.56}$$

If every ϕ_i is replaced by $\phi_i^c + \phi_i^a$, it can be expanded into a linear combination of terms which are all pure products of creation and annihilation operators.

ii) **Normal Ordering**: The normal product reorders all products arising from a complete linear expansion of all fields according to i) in such a way that all annihilators stand to the right of all creators. If the operators ϕ_i describe fermions, the definition requires a factor -1 to be inserted for every transmutation of the order of two operators.

For example, let ϕ_1, ϕ_2, ϕ_3 be scalar fields, then with two field operators normal ordering produces:

$$\begin{aligned}
\hat{N}(\phi_1^c \phi_2^c) &= \phi_1^c \phi_2^c = \phi_2^c \phi_1^c, \\
\hat{N}(\phi_1^c \phi_2^a) &= \phi_1^c \phi_2^a, \\
\hat{N}(\phi_1^a \phi_2^c) &= \phi_2^c \phi_1^a, \\
\hat{N}(\phi_1^a \phi_2^a) &= \phi_1^a \phi_2^a = \phi_2^a \phi_1^a,
\end{aligned} \tag{1.57}$$

and with three field operators:

$$\begin{aligned}
\hat{N}(\phi_1^c \phi_2^c \phi_3^a) &= \phi_1^c \phi_2^c \phi_3^a = \phi_2^c \phi_1^c \phi_3^a, \\
\hat{N}(\phi_1^c \phi_2^a \phi_3^c) &= \phi_1^c \phi_3^c \phi_2^a = \phi_3^c \phi_1^c \phi_2^a, \\
\hat{N}(\phi_1^a \phi_2^c \phi_3^c) &= \phi_2^c \phi_3^c \phi_1^a = \phi_3^c \phi_2^c \phi_1^a.
\end{aligned} \tag{1.58}$$

If the operators ϕ_i are fermions, the effect is

$$\begin{aligned}
\hat{N}(\phi_1^c \phi_2^c) &= \phi_1^c \phi_2^c = -\phi_2^c \phi_1^c, \\
\hat{N}(\phi_1^c \phi_2^a) &\equiv \phi_1^c \phi_2^a, \\
\hat{N}(\phi_1^a \phi_2^c) &= -\phi_2^c \phi_1^a, \\
\hat{N}(\phi_1^a \phi_2^a) &= \phi_1^a \phi_2^a = -\phi_2^a \phi_1^a,
\end{aligned} \tag{1.59}$$

and

$$\begin{aligned}
\hat{N}(\phi_1^c \phi_2^c \phi_3^a) &= \phi_1^c \phi_2^c \phi_3^a = -\phi_2^c \phi_1^c \phi_3^a, \\
\hat{N}(\phi_1^c \phi_2^a \phi_3^c) &= -\phi_1^c \phi_3^c \phi_2^a = \phi_3^c \phi_1^c \phi_2^a, \\
\hat{N}(\phi_1^a \phi_2^c \phi_3^c) &= \phi_2^c \phi_3^c \phi_1^a = -\phi_3^c \phi_2^c \phi_1^a.
\end{aligned} \tag{1.60}$$

The normal product is uniquely defined. The remaining order of creation or annihilation parts among themselves is irrelevant, since these commute or anticommute with each other by virtue of the canonical free-field commutation rules. In the following, the fields ϕ may be Bose or Fermi fields and the sign of the Fermi case is recorded underneath the Bose sign.

The advantage of defining normal products is their important property that they have no vacuum expectation values. There is always an annihilator on the right-hand side or a creator on the left-hand side which produce 0 when matched between vacuum states. The method of calculating all n-point functions consists in expanding all time ordered products of n field operators completely into normal products. Then only the terms with no operators survive between vacuum states. This is the desired value of the n-point function.

Let us see how this works for the simplest case of a time-ordered product of two identical field operators

$$\hat{T}(\phi(x_1)\phi(x_2)) \equiv \Theta(x_1^0 - x_2^0)\phi(x_1)\phi(x_2) \pm \Theta(x_2^0 - x_1^0)\phi(x_2)\phi(x_1). \quad (1.61)$$

The basic expansion formula is

$$\hat{T}(\phi(x_1)\phi(x_2)) = \hat{N}(\phi(x_1)\phi(x_2)) + \langle 0|\hat{T}(\phi(x_1)\phi(x_2))|0\rangle. \quad (1.62)$$

For brevity, we shall denote the propagator of two fields as follows:

$$\langle 0|T(\phi(x_1)\phi(x_2))|0\rangle = \overline{\phi(x_1)\phi}(x_2) = G(x_1 - x_2). \quad (1.63)$$

The hook which connects the two fields on the top are referred to as a *contraction* of the fields.

We shall prove the basic expansion formula (1.62) by considering it separately for the creation and annihilation parts ϕ^c and ϕ^a. This will be sufficient since the time ordered product is linear in each field just as the normal product. Now, in both cases $x_1^0 \gtrless x_2^0$ we have

$$\hat{T}(\phi^c(x_1)\phi^c(x_2)) = \left\{ \begin{array}{c} \phi^c(x_1)\phi^c(x_2) \\ \pm\, \phi^c(x_2)\phi^c(x_1) \end{array} \right\}$$
$$= \phi^c(x_1)\phi^c(x_2) + \langle 0| \left\{ \begin{array}{c} \phi^c(x_1)\phi^c(x_2) \\ \pm\, \phi^c(x_2)\phi^c(x_1) \end{array} \right\} |0\rangle, \quad (1.64)$$

which is true since $\phi^c(x_1)\phi^c(x_2)$ commute or anticommute with each other, and annihilate the vacuum state $|0\rangle$. The same equation holds for $\phi^a(x_1)\phi^a(x_2)$. The only nontrivial cases are those with a time-ordered product of $\phi^c(x_1)\phi^a(x_2)$ and $\phi^a(x_1)\phi^c(x_2)$. The first becomes for $x_1^0 \gtrless x_2^0$:

$$T(\phi^c(x_1)\phi^a(x_2)) = \left\{ \begin{array}{c} \phi^c(x_1)\phi^a(x_2) \\ \pm\, \phi^a(x_2)\phi^c(x_1) \end{array} \right\}$$
$$= \phi^c(x_1)\phi^a(x_2) + \langle 0| \left\{ \begin{array}{c} \phi^c(x_1)\phi^a(x_2) \\ \pm\, \phi^a(x_2)\phi^c(x_1) \end{array} \right\} |0\rangle. \quad (1.65)$$

For $x_1^0 > x_2^0$, this equation is obviously true. For $x_1^0 < x_2^0$, the normal ordering produces an additional term

$$\pm(\phi^a(x_2)\phi^c(x_1) \mp \phi^c(x_1)\phi^a(x_2)) = \pm[\phi^a(x_2), \phi^c(x_1)]_{\mp}. \quad (1.66)$$

As the commutator or anticommutator of free fields is a c-number, they may equally well be evaluated between vacuum states, so that we may replace (1.66) by

$$\pm\langle 0| [\phi^a(x_2), \phi^c(x_1)]_{\mp} |0\rangle. \quad (1.67)$$

Moreover, since ϕ^a annihilates the vacuum, this reduces to

$$\pm\langle 0|\phi^a(x_2), \phi^c(x_1)|0\rangle. \quad (1.68)$$

The oppositely ordered operators $\phi^a(x_1)\phi^c(x_2)$ can be processed by complete analogy.

We shall now generalize this basic result to an arbitrary number of field operators. In order to abbreviate the expressions let us define the concept of a contraction inside a normal product

$$\hat{N}\left(\phi_1 \cdots \phi_{i-1}\overbracket{\phi_i\phi_{i+1}\cdots\phi_{j-1}\phi_j}\phi_{j+1}\cdots\phi_n\right)$$

$$\equiv \eta\,\overbracket{\phi_i\phi_j}\,\hat{N}\left(\phi_1\cdots\phi_{i-1}\phi_{i+1}\cdots\phi_{j-1}\phi_{j+1}\cdots\phi_n\right). \tag{1.69}$$

Here $\eta = 1$ for bosons and $\eta = (-1)^{j-i-1}$ for fermions, each minus-sign counting the number of fermion transmutations which is necessary to reach the final order. A normal product with several contractions is defined by the successive execution of each of them. If only one field is left inside the normal ordering symbol, it is automatically normally ordered so that

$$\hat{N}(\phi) = \phi. \tag{1.70}$$

Similarly, if all fields inside a normal product are contracted, the result is no longer an operator and the symbol $\hat{N}$ may be dropped using linearity and the trivial property

$$\hat{N}(1) \equiv 1. \tag{1.71}$$

The fully contracted normal product is the relevant one in determining the n-particle propagator. With these preliminaries we are now ready to prove Wick's theorem for the expansion of a time-ordered product in terms of normally ordered products.[1]

The formula for an arbitrary functional of free fields ψ, ψ^* is

$$TF[\psi^*, \psi] = e^{\int d^3x dt d^3x' dt'\,\frac{\delta}{\delta\psi(\mathbf{x},t)}G_0(\mathbf{x},t;\mathbf{x}',t')\frac{\delta}{\delta\psi^*(\mathbf{x},t')}}\,\hat{N}\left(F[\psi^*,\psi]\right). \tag{1.72}$$

Applying this to

$$\langle 0|TF[\psi^*,\psi]|0\rangle = \langle 0|T\exp\left[i\int dx dt(\psi^*\eta + \eta^*\psi)\right]|0\rangle \tag{1.73}$$

one finds:

$$Z_0[\eta^*,\eta] = \exp\left[-\int dx dt dx' dt'\,\eta^*(\mathbf{x},t)G_0(\mathbf{x},t;\mathbf{x}',t')\eta(\mathbf{x}',t')\right]$$

$$\times \langle 0|\hat{N}\left(\exp\left[i\int dx dt(\psi^*\eta + \eta^*\psi)\right]\right)|0\rangle. \tag{1.74}$$

Each term can be pictured graphically by so-called Feynman diagrams. They have the physical interpretation as a virtual process.

The perturbation expansion of (1.51) may be used to *define* an interacting theory. In praxis, however, it can only be carried up to a certain finite order in n.

---

[1]G.C. Wick, Phys. Rev. *80*, 268 (1950); F. Dyson, Phys. Rev. *82*, 428 (1951).

As such it is unable to describe many important physical phenomena. Examples are bound states living in the vacuum, or collective excitations of many-body systems. These require the summation of infinite subsets of Feynman diagrams to all orders. In many situations it is well known which subsets have to be taken if we want to account approximately for a specific effect. What is not clear is how such approximations can be improved in a systematic manner. The point is that, as soon as a selective summation of Feynman diagrams is performed, the original coupling constant has lost its meaning as an organizer of the expansion and there is need for a new systematics of diagrams. Such a systematic approach will be presented in what follows.

As soon as bound states or other collective excitations are formed, it is suggestive to construct a quantum field theory for *these* and continue working with the new fields rather than the original fundamental fields $\psi(\mathbf{x}, t)$. The goal would then to rewrite the expression (1.51) for $Z[\eta^*, \eta]$ in terms of new fields whose unperturbed propagator has the free energy spectrum of the *bound states* or of the other collective excitations. It would also display their mutual interactions. In the operator form (1.51), such changes of fields are not so easy to achieve.

The ideal theoretical framework for describing the generating functional $Z[\eta^*, \eta]$ of a physical system in terms of the new quantum fields is offered by the above-introduced functional integral techniques [4, 5, 6]. In these, changes of fields amount to changes of integration variables, as we shall see in the sequel.

1.1.5   Functional Formulation

In the functional integral approach, the generating functional (1.49) is given by

$$Z[\eta^*, \eta] = N \int \mathcal{D}\psi^*(\mathbf{x}, t) \mathcal{D}\psi(\mathbf{x}, t) \exp\left\{ i\mathcal{A}[\psi^*, \psi] + i \int d^3x \, dt \, [\psi^*(\mathbf{x}, t)\eta(\mathbf{x}, t) + \text{c.c.}] \right\}.$$
(1.75)

Note that in contrast to the expression (1.49), the field $\psi(\mathbf{x}, t)$ is now a complex-valued field, *not* an operator. All quantum effects are accounted for by the fluctuations in the functional integral. This does not only include the classical field configurations, but all possible field configurations, also those which are classically forbidden, i.e., all those which do not run through the valley of extremal actions in the exponent.

In order to evaluate functional integrals of the type (1.75) involving source terms, we must extend the Gaussian formulas (1.24), (1.25) and (1.43), (1.44) to include linear terms. This complicates the integrals only slightly. We simply eliminate the linear terms by a quadratic completion. If this is done in (1.24) and (1.25), we obtain for both bosons and fermions (dropping product and summation symbols):

$$\int_{-\infty}^{\infty} \frac{d\xi}{\sqrt{2\pi i}} \exp\left(\frac{1}{2}\xi A\xi + ij\xi \right)$$
$$= \int_{-\infty}^{\infty} \frac{d\xi}{\sqrt{2\pi i}} \exp\left[\frac{i}{2} \left(\xi + jA^{-1} \right) A \left(\xi + A^{-1}j \right) - \frac{i}{2} jA^{-1}j \right],$$
(1.76)

$$\int \frac{d\xi^* d\xi}{\sqrt{2\pi i}\sqrt{2\pi i}} \exp(i\xi^* A\xi + ij^*\xi + i\xi^* j)$$

$$= \int \frac{d\xi^* d\xi}{\sqrt{2\pi i}\sqrt{2\pi i}} \exp\left[i\left(\xi^* + j^* A^{-1}\right) A \left(\xi + A^{-1} j\right) - ij^* A^{-1} j\right]. \quad (1.77)$$

The shift in the integral $\xi \to \xi + A^{-1}\xi$ gives no change due to the infinite range of integration. Hence

$$\int_{-\infty}^{\infty} \frac{d\xi}{\sqrt{2\pi i}} \exp\left(\frac{i}{2}\xi A\xi + ij\xi\right) = \left\{\begin{array}{c} 1 \\ \frac{1}{i^{1/2}} \end{array}\right\} A^{\mp 1/2} \exp\left(-\frac{i}{2}jA^{-1}j\right),$$

$$\int_{-\infty}^{\infty} \frac{d\xi^* d\xi}{\sqrt{2\pi i}\sqrt{2\pi i}} \exp(i\xi^* A\xi + ij^*\xi + i\xi^* j) = A^{\mp 1} \exp(-ij^* A^{-1} j). \quad (1.78)$$

A corresponding operation on the functional formulas (1.43) and (1.44) leads to the so-called Hubbard-Stratonovich transformations:

$$\int \mathcal{D}\varphi(\mathbf{x}, t) e^{\frac{i}{2}\int d^3x dt d^3x' dt'\left[\varphi(\mathbf{x},t)A(\mathbf{x},t;\mathbf{x}',t')\varphi(\mathbf{x}',t') + 2j(\mathbf{x},t)\varphi(\mathbf{x},t)\delta^3(\mathbf{x}-\mathbf{x}',t)\delta(t-t')\right]}$$

$$= e^{i\left(\pm\frac{i}{2}\text{Trlog}\left\{\frac{1}{i}\right\}A\right) - \frac{i}{2}\int d^3x dt d^3x' dt' \, j(\mathbf{x},t)A^{-1}(\mathbf{x},t;\mathbf{x}',t')j(\mathbf{x}',t')}, \quad (1.79)$$

or

$$\int \mathcal{D}\psi^*(\mathbf{x}, t)\mathcal{D}\psi(\mathbf{x}, t) e^{i\int d^3x dt d^3x' dt' \left\{\psi^*(\mathbf{x},t)A(\mathbf{x},t;\mathbf{x}',t')\psi(\mathbf{x},t') + [\eta^*(\mathbf{x},t)\psi(\mathbf{x})\delta^3(\mathbf{x}-\mathbf{x}')\delta(t-t') + \text{c.c.}]\right\}}$$

$$= e^{i(\pm i\text{Trlog}A) - i\int d^3x dt d^3x' dt\eta^*(\mathbf{x},t)A^{-1}(\mathbf{x},t;\mathbf{x}',t')\eta(\mathbf{x}',t')}. \quad (1.80)$$

These integration formulas will be needed repeatedly in the remainder of this text. They have been applied frequently in many-body theory, ever since the work of Hubbard and Stratonovic [10], and for this reason they have been named in many publications after these authors. They are the basis for the treatment of any interacting quantum field theory in terms of *collective quantum fields*.

Although this transformation is mathematically exact, it may be of little use in applications in which various collective effects compete with each other. This can be understood only after treating a few important phenomena using this transformation. A way out of the difficulties will be shown in Section 3.9. The improved treatment will allow us to study competing mechanisms in terms of *collective classical fields*.

1.1.6 Equivalence of Functional and Operator Methods

As an exercise we shall apply (1.79) and (1.80) and demonstrate the equivalence betweeen the operator expression (1.51) for the generating functional $Z[\eta^*, \eta]$ with Feynman's functional integral formula (1.75).

First we note that the interaction can be taken outside the integral or the vacuum expectation value in either formula as

$$Z[\eta^*, \eta] = \exp\left\{i\mathcal{A}_{\text{int}}\left[\frac{1}{i}\frac{\delta}{\delta\eta}, \frac{1}{i}\frac{\delta}{\delta\eta^*}\right]\right\} Z_0[\eta^*, \eta], \quad (1.81)$$

where $Z_0[\eta^*, \eta]$ is the generating functional for the free fields. Thus Eq. (1.75) contains only $\mathcal{A}_0$ of (1.45) in the exponent, i.e.

$$\mathcal{A}_0[\psi^*, \psi] = \int dx dt \psi^*(\mathbf{x}, t) \left[i\partial_t - \epsilon(-i\boldsymbol{\nabla}) \right] \psi(\mathbf{x}, t). \tag{1.82}$$

The functional integral in $Z_0[\eta^*, \eta]$ is of the type (1.80), where $A(\mathbf{x}, t; \mathbf{x}', t')$ is the functional matrix

$$A(\mathbf{x}, t; \mathbf{x}', t') = \left[i\partial_t - \epsilon(-i\boldsymbol{\nabla}) \right] \delta^{(3)}(\mathbf{x} - \mathbf{x}') \delta(t - t'). \tag{1.83}$$

The inverse of this functional matrix yields the so-called *propagator* of the free particle:

$$G_0(\mathbf{x}, t; \mathbf{x}', t') = iA^{-1}(\mathbf{x}, t; \mathbf{x}', t'). \tag{1.84}$$

It can be calculated explicitly in the spectral representation which, for translationally invariant operators, is a Fourier representation:

$$G_0(\mathbf{x}, t; \mathbf{x}', t') = \int \frac{dE}{2\pi} \int \frac{d^3 p}{(2\pi)^4} e^{-i[E(t-t') - \mathbf{p}(\mathbf{x}-\mathbf{x}')]} \frac{i}{E - \epsilon(\mathbf{p}) + i\eta}. \tag{1.85}$$

Inserting this into (1.80), we see that

$$Z_0[\eta^*, \eta] = N \exp\left[i \left(\pm i \mathrm{Tr} \log iG_0^{-1} \right) - \int d^3 x dt d^3 x' dt' \, \eta^*(\mathbf{x}, t) G_0(\mathbf{x}', t') \eta(\mathbf{x}', t') \right]. \tag{1.86}$$

We now fix the normalization constant N to satisfy the condition (1.52):

$$N = \exp\left[i \left(\pm i \mathrm{Tr} \log iG_0 \right) \right], \tag{1.87}$$

and arrive at

$$Z_0[\eta^*, \eta] = \exp\left[- \int d^3 x dt d^3 x' dt' \, \eta^*(\mathbf{x}, t) G_0(\mathbf{x}, t; \mathbf{x}', t') \eta(\mathbf{x}', t') \right]. \tag{1.88}$$

This coincides exactly with what would have been obtained from the operator expression (1.51) for $Z_0[\eta^*, \eta]$ (i.e., without $\mathcal{A}_{\mathrm{int}}$).

1.1.7 Grand-Canonical Ensembles at Zero Temperature

All these results are easily generalized from vacuum expectation values to thermodynamic averages at fixed temperatures T and chemical potential μ. The change at $T = 0$ is trivial: The single particle energies in the action (1.45) have to be replaced by

$$\xi(-i\boldsymbol{\nabla}) = \epsilon(-i\boldsymbol{\nabla}) - \mu, \tag{1.89}$$

and new boundary conditions have to be imposed upon all Green functions via an appropriate $i\epsilon$ prescription in $G_0(\mathbf{x}, t; \mathbf{x}', t')$ of (1.85) [see [5, 8]]:

$$^{T=0}G_0(\mathbf{x}, t; \mathbf{x}', t') = \int \frac{dE d^3p}{(2\pi)^4} e^{-iE(t-t')+i\mathbf{p}(\mathbf{x}-\mathbf{x}')} \frac{i}{E - \xi(\mathbf{p}) + i\eta \operatorname{sgn}\xi(\mathbf{p})}. \quad (1.90)$$

As a consequence of the chemical potential, fermions with $\xi < 0$ inside the Fermi sea propagate backwards in time. Bosons, on the other hand, have in general $\xi > 0$ and, hence, always propagate forward in time.

In order to simplify the notation we shall often use the four-vectors $p = (p^0, \mathbf{p})$ and write the measure of integration in (1.90) as

$$\int \frac{dE d^3p}{(2\pi)^4} = \int \frac{d^4p}{(2\pi)^4}. \quad (1.91)$$

Note that in a solid, the momentum integration is really restricted to a Brillouin zone. If the solid has a finite volume V, the integral over spacial momenta becomes a sum over momentum vectors,

$$\int \frac{d^3p}{(2\pi)^3} = \frac{1}{V} \sum_{\mathbf{p}}, \quad (1.92)$$

and the Green function (1.90) reads

$$^{T=0}G_0(\mathbf{x}, t; \mathbf{x}', t') \equiv \int \frac{dE}{2\pi} \frac{1}{V} \sum_{\mathbf{p}} e^{-ip(x-x')} \frac{i}{p^0 - \xi(\mathbf{p}) + i\eta \operatorname{sgn}\xi(\mathbf{p})}. \quad (1.93)$$

The resulting formulas for $^{T=0}Z[\eta^*, \eta]$ can be brought to a conventional form by performing a Wick rotation in the complex energy plane in all energy integrals (1.90), implied by formulas (1.51) and (1.74). For this, one sets $E = p^0 \equiv i\omega$ and replaces

$$\int_{-\infty}^{\infty} \frac{dE}{2\pi} \to i \int_{-\infty}^{\infty} \frac{d\omega}{2\pi}. \quad (1.94)$$

Then the Green function (1.90) becomes

$$^{T=0}G_0(\mathbf{x}, t; \mathbf{x}', t') = -\int \frac{d\omega}{2\pi} \frac{d^3p}{(2\pi)^3} e^{\omega(t-t')+i\mathbf{p}(\mathbf{x}-\mathbf{x}')} \frac{1}{i\omega - \xi(\mathbf{p})}. \quad (1.95)$$

Note that with formulas (1.88) and (1.81), the generating functional $^{T=0}Z[\eta^*, \eta]$ is the grand-canonical partition function in the presence of sources [8].

Finally, we have to introduce arbitrary temperatures T. According to the standard rules of quantum field theory (for an elementary introduction see Chapter 2 in Ref. [5]), we must continue all times to imaginary values $t = i\tau$, restrict the imaginary time interval to the inverse temperature[2] $\beta \equiv 1/T$, and impose periodic or

---

[2]Throughout this chapter we use natural units so that $k_B = 1, \hbar = 1$.

antiperiodic boundary conditions upon the fields $\psi(\mathbf{x}, -i\tau)$ of bosons and fermions, respectively [5, 8]:

$$\psi(\mathbf{x}, -i\tau) = \pm\psi(\mathbf{x}, -i(\tau + 1/T)). \tag{1.96}$$

When there is no danger of confusion, we shall usually drop the factor $-i$ in front of the imaginary times in all field arguments, for brevity. The same thing will be done in the Green functions.

By virtue of (1.81) and (1.88), also the Green functions satisfy these boundary conditions, implying that

$${}^{T}G_0\left(\mathbf{x}, \tau + 1/T; \mathbf{x}', \tau'\right) \equiv \pm {}^{T}G_0(\mathbf{x}, -i\tau; \mathbf{x}', -i\tau'). \tag{1.97}$$

This property is enforced automatically by replacing the energy integrations $\int_{-\infty}^{\infty} d\omega/2\pi$ in (1.95) by a summation over the discrete Matsubara frequencies [in analogy to the momentum sum (1.92), the temporal "volume" being $\beta = 1/T$]:

$$\int_{-\infty}^{\infty} \frac{d\omega}{2\pi} \to T\sum_{\omega_n}, \tag{1.98}$$

which are even or odd multiples of πT

$$\omega_n = \left\{ \begin{array}{c} 2n \\ 2n+1 \end{array} \right\} \pi T \quad \text{for} \quad \left\{ \begin{array}{c} \text{bosons} \\ \text{fermions} \end{array} \right\}. \tag{1.99}$$

The prefactor T of the sum over the discrete Matsubara frequencies accounts for the density of these frequencies yielding the correct $T \to 0$-limit.

Thus, for the imaginary-time Green function of a free nonrelativistic field at finite temperature (the so-called *free thermal Green function*), we obtain the following expression:

$${}^{T}G_0(\mathbf{x}, \tau, \mathbf{x}', \tau') = -T\sum_{\omega_n} \int \frac{d^3 p}{(2\pi)^3} e^{-i\omega_n(\tau-\tau')+i\mathbf{p}(\mathbf{x}-\mathbf{x}')} \frac{1}{i\omega_n - \xi(\mathbf{p})}. \tag{1.100}$$

Incorporating the Wick rotation in the sum notation we may write

$$T\sum_{p_0} = -iT\sum_{\omega_n} = -iT\sum_{p_4}, \tag{1.101}$$

where $p_4 = -ip_0 = \omega$. If both temperature and volume are finite, the Green function is written as

$${}^{T}G_0(\mathbf{x}, \tau, \mathbf{x}', \tau') = -\frac{T}{V}\sum_{p_0}\sum_{\mathbf{p}} e^{-i\omega_n(\tau-\tau')+i\mathbf{p}(\mathbf{x}-\mathbf{x}')} \frac{1}{i\omega_n - \xi(\mathbf{p})}. \tag{1.102}$$

At equal space points and equal imaginary times, the sum can easily be evaluated. One must, however, specify the order in which $\tau \to \tau'$. Let η denote an infinitesimal positive number and consider the case $\tau' = \tau + \eta$, i.e., the Green function

$${}^{T}G_0(\mathbf{x}, \tau, \mathbf{x}, \tau + \eta) = -T\sum_{\omega_n} \int \frac{d^3 p}{(2\pi)^3} e^{i\omega_n \eta} \frac{1}{i\omega_n - \xi(\mathbf{p})}.$$

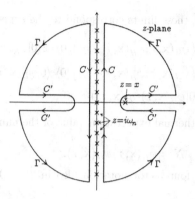

FIGURE 1.1 Contour C in the complex z-plane for evaluating the Matsubara sum (1.104).

The sum is now found by changing it into a contour integral

$$T \sum_{\omega_n} e^{i\omega_n \eta} \frac{1}{i\omega_n - \xi(\mathbf{p})} = \frac{T}{2\pi i} \int_C dz \frac{e^{\eta z}}{e^{z/T} \mp 1} \frac{1}{z - \xi}. \qquad (1.103)$$

The upper sign holds for bosons, the lower for fermions. The contour of integration C encircles the imaginary z axis in the positive sense, thereby enclosing all integer or half-integer valued poles of the integrand at the Matsubara frequencies $z = i\omega_m$ (see Fig. 1.1). The factor $e^{\eta z}$ ensures that the contour in the left half-plane does not contribute.

By deforming the contour C into C' and by contracting C' to zero we pick up the pole at $z = \xi$ and find

$$T \sum_{\omega_n} e^{i\omega_n \eta} \frac{1}{i\omega_n - \xi(\mathbf{p})} = \mp \frac{1}{e^{\xi(\mathbf{p})/T} \mp 1} = \mp \frac{1}{e^{\xi(\mathbf{p})/T} \mp 1} = \mp n(\xi(\mathbf{p})). \qquad (1.104)$$

The function on the right is known as the *Bose* or *Fermi distribution function*.

By subtracting from (1.104) the sum with ξ replaced by $-\xi$, we obtain the important sum formula

$$T \sum_{\omega_n} \frac{1}{\omega_n^2 + \xi^2(\mathbf{p})} = \frac{1}{2\xi(\mathbf{p})} \coth^{\pm 1} \frac{\xi(\mathbf{p})}{T}. \qquad (1.105)$$

In the opposite limit $\tau' = \tau - \eta$, the phase factor in the sum would be $e^{-i\omega_m \eta}$, and the Matsubara sum would be converted into a contour integral

$$-k_B T \sum_{\omega_m} e^{i\omega_m \eta} \frac{1}{i\omega_m - \xi(\mathbf{p})} = \pm \frac{k_B T}{2\pi i} \int_C dz \frac{e^{-\eta z}}{e^{-z/k_B T} \mp 1} \frac{1}{z - \xi}, \qquad (1.106)$$

yielding $1 \pm n_{\xi(\mathbf{p})}$.

In the operator language, these limits correspond to the expectation values

$$^T G\left(\mathbf{x}, \tau; \mathbf{x}, \tau + \eta\right) = \langle 0|\hat{T}\left(\hat{\psi}_H(\mathbf{x}, \tau)\hat{\psi}_H^\dagger(\mathbf{x}, \tau + \eta)\right)|0\rangle \pm \langle 0|\hat{\psi}_H^\dagger(\mathbf{x}, \tau)\hat{\psi}_H(\mathbf{x}, \tau)|0\rangle,$$

$$^T G\left(\mathbf{x}, \tau; \mathbf{x}, \tau - \eta\right) = \langle 0|\hat{T}\left(\hat{\psi}_H(\mathbf{x}, \tau)\hat{\psi}_H^\dagger(\mathbf{x}, \tau - \eta)\right)|0\rangle \langle 0|\hat{\psi}_H(\mathbf{x}, \tau)\hat{\psi}_H^\dagger(\mathbf{x}, \tau)|0\rangle$$

$$= 1 \pm \langle 0|\hat{\psi}_H^\dagger(\mathbf{x}, \tau)\hat{\psi}_H(\mathbf{x}, \tau \mp \eta)|0\rangle.$$

The function $n(\xi(\mathbf{p}))$ is the thermal expectation value of the number operator

$$\hat{N} = \hat{\psi}_H^\dagger(\mathbf{x}, \tau)\hat{\psi}_H(\mathbf{x}, \tau). \tag{1.107}$$

It is useful to employ a four-vector notation also in $T \neq 0$ -ensembles. The four-vector

$$p_E \equiv (p_4, \mathbf{p}) = (\omega, \mathbf{p}) \tag{1.108}$$

is called the *Euclidean four-momentum*. Correspondingly, we define the *Euclidean spacetime coordinate*

$$x_E \equiv (-\tau, \mathbf{x}). \tag{1.109}$$

The exponential in (1.100) can be written as

$$p_E x_E = -\omega\tau + \mathbf{p}\mathbf{x}. \tag{1.110}$$

Collecting integral and sum in a single four-summation symbol, we shall write (1.100) as

$$^T G_0(x_E - x') \equiv -\frac{T}{V}\sum_{p_E}\exp\left[-ip_E(x_E - x'_E)\right]\frac{1}{ip_4 - \xi(\mathbf{p})}. \tag{1.111}$$

It is quite straightforward to derive the general $T \neq 0$ Green function from a path integral formulation analogous to (1.75). For this we consider classical fields $\psi(\mathbf{x}, \tau)$ with the periodicity or anti-periodicity

$$\psi(\mathbf{x}, \tau) = \pm\psi\left(\mathbf{x}, \tau + 1/T\right). \tag{1.112}$$

They can be Fourier-decomposed as

$$\psi(\mathbf{x}, \tau) = \frac{T}{V}\sum_{\omega_n}\sum_{\mathbf{p}}e^{-i\omega_n\tau + i\mathbf{p}\mathbf{x}}a(\omega_n, \mathbf{p}) \equiv \frac{T}{V}\sum_{p_E}e^{-ip_E x_E}a(p_E), \tag{1.113}$$

with a sum over even or odd Matsubara frequencies ω_n. If now a free action is defined as

$$\mathcal{A}_0[\psi^*, \psi] = -i\int_{-1/2T}^{1/2T}d\tau\int d^3x\,\psi^*(\mathbf{x}, \tau)\left[-\partial_\tau - \xi\left(-i\boldsymbol{\nabla}\right)\right]\psi(\mathbf{x}, \tau),$$

$$\tag{1.114}$$

Eq. (1.80) renders [4, 10]

$$^T Z_0[\eta^*,\eta] = e^{\mp \operatorname{Tr}\log A + \int \int_{-1/2T}^{1/2T} d\tau d\tau' \int d^3x d^3x' \eta^*(\mathbf{x},\tau) A^{-1}(\mathbf{x},\tau,\mathbf{x}',\tau') \eta(\mathbf{x}',\tau')}, \qquad (1.115)$$

with

$$A(\mathbf{x},\tau;\mathbf{x}',\tau') = [\partial_\tau + \xi(-i\boldsymbol{\nabla})]\delta^{(3)}(\mathbf{x}-\mathbf{x}')\delta(\tau-\tau'). \qquad (1.116)$$

Henceforth A^{-1} is equal to the propagator (1.100). The Matsubara frequencies arise from the finite τ interval of Euclidean space together with the periodic boundary condition (1.112).

Again, interactions are taken care of by multiplying $^T Z_0[\eta^*\eta]$ with the factor (1.81). In terms of the fields $\psi(\mathbf{x},\tau)$, the exponent has the form

$$\mathcal{A}_{\text{int}} = \frac{1}{2}\int\int_{-1/2T}^{1/2T} d\tau d\tau'$$
$$\times \int d^3x d^3x' \psi^*(\mathbf{x},\tau)\psi^*(\mathbf{x}',\tau')\psi(\mathbf{x}',\tau')\psi(\mathbf{x},\tau)V(\mathbf{x},-i\tau;\mathbf{x}',-i\tau'). \quad (1.117)$$

In the case of a potential of type (1.47), which becomes instantaneous in τ:

$$V(\mathbf{x},-i\tau;\mathbf{x}',-i\tau') = V(\mathbf{x}-\mathbf{x}')\, i\delta(\tau-\tau'). \qquad (1.118)$$

Then $\mathcal{A}_{\text{int}}$ can be written in terms of the interaction Hamiltonian as

$$\mathcal{A}_{\text{int}} = i\int_{-1/2T}^{1/2T} d\tau H_{\text{int}}(\tau). \qquad (1.119)$$

Thus the grand canonical partition function in the presence of external sources may be calculated from the path integral [10]:

$$^T Z[\eta^*,\eta] = \int \mathcal{D}\psi^*(\mathbf{x},\tau)\mathcal{D}\psi(\mathbf{x},\tau) e^{i\,^T A + \int_{-1/2T}^{1/2T} d\tau \int d^3x [\psi^*(\mathbf{x},\tau)\eta(\mathbf{x},\tau)+\text{c.c.}]}, \qquad (1.120)$$

where the grand-canonical action is

$$i\,^T\mathcal{A}[\psi^*,\psi] = -\int_{-1/2T}^{1/2T} d\tau \int d^3x \psi^*(\mathbf{x},\tau)[\partial_\tau + \xi(-i\boldsymbol{\nabla})]\psi(\mathbf{x},\tau) \qquad (1.121)$$
$$+\frac{i}{2}\int_{-1/2T}^{1/2T} d\tau d\tau' \int d^3x d^3x' \psi^*(\mathbf{x},\tau)\psi^*(\mathbf{x}',\tau')\psi(\mathbf{x},\tau')\psi(\mathbf{x},\tau)V(\mathbf{x},-i\tau;\mathbf{x},-i\tau').$$

The Green functions are obtained from the functional derivatives

$$G(\mathbf{x}_1,\tau_1,\ldots,\mathbf{x}_n,\tau_n;\mathbf{x}_{n'},\tau_{n'},\ldots,\mathbf{x}_{1'},\tau_{1'}) \qquad (1.122)$$
$$= (-i)^{n+n'} \frac{\delta^{n+n'} Z[\eta^*,\eta]}{\delta\eta^*(\mathbf{x}_1,\tau_1)\cdots\delta\eta^*(\mathbf{x}_n,\tau_n)\delta\eta(\mathbf{x}_{n'},\tau_{n'})\cdots\delta\eta(\mathbf{x}_{1'},\tau_{n'})}\bigg|_{\eta=\eta^*\equiv 0}.$$

The right-hand side consists of the functional integrals

$$N\int \mathcal{D}\psi^*(\mathbf{x},t)\mathcal{D}\psi(\mathbf{x},t)\,\hat{\psi}(\mathbf{x}_1,\tau_1)\cdots\hat{\psi}(\mathbf{x}_n,\tau_n)\hat{\psi}^*(\mathbf{x}_{n'},\tau_{n'})\cdots\hat{\psi}^*(\mathbf{x}_{1'},\tau_{1'})e^{i\mathcal{A}[\psi^*,\psi]}. \quad (1.123)$$

In the sequel, we shall always assume the normalization factor to be chosen in such a way that $Z[0,0]$ is normalized to unity. Then the functional integrals (1.123) are obviously the *correlation functions* of the fields, commonly written in the form

$$\langle\hat{\psi}(\mathbf{x}_1,\tau_1)\cdots\hat{\psi}(\mathbf{x}_n,\tau_n)\hat{\psi}^*(\mathbf{x}_{n'},\tau_{n'})\cdots\hat{\psi}^*(\mathbf{x}_{1'},\tau_{1'})\rangle. \quad (1.124)$$

In contrast to Section 1.2, the bra and ket symbols denote now a thermal average of the classical fields.

The functional integral expression (1.120) for the generating functional offers the advantageous flexibility with respect to changes in the field variables.

Summarizing we have seen that the functional (1.120) defines the most general type of theory involving two-body forces. It contains all information on the physical system in the vacuum as well as in thermodynamic ensembles. The vacuum theory is obtained by setting $T = 0$, $\mu = 0$, and continuing the result back from T to physical times. Conversely, the functional (1.75) in the vacuum can be generalized to ensembles in the straightforward manner by first continuing the times t to imaginary values $-i\tau$ via a Wick rotation in all energy integrals and then going to periodic functions in τ.

There is a complete correspondence between the real-time generating functional (1.75) and the thermodynamic imaginary-time expression (1.120). For this reason it is sufficient to exhibit all techniques only in one version for which we shall choose (1.75). Note, however, that due to the singular nature of the propagators (1.85) in real energy-momentum, the thermodynamic formulation specifies the way how to avoid singularities.

1.2 Relativistic Fields

We shall also study collective phenomena in relativistic fermion systems. For this we shall need fields describing relativistic particles of spin zero, 1/2, and 1. Their properties will now be briefly reviewed.

1.2.1 Lorentz and Poincaré Invariance

For relativistic particles, the relation between the physical laws in two coordinate frames which move with a constant velocity with respect to each other are different from the nonrelativistic case. Suppose a frame moves with velocity v into the $-z$-direction of another fixed frame. Then in the moving frame, the z-momentum of the particle will be increased. The particle appears *boosted* in the z-direction with respect to the original observer. The momenta in x- and y-directions are unaffected. Now, the total four momentum still satisfies the energy momentum relation

$$E(\mathbf{p}) = \sqrt{\mathbf{p}^2 + M^2}. \quad (1.125)$$

Introducing the four-vector notation

$$p^\mu \equiv (p^0, p^i) \quad \text{with} \quad p0 \equiv (\mathbf{p})/c, \tag{1.126}$$

we see that the four-vector satisfies the mass shell condition

$$p^{0^2} - \mathbf{p}^2 = M^2. \tag{1.127}$$

For the particle moving in the z-direction, the combination $p^{0^2} - p^{3^2}$ remains invariant. This implies that there must be a hyperbolic transformation mixing p^0 and p^3, which may be parametrized by a hyperbolic angle ζ called *rapidity*:

$$\begin{aligned} p'^0 &= \cosh\zeta\, p^0 + \sinh\zeta\, p^3, \\ p'^3 &= \sinh\zeta\, p^0 + \cosh\zeta\, p^3. \end{aligned} \tag{1.128}$$

This is called a *pure Lorentz transformation*. We may write it in a 4×4 matrix form as

$$p'^\mu = \begin{pmatrix} \cosh\zeta & 0 & 0 & \sinh\zeta \\ 0 & 1 & 0 & 0 \\ 0 & 0 & 1 & 0 \\ \sinh\zeta & 0 & 0 & \cosh\zeta \end{pmatrix}^\mu_{\ \nu} p^\nu \equiv B_3(\zeta)^\mu_{\ \nu} p^\nu. \tag{1.129}$$

The subscript 3 of B_3 indicates that the particle is boosted into the z-direction. A similar matrix can be written down for x and y-directions. In an arbitrary direction $\hat{\mathbf{p}}$, the matrix elements are

$$B_{\hat{\mathbf{p}}}(\zeta) = \left(\begin{array}{c|c} \cosh\zeta & \hat{p}^i \sinh\zeta \\ \hline \hat{p}^i \sinh\zeta & \delta^{ij} + \hat{p}^i\hat{p}^j(\cosh\zeta - 1) \end{array} \right). \tag{1.130}$$

By combining rotations and boosts one obtains a 6-parameter manifold of matrices

$$\Lambda = B_{\hat{\mathbf{p}}}(\zeta) R_{\hat{\varphi}}(\varphi), \tag{1.131}$$

called *proper Lorentz transformations*. For all these

$$p'^{0^2} - \mathbf{p}'^2 = p^{0^2} - \mathbf{p}^2 = M^2 c^2 \tag{1.132}$$

is an invariant. These matrices form a group, the *proper Lorentz group*. We can easily see that the Lorentz group allows reaching *every* momentum p^μ on the mass shell by applying an appropriate group element to some *fixed* reference momentum p_R^μ. For example, if the particle has a mass M we may choose for p_R^μ the so-called *rest momentum*

$$p_R^\mu = (M, 0, 0, 0), \tag{1.133}$$

and apply the boost in the $\hat{\mathbf{p}}$ direction

$$\Lambda = B_{\hat{\mathbf{p}}}(\zeta), \tag{1.134}$$

with the rapidity given by

$$\cosh \zeta = \frac{p^0}{M}, \quad \sinh \zeta = \frac{|\mathbf{p}|}{M}. \tag{1.135}$$

But we may also choose $\Lambda(p) = B_{\hat{\mathbf{p}}}(\zeta) R_{\hat{\varphi}}(\varphi)$ where $R_{\hat{\varphi}}$ is an arbitrary rotation, since any of these leaves the rest momentum p_R^μ invariant. In fact, the rotations form the largest subgroup of the group of all proper Lorentz transformations, which leaves the rest momentum p_R^μ invariant. It is referred to as the *little group* or *Wigner group* of a massive particle. It has an important physical significance since it serves to specify the intrinsic rotational degrees of freedom of the particle. If the particle is at rest it carries no orbital angular momentum. If its quantum mechanical state remains completely invariant under any member $R_{\hat{\varphi}}$ of the little group, the particle must also have zero intrinsic angular momentum or zero *spin*. Besides this trivial representation, the little group being a rotation group can have representations of any angular momentum $s = \frac{1}{2}, 1, \frac{3}{2}, \dots$. In these cases, the state at rest has $2s + 1$ components which are mixed with each other upon rotations.

The situation is quite different in the case of massless particles. They move with the speed of light and p^μ cannot be brought by a Lorentz transformation from the light cone to a rest frame. There is, however, another standard reference momentum from which one can generate all other momenta on the light cone. It is given by

$$p_R^\mu = (1, 0, 0, 1)p, \tag{1.136}$$

and it remains invariant under a different little group, which is again a three-parameter subgroup of the Lorentz group. This will be discussed later.

It is useful to write the invariant expression (1.132) as a square of a four vector p^μ formed with the metric

$$g_{\mu\nu} = \begin{pmatrix} 1 & & & \\ & -1 & & \\ & & -1 & \\ & & & -1 \end{pmatrix}, \tag{1.137}$$

namely

$$p^2 = g_{\mu\nu} p^\mu p^\nu. \tag{1.138}$$

In general, we define a scalar product between any two vectors as

$$pp' \equiv g_{\mu\nu} p^\mu p'^\nu = p^0 p'^0 - \mathbf{p} \mathbf{p}'. \tag{1.139}$$

A space with this scalar product is called *Minkowski space*. . It is useful to introduce the covariant components of any vector v^μ as

$$v_\mu \equiv g_{\mu\nu} v^\nu. \tag{1.140}$$

Then the scalar product can also be written as

$$pp' = p_\mu p'^\mu. \tag{1.141}$$

With this notation the mass shell condition for a particle before and after a Lorentz transformation reads simply

$$p'^2 = p^2 = M^2c^2. \tag{1.142}$$

Note that, apart from the minus signs in the metric (1.137), the mass shell condition $p^2 = p^{0^2} - p^{1^2} - p^{2^2} - p^{3^2} = M^2c^2$, left invariant by the Lorentz group, is completely analogous to the spherical condition $p^{4^2} + p^{1^2} + p^{2^2} + p^{3^2} = M^2c^2$ which is left invariant by the rotation group in a four-dimensional euclidean space. Both groups are parametrized by six parameters which are associated with linear transformations in the six planes $12, 23, 31; 10, 20, 30$ or $12, 23, 31; 14, 24, 34$, respectively. In the case of the four-dimensional euclidean space these are all rotations which form the group of *special orthogonal matrices* called O(4). The letter S indicates the property *special*. A group is called *special* if all its transformation matrices have a unit determinant. In the case of the proper Lorentz group one uses by analogy the notation SO(1,3). The numbers indicate the fact that in the metric (1.137), one diagonal element is equal to $+1$ and three are equal to -1.

The fact that all group elements are *special* follows from a direct calculation of the determinant of (1.130), (1.131).

How do we have to describe the quantum mechanics of a free relativistic particle in Minkowski space? The energy and momenta $p^0, \mathbf{p}$ must be related to the time and space derivatives of particle waves in the usual way

$$p^0 = \frac{\epsilon}{c} = i\hbar \frac{\partial}{\partial ct} \equiv i\hbar \frac{\partial}{\partial x^0},$$
$$p^i = -i\hbar \frac{\partial}{\partial x^i}. \tag{1.143}$$

They satisfy the canonical commutation rules

$$[p^\mu, p^\nu] = 0,$$
$$[x^\mu, x^\nu] = 0,$$
$$[p^\mu, x^\nu] = -i\hbar g^{\mu\nu}. \tag{1.144}$$

We expect that associated with the pure momentum state $\mathbf{p}$ there will be some wave function

$$f_{\mathbf{p}}(x) = e^{-i(p^0 x^0 - p^i x^i)/\hbar} \equiv e^{-ipx/\hbar}. \tag{1.145}$$

At this point we do not yet know the proper scalar product necessary to extract physical information from such wave functions.

We have stated previously that permissible energy momentum states of a free particle can be realized by considering *one and the same* particle in different coordinate frames connected by the transformation $\Lambda(p)$. Suppose that we change the coordinates of the same space time point as follows:

$$x \rightarrow x' = \Lambda x. \tag{1.146}$$

Under this transformation the scalar product of any two vectors remains invariant:

$$x'y' = xy. \tag{1.147}$$

This holds also for scalar products between momentum and coordinate vectors

$$p'x' = px. \tag{1.148}$$

For the transformation matrix Λ, it implies that

$$(\Lambda p)(\Lambda x) = px. \tag{1.149}$$

If the scalar products are written out explicitly in terms of the metric $g_{\mu\nu}$ this amounts to

$$g_{\mu\nu}\Lambda^\mu{}_\lambda p^\lambda \Lambda^\nu{}_\kappa x^\kappa = g_{\lambda\kappa}p^\lambda x^\kappa, \tag{1.150}$$

for all p, x. The Lorentz matrices Λ satisfy therefore the identity

$$g_{\mu\nu}\Lambda^\mu{}_\lambda \Lambda^\nu{}_\kappa = g_{\lambda\kappa}, \tag{1.151}$$

or, written without indices,

$$\Lambda^T g\Lambda = g. \tag{1.152}$$

If the metric is Euclidean, this would be the definition of orthogonal matrices. In fact, in the notation of scalar products in which the metric is suppressed as in (1.153), there is no difference between the manipulation of orthogonal and Lorentz matrices. In both cases one has

$$(\Lambda p)(\Lambda x) = p\Lambda^{-1}\Lambda x = px. \tag{1.153}$$

When changing the coordinates, the same particle wave in space behaves like

$$f_p(x) = e^{-ip\Lambda^{-1}x'/\hbar}$$
$$= e^{-i(\Lambda p)x'/\hbar} = f_{\Lambda p}(x') = f_{p'}(x'). \tag{1.154}$$

This shows that, in the new coordinates, the same particle appears with different momentum components

$$p' = \Lambda p. \tag{1.155}$$

Consider a wave $\psi(x)$ which is an arbitrary superposition of different momentum states. After a coordinate transformation it will still have the same value at the same space time point. Thus $\psi'(x')$, as seen in the new frame, must be equal to $\psi(x)$ in the old frame

$$\psi'(x') = \psi(x). \tag{1.156}$$

At this place one defines the *substantial change* under the Lorentz transformation Λ as the change at the same values of the coordinates x (which corresponds to a transformed point in space)

$$\psi(x) \xrightarrow{\ \Lambda\ } \psi'_\Lambda(x) = \psi(\Lambda^{-1}x). \tag{1.157}$$

We have marked by a subscript the transformation under which $\psi'(x)$ arises. Clearly, this transformation property is valid only if the particle does not possess any intrinsic orientational degree of freedom, i.e., no spin. A field with this property is called a *scalar field* or, for historical reasons, a *Klein-Gordon field*.

If a particle has spin degrees of freedom, the situation is quite different. Then the wave function has several components to account for the spin orientations. The transformation law must be such that the spin orientation in space remains unchanged at the same space point. This implies that the field components which specify the orientation with respect to the different coordinate axes will have to be transformed by certain matrices. It is well-known how this is done in the case of electromagnetic and gravitational fields, whose vector and tensor transformation properties follow standard rules. In the next sections these will be recalled. Afterwards it will be easy to generalize everything to the case of arbitrary spin.

Before coming to this, however, let us conclude this section by mentioning that there are other space transformations which leave the scalar products $p_\mu x^\mu$ invariant but which are not contained in the group SO(1,3): Most importantly there is the *space inversion*, also called *mirror reflection* or *parity transformation*:

$$P = \begin{pmatrix} 1 & & & \\ & -1 & & \\ & & -1 & \\ & & & -1 \end{pmatrix}, \tag{1.158}$$

which reverses the direction of the spatial vectors, $\mathbf{x} \to -\mathbf{x}$. There is further the *time inversion*

$$T = \begin{pmatrix} -1 & & & \\ & 1 & & \\ & & 1 & \\ & & & 1 \end{pmatrix}, \tag{1.159}$$

which changes the sign of x^0. If P and T are incorporated into the special Lorentz group SO(1,3), one deals with the *full Lorentz group*.

Note that the determinants of both (1.158) and (1.159) are negative, so that the full Lorentz group no longer deserves the letter S in its name. It is then called O(1,3).

1.2.2   Relativistic Free Scalar Fields

From all this it is obvious how the non-relativistic free field action

$$\mathcal{A} = \int dt dx \psi^*(\mathbf{x}, t) \left[i\hbar \partial_t + \hbar^2 \frac{\partial_\mathbf{x}^2}{2M} \right] \psi(\mathbf{x}, t) \tag{1.160}$$

must be modified to describe relativistic n-particle states. In order to accommodate the kinematic features, discussed in the last section, we require the action to be invariant under Lorentz transformations. Depending on the possible internal spin degrees of freedom there are different ways of making the action relativistic. These will now be discussed separately.

Scalar Fields

If the field $\psi(\mathbf{x}, t)$ carries no spin degree of freedom which varies under space rotations, the spatial derivative $\partial_{\mathbf{x}}$ always has to appear squared in the action to guarantee rotational invariance. With the Lorentz symmetry between ∂_0 and $\partial_{\mathbf{x}}$ we are led to a classical action

$$\mathcal{A} = \int dx^0 L = \int dx^0 d^3 x \psi^*(\mathbf{x}, t) \left[c_1 \partial^\mu \partial_\mu + c_2 \right] \psi(\mathbf{x}, t), \tag{1.161}$$

where c_1, c_2 are two arbitrary real constants. It is now easy to see that this action is indeed Lorentz invariant: Under the transformation (1.146), the four-volume element does not change

$$dx^0 d^3 x \equiv d^4 x \to d^4 x' = d^4 x. \tag{1.162}$$

If we therefore take the action in the new frame

$$\mathcal{A} = \int d^4 x' \psi^{*\prime}(x') \left[c_1 \partial^{\prime\mu} \partial'_\mu + c_2 \right] \psi'(x'), \tag{1.163}$$

we can use (1.161) and (1.156) to rewrite

$$\mathcal{A} = \int d^4 x \psi^*(x) \left[c_1 \partial^{\prime\mu} \partial'_\mu + c_2 \right] \psi(x). \tag{1.164}$$

But since

$$\partial'_\mu = \Lambda_\mu{}^\nu \partial_\nu, \quad \partial^{\mu\prime} = \Lambda^\mu{}_\nu \partial^\nu \tag{1.165}$$

with $\Lambda_\mu{}^\nu \equiv g_{\mu\lambda} g^{\nu\kappa} \Lambda^\lambda{}_\kappa$, we see that

$$\partial'^2 = \partial^2, \tag{1.166}$$

and the transformed action becomes

$$\mathcal{A} = \int dx^0 d^3 x \psi^*(\mathbf{x}, t) \left[c_1 \partial^\mu \partial_\mu + c_2 \right] \psi(\mathbf{x}, t), \tag{1.167}$$

which is the same as (1.161).

It is useful to introduce the integrand of the action as the so-called *Lagrangian density*

$$\mathcal{L}(\mathbf{x}, t) = \psi^*(\mathbf{x}, t) \left[c_1 (\partial^{0^2} - \partial_{\mathbf{x}}{}^2) + c_2 \right] \psi(\mathbf{x}, t). \tag{1.168}$$

Then the invariance of the action under Lorentz transformation is a direct consequence of the Lagrangian density being a scalar field, satisfying the transformation law (1.156),

$$\mathcal{L}'(x') = \mathcal{L}(x), \tag{1.169}$$

as implied by (1.163), (1.164), and (1.166).

The free-field equations of motion are derived from (1.161) as follows. We write

$$A = \int dx^0 L = \int dx^0 \int d^3x \psi^*(\mathbf{x}, t) \left[c_1(\partial^{0^2} - \partial_{\mathbf{x}}^2) + c_2 \right] \psi(\mathbf{x}, t), \qquad (1.170)$$

and vary this independently with respect to the fields $\psi(x)$ and $\psi^*(x)$. The independence of these variables is expressed by the functional differentiation rules

$$\frac{\delta\psi(x)}{\delta\psi(x')} = \delta^{(4)}(x - x'), \quad \frac{\delta\psi^*(x)}{\delta\psi^*(x')} = \delta^{(4)}(x - x'),$$

$$\frac{\delta\psi(x)}{\delta\psi^*(x')} = 0, \qquad \frac{\delta\psi^*(x)}{\delta\psi(x')} = 0. \qquad (1.171)$$

Applying these rules to (1.170) we obtain directly

$$\frac{\delta A}{\delta\psi^*(x)} = \int d^4x' \delta^{(4)}(x' - x)(c_1\partial^2 + c_2)\psi(x)$$

$$= (c_1\partial^2 + c_2)\psi(x) = 0. \qquad (1.172)$$

Similarly,

$$\frac{\delta A}{\delta\psi(x)} = \int d^4x' \psi^*(x')(c_1\partial^2 + c_2)\delta(x' - x)$$

$$= \psi^*(x)(c_1 \overleftarrow{\partial^2} + c_2), \qquad (1.173)$$

where the arrow on top of the last derivative indicates that it acts on the field to the left. The second equation is just the complex conjugate of the previous one. Then the functional derivative with respect to $\psi^*(x)$ is simple. In terms of the Lagrangian density, the extremality condition can be expanded in terms of partial derivatives with respect to increasing partial derivatives of all fields in $\mathcal{L}$,

$$\frac{\delta A}{\delta\psi(x)} = \frac{\partial\mathcal{L}(x)}{\partial\psi(x)} - \partial_\mu \frac{\partial\mathcal{L}(x)}{\partial\partial_\mu\psi(x)} + \partial_\mu\partial_\nu \frac{\partial\mathcal{L}(x)}{\partial\partial_\mu\partial_\nu\psi(x)} + \dots, \qquad (1.174)$$

with the same equation for $\psi^*(x)$. This follows directly from the defining relations in (1.171). The field equation for $\psi(x)$ is particularly simple:

$$\frac{\delta A}{\delta\psi^*(x)} = \frac{\partial\mathcal{L}(x)}{\partial\psi^*(x)}. \qquad (1.175)$$

For $\psi^*(x)$, on the other hand, all derivatives written out in (1.174) have to be evaluated.

Both field equations (1.172) and (1.173) are solved by the quantum mechanical plane wave (1.145),

$$f_p(x) = e^{-ipx/\hbar}, \qquad (1.176)$$

if the momentum satisfies the condition

$$-c_1 p^\mu p_\mu + c_2 = 0. \tag{1.177}$$

This has precisely the form of the mass shell relation (1.142) if we choose

$$c_2 \hbar^2 / c_1 = M^2 c^2. \tag{1.178}$$

It is customary to normalize c_1 to

$$c_1 = -\hbar^2. \tag{1.179}$$

The sign is necessary to have stable field fluctuations. The size can be brought to this value by a multiplicative renormalization of the field. Then the mass shell condition fixes the free field action to the standard form

$$\mathcal{A} = \int dx^0 d^3 x \psi^*(\mathbf{x}, t) \left[-\hbar^2 \partial^\mu \partial_\mu - M^2 c^2 \right] \psi(\mathbf{x}, t). \tag{1.180}$$

The appearance of the constants $\hbar$ and c in all future formulas can be avoided if we agree to work with *natural units* l_0, m_0, t_0, E_0 different from the ordinary cgs units. They are chosen to give $\hbar$ and c the value 1. Expressed in terms of the conventional length, time, mass, and energy, these new *natural units* are given by

$$l_0 = \frac{\hbar}{Mc} = \frac{\hbar}{E_0} c, \qquad t_0 = \frac{\hbar}{Mc^2}, \tag{1.181}$$

$$m_0 = M, \quad E_0 = Mc^2. \tag{1.182}$$

If, for example, the particle is a proton with mass m_p, these units are

$$l_0 = 2.103138 \times 10^{-11} \text{cm} \tag{1.183}$$
$$- \text{ Compton wavelength of proton,}$$

$$t_0 = l_0/c = 7.0153141 \times 10^{-22} \text{sec} \tag{1.184}$$
$$= \text{ time taken by light running along Compton wavelength,}$$

$$m_0 = m_p = 1.6726141 \times 10^{-24} \text{g}, \tag{1.185}$$

$$E_0 = 938.2592 \text{ MeV}. \tag{1.186}$$

For any other mass, they can easily be rescaled.

With these natural units we can drop c and $\hbar$ in all formulas and write the action simply as

$$\mathcal{A} = \int d^4 x \mathcal{L}(x) = \int d^4 x \psi^*(x)(-\partial^2 - M^2)\psi(x). \tag{1.187}$$

Actually, since we are dealing with relativistic particles there is no fundamental reason to assume $\psi(x)$ to be a complex field. In the non-relativistic theory this was necessary in order to construct a term linear in the time derivative

$$\int dt \psi^* i \partial_t \psi. \tag{1.188}$$

For a real field $\psi(x)$ this would have been a pure surface term that does not influence the dynamics of the system. For second-order time derivatives as in (1.187) this is no longer necessary.

Thus we shall also study the real scalar field with an action

$$\mathcal{A} = \int d^4x \mathcal{L}(x) = \frac{1}{2} \int d^4x \phi(x)(-\partial^2 - M^2)\phi(x). \tag{1.189}$$

In this case, a prefactor $\frac{1}{2}$ is the normalization convention for the field. We have also here used the letter $\phi(x)$ to denote the real field, as is commonly done.

1.2.3 Electromagnetic Fields

Electromagnetic fields move with light velocity and have no mass term.[3] The fields have two polarization degrees of freedom (right and left polarized) and are described by the usual electromagnetic action. Historically, this was the very first example of a relativistic classical field theory. Thus it could also have served as a guideline for the previous construction of the action of the scalar field $\phi(\mathbf{x})$.

The action may be given in terms of a real auxiliary four-vector potential $A_\mu(x)$ from which the physical electric and magnetic fields can be derived as follows

$$E^i = -(\partial^0 A^i - \partial^i A^0) = -\partial_t A^i - \partial_i A^0, \tag{1.190}$$

$$H^i = -\frac{1}{2}\epsilon_{ijk}(\partial^j A^k - \partial^k A^i) = \frac{1}{2}\epsilon_{ijk}(\partial_j A^k - \partial_k A^j). \tag{1.191}$$

Here ϵ_{ijk} is the completely antisymmetric *Levi-Cività tensor* with $\epsilon_{123} = 1$. It is useful to introduce the so-called *four-curl* of the vector potential

$$F_{\mu\nu} = \partial_\mu A_\nu - \partial_\nu A_\mu. \tag{1.192}$$

Its six components are directly the field strengths

$$E^i = -F^{0i} = F_{0i}, \quad H^i = -F^{jk} = -F_{jk}; \quad ijk = \text{cyclic}. \tag{1.193}$$

For this reason $F_{\mu\nu}$ is also called the field tensor. The electromagnetic action reads

$$\mathcal{A} = \int d^4x \mathcal{L}(x) = \int d^4x \frac{1}{2}(\mathbf{H}^2 - \mathbf{E}^2) = -\frac{1}{4} \int d^4x F_{\mu\nu}^{\ 2}. \tag{1.194}$$

The four-curl $F_{\mu\nu}$ satisfies the so-called *Bianchi identity* for any smooth A_μ [which satisfies the Schwartz integrability condition $(\partial_\lambda \partial_\kappa - \partial_\kappa \partial_\lambda)A_\mu = 0$]:

$$\partial_\mu \tilde{F}^{\mu\nu} = 0, \tag{1.195}$$

---

[3]The best upper limit for the mass of the electromagnetic field M_γ deduced under terrestrial conditions, from the shape of the earth's magnetic field, is $M_\gamma < 4 \cdot 10^{-48}$g corresponding to a Compton wavelength $\lambda_\gamma = \hbar/M_\gamma c > 10^{10}$cm (= larger than the diameter of the sun). Astrophysical considerations ("whisps" in the crab nebula) give $\lambda_\gamma > 10^{16}$cm. If metagalactic magnetic fields are discovered, the Compton wavelength would be larger than $10^{24} - 10^{25}$cm, quite close to the ultimate limit set by the horizon of the universe = c× age of the universe $\sim 10^{28}$cm. See G.V. Chibisov, Sov. Phys. Usp. *19*, 624 (1976).

where

$$\tilde{F}^{\mu\nu} = \epsilon^{\mu\nu\lambda\kappa} F_{\lambda\kappa} \tag{1.196}$$

is the so called *dual field tensor*, with $\epsilon^{\mu\nu\lambda\kappa}$ being the four-dimensional Levi-Cività tensor with $\epsilon_{0123} = 1$.

The equations of motion which extremize the action are

$$\frac{\delta\mathcal{A}}{\delta A^\mu(x)} = -\partial_\mu \frac{\partial\mathcal{L}(x)}{\partial_\mu A_\nu(x)} = \partial_\mu F^{\mu\nu}(x) = 0. \tag{1.197}$$

Separating the equations (1.196) and (1.197) into space and time components they are seen to coincide with Maxwell's equation in empty space

$$\partial_\mu F^{\mu\nu} = 0: \quad \nabla \cdot \mathbf{E} = 0, \quad \nabla \times \mathbf{B} - \partial_t \mathbf{E} = 0, \tag{1.198}$$

$$\partial_\mu \tilde{F}^{\mu\nu} = 0: \quad \nabla \cdot \mathbf{B} = 0, \quad \nabla \times \mathbf{E} + \partial_t \mathbf{B} = 0. \tag{1.199}$$

The field tensor is invariant under local gauge transformations

$$A_\mu(x) \longrightarrow A_\mu(x) + \partial_\mu \Lambda(x), \tag{1.200}$$

where $\Lambda(x)$ is any smooth field which satisfies the integrability condition $(\partial_\mu\partial_\nu - \partial_\nu\partial_\mu)\Lambda = 0$. In terms of the vector field A^μ, the action reads explicitly

$$\begin{aligned}
\mathcal{A} = \int d^4x \mathcal{L}(x) &= -\frac{1}{2} \int d^4x [\partial^\mu A^\nu(x)\partial_\mu A_\nu(x) - \partial^\nu A_\nu(x)\partial^\mu A_\mu(x)] \\
&= \frac{1}{2} \int dx A_\mu(x)(g^{\mu\nu}\partial^2 - \partial^\mu\partial^\nu)A_\nu(x).
\end{aligned} \tag{1.201}$$

The latter form is very similar to the scalar action (1.160). The first piece is the same as (1.161) for each of the spatial components A^1, A^2, A^3. The time component A^0, however, appears with an opposite sign. A field with this property is called a *ghost field*. When trying to quantize such a field, the associated particle states turn out to have a negative norm. In order for the theory to be physically consistent it will be necessary to make sure that such states can never appear in any scattering process. The second piece in the action $\partial^\nu A_\nu\partial^\mu A_\mu$ is novel with respect to the scalar case. It exists here as an additional Lorentz invariant since A_μ is a vector field under Lorentz transformation.

In order to see the Lorentz transformation properties, let us remember that in electrodynamics the Lorentz forces on a moving particle carrying a charge and a classical magnetic pole are obtained from the field transformation

$$E_\parallel{}' = E_\parallel, \quad E_\perp{}' = \gamma(E_\perp + \mathbf{v} \times \mathbf{B}), \tag{1.202}$$

$$B_\parallel{}' = B_\parallel, \quad B_\perp{}' = \gamma(B_\perp - \mathbf{v} \times \mathbf{E}), \tag{1.203}$$

with $\mathbf{v}$ being the velocity of the particle and $\gamma \equiv \sqrt{1 - \mathbf{v}^2/c^2}$. Here $\mathbf{E}$ and $\mathbf{B}$ are the fields in the laboratory, whereas $\mathbf{E}'$ and $\mathbf{B}'$ are the corresponding fields in the frame of the moving particle. They exert electric and magnetic forces $e\mathbf{E}' + g\mathbf{B}'$. The subscripts $\parallel$ and $\perp$ denote the components parallel and orthogonal to $\mathbf{v}$.

From this experimental fact we can derive the transformation law of the vector field A_μ under Lorentz transformations. The frame in which the moving particle is at rest is related to the laboratory frame by

$$x' = B_{\hat{\mathbf{v}}}(\zeta)x, \tag{1.204}$$

where $B_{\hat{\mathbf{v}}}(\zeta)$ is a boost in the $\mathbf{v}$-direction with the rapidity

$$\cosh\zeta = \gamma, \quad \sinh\zeta = \gamma\frac{v}{c}, \quad \tanh\zeta = \frac{v}{c}. \tag{1.205}$$

The transformation law (1.202) is equivalent to

$$A'^\mu(x') = B_{\hat{\mathbf{v}}}(\zeta)^\mu{}_\nu A^\nu(x). \tag{1.206}$$

An analogous transformation law holds for rotations so that we can write, in general,

$$A'^\mu(x') = \Lambda^\mu{}_\nu A^\nu(x). \tag{1.207}$$

This transformation law differs from that of a scalar field (1.156) in the way envisaged above for particles with non-zero intrinsic angular momentum. The field has several components. It points in the same spatial direction before and after the coordination change. This is ensured by its components changing in the same way as the coordination of the point x_μ. Notice that as a consequence, $\partial^\mu A_\mu(x)$ is a scalar field in the sense defined in (1.156). Indeed

$$\partial'^\mu A'_\mu(x') = (\Lambda^\mu{}_\nu\partial^\nu)\Lambda_\mu{}^\lambda A_\lambda(x) = \partial^\nu A_\nu(x). \tag{1.208}$$

For this reason the second term in the action (1.218) is Lorentz invariant, just as the mass term in (1.189). The invariance of the first term is shown similarly

$$\begin{aligned}
A'^\nu(x')\partial'^2 A'_\nu(x') &= \Lambda^\nu{}_\lambda A^\nu(x)\partial'^2\Lambda_\nu{}^\kappa A_\kappa(x') \\
&= A^\nu(x)\partial'^2 A_\nu(x) = A^\nu(x)\partial^2 A_\nu(x).
\end{aligned} \tag{1.209}$$

Hence the action (1.218) does not change under Lorentz transformations, as it should.

Just as the scalar action, also the electromagnetic action (1.194) is invariant under a Lorentz group extended by translations (the so-called *Poincaré group*):

$$A'^\mu(x') = A^\mu(x) \tag{1.210}$$

where

$$x'^\mu = \Lambda^\mu{}_\nu x^\nu + a^\mu. \tag{1.211}$$

Similarly, we have under parity

$$A^\mu \xrightarrow{\;P\;} A'^\mu_P(x) = \tilde{A}^\mu(\tilde{x}), \tag{1.212}$$

and under time reversal,

$$A \xrightarrow{\quad T \quad} A'^{\mu}_T(x) = \tilde{A}^{\mu}(x_T) \qquad (1.213)$$

where the tilde inverts the spacial components

$$\tilde{A}^{\mu} = (A^0, -A^i). \qquad (1.214)$$

In principle, there would have been the possibility of a parity transformation

$$A^{\mu} \xrightarrow{\quad P \quad} A'^{\mu}_P(x) = \eta_P \tilde{A}^{\mu}(\tilde{x}), \qquad (1.215)$$

with $\eta_P = \pm 1$, and in the case $\eta_P = -1$ the field A^{μ} would have been called an *axial vector field*. The electromagnetic gauge field A^{μ}, however, is definitely a vector field. This follows from the vector nature of the electric and the axial vector nature of the magnetic field, which are observed in the laboratory. Similarly, the phase under time reversal of A^{μ}, which in principle could have been

$$A^{\mu} \xrightarrow{\quad T \quad} A'^{\mu}_T(x) = \eta_T \tilde{A}^{\mu}(x_T) \qquad (1.216)$$

with $\eta_T = \pm 1$, is given by (1.213). This is due to time reversal, under which all currents change their direction. This reverses the direction of the **B**-field but has no influence on the **E**-field.

It is also possible to perform the operation of charge conjugation by exchanging the sign of all charges without changing their direction of flow. Then **E** and **B** change directions. Hence

$$A^{\mu} \xrightarrow{\quad C \quad} A'^{\mu}_C(x) = -A^{\mu}(x). \qquad (1.217)$$

In general, the vector field could have been transformed as

$$A^{\mu} \xrightarrow{\quad C \quad} A'^{\mu}_C(x) = \eta_C A^{\mu}(x) \qquad (1.218)$$

with $\eta_C = \pm 1$. The phase factor $\eta_C = -1$ expresses the experimental fact that the electromagnetic field is odd under charge conjugation.

1.2.4 Relativistic Free Fermi Fields

For Fermi fields, the situation is technically more involved. Experimentally, fermions always have an even number of spin degrees of freedom. In order to describe these we give the field ψ a spin index α running through $(2s+1)$ components. Under rotations, these spin components are mixed with each other as observed experimentally in the *Stern-Gerlach experiment*. Lorentz transformations lead to certain well defined mixtures of different spin components.

The question arises whether we can construct a Lorentz invariant action involving $(2s + 1)$ spinor field components. To see the basic construction principle we use the known transformation law (1.207) for the 4-vector field A^μ as a guide. For an arbitrary *spinor field* we postulate the transformation law

$$\psi(x)_\alpha \xrightarrow{\Lambda} \psi'_\alpha(x') = D_\alpha{}^\beta(\Lambda)\psi_\beta(x), \qquad (1.219)$$

with an appropriate $(2s+1) \times (2s+1)$ spinor transformation matrix $D_\alpha{}^\beta(\Lambda)$ which we have to construct. This can be done by purely mathematical arguments. The construction is the subject of the so-called *group representation theory*. First of all, we perform two successive Lorentz transformations,

$$x'' = \Lambda x = \Lambda_2 x' = \Lambda_2\Lambda_1 x. \qquad (1.220)$$

Since the Lorentz transformations Λ_1, Λ_2 are elements of a group, the product $\Lambda \equiv \Lambda_2\Lambda_1$ is again a Lorentz transformation. Under the individual factors Λ_2 and Λ_1, the field transforms as

$$\Psi(x) \xrightarrow{\Lambda_1} \Psi'(x') = D(\Lambda_1)\Psi(x), \qquad (1.221)$$

$$\Psi'(x) \xrightarrow{\Lambda_2} \Psi''(x'') = D(\Lambda_2)\Psi'(x'), \qquad (1.222)$$

so that under $\Lambda = \Lambda_2\Lambda_1$,

$$\Psi(x) \xrightarrow{\Lambda_2\Lambda_1} \Psi''(x'') = D(\Lambda_2)D(\Lambda_1)\Psi(x). \qquad (1.223)$$

But for Λ itself, the transformation matrix is $D(\Lambda)$ and

$$\Psi''(x'') = D(\Lambda_2\Lambda_1)\Psi(x). \qquad (1.224)$$

Comparison of this with (1.223) shows that the matrices $D(\Lambda)$ which mix the spinor field components under the Lorentz group must follow a group multiplication law which has to be compatible with that of the group itself. The mapping

$$\Lambda \longrightarrow D(\Lambda) \qquad (1.225)$$

is a homomorphism and the $D(\Lambda)$'s form a *matrix representation* of the group.

Notice that the transformation law (1.207) for A_μ follows the same rule, with

$$D(\Lambda) \equiv \Lambda \qquad (1.226)$$

being the defining 4×4 representation of the Lorentz group.

The group laws for Λ and $D(\Lambda)$ are sufficiently stringent to allow only a countable set of fundamental[4] finite dimensional transformation laws $D(\Lambda)$. They are characterized by two quantum numbers, s_1 and s_2, with either one taking the possible half-integer or integer values $0, \frac{1}{2}, 1, \frac{3}{2}, \ldots$.

---

[4]Mathematically, "fundamental" means that the representation is irreducible. Any arbitrary representation is equivalent to a direct sum of irreducible ones.

A representation $D^{(s_1,s_2)}(\Lambda)$ will turn out to harbor particles of spin $|s_1 - s_2|$ to $s_1 + s_2$. Hence, particles with a single fixed spin s can only follow the $D^{(s,0)}(\Lambda)$ or $D^{(0,s)}(\Lambda)$ transformation laws. For spin $1/2$, the relativistic free-field which is invariant under parity has four components and is called the *Dirac field*. It is described by the action

$$\mathcal{A} = \int d^4x \mathcal{L}(x) = \int d^4x \bar{\psi}(x)\, (i\gamma^\mu \partial_\mu - M)\, \psi(x), \qquad (1.227)$$

where M is the mass of the spin-$1/2$ -particles described by $\psi(x)$. The quantities γ^μ are the *Dirac matrices*. , defined by defined by

$$\gamma^\mu = \begin{pmatrix} 0 & \sigma^\mu \\ \tilde{\sigma}^\mu & 0 \end{pmatrix}, \qquad (1.228)$$

where σ^μ is a four-vector formed from the 2×2-dimensional Pauli matrices as follows:

$$\sigma^\mu \equiv (\mathbf{1}, \sigma^i), \qquad (1.229)$$

and

$$\tilde{\sigma}^\mu \equiv (\mathbf{1}, -\sigma^i). \qquad (1.230)$$

The symbol $\bar{\psi}(x)$ is short for

$$\bar{\psi} \equiv \psi^\dagger \gamma^0. \qquad (1.231)$$

As a historical note we mention that Dirac did not find his equation by invoking group-theoretic arguments. Instead, he was searching for an alternative solution to the relativistic time-independent Schrödinger equation of an electron

$$\hat{H}\psi(\mathbf{x}) = \sqrt{\hat{\mathbf{p}}^2 + M^2}\,\psi(\mathbf{x}) = E\psi(\mathbf{x}). \qquad (1.232)$$

He observed that a square root linear in the momentum operator exists if the equation is considered as a matrix equation acting on several components of $\psi(\mathbf{x},t)$. These would indeed be necessary to represent the spin degrees of freedom of the electron. So he made the ansatz

$$\hat{H}_D\psi(\mathbf{x}) = (-i\alpha_i \hat{p}_i + \beta M)\psi(\mathbf{x}) = E\psi(\mathbf{x}), \qquad (1.233)$$

with unknown matrices α_i, β. Being the square root of $\hat{H}$, the operator $\hat{H}_D$ has to fulfill the equation $\hat{H}_D^2 = \hat{\mathbf{p}}^2 + M^2$. This implies that the matrices satisfy the algebraic relations:

$$\{\alpha_i, \alpha_j\} = \delta_{ij}, \qquad \{\alpha_i, \beta\} = 0, \qquad \beta^2 = 1. \qquad (1.234)$$

By multiplying Eq. (1.235) with β and going over to a time-dependent equation by replacing E by $i\partial_{x^0}$, he obtained the *Dirac equation*

$$(i\gamma^\mu \hat{p}_\mu - M)\psi(x) = 0, \qquad (1.235)$$

with the matrices

$$\gamma^0 \equiv \beta, \quad \gamma^i \equiv \beta\alpha_i. \tag{1.236}$$

These satisfy the anticommutation rules

$$\{\gamma^\mu, \gamma^\nu\} = 2g^{\mu\nu}, \tag{1.237}$$

which are indeed solved by the Dirac matrices (1.228).

It has become customary to abbreviate the contraction of γ^μ with any vector v^μ by

$$\slashed{v} \equiv \gamma^\mu v_\mu, \tag{1.238}$$

and write the Dirac equation as

$$(\slashed{p} - M)\psi(x) = 0, \tag{1.239}$$

or

$$(i\slashed{\partial} - M)\psi(x) = 0. \tag{1.240}$$

1.2.5 Perturbation Theory of Relativistic Fields

If interactions are present, the Lagrangian consists of a sum

$$\mathcal{L}\left(\psi, \bar{\psi}, \varphi\right) = \mathcal{L}_0 + \mathcal{L}_{\text{int}}. \tag{1.241}$$

As in the case of nonrelativistic fields, all time ordered Green's functions can be obtained from the derivatives with respect to the external sources of the generating functional

$$Z\left[\eta, \bar{\eta}, j\right] = \text{const} \times \langle 0|Te^{i\int dx\left(\mathcal{L}_{\text{int}} + \bar{\eta}\psi + \bar{\psi}\eta + j\varphi\right)}|0\rangle. \tag{1.242}$$

The fields in the exponent follow free equations of motion and $|0\rangle$ is the free-field vacuum. The constant is conventionally chosen to make $Z[0, 0, 0] = 1$, i. e.

$$\text{const} = \left[\langle 0|Te^{i\int dx\mathcal{L}_{\text{int}}\left(\psi, \bar{\psi}, \varphi\right)}|0\rangle\right]^{-1}. \tag{1.243}$$

This normalization may always be enforced at the very end of any calculation such that $Z\left[\eta, \bar{\eta}, j\right]$ is only interesting as far as its functional dependence is concerned, modulo the irrelevant constant in front.

It is then straightforward to show that $Z\left[\eta, \bar{\eta}, j\right]$ can alternatively be computed via the Feynman path integral formula

$$Z\left[\eta, \bar{\eta}, j\right] = \text{const} \times \int \mathcal{D}\psi \mathcal{D}\bar{\psi} \mathcal{D}\varphi \, e^{i\int dx\left[\mathcal{L}_0\left(\psi, \bar{\psi}, \varphi\right) + \mathcal{L}_{\text{int}} + \bar{\eta}\psi + \bar{\psi}\eta + j\varphi\right]}. \tag{1.244}$$

Here the fields are no more operators but classical functions (with the mental reservation that classical Fermi fields are anticommuting objects). Notice that contrary to the operator formula (1.242) the *full* action appears in the exponent.

For simplicity, we demonstrate the equivalence only for one real scalar field $\varphi(x)$. The extension to other fields is immediate [1, 3]. First note that it is sufficient to give the proof for the free field case, i. e.,

$$
\begin{aligned}
Z_0[j] &= \langle 0|Te^{i\int dx j(x)\varphi(x)}|0\rangle \\
&= \text{const} \times \int \mathcal{D}\varphi e^{i\int dx\left[\frac{1}{2}\varphi(x)\left(-\Box_x - \mu^2\right)\varphi(x) + j(x)\varphi(x)\right]}.
\end{aligned}
\tag{1.245}
$$

Indeed, if it holds there, a simple multiplication on both sides of (1.245) by the differential operator

$$
e^{i\int dx \mathcal{L}_{\text{int}}\left(\frac{1}{i}\frac{\delta}{\delta j(x)}\right)}
\tag{1.246}
$$

would extend it to the interacting functionals (1.242) or (1.244). Equation (1.245) follows directly from Wick's theorem according to which any time ordered product of a free field can be expanded into a sum of normal products with all possible time ordered contractions. This statement can be summarized in an operator form valid for any functional $F[\varphi]$ of a free field $\varphi(x)$:

$$
TF[\varphi] = e^{\frac{1}{2}\int dx dy \frac{\delta}{\delta\varphi(x)}D(x-y)\frac{\delta}{\delta\varphi(y)}}\hat{N}(F[\varphi]),
\tag{1.247}
$$

where $D(x-y)$ is the free-field propagator

$$
D(x-y) = \frac{i}{-\Box_x - \mu^2 + i\epsilon}\delta(x-y) = \int \frac{d^4q}{(2\pi)^4}e^{-iq(x-y)}\frac{i}{q^2 - \mu^2 + i\epsilon}.
\tag{1.248}
$$

Applying this to (1.247) gives

$$
\begin{aligned}
Z_0 &= e^{\frac{1}{2}\int dx dy \frac{\delta}{\delta\hat{\varphi}(x)}D(x-y)\frac{\delta}{\delta\hat{\varphi}(y)}} \langle 0|\hat{N}(e^{i\int dx j(x)\hat{\varphi}(x)})|0\rangle \\
&= e^{-\frac{1}{2}\int dx dy j(x)D(x-y)j(y)} \langle 0|\hat{N}(e^{i\int dx j(x)\hat{\varphi}(x)})|0\rangle \\
&= e^{-\frac{1}{2}\int dx dy j(x)D(x-y)j(y)}.
\end{aligned}
\tag{1.249}
$$

The last part of the equation follows from the vanishing of all normal products of $\varphi(x)$ between vacuum states.

Exactly the same result is obtained by performing the functional integral in (1.245) and using the functional integral formula (1.79). The matrix A is equal to $A(x,y) = (-\Box_x - \mu^2)\,\delta(x-y)$, and its inverse yields the propagator $D(x-y)$:

$$
A^{-1}(x,y) = \frac{1}{-\Box_x - \mu^2 + i\epsilon}\delta(x-y) = -iD(x-y)
\tag{1.250}
$$

thus reproducing once more (1.249).

For the generating functional of a free Dirac field theory

$$
\begin{aligned}
Z_0[\eta,\bar{\eta}] &= \langle 0|Te^{i\int(\bar{\eta}\hat{\psi}+\hat{\bar{\psi}}\eta)dx}|0\rangle \\
&= \text{const} \times \int \mathcal{D}\psi \mathcal{D}\bar{\psi} e^{i\int dx[\mathcal{L}_0(\psi,\bar{\psi})+\bar{\eta}\psi+\bar{\psi}\eta]},
\end{aligned}
\tag{1.251}
$$

with the free-field Lagrangian

$$\mathcal{L}_0(x) = \bar{\psi}(x)\left(i\gamma^\mu\partial_\mu - M\right)\psi(x),$$ (1.252)

we obtain, similarly,

$$
\begin{aligned}
Z_0[\bar{\eta}, \eta] &= e^{\frac{1}{2}\int dxdy\frac{\delta}{\delta\psi(x)}G_0(x-y)\frac{\delta}{\delta\bar{\psi}(y)}}\langle 0|\hat{N}(e^{i\int dx(\bar{\eta}\hat{\psi}+\hat{\bar{\psi}}\eta)})|0\rangle \\
&= e^{-\frac{1}{2}\int dxdy\bar{\eta}(x)G_0(x-y)\eta(y)}\langle 0|\hat{N}(e^{i\int dx(\bar{\eta}\hat{\psi}+\hat{\bar{\psi}}\eta)})|0\rangle \\
&= e^{-\frac{1}{2}\int dxdy\bar{\eta}(x)G_0(x-y)\eta(y)}.
\end{aligned}
$$ (1.253)

Now,

$$A(x, y) = (i\gamma^\mu\partial_\mu - M)\,\delta(x - y),$$ (1.254)

and its inverse yields the fermion propagator $G_0(x - y)$:

$$A^{-1}(x, y) = \frac{1}{i\gamma^\mu\partial_\mu - M + i\epsilon}\delta(x - y) = -iG_0(x - y).$$ (1.255)

Note that it is Wick's expansion which supplies the free part of the Lagrangian when going from the operator form (1.247) to the functional version (1.244).

Notes and References

[1] For the introduction of collective bilocal fields in particle physics and applications see:
H. Kleinert, *On the Hadronization of Quark Theories*, Erice Lectures 1976, publ. in *Understanding the Fundamental Constituents of Matter*, Plenum Press, 1978, (ed. by A. Zichichi).
See also:
H. Kleinert, Phys. Letters B **62**, 429 (1976), B **59**, 163 (1975).

[2] H. Kleinert, *Particles and Quantum Fields*, World Scientific, Singapore, 2016 (klnrt.de/b6).

[3] H.E. Stanley, *Phase Transitions and Critical Phenomena*, Clarendon Press, Oxford, 1971; F.J. Wegner, *Phase Transitions and Critical Phenomena*, ed. by C. Domb and M. S. Green, Academic Press, 1976, p. 7; E. Brézin, J. C. Le Guillou, and J. Zinn-Justin, ibid., p. 125, see also L. Kadanoff, Rev. Mod. Phys. **49**, 267 (1977).

[4] R.P. Feynman, Rev. Mod. Phys. **20**, 367 (1948);
R.P. Feynman and A.R. Hibbs, *Path Integrals and Quantum Mechanics*, McGraw-Hill, New York (1968).

[5] H. Kleinert, *Path Integrals in Quantum Mechanics, Statistics, Polymer Physics, and Financial Markets*, 5th ed., World Scientific, Singapore, 2006, pp. 1-1547 (`klnrt.de/b5`).

[6] J. Rzewuski, *Quantum Field Theory II*, Hefner, New York (1968).

[7] S. Coleman, Erice Lectures 1974, in *Laws of Hadronic Matter*, ed. by A. Zichichi, p. 172.

[8] See for example:
A.A. Abrikosov, L.P. Gorkov, and I.E. Dzyaloshinski, *Methods of Quantum Field Theory in Statistical Physics*, Dover, New York (1975);
L.P. Kadanoff, G. Baym, *Quantum Statistical Mechanics*, Benjamin, New York (1962);
A.L. Fetter and J.D. Walecka, *Quantum Theory of Many-Particle Systems*, McGraw-Hill, New York (1971).

[9] The first authors to employ such identities were
P.T. Mathews and A. Salam, Nuovo Cimento **12**, 563 (1954); **2**, 120 (1955).

[10] R.L. Stratonovich, Sov. Phys. Dokl. **2**, 416 (1958);
J. Hubbard, Phys. Rev. Letters **3**, 77 (1959);
B. Muehlschlegel, J. Math. Phys. **3**, 522 (1962);
J. Langer, Phys. Rev. A **134**, 553 (1964);
T.M. Rice, Phys. Rev. A **140**, 1889 (1965); J. Math. Phys. **8**, 1581 (1967);
A.V. Svidzinskij, Teor. Mat. Fiz. **9**, 273 (1971);
D. Sherrington, J. Phys. C **4** 401 (1971).

Our observation of nature must be diligent,
our reflection profound, and our experiments exact.
DENIS DIDEROT (1713–1784)

2

Plasma Oscillations

In this chapter we develop a collective quantum field theory for a gas of many electrons which interact only via long-range Coulomb forces. The Coulomb forces give rise to collective modes called plasmons.

2.1 General Formalism

The simplest application of the functional method transforms the grand-canonicel partition function (1.120) from the defining formulation in terms of a fundamental field to a re-formulation in terms of a collective quantum field. The new formulation describes the phenomeana directly by means of its fundamental excitations called plasmons. For this, we make use of the Hubbard-Stratonovich transformation in the form (1.79) and observe that a two-body interaction (1.45) in the generating functional can be created by a fluctuating auxiliary field $\varphi(x)$ as follows:

$$\exp\left[-\frac{i}{2}\int dxdx'\psi^*(x)\psi^*(x')\psi(x)\psi(x')V(x,x')\right] \tag{2.1}$$
$$= \text{const} \times \int \mathcal{D}\varphi \left\{\frac{i}{2}\int dxdx'\left[\varphi(x)V^{-1}(x,x')\varphi(x') - 2\varphi(x)\psi^*(x)\psi(x)\delta(x-x')\right]\right\}.$$

To abbreviate the notation, we have used a four-vector notation with

$$x \equiv (\mathbf{x},t), \quad dx \equiv d^3xdt, \quad \delta(x) \equiv \delta^3(\mathbf{x})\delta(t).$$

The symbol $V^{-1}(x,x')$ denotes the functional inverse of the matrix $V(x,x')$, which is the solution of the equation

$$\int dx'V^{-1}(x,x')V(x',x'') = \delta(x-x''). \tag{2.2}$$

The constant prefactor in (2.1) is $[\det V]^{-1/2}$. Absorbing this in the always omitted normalization factor N of the functional integral, the grand-canonical partition function $\Omega = Z$ becomes

$$Z[\eta^*,\eta] = \int \mathcal{D}\psi^*\mathcal{D}\psi\mathcal{D}\varphi \exp\left[i\mathcal{A} + i\int dx\left(\eta^*(x)\psi(x) + \psi^*(x)\eta(x)\right)\right], \tag{2.3}$$

41

where the new action is

$$\mathcal{A}[\psi^*, \psi, \varphi] = \int dx dx' \Big\{ \psi^*(x) \left[i \partial_t - \xi(-i\nabla) - \varphi(x) \right] \delta(x - x')\psi(x') \tag{2.4}$$
$$+ \frac{1}{2}\varphi(x)V^{-1}(x, x')\varphi(x') \Big\}.$$

Note that the effect of using formula (1.79) in the generating functional amounts to the addition of the complete square involving the field φ in the exponent:

$$\frac{1}{2}\int dx dx' \Big[\varphi(x) - \int dy V(x, y)\psi^*(y)\psi(y) \Big] V^{-1}(x, x') \Big[\varphi(x') - \int dy' V(x', y')\psi^*(y')\psi(y') \Big],$$
$$\tag{2.5}$$

followed by a functional integration over $\varphi(x)$. The addition of (2.5) to the action (2.12) can be generated from a source term

$$\mathcal{A}_I = \int \mathcal{D}I(x) \, e^{-\frac{1}{2}\int dx I(x)V(x.x')I(x')+I(x)[\varphi(x)-\int dy \, V(x,y) \, \psi^*(y)\psi(y)]}. \tag{2.6}$$

The generating functional Z remains unchanged by the two successive manipulations as follows from the observation that the integral $\mathcal{D}\varphi$ produces the irrelevant constant $[\det V]^{1/2}$, which is precisely cancelled by the functional over $I(x)$. This procedure of going from (1.45) to (2.4) is probably simpler mnemonically than formula (1.79).

The physical significance of the new field $\varphi(x)$ is easy to understand: $\varphi(x)$ is directly related to the particle density. At the classical level this is seen immediately by extremizing the action (2.4) with respect to variations $\delta\varphi(x)$, which yield:

$$\frac{\delta\mathcal{A}}{\delta\varphi(x)} = \varphi(x) - \int dy \, V(x, y) \, \psi^*(y)\psi(y) = 0. \tag{2.7}$$

Quantum mechanically, there are fluctuations around the field configuration $\varphi(x)$ that is determined by Eq. (2.7). These make the field $\varphi(x)$ different from the composite operator $\mathcal{O}(x) \equiv \int dy V(x, y)\psi^*(y)\psi(y)$. But due to the Gaussian nature of the functional integral over $\varphi(x)$, the fluctuations are rather trivial. Thus we can easily see that the propagators of the two fields $\varphi(x)$ and $\mathcal{O}(x)$ differ only by the direct interaction, i.e.,

$$\langle T(\varphi(x)\varphi(x'))\rangle = V(x - x') + \langle T(\mathcal{O}(x)\mathcal{O}(x'))\rangle.$$

Note that if a potential $V(x, y)$ is dominantly caused by a single fundamental-particle exchange, the field $\varphi(x)$ coincides with the field of this particle: If, for example, $V(x, y)$ represents the Coulomb interaction

$$V(x, x') = \frac{e^2}{|\mathbf{x} - \mathbf{x}'|}\delta(t - t'), \tag{2.8}$$

then Eq. (2.7) amounts to

$$\varphi(\mathbf{x}, t) = -\frac{4\pi e^2}{\nabla^2}\psi^*(\mathbf{x}, t)\psi(\mathbf{x}, t). \tag{2.9}$$

This reveals that the auxiliary field $\varphi(\mathbf{x}, t)$ is the electric potential of the system.

If the particles $\psi(x)$ have spin indices, the potential will, in this example, be spin conserving at every vertex, and Eq. (2.7) must be read as spin-contracted:

$$\frac{\delta \mathcal{A}}{\partial \varphi(x)} = \varphi(x) - \mathcal{O}(x) \equiv \varphi(x) - \int d^4 y V(x, y) \psi^{*\alpha}(y) \psi_\alpha(y) = 0. \qquad (2.10)$$

This restriction is just for convenience and can easily be lifted later. Nothing in our procedure depends on this particular form of $V(x, y)$ and $\mathcal{O}$. In fact, V could arise from the exchange of one or many different fundamental particles and their multiparticle configurations (for example, $\pi, \pi\pi, \sigma, \varphi$, etc. in nuclei [1]) so that the spin dependence is the rule rather than the exception.

The important point is now that the auxiliary field $\varphi(x)$ can be made the *only* field of the theory by integrating out ψ^*, ψ in Eq. (2.3), using formula (1.80). Thus one obtains

$$Z[\eta^*, \eta] \equiv \Omega[\eta^*, \eta] = N e^{i\mathcal{A}}, \qquad (2.11)$$

where the new action is

$$\mathcal{A}[\varphi] = \pm \text{Tr} \log\left(iG_\varphi^{-1}\right) + \frac{1}{2} \int dx dx' \eta^*(x) G_\varphi(x, x') \eta(x'), \qquad (2.12)$$

with $G_\varphi(x, x')$ being the Green function of the fundamental particles in an external classical field $\varphi(x)$:

$$[i\partial_t - \chi(-i\boldsymbol{\nabla}) - \varphi(x)] G_\varphi(x, x') = i\delta(x - x'). \qquad (2.13)$$

The field $\varphi(x)$ is called a *plasmon* field. The new plasmon action can easily be interpreted graphically. For this, one expands $G_\varphi(x, x')$ in powers of φ:

$$G_\varphi(x, x') = G_0(x - x') - i \int dx_1 G_0(x - x_1) \varphi(x_1 - x') + \dots \qquad (2.14)$$

Hence the couplings to the external currents η^*, η in (2.12) amount to radiating one, two, etc. φ fields from every external line of fundamental particles (see Fig. 2.1). An expansion of the expression $\text{Tr} \log(iG_\varphi^{-1})$ in powers of φ gives

$$\begin{aligned}
\pm i \text{Tr} \log(iG_\varphi^{-1}) &= \pm i \text{Tr} \log(iG_0^{-1}) \pm i \text{Tr} \log(1 + iG_0\varphi) \\
&= \pm i \text{Tr} \log(iG_0^{-1}) \mp i \text{Tr} \sum_{n=1}^{\infty} (-iG_0\varphi)^n \frac{1}{n}. \qquad (2.15)
\end{aligned}$$

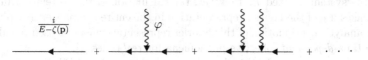

FIGURE 2.1 This diagram displays the piece of the collective action (2.12). The original fundamental particle (fat line) can enter and leave the diagrams only via external currents. It emits an arbitrary number of plasmonson its way (wiggly lines).

The nth term corresponds to a loop of the original fundamental particle emitting $n\varphi$ lines (see Fig. 2.2).

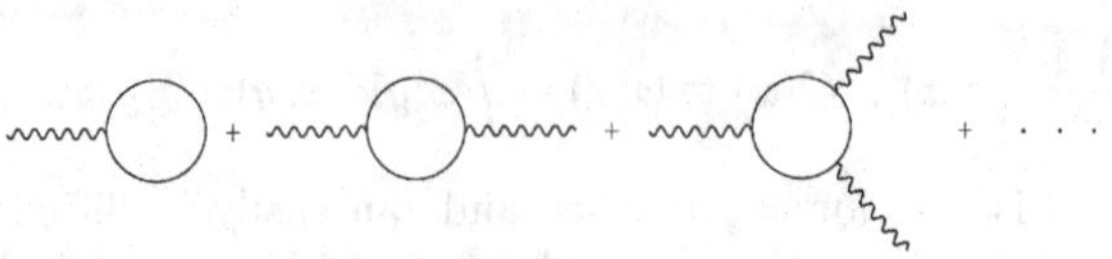

FIGURE 2.2 The non-polynomial self-interaction terms of plasmons arising from the Tr log in (2.12) are equal to the single loop diagrams emitting n plasmons.

Let us now use the action (2.12) to construct a quantum field theory of plasmons. For this we may include the quadratic term

$$\pm i \text{Tr}(G_0\varphi)^2 \frac{1}{2} \tag{2.16}$$

into the free part of φ in (2.12) and treat the remainder perturbatively. The free propagator of the plasmon becomes

$$\{0|T\varphi(x)\varphi(x')|0\} \equiv (2s+1)G_0(x',x). \tag{2.17}$$

This corresponds to an inclusion of all ring graphs into the V-propagator (see Fig. 2.3).

FIGURE 2.3 Free plasmon propagator containing an infinite sequence of single loop corrections ("bubblewise summation")

It is worth pointing out that the propagator in momentum space $G^{\text{pl}}(k)$ contains actually two important physical informations. From the derivation at fixed temperature it appears in the transformed action (2.12) as a function of discrete Euclidean frequencies $\nu_n = 2\pi nT$ only. In this way it serves for the time-independent description of the system at fixed T. However, the calculation of the correlation function (2.17) makes use of the Fourier representation in the entire complex energy plane. A suitable analytic continuation of this Fourier representation can be used to calculate also the *time-dependent* collective phenomena for *real* times [2].

With the propagator (2.17) and the interactions given by (2.15), the original theory of fundamental fields ψ^*, ψ has been transformed into a theory of φ-fields whose bare propagator accounts for the original potential which has absorbed ringwise an infinite sequence of fundamental loops.

This transformation is exact. Nothing in our procedure depends on the statistics of the fundamental particles nor on the shape of the potential. Such properties are important when it comes to *solving* the theory perturbatively. Only under appropriate physical circumstances will the field φ represent important collective excitations with weak residual interactions. Then the new formulation is of great use in understanding the dynamics of the system. As an illustration consider a dilute fermion gas of very low temperature. Then the function $\xi(-i\nabla)$ is $\epsilon(-i\nabla) - \mu$ with $\epsilon(-i\nabla) = -\nabla^2/2m$.

2.2 Physical Consequences

Let the potential be translationally invariant and instantaneous:

$$V(x, x') = \delta(t - t')V(\mathbf{x} - \mathbf{x'}). \tag{2.18}$$

Then the plasmon propagator (2.17) reads in momentum space

$$G_{\text{pl}}(\nu, \mathbf{k}) = V(\mathbf{k}) \frac{1}{1 - V(\mathbf{k})\pi(\nu, \mathbf{k})}, \tag{2.19}$$

where the single electron loop symbolizes the analytic expression[1]

$$\pi(\nu, \mathbf{k}) = 2\frac{T}{V} \sum_p \frac{1}{i\omega - \mathbf{p}^2/2m + \mu} \frac{1}{i(\omega + \nu) - (\mathbf{p} + \mathbf{k})^2/2m + \mu}. \tag{2.20}$$

The frequencies ω and ν are odd and even multiples of πT, respectively. In order to calculate the sum we introduce a convergence-enforcing factor $e^{i\omega\eta}$, and rewrite (2.20) as [3]

$$\pi(\nu, \mathbf{k}) = 2 \int \frac{d^3p}{(2\pi)^3} \frac{1}{\xi(\mathbf{p} + \mathbf{k}) - \xi(\mathbf{p}) - i\nu}$$
$$\times T \sum_{\omega_n} e^{i\omega_n\eta} \left[\frac{1}{i(\omega_n + \nu) - \xi(\mathbf{p} + \mathbf{k})} - \frac{1}{i\omega_n - \xi(\mathbf{p})} \right]. \tag{2.21}$$

Using the summation formula (1.104), this becomes

$$\pi(\nu, \mathbf{k}) = 2 \int \frac{d^3p}{(2\pi)^3} \frac{n(\mathbf{p} + \mathbf{k}) - n(\mathbf{p})}{\epsilon(\mathbf{p} + \mathbf{k}) - \epsilon(\mathbf{p}) - i\nu}, \tag{2.22}$$

or, after some rearrangement,

$$\pi(\nu, \mathbf{k}) = -2 \int \frac{d^3p}{(2\pi)^3} n(\mathbf{p}) \left[\frac{1}{\epsilon(\mathbf{p} + \mathbf{k}) - \epsilon(\mathbf{p}) - i\nu} + \frac{1}{\epsilon(\mathbf{p} - \mathbf{k}) - \epsilon(\mathbf{p}) + i\nu} \right]. \tag{2.23}$$

---

[1]The factor 2 stems from the trace over the electron spin.

Let us study this function for real physical frequencies $\omega = i\nu$ where we rewrite it as

$$\pi(\omega, \mathbf{k}) = -2 \int \frac{d^3p}{(2\pi)^3} n(\mathbf{p}) \left[\frac{1}{\epsilon(\mathbf{p}+\mathbf{k}) - \epsilon(\mathbf{p}) - \omega} + \frac{1}{\epsilon(\mathbf{p}-\mathbf{k}) - \epsilon(\mathbf{p}) + \omega} \right], \quad (2.24)$$

which can be brought to the form

$$\pi(\omega, \mathbf{k}) = 2 \frac{k^2}{m\omega^2} \int \frac{d^3p}{(2\pi)^3} n(\mathbf{p}) \frac{1}{(\omega - \mathbf{p}\cdot\mathbf{k}/m + i\eta)^2 - (k^2/2M)^2}. \quad (2.25)$$

For $|\omega| > p_F k/m + k^2/2m$, the integrand is real and we can expand

$$\pi(\omega, \mathbf{k}) = 2 \frac{k^2}{m\omega^2} \int \frac{d^3p}{(2\pi)^3} n(\mathbf{p}) \left[1 + \frac{2\mathbf{p}\cdot\mathbf{k}}{m\omega} + 3 \left(\frac{\mathbf{p}\cdot\mathbf{k}}{m\omega} \right)^2 \right.$$
$$\left. + \left(\frac{\mathbf{p}\cdot\mathbf{k}}{m\omega} \right)^3 + \frac{80(\mathbf{p}\cdot\mathbf{k})^4 + m^2\omega^2 k^4}{16 m^2 \omega^4} + \dots \right]. \quad (2.26)$$

2.2.1 Zero Temperature

For zero temperature, the chemical potential μ is equal to the Fermi energy $\varepsilon_F = p_F^2/2m$, and all states below the Fermi momentum p_F are occupied so that the occupation number is given by a Heaviside function $n(\mathbf{p}) = \Theta(p - p_F)$. Then the integral in (2.26) can be performed trivially as

$$\frac{N}{V} = n = 2 \int \frac{d^3p}{(2\pi)^3} n_{T=0}(\mathbf{p}) = \frac{p_F^3}{3\pi^2}, \quad (2.27)$$

and we obtain the expansion

$$\pi(\omega, \mathbf{k}) = \frac{k^2}{\omega^2} \frac{n}{m} \left[1 + \frac{3}{5} \left(\frac{p_F k}{m\omega} \right)^2 + \frac{1}{5} \left(\frac{p_F k}{m\omega} \right)^4 + \frac{1}{16} \frac{k^4}{m^2\omega^2} + \dots \right]. \quad (2.28)$$

Inserting this into (2.19) we find, for long wavelengths, the Green function

$$G^{\mathrm{pl}}(\nu, \mathbf{k}) \approx V(\mathbf{k}) \left[1 - \frac{V(\mathbf{k})}{\omega^2} \frac{n}{m} + \dots \right]^{-1}. \quad (2.29)$$

Thus the original propagator is modified by a factor

$$\epsilon(\omega, \mathbf{k}) = 1 - \frac{4\pi e^2}{\omega^2} \frac{n}{m} + \dots . \quad (2.30)$$

The dielectric constant vanishes at the frequency

$$\omega = \omega_{\mathrm{pl}} = \sqrt{\frac{4\pi e^2}{m}}, \quad (2.31)$$

which is the famous plasma frequency of the electron gas. At this frequency, the plasma propagator (2.19) has a pole on the real-ω axis, implying the existence of an undamped excitation of the system.

For an electron gas, we insert the Coulomb interaction (2.9) and obtain

$$G^{\mathrm{pl}}(\nu, \mathbf{k}) \approx \frac{4\pi e^2}{\mathbf{k}^2} \left[1 - \frac{4\pi e^2}{m\omega^2} n + \ldots \right]^{-1}. \tag{2.32}$$

Thus the original Coulomb propagator is modified by a factor

$$\epsilon(\omega, \mathbf{k}) = 1 - \frac{4\pi e^2}{m\omega^2} n + \ldots, \tag{2.33}$$

which is simply the *dielectric* constant.

The zero temperature limit can also be calculated exactly starting from the expression (2.26), written in the form

$$\pi(\omega, \mathbf{k}) = -2 \int \frac{d^3 p}{(2\pi)^3} \Theta(p - p_F) \left[\frac{1}{\mathbf{p} \cdot \mathbf{k} + k^2/2m - \omega} + (\omega \to -\omega) \right]. \tag{2.34}$$

Performing the integral yields

$$
\begin{aligned}
\pi(\omega, \mathbf{k}) = & -\frac{m p_F}{2\pi^2} \left\{ 1 - \frac{1}{2k p_F} \left[p_F^2 - \left(\frac{k}{2} + \frac{m\omega}{k} \right)^2 + p_F^2 \right] \log \frac{k^2 + 2m\omega - 2k p_F}{k^2 + 2m\omega + 2k p_F} \right\} \\
& + (\omega \to -\omega).
\end{aligned} \tag{2.35}
$$

The lowest terms of a Taylor expansion in powers of k agree with (2.28).

2.2.2   Short-Range Potential

Let us also find the real poles of $G_{\mathrm{pl}}(\nu, \mathbf{k})$ for a short-range potential where the singularity at $\mathbf{k} = 0$ is absent. Then a rotationally invariant $[V(\mathbf{k})]^{-1}$ has the long-wavelength expansion

$$[V(\mathbf{k})]^{-1} = [V(0)]^{-1} + a\mathbf{k}^2 + \ldots, \tag{2.36}$$

as long as $[V(0)]^{-1}$ is finite and positive, i.e., for a well behaved overall repulsive potential satisfying $V(0) = \int d^3 x V(\mathbf{x}) > 0$. Then the Green function (2.19) becomes

$$G_{\mathrm{pl}}(\omega, \mathbf{k}) = \omega^2 \left\{ \omega^2 [V(0)]^{-1} + a\omega^2 k^2 - k^2 \frac{n}{m} \left[1 + \frac{3}{5} \left(\frac{p_F k}{m\omega} \right)^2 + \ldots \right] \right\}^{-1}. \tag{2.37}$$

There is a pole at $\omega = \pm c_0 k$, where

$$c_0 = V(0) \frac{n}{m} \tag{2.38}$$

is the velocity of zero sound.

In the neighborhood of the positive-energy pole, the propagator has the form

$$G^{\mathrm{pl}}(k_0, k) \approx V(0) \times \frac{|\mathbf{k}|}{\omega - c_0 |\mathbf{k}|}. \tag{2.39}$$

More details can be studied in the textbook [4].

Appendix 2A   Fluctuations around the Plasmon Field

Here we derive the quantum mechanical fluctuations around the *classical* equation of motion [recall (2.7)]

$$\varphi(x) = \int dy \, V(x, y) \, \psi^\dagger(y)\psi(y). \tag{2A.1}$$

They are quite simple to calculate. Let us compare the Green function of $\varphi(x)$ with that of the composite operators on the right-hand side of Eq. (2A.1). The Green functions of φ are generated by adding external currents $\int dx \varphi(x) I(x)$ to the final action (2.12) respectively, and by forming functional derivatives $\delta/\delta I$. The Green functions of the composite operators, on the other hand, are obtained by adding

$$\int dx \left(\int dy V(x, y)\psi^\dagger(y)\psi(y) \right) K(x)$$

to the original actions (2.4) and by forming functional derivatives $\delta/\delta K$. It is obvious that the sources $K(x)$ can be included in the final action (2.12) by simply replacing:

$$\varphi(x) \to \varphi'(x) = \varphi(x) - \int dx' K(x') V(x', x).$$

If one now shifts the functional integrations to these new translated variables and drops the irrelevant superscript "prime", the actions can be rewritten as

$$\mathcal{A}[\varphi] = \pm i \mathrm{Tr} \log(i G_\varphi^{-1}) + \frac{1}{2} \int dx dx' \varphi(x) V^{-1}(x, x')\varphi(x') + i \int dx dx' \eta^\dagger(x) G_\varphi(x, x')\eta(x)$$

$$+ \int dx \varphi(x) I(x) - \frac{1}{2} \int dx dx' I(x) V(x, x') I(x'). \tag{2A.2}$$

In this form, the action display clearly the fact that derivatives of the partition function with respect to the source $I(x)$ coincide exactly with the the right-hand side of (2.1). Thus the propagators of the plasmon field $\varphi(x)$ and of the composite operator $\int dy V(x, y)\psi^\dagger(y)\psi(y)$ are related by

$$\overset{\frown}{\varphi(x)\varphi(x')} \;=\; -\frac{\delta^{(2)} Z}{\delta I(x)\delta I(x')} = V^{-1}(x, x') - \frac{\delta^{(2)} Z}{\delta K(x)\delta K(x')} \tag{2A.3}$$

$$= V^{-1}(x, x') + \langle 0| \hat{T} \left(\int dy V(x, y)\psi^\dagger(y)\psi(y) \right) \left(\int dy' V(x', y')\psi^\dagger(y')\psi(y') \right) |0\rangle,$$

where $\hat{T}$ is the time-ordering operator.

Notes and References

[1] H. Kleinert, *Particles and Quantum Fields*, World Scientific, Singapore, 2016.

[2] L.P. Kadanoff and G. Baym, *Quantum Statistical Mechanics*, Benjamin, New York (1962);
and the review papers
G. Baym and N.D. Mermin, J. Math. Phys. **2**, 232 (1961);
L.P. Kadanoff, Rev. Mod. Phys. **49**, 267 (1977);
and Chapter 14 of the textbook [1].

[3] A.A. Abrikosov, L.P. Gorkov, and I.E. Dzyaloshinski, *Methods of Quantum Field Theory in Statistical Physics*, Dover, New York (1975);
A. Fetter and J.D. Walecka, *Quantum Theory of Many-Particle Systems*, McGraw-Hill, New York (1971). See in particular Chapter 9.

[4] H.L. Pècseli, *Waves and Oscillations in Plasmas*, CRC Press, New York, 2013.

*What we wish, we readily believe,
and what we ourselves think, we imagine others think also.*
JULIUS CAESAR (100 B.C.–44 B.C.)

3

Superconductors

Superconductors are made from materials which do not pose any resistance to the flow of electricity. The phenomenon was first observed in 1911 by the Dutch physicist Heike Kamerlingh Onnes at Leiden University. When he cooled mercury down to the temperature of liquid helium, which appears at about 4 degree Kelvin (1 degree Kelvin $-273.15\,°C$), its resistance suddenly disappeared. For this discovery, Onnes won the Nobel Prize in physics in 1913.

Superconductors have an important property, which distinguishes them from ordinary conductors of extremely low resistance: They are perfect diamagnets. This implies that they do not tolerate, in their inside, any magnetic fields. This is the so-called *Meissner-Ochsenfeld effect* discovered in 1933. This effect causes superconductive materials to hover over a sufficiently strong magnetic field. They are lifted as soon as they are cooled below the critical temperature (levitation). A perfect conductor would only hover above the magnet if brought in from the outside due to induction, generating a current with a magnetic moment opposite to the external field.

For the purpose of energy conservation, it is a challenge to find superconductive materials which can transport high currents without loss at room temperature. Since 1941, the record was held for a long time by niobium-nitride, which becomes superconductive at 16 K, surpassed in 1953 by vanadium-silicon with a critical temperature of 17.5 K. In 1962, a first commercial superconducting wire was manufactured from an alloy of niobium and titanium. First applications were made in 1987 in the Fermilab high-energy particle-accelerator Tevatron where the necessary strong magnetic fields were produced by supercurrents in copper-clad niobium-titanium. The magnets had been developed in 1960 at the Rutherford Appleton Laboratory in the UK.

The first satisfactory theory of superconductivity was developed in 1957 by J. Bardeen, L.N. Cooper, and J.R. Schrieffer [1], now called BCS theory, which won them the Nobel prize in 1972. The theory made essential use of a fermionic version of a canonical transformation that had been invented ten years earlier for bosons by N.N. Bogoliubov, to expain the phenomenon of superfluidity in a dilute gas of bosons [2]. The BCS theory explains the early forms of superconductivity observed for elements and simple alloys at temperatures close to absolute zero.

New advances were made in the 1980 when the first organic superconductor was synthesized by the Danish researcher Klaus Bechgaard of the University of Copenhagen and his group [3]. The new material turned out to become superconductive at a transition temperature of about 1.2 K. The possibility that this could happen had been pointed out in 1964 by Bill Little at Stanford University [4].

A more recent major breakthrough was made in 1986 by Alex Müller and Georg Bednorz at the IBM Research Laboratory in Rüschlikon, Switzerland [5]. They synthesized brittle ceramic compound that became superconducting at the record temperature of 30 K. What made this discovery so remarkable was that ceramics are normally insulators, and do not conduct electricity at all. So, researchers had not considered them as possible high-temperature superconductor candidates. The compound that Müller and Bednorz synthesized of a mixture of Lanthanum, Barium, Copper, and Oxygen behaved in a not-as-yet-understood way. Their discovery won them the Nobel Prize in 1987. It was later found that tiny amounts of this material were actually superconducting at 58 K. Since then there has been a great deal of activity trying to find ceramics of many combinations with higher and higher critical temperatures. In 1987 superconductivity was reached in a material called YBCO (Yttrium Barium Copper Oxide) at 92 K, a temperature which can simply be reached by cooling the material with liquid nitrogen.

The present world record was reached in 2015 at $T_c = 203$ K in a sulfur hydride system. Under extreme pressure of 300,000 atmospheres, this critical temperature can be raised by 25 to 30 more degrees (see Fig. 3.1). For more details see [6, 7, 8].

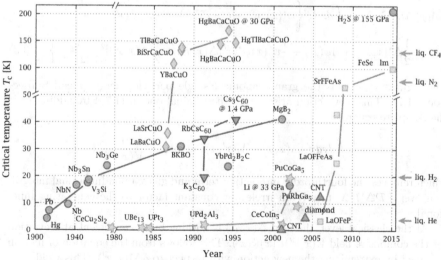

FIGURE 3.1 Time evolution of critical temperatures at which superconductivity sets in (in units of Kelvin). The right-hand margins indicate the liquid by which the temperatures can be reached. From Ref. [6].

3.1 General Formulation

The theoretical description of low-temperature superconductivity is based on a collective field complementary to the plasmon field. The complementary field is a *pair field* which describes the dominant low-energy collective systems such as the superconductors to be discussed in this Chapter. The pair field is in general bilocal and will be denoted by $\Delta(\mathbf{x}\,t; \mathbf{x}'t')$, with two space and two time arguments. It is introduced into the generating functional by performing a Hubbard-Stratonovich transformation of the type (1.80), according to which one rewrites the exponential of the interaction term in (2.1) in the partition function (1.75) as [9, 10]:

$$
\exp\left[-\frac{i}{2}\int dx dx'\, \psi^*(x)\psi^*(x')\psi(x')\psi(x)V(x,x')\right] = \text{const} \times \int \mathcal{D}\Delta(x,x')\mathcal{D}\Delta^*(x,x')
$$

$$
\times \exp\left[\frac{i}{2}\int dx dx'\left\{|\Delta(x,x')|^2\frac{1}{V(x,x')} - \Delta^*(x,x')\psi(x)\psi(x') - \psi^*(x)\psi^*(x')\Delta(x,x')\right\}\right].
$$
(3.1)

In contrast to the similar-looking plasmon expression (2.1), the inverse $1/V(x,x')$ in (3.1) is understood as a numeric division for each x,y, not as a functional inversion. Hence the grand-canonical potential becomes

$$
Z[\eta,\eta^*] = \int \mathcal{D}\psi^*\mathcal{D}\psi\mathcal{D}\Delta^*\mathcal{D}\Delta\, e^{i\mathcal{A}[\psi^*,\psi,\Delta^*,\Delta]+i\int dx(\psi^*(x)\eta(x)+\text{c.c.})}, \tag{3.2}
$$

with the action

$$
\mathcal{A}[\psi^*,\psi,\Delta^*,\Delta] = \int dx dx'\left\{\psi^*(x)\left[i\partial_t - \xi(-i\boldsymbol{\nabla})\right]\delta(x-x')\psi(x')\right.
$$

$$
\left. -\frac{1}{2}\Delta^*(x,x')\psi(x)\psi(x') - \frac{1}{2}\psi^*(x)\psi^*(x')\Delta(x,x') + \frac{1}{2}|\Delta(x,x')|^2\frac{1}{V(x,x')}\right\}, \tag{3.3}
$$

where $\xi_{\mathbf{p}} \equiv \varepsilon_{\mathbf{p}} - \mu$ is the grand-canonical single particle energy (recall Subsection 1.1.7). This new action arises from the original one in (1.75) by adding to it the complete square

$$
\frac{i}{2}\int dx dx'\,|\Delta(x,x) - V(x',x)\psi(x')\psi(x)|^2\frac{1}{V(x,x')},
$$

which removes the fourth-order interaction term and gives, upon functional integration over $\int \mathcal{D}\Delta^*\mathcal{D}\Delta$, merely an irrelevant constant factor to the generating functional.

At the classical level, the field $\Delta(x,x')$ is nothing but a convenient abbreviation for the composite field $V(x,x')\psi(x)\psi(x')$. This follows from the equation of motion obtained by extremizing the new action with respect to $\delta\Delta^*(x,x')$. This yields

$$
\frac{\delta\mathcal{A}}{\delta\Delta^*(x,x')} = \frac{1}{2V(x,x')}\left[\Delta(x,x') - V(x,x')\psi(x)\psi(x')\right] \equiv 0. \tag{3.4}
$$

Quantum mechanically, there are Gaussian fluctuations around this solution which are discussed in Appendix 3C.

Expression (3.3) is quadratic in the fundamental fields $\psi(x)$ and can be rewritten in a matrix form as

$$\frac{1}{2}f^*(x)A(x,x')f(x')$$

$$= \frac{1}{2}f^\dagger(x) \left(\begin{array}{cc} [i\partial_t - \xi(-i\boldsymbol{\nabla})]\,\delta(x-x') & -\Delta(x,x') \\ -\Delta^*(x,x') & \mp [i\partial_t + \xi(i\boldsymbol{\nabla})]\,\delta(x-x') \end{array} \right) f(x'), \quad (3.5)$$

where $f(x)$ denotes the fundamental field doublets $f(x) = \left(\begin{array}{c} \psi(x) \\ \psi^*(x) \end{array} \right)$ and $f^\dagger \equiv f^{*T}$. The field $f^*(x)$ is not independent of $f(x)$. Indeed, there is an identity

$$f^\dagger A f = f^T \left(\begin{array}{cc} 0 & 1 \\ 1 & 0 \end{array} \right) A f. \quad (3.6)$$

Therefore, the real-field formula (1.79) must be used to evaluate the functional integral for the generating functional

$$Z[\eta^*,\eta] = \int \mathcal{D}\Delta^* \mathcal{D}\Delta \; e^{i\mathcal{A}[\Delta^*,\Delta] - \frac{1}{2}\int dx \int dx' j^\dagger(x) G_\Delta(x,x') j(x')}, \quad (3.7)$$

where $j(x)$ collects the external source $\eta(x)$ and its complex conjugate, $j(x) \equiv \left(\begin{array}{c} \eta(x) \\ \eta^*(x) \end{array} \right)$. Then the collective action reads

$$\mathcal{A}[\Delta^*,\Delta] = \pm \frac{i}{2} \mathrm{Tr} \log \left[i\mathbf{G}_\Delta^{-1}(x,x') \right] + \frac{1}{2}\int dx dx' |\Delta(x,x')|^2 \frac{1}{V(x,x')}. \quad (3.8)$$

The 2×2 matrix $\mathbf{G}_\Delta$ denotes the propagator iA^{-1} which satisfies the functional equation

$$\int dx'' \left(\begin{array}{cc} [i\partial_t - \xi(-i\boldsymbol{\nabla})]\,\delta(x-x'') & -\Delta(x,x'') \\ -\Delta^*(x,x'') & \mp [i\partial_t + \xi(i\boldsymbol{\nabla})]\,\delta(x-x'') \end{array} \right) \mathbf{G}_\Delta(x'',x') = i\delta(x-x'). \quad (3.9)$$

Writing $\mathbf{G}_\Delta$ as a matrix $\left(\begin{array}{cc} G & G_\Delta \\ G_\Delta^\dagger & \tilde{G} \end{array} \right)$, the mean-field equations associated with this action are precisely the equations used by Gorkov to study the behavior of type II superconductors.[1] With $Z[\eta^*,\eta]$ being the *full* partition function of the system, the fluctuations of the collective field $\Delta(x,x')$ can now be incorporated, at least in principle, thereby yielding corrections to these equations.

Let us set the sources in the generating functional $Z[\eta^*,\eta]$ equal to zero and investigate the behavior of the collective quantum field Δ. In particular, we want to

---

[1]As an example see p. 444 in Ref. [11].

develop Feynman rules for a perturbative treatment of the fluctuations of $\Delta(x, x')$. As a first step we expand the Green function $\mathbf{G}_\Delta$ in powers of Δ as

$$\mathbf{G}_\Delta = \mathbf{G}_0 - i\mathbf{G}_0 \begin{pmatrix} 0 & \Delta \\ \Delta^* & 0 \end{pmatrix} \mathbf{G}_0 - \mathbf{G}_0 \begin{pmatrix} 0 & \Delta \\ \Delta^* & 0 \end{pmatrix} \mathbf{G}_0 \begin{pmatrix} 0 & \Delta \\ \Delta^* & 0 \end{pmatrix} \mathbf{G}_0 + \ldots \quad (3.10)$$

with

$$\mathbf{G}_0(x, x') = \begin{pmatrix} \dfrac{i}{i\partial_t - \xi(-i\boldsymbol{\nabla})} \delta(x - x') & 0 \\ 0 & \mp\dfrac{i}{i\partial_t + \xi(i\boldsymbol{\nabla})} \delta(x - x') \end{pmatrix}. \quad (3.11)$$

We shall see later that this expansion is applicable only close to the critical temperature T_c. Inserting this expansion into (3.7), the source term can be interpreted graphically by the absorption and emission of lines $\Delta(k)$ and $\Delta^*(k)$, respectively, from virtual zig-zag configurations of the underlying particles $\psi(k)$, $\psi^*(k)$ (see Fig. 3.2).

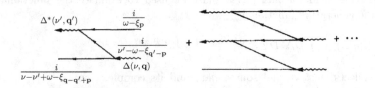

FIGURE 3.2 Fundamental particles (fat lines) entering any diagram only via the external currents in the last term of (3.7), absorbing n pairs from the right (the past) and emitting the same number from the left (the future).

The functional submatrices in $\mathbf{G}_0$ have the Fourier representation

$$G_0(x, x') = \frac{T}{V} \sum_p \frac{i}{p^0 - \xi_{\mathbf{p}}} e^{-i(p^0 t - \mathbf{p}\mathbf{x})}, \quad (3.12)$$

$$\tilde{G}_0(x, x') = \pm\frac{T}{V} \sum_p \frac{i}{-p^0 - \xi_{-\mathbf{p}}} e^{-i(p^0 t - \mathbf{p}\mathbf{x})}, \quad (3.13)$$

where we have used the notation $\xi_{\mathbf{p}}$ for the Fourier components $\xi(\mathbf{p})$ of $\xi(-i\boldsymbol{\nabla})$.

The first matrix coincides with the operator Green function

$$G_0(x - x') = \langle 0|T\psi(x)\psi^\dagger(x')|0\rangle. \quad (3.14)$$

The second one corresponds to

$$\tilde{G}_0(x - x') = \langle 0|T\psi^\dagger(x)\psi(x')|0\rangle = \pm\langle 0|T\left(\psi(x')\psi^\dagger(x)\right)|0\rangle$$
$$= \pm G_0(x' - x) \equiv \pm[G_0(x, x')]^T, \quad (3.15)$$

where T denotes the transposition in the functional sense (i.e., x and x' are interchanged). After a Wick rotation of the energy integration contour, the Fourier components of the Green functions at fixed energy read

$$G_0(\mathbf{x} - \mathbf{x}', \omega) = -\sum_{\mathbf{p}} \frac{1}{i\omega - \xi_{\mathbf{p}}} e^{i\mathbf{p}(\mathbf{x}-\mathbf{x}')} \tag{3.16}$$

$$\tilde{G}_0(\mathbf{x} - \mathbf{x}', \omega) = \mp\sum_{\mathbf{p}} \frac{1}{-i\omega - \xi_{-\mathbf{p}}} e^{i\mathbf{p}(\mathbf{x}-\mathbf{x}')} = \mp G_0(\mathbf{x}' - \mathbf{x}, -\omega). \tag{3.17}$$

The Tr log term in Eq. (3.8) can be interpreted graphically just as easily by expanding as in (3.134):

$$\pm\frac{i}{2}\mathrm{Tr}\log\left(i\mathbf{G}_\Delta^{-1}\right) = \pm\frac{i}{2}\mathrm{Tr}\log\left(i\mathbf{G}_0^{-1}\right) \mp \frac{i}{2}\mathrm{Tr}\left[-i\mathbf{G}_0\begin{pmatrix} 0 & \Delta \\ \Delta^* & 0 \end{pmatrix}\Delta^*\right]^n\frac{1}{n}. \tag{3.18}$$

The first term only changes the irrelevant normalization N of Z. To the remaining sum only even powers can contribute so that we can rewrite

$$\begin{aligned}
\mathcal{A}[\Delta^*, \Delta] &= \mp i\sum_{n=1}^{\infty}\frac{(-)^n}{2n}\mathrm{Tr}\left[\left(\frac{i}{i\partial_t - \xi(-i\boldsymbol{\nabla})}\delta\right)\Delta\left(\frac{\mp i}{i\partial_t + \xi(i\boldsymbol{\nabla})}\delta\right)\Delta^*\right]^n \\
&\quad + \frac{1}{2}\int dx dx' |\Delta(x, x')|^2 \frac{1}{V(x, x')} \\
&= \sum_{n=1}^{\infty}\mathcal{A}_n[\Delta^*, \Delta] + \frac{1}{2}\int dx dx' |\Delta(x, x')|^2 \frac{1}{V(x, x')}. \tag{3.19}
\end{aligned}$$

This form of the action allows an immediate quantization of the collective field Δ. The graphical rules are slightly more involved technically than in the plasmon case since the pair field is bilocal. Consider at first the *free* collective fields which can be obtained from the quadratic part of the action:

$$\mathcal{A}_2[\Delta^*, \Delta] = -\frac{i}{2}\mathrm{Tr}\left[\left(\frac{i}{i\partial_t - \xi(-i\boldsymbol{\nabla})}\delta\right)\Delta\left(\frac{i}{i\partial_t + \xi(i\boldsymbol{\nabla})}\delta\right)\Delta^*\right]. \tag{3.20}$$

Variation with respect to Δ displays the equations of motion

$$\Delta(x, x') = iV(x, x')\left[\left(\frac{i}{i\partial_t - \xi(-i\boldsymbol{\nabla})}\delta\right)\Delta\left(\frac{i}{i\partial_t + \xi(i\boldsymbol{\nabla})}\delta\right)\right]. \tag{3.21}$$

This equation coincides exactly with the Bethe-Salpeter equation [18], in ladder approximation, for two-body bound-state vertex functions, usually denoted in momentum space by

$$\Gamma(p, p') = \int dx dx' \exp[i(px + p'x')]\Delta(x, x'). \tag{3.22}$$

Thus the free excitations of the field $\Delta(x, x')$ consist of bound pairs of the original fundamental particles. The field $\Delta(x, x')$ will consequently be called pair field. If we introduce total and relative momenta q and $P = (p - p')/2$, then (3.21) can be written as[2]

$$\Gamma(P|q) \;=\; -i \int \frac{d^4 P'}{(2\pi)^4} V(P - P') \frac{i}{q_0/2 + P'_0 - \xi_{\mathbf{q}/2+\mathbf{P}'} + i\eta \operatorname{sgn}\xi}$$
$$\times \Gamma(P'|q) \frac{i}{q_0/2 - P'_0 - \xi_{\mathbf{q}/2-\mathbf{P}'} + i\eta \operatorname{sgn}\xi}. \qquad (3.23)$$

Graphically this formula can be represented shown in Fig. 3.3. The vertex $\Gamma(P|q)$

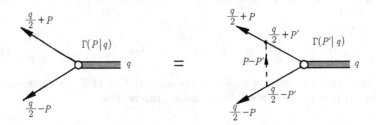

FIGURE 3.3 Free pair field following the Bethe-Salpeter equation as pictured in this diagram.

produces a Bethe-Salpeter *wave function*:

$$\Phi(P|q) \;=\; N \frac{i}{q_0/2 + P_0 - \xi_{\mathbf{q}/2+\mathbf{P}} + i\eta \operatorname{sgn}\xi}$$
$$\times \frac{i}{q_0/2 + P_0 - \xi_{\mathbf{q}/2+\mathbf{P}} + i\eta \operatorname{sgn}\xi} \Gamma(P|q). \qquad (3.24)$$

It satisfies

$$G_0\left(q/2 + P\right) G_0\left(q/2 - P\right) \Phi(P|q) = -i \int \frac{dP'}{(2\pi)^4} V(P, P') \Phi(P'|q), \qquad (3.25)$$

thus coinciding, up to a normalization, with the Fourier transform of the two-body state wave functions

$$\psi(\mathbf{x}, t; \mathbf{x}', t') = \langle 0 | T\left(\psi(\mathbf{x}, t)\psi(\mathbf{x}', t')\right) | B(q) \rangle. \qquad (3.26)$$

If the potential is instantaneous, then (3.21) shows $\Delta(x, x')$ to be factorizable according to

$$\Delta(x, x') = \delta(t - t')\Delta(\mathbf{x}, \mathbf{x}'; t), \qquad (3.27)$$

so that $\Gamma(P|q)$ becomes independent of P_0.

---

[2]Here q is short for the four-vector $q^\mu = (q^0, \mathbf{q})$ with $q_0 = E$.

Consider now the system at $T = 0$ in the vacuum. Then $\mu = 0$ and $\xi_\mathbf{p} = \varepsilon_\mathbf{p} > 0$. One can perform the P_0 integral in (3.23) with the result

$$\Gamma(\mathbf{P}|q) = \int \frac{d^3P'}{(2\pi)^4} V(\mathbf{P} - \mathbf{P}') \frac{1}{q_0 - \varepsilon_{\mathbf{q}/2+\mathbf{P}'} - \varepsilon_{\mathbf{q}/2-\mathbf{P}'} + i\eta} \Gamma(\mathbf{P}'|q). \tag{3.28}$$

Now the equal-time Bethe-Salpeter wave function

$$\psi(\mathbf{x}, \mathbf{x}'; t) \equiv N \int \frac{d^3P\, dq_0 d^3q}{(2\pi)^7} \exp\left[-i\left(q_0 t - \mathbf{q} \cdot \frac{\mathbf{x} + \mathbf{x}'}{2} - \mathbf{P} \cdot (\mathbf{x} - \mathbf{x}')\right)\right]$$
$$\times \frac{1}{q_0 - \varepsilon_{\mathbf{q}/2+\mathbf{P}} - \varepsilon_{\mathbf{q}/2-\mathbf{P}} + i\eta} \tag{3.29}$$

satisfies

$$\left[i\partial_t - \epsilon_{(-i\boldsymbol{\nabla})} - \epsilon_{(-i\boldsymbol{\nabla}')}\right] \psi(\mathbf{x}, \mathbf{x}'; t) = V(\mathbf{x} - \mathbf{x})\psi(\mathbf{x}, \mathbf{x}'; t), \tag{3.30}$$

which is simply the Schrödinger equation of the two-body system. Thus, in the instantaneous case, the free collective excitations in $\Delta(x, x')$ are the bound states derived from the Schrödinger equation.

In a thermal ensemble, the continuous integrals over the energies P'^0 in (3.23) are restricted to sums over the Matsubara frequencies. First, we write the Schrödinger equation as

$$\Gamma(\mathbf{P}|q) = -\int \frac{d^3P'}{(2\pi)^3} V(\mathbf{P} - \mathbf{P}') l(\mathbf{P}'|q) \Gamma(\mathbf{P}'|q) \tag{3.31}$$

with

$$l(\mathbf{P}|q) = -i \sum_{P_0} G_0(q/2 + P)\, \tilde{G}_0(P - q/2)$$
$$= -i \sum_{P_0} \frac{i}{q_0/2 + P_0 - \xi_{\mathbf{q}/2+\mathbf{P}} + i\eta \operatorname{sgn}\xi} \frac{i}{q_0/2 - P_0 - \xi_{\mathbf{q}/2-\mathbf{P}} + i\eta \operatorname{sgn}\xi}. \tag{3.32}$$

After a Wick rotation and setting $q_0 \equiv i\nu$, the replacement of the energy integration by a Matsubara sum leads to

$$l(\mathbf{P}|q) = -T \sum_{\omega_n} \frac{1}{i(\omega_n + \nu/2) - \xi_{\mathbf{q}/2+\mathbf{P}}} \frac{1}{i(\omega_n - \nu/2) + \xi_{\mathbf{q}/2-\mathbf{P}}}$$
$$= T \sum_{\omega_n} \frac{1}{i\nu - \xi_{\mathbf{q}/2+\mathbf{P}} - \xi_{\mathbf{q}/2-\mathbf{P}}}$$
$$\times \left[\frac{1}{i(\omega_n + \nu/2) - \xi_{\mathbf{q}/2+\mathbf{P}}} - \frac{1}{i(\omega_n - \nu/2) + \xi_{\mathbf{q}/2-\mathbf{P}}}\right]$$
$$= -\frac{\pm\left[n_{\mathbf{q}/2+\mathbf{P}} + n_{\mathbf{q}/2-\mathbf{P}}\right]}{i\nu - \xi_{\mathbf{q}/2+\mathbf{P}} - \xi_{\mathbf{q}/2-\mathbf{P}}}. \tag{3.33}$$

Here we have used the frequency sum [see (1.104)]

$$T \sum_{\omega_n} \frac{1}{i\omega_n - \xi_{\mathbf{p}}} = \mp \frac{1}{e^{\xi_{\mathbf{p}}/T} \mp 1} \equiv \mp n_{\mathbf{p}}, \tag{3.34}$$

with $n_{\mathbf{p}}$ being the occupation numbers of the state of energy $\xi_{\mathbf{p}}$. In Chapter 1 we have used the generic notation $n(\mathbf{p})$ for the occupation numbers of fermions and bosons. In this Chapter we shall save some space by placing the momentum argument into a subscript.

The expression in brackets is antisymmetric under the exchange $\xi \to -\xi$, since under this substitution $n_{(\mathbf{p})} \to \mp 1 - n_{(\mathbf{p})}$. In fact, one can write it in the form $-N(\mathbf{P}, \mathbf{q})$ with

$$\begin{aligned} N(\mathbf{P}|q) &\equiv 1 \pm \left(n_{\mathbf{q}/2+\mathbf{P}} + n_{\mathbf{q}/2-\mathbf{P}} \right) \\ &= \frac{1}{2} \left(\tanh^{\mp 1} \frac{\xi_{\mathbf{q}/2+\mathbf{P}}}{2T} + \tanh^{\mp 1} \frac{\xi_{\mathbf{q}/2-\mathbf{P}}}{2T} \right), \end{aligned} \tag{3.35}$$

so that

$$l(\mathbf{P}|q) = -\frac{N(\mathbf{P}|\mathbf{q})}{i\nu - \xi_{\mathbf{q}/2+\mathbf{P}} - \xi_{(\mathbf{q}/2-\mathbf{P}}}. \tag{3.36}$$

Defining again a Schrödinger type wave function as in (3.29), the bound-state problem can be brought to the form (3.28) but with a momentum dependent potential $V(\mathbf{P} - \mathbf{P}') \times N(\mathbf{P}'|\mathbf{q})$. We are now ready to construct the propagator of the pair field $\Delta(x, x')$ for $T = 0$. This is most simply done by considering Eq. (3.23) with a potential $\lambda V(P, P')$ rather than V, and asking for all eigenvalues λ_n at *fixed* q. Let $\Gamma_n(P|q)$ be a complete set of vertex functions for this q. Then one can write the propagator as

$$\overline{\Delta(P|q)\Delta^{\dagger}(P'|q')} = -i \sum_n \left. \frac{\Gamma_n(P|q)\Gamma_n^*(P'|q)}{\lambda - \lambda_n(q)} \right|_{\lambda=1} (2\pi)^4 \delta^{(4)}(q - q'), \tag{3.37}$$

where a hook denotes, as usual, the Wick contraction of the fields. Obviously the vertex functions have to be normalized in a specific way, as discussed in Appendix 3B.

An expansion of (3.37) in powers of $[\lambda/\lambda_n(q)]^n$ exhibits the propagator of Δ as a ladder sum of exchanges as shown in Fig. 3.4 (see also Appendix 3B).

FIGURE 3.4 Free pair propagator, amounting to a sum of all ladders of fundamental potential exchanges. This is revealed explicitly by the expansion of (3.37) in powers of $[\lambda/\lambda_n(q)]$.

For an instantaneous interaction, either side is independent of P_0, P_0'. Then the propagator can be shown to coincide directly with the scattering matrix T of the Schrödinger equation (3.30) and the associated integral equation in momentum space (3.28) [see Eq. (3B.13)].

$$\overset{\square}{\Delta\Delta^{\dagger}} = iT \equiv iV + iV\frac{1}{E-H}V. \tag{3.38}$$

Consider now the higher interactions $\mathcal{A}_n, n \geq 3$ of Eq. (3.19). They correspond to zig-zag loops shown in Fig. 3.5. These have to be calculated with every possible $\Gamma_n(P|q), \Gamma_m^*(P|q)$ entering or leaving, respectively.

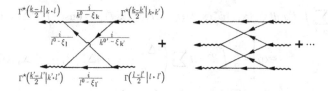

FIGURE 3.5 Self-interaction terms of the non-polynomial pair Lagrangian amounting to the calculation of all single zig-zag loop diagrams absorbing and emitting n pair fields.

Due to the P dependence at every vertex, the loop integrals become very involved. A slight simplification arises for an instantaneous potential where at least the frequency sums can be performed immediately. Only in the special case of a completely local action the full P-dependence disappears and the integrals can be calculated at least approximately. This will be done in the following section.

3.2 Local Interaction and Ginzburg-Landau Equations

Let us study the case of a completely local potential in detail. For the electrons in a crystal, such a local potential is only an approximation which, however, happens to be quite reliable. In a crystal, the interaction between the electrons is mediated by phonon exchange. An electron moving through the lattice attracts the positive ions in its neighborhood and thus creates a cloud of positive charge around its path. This cloud, in turn, attracts other electrons and this is the origin of pair formation. The size of the cloud is of the order of the lattice spacing, i.e., a few Å. Although this can hardly be called local, it is effectively so, as far as the formation of bound states is concerned. The reason is that the strength of the interaction is quite small. This leads to a rather wide bound-state wave function. Its radius will be seen to extend over many lattice spacings. Thus, as far as the bound-states are concerned, the potential may just as well be considered as local. This is what justifies the theoretical treatment to be developed in the sequel.

We assume the fundamental action to be a sum

$$\mathcal{A} = \mathcal{A}_0 + \mathcal{A}_{\text{int}}, \tag{3.39}$$

with a free-particle term

$$\mathcal{A}_0 = \int d^3x dt \psi_\beta^*(\mathbf{x},t)[i\partial_t - \xi(-i\boldsymbol{\nabla})]\psi_\beta(\mathbf{x},t), \tag{3.40}$$

and a δ-function interaction

$$\mathcal{A}_{\text{int}} = \frac{g}{2}\sum_{\alpha,\beta}\int d^3x dt \psi_\alpha^*(\mathbf{x},t)\psi_\beta^*(\mathbf{x},t)\psi_\beta(\mathbf{x},t)\psi_\alpha(\mathbf{x},t), \tag{3.41}$$

with $g > 0$, for an attractive potential. Following the general arguments leading to formula (3.1), we rewrite the exponential of this interaction as[3]

$$\exp\left[\frac{i}{2}g\sum_{\alpha,\beta}\int d^3x\,dt\,\psi_\alpha^*(\mathbf{x},t)\psi_\beta^*(\mathbf{x},t)\psi_\beta(\mathbf{x},t)\psi_\alpha(\mathbf{x},t)\right] = \text{const} \times \int \mathcal{D}\Delta(\mathbf{x},t)\mathcal{D}\Delta^*(\mathbf{x},t)$$

$$\times \exp\left[-\frac{i}{2}\int d^3x dt \sum_{\alpha\beta}\left(\frac{|\Delta_{\alpha\beta}|^2}{g} - \psi_\beta\Delta_{\beta\alpha}^*\psi_\alpha - \psi_\alpha^*\Delta_{\alpha\beta}\psi_\beta^*\right)\right], \tag{3.42}$$

where the new auxiliary field is a $(2s+1) \times (2s+1)$ non-hermitian matrix which satisfies the equation of constraint:

$$\Delta_{\alpha\beta}(\mathbf{x},t) = g\psi_\alpha(\mathbf{x},t)\psi_\beta(\mathbf{x},t). \tag{3.43}$$

Observe the hermiticity property

$$\Delta_{\alpha\beta}(\mathbf{x},t)^* = \Delta_{\beta\alpha}^*(\mathbf{x},t). \tag{3.44}$$

The free part of the action is now written in a 2×2 matrix form analogous to that in (3.5):

$$\mathcal{A}_0 = \int d^3x\,dt\,d^3x'\,dt'\,f^*(x)A(\mathbf{x},t;\mathbf{x}',t')f(x'), \tag{3.45}$$

where $f^T(x)$ denotes here the doubled fundamental field

$$f^T(x) = \left(\psi_\alpha(x), \psi_\beta^*(x)\right), \tag{3.46}$$

and $A(\mathbf{x},t;\mathbf{x}',t')$ is the functional matrix

$$A(\mathbf{x},t;\mathbf{x}',t') = \begin{pmatrix} [i\partial_t - \xi(-i\boldsymbol{\nabla})]\,\delta(x-x')\delta_{\alpha\beta} & -\Delta_{\alpha\beta}(x)\delta(x-x') \\ -\Delta_{\alpha\beta}^*(x)\delta(x-x') & \mp[i\partial_t + \xi(i\boldsymbol{\nabla})]\,\delta(x-x')\delta_{\alpha\beta} \end{pmatrix}. \tag{3.47}$$

Then the action (3.19) becomes

$$\mathcal{A}[\Delta^*,\Delta] = \mp i\sum_{n=1}^{\infty}\frac{(-)^n}{2n}\text{Tr}\,\text{tr}_{\text{spin}}\left[\left(\frac{i}{i\partial_t - \xi(-i\boldsymbol{\nabla})}\delta\right)(\Delta\,\delta)\left(\frac{\mp i}{i\partial_t + \xi(i\boldsymbol{\nabla})}\delta\right)(\Delta^*\,\delta)\right]^n, \tag{3.48}$$

---

[3]Note that the hermitian adjoint $\Delta_{\uparrow\downarrow}^*(x)$ comprises transposition of the spin indices, i.e., $\Delta_{\uparrow\downarrow}^*(x) = [\Delta_{\downarrow\uparrow}(x)]^*$.

where tr_{spin} indicates the trace over the spin indices, and Tr refers to the trace in the functional matrix space. The different terms on the right-hand side will be denoted by $\mathcal{A}_n[\Delta^*, \Delta]$.

Consider fermions of spin $1/2$ close to a critical region, i.e., for $T \approx T_c$. There only long-range properties of the system dominate. As far as such questions are concerned, the expansion

$$\mathcal{A}[\Delta^*, \Delta] = \sum_2^\infty \mathcal{A}_n[\Delta^*, \Delta] \tag{3.49}$$

may be truncated after the fourth term without much loss of information. The dimensions of the neglected terms are so high that they become irrelevant at long distances. The free part of the action $\mathcal{A}_2[\Delta^*, \Delta]$ is given by

$$\begin{aligned}
\mathcal{A}_2[\Delta^*, \Delta] = &\pm i \text{Tr} \, \text{tr}_{\text{spin}} \left[\left(\frac{i}{i\partial_t - \xi(-i\boldsymbol{\nabla})} \delta \right) (\Delta\delta) \left(\frac{\mp i}{i\partial_t + \xi(i\boldsymbol{\nabla})} \delta \right) (\Delta^*\delta) \right] \\
&- \frac{1}{2} \text{tr}_{\text{spin}} \int dx \Delta^*(x) \Delta(x) \frac{1}{g}.
\end{aligned} \tag{3.50}$$

The spin traces can be performed by noting that due to Fermi statistics, the square of the field at a point vanishes, $\psi_\downarrow^2(x) = 0$, $\psi_\uparrow^2(x) = 0$, so that there is really only one independent pair field:

$$\Delta(x) \equiv \Delta_{\downarrow\uparrow}(x) = g\psi_\downarrow(x)\psi_\uparrow(x) = -\Delta_{\uparrow\downarrow}(x). \tag{3.51}$$

Thus $\mathcal{A}_2$ becomes:

$$\mathcal{A}_2[\Delta^*\Delta] = -i \int dx dx' G_0(x, x') \tilde{G}_0(x', x) \Delta^*(x)\Delta(x') - \frac{1}{g} \int dx |\Delta(x)|^2. \tag{3.52}$$

Let us expand the pair field into its Fourier components

$$\Delta(\tau, \mathbf{x}) = T \sum_{\nu_n} \int \frac{d^3k}{(2\pi)^3} e^{-i(\tau\nu_n - \mathbf{kx})} \Delta(\nu_n, \mathbf{k}), \tag{3.53}$$

with the bosonic Matsubara frequencies

$$\nu_n = 2n\pi T. \tag{3.54}$$

Using the short notation

$$T \sum_{\nu_n} \int \frac{d^3k}{(2\pi)^3} f(\nu_n, \mathbf{k}) = T \sum_k f(k), \tag{3.55}$$

the quadratic action $\mathcal{A}_2[\Delta^*\Delta]$ can be written in momentum space as

$$\mathcal{A}_2[\Delta^*, \Delta] = \frac{T}{V} \sum_k \Delta^*(k) L(k) \Delta(k), \tag{3.56}$$

where

$$L(k) \equiv -i\frac{T}{V} \sum_p \frac{i}{p^0 + k^0 - \xi_{p+k} + i\eta \operatorname{sgn} \xi_{p+k}} \frac{i}{p^0 + \xi_p - i\eta \operatorname{sgn} \xi_p} - \frac{1}{g}$$

$$= \frac{T}{V} \sum_p l(\mathbf{p}|k) - \frac{1}{g}. \tag{3.57}$$

This is pictured by a Feynman diagram in Fig. 3.6.

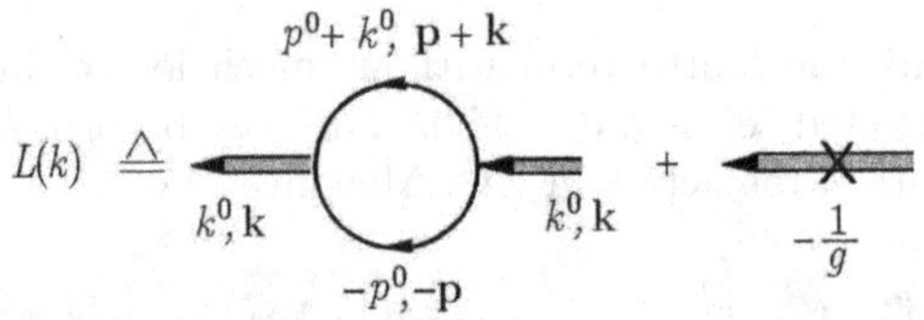

FIGURE 3.6 Free part of the Δ-Lagrangian containing the direct term plus the one loop diagram. As a consequence, the free Δ-propagator sums up an infinite sequence of such loops.

The expression $l(\mathbf{p}|k)$ appeared in the general discussion in Eq. (3.31), where it was brought to the form (3.33). In the present case of Fermi statistics this leads to

$$L(\nu, \mathbf{k}) = \frac{1}{2}\frac{1}{V} \sum_{\mathbf{p}} \frac{1}{\xi_{p+k} + \xi_p - i\nu} \left[\tanh\frac{\xi_{p+k}}{2T} + \tanh\frac{\xi_p}{2T} \right] - \frac{1}{g}. \tag{3.58}$$

At $k = 0$, one has

$$L(0) = \frac{1}{2}\frac{1}{V} \sum_{\mathbf{p}} \frac{1}{\xi_p} \tanh\frac{\xi_p}{2T} - \frac{1}{g}$$

$$\approx \mathcal{N}(0) \int_0^\infty \frac{d\xi}{\xi} \tanh\frac{\xi}{2T} - \frac{1}{g}. \tag{3.59}$$

When going from the first to the second line we have used the equality in a large volume for rotationally symmetric integrands

$$\frac{1}{V} \sum_{\mathbf{p}} \equiv \int \frac{d^3p}{(2\pi\hbar)^3} = \frac{1}{(2\pi\hbar)^3} \int d\hat{\mathbf{p}} \int dp \, p^2 = \frac{4\pi}{(2\pi\hbar)^3} \int p^2 \frac{dp}{d\xi} d\xi. \tag{3.60}$$

In a further approximation of a weak attraction between the electrons caused by phonons, we include only momenta near the surface of the Fermi sphere in momentum space which cover the neighborhood of $|\mathbf{p}| \approx p_F$. There we can approximate the sum over states by the integrals

$$\frac{1}{V} \sum_{\mathbf{p}} \equiv \frac{4\pi}{(2\pi\hbar)^3} \int p^2 \frac{dp}{d\xi} d\xi \approx \mathcal{N}(0) \int d\xi, \tag{3.61}$$

where

$$\mathcal{N}(0) = \frac{mp_F}{2\pi^2\hbar^3} = \frac{3}{4\hbar^3}\frac{\rho}{p_F^2} \tag{3.62}$$

is the density of states on the Fermi surface at zero temperature. Here ρ is the mass density which is related to the particle density N/V by

$$\rho = m\frac{N}{V}. \tag{3.63}$$

In (3.62), we have expressed the *Fermi momentum* of free spin-1/2 particles in terms of ρ by

$$p_F = \left(3\pi^2\right)^{1/3}\rho^{1/3}\hbar \approx g \times 10^{-20}\text{g cm/sec.} \tag{3.64}$$

It is only slightly pressure dependent. The associated *Fermi temperature* defined by $T_F \equiv (1/k_B)p_F^2/2m$ corresponds in most materials to around 10 000 times the *Fermi momentum* of free spin-1/2 particles.

The ξ-integral in (3.59) is logarithmically divergent. This is a consequence of the local approximation to the attractive interaction between the electrons assumed in Eq. (3.41). As explained earlier, the attraction between electrons is caused by phonon exchange. Phonons, however, have frequencies which are at most of the order of the *Debye frequency* ω_D. This may be used as a cutoff to all energy integrals of the type $\int d\xi$, which will be restricted to the interval $\xi \in (-\omega_D, \omega_D)$. The associated *Debye temperature* $T_D \equiv \hbar\omega_D/k_B$ is of the order of 1000 K and thus quite large compared to the characteristic temperature T_c where superconductivity sets in, the so-called *critical temperature*.

The Debye temperature T_D, although being much larger than T_c, is an order of magnitude smaller than T_F. As a consequence, the attraction between electrons is active only between states within a thin layer in the neighborhood of the surface of the Fermi sphere. Using the cutoff energy $\hbar\omega_D$ in Eq. (3.59) yields (from now on in natural units with $\hbar = k_B = 1$)

$$L(0) \approx \mathcal{N}(0)\int_0^{\omega_D}\frac{d\xi}{\xi}\tanh\frac{\xi}{2T} - \frac{1}{g} = \mathcal{N}(0)\log\left(\frac{\omega_D}{T}\frac{2e^\gamma}{\pi}\right) - \frac{1}{g}, \tag{3.65}$$

where γ is Euler's constant

$$\gamma = -\Gamma'(1)/\Gamma(1) \approx 0.577, \tag{3.66}$$

implying that $e^\gamma/\pi \approx 1.13$.

The integral in (3.65) is evaluated as follows: First there is an integration by parts, yielding

$$\int_0^{\omega_D}\frac{d\xi}{\xi}\tanh\frac{\xi}{2T} = \log\frac{\xi}{T}\tanh\frac{\xi}{2T}\bigg|_0^{\omega_D} - \frac{1}{2}\int_0^\infty d\frac{\xi}{T}\log\frac{\xi}{T}\frac{1}{\cosh^2\frac{\xi}{2T}}. \tag{3.67}$$

Since $\omega_D/\pi T \gg 1$, the first term is equal to $\log(\omega_D/2T)$, with exponentially small corrections which can be ignored. In the second integral, we have taken the upper limit of integration to infinity since it converges. We may use the integral formula[4]

$$\int_0^\infty dx \frac{x^{\mu-1}}{\cosh^2(ax)} = \frac{4}{(2a)^\mu}\left(1 - 2^{2-\mu}\right)\Gamma(\mu)\zeta(\mu-1), \tag{3.68}$$

set $\mu = 1 + \delta$, expand the formula to order δ, and insert the special values

$$\Gamma'(1) = -\gamma, \quad \zeta'(0) = -\frac{1}{2}\log(2\pi)\log(4e^\gamma/\pi), \tag{3.69}$$

to find from the linear terms in δ:

$$\int_0^\infty dx \frac{\log x}{\cosh^2(x/2)} = -2\log(2e^\gamma/\pi). \tag{3.70}$$

Hence we obtain

$$\int_0^{\omega_D} \frac{d\xi}{\xi} \tanh\frac{\xi}{2T} = \log\left(\frac{\omega_D}{T}\frac{2e^\gamma}{\pi}\right). \tag{3.71}$$

The value $L(0)$ of Eq. (3.65) vanishes at a critical temperature determined by

$$T_c \equiv \frac{2e^\gamma}{\pi}\omega_D e^{-1/\mathcal{N}(0)g}. \tag{3.72}$$

Using this, we can rewrite Eq. (3.65) as

$$L(0) = \mathcal{N}(0)\log\frac{T_c}{T} \approx \mathcal{N}(0)\left(1 - \frac{T}{T_c}\right). \tag{3.73}$$

The constant $L(0)$ obviously plays the role of the chemical potential of the pair field. Its vanishing at $T = T_c$ implies that, at this temperature, the field propagates over a long range (with a power law) in the system. Critical phenomena are observed [19]. For $T < T_c$, the chemical potential becomes positive indicating the appearance of a Bose condensate. If $\nu \neq 0$ and $\mathbf{k} = 0$, one can write (3.52) in the subtracted form

$$L(\nu, \mathbf{0}) - L(0, \mathbf{0}) = \frac{T}{V}\sum_p \left[\frac{1}{2\xi_\mathbf{p} - i\nu} - \frac{1}{2\xi(\mathbf{p})}\right]\tanh\frac{\xi_\mathbf{p}}{2T} \tag{3.74}$$

$$\approx i\nu\mathcal{N}(0)\int_{-\omega_D}^{\omega_D}\frac{d\xi}{2\xi - i\nu}\frac{1}{2\xi}\tanh\frac{\xi}{2T}. \tag{3.75}$$

Since the subtracted integral converges fast it can be performed over the entire ξ-axis with the small error of relative order $T/\omega_D \ll 1$. For $\nu < 0$, the contour may be closed above, picking up poles exactly at the Matsubara frequencies $\xi = i(2n+1)\pi T = i\omega_n$. Hence

$$L(\nu, \mathbf{0}) - L(0, \mathbf{0}) \approx \nu\mathcal{N}(0)\pi T\sum_{\omega_n > 0}\frac{1}{\omega_n - \nu/2}\frac{1}{\omega_n}. \tag{3.76}$$

---

[4]See, for instance, I.S. Gradshteyn and I.M. Ryzhik, op. cit., Formula 3.527.3.

The sum can be expressed in terms of Digamma functions: For $|\nu| \ll T$, one expands

$$\sum_{\omega_n > 0} \left[\frac{1}{\omega_n^2} + \frac{\nu}{2} \frac{1}{\omega_n^3} + \frac{\nu^2}{4} \frac{1}{\omega_n^4} + \dots \right], \tag{3.77}$$

and applies the formula

$$\sum_{\omega_n > 0} \frac{1}{\omega_n^k} = \frac{1}{\pi^k T^k} [1 - 2^{-k}] \zeta(k), \tag{3.78}$$

to express everything in terms of the Riemann zeta function

$$\zeta(z) = \sum_{n=1}^{\infty} n^{-z}. \tag{3.79}$$

Some of its values are

$$\zeta(2) = \frac{\pi^2}{6}, \quad \zeta(3) = 1.202057,$$
$$\zeta(4) = \frac{\pi^4}{90}, \quad \zeta(5) = 1.036928, \tag{3.80}$$
$$\vdots$$

implying the Matsubara sums

$$\sum_{\omega_n > 0} \frac{1}{\omega_n^2} = \frac{1}{\pi^2 T^2} \frac{3}{4} \frac{\pi^2}{6} = \frac{1}{8T^2}, \tag{3.81}$$

$$\sum_{\omega_n > 0} \frac{1}{\omega_n^3} = \frac{1}{\pi^3 T^3} \frac{7}{8} \zeta(3), \tag{3.82}$$

$$\sum_{\omega_n > 0} \frac{1}{\omega_n^4} = \frac{1}{\pi^4 T^4} \frac{15}{16} \frac{\pi^4}{90} = \frac{1}{96T^4}, \tag{3.83}$$

$$\vdots \tag{3.84}$$

Using the power series for the Digamma function

$$\psi(1 - x) = -\gamma - \sum_{k=2}^{\infty} \zeta(k) x^{k-1}, \tag{3.85}$$

the sum is

$$-\frac{2}{\nu \pi T} \left[\psi \left(1 - \frac{\nu}{2\pi T} \right) - \psi \left(1 - \frac{\nu}{4\pi T} \right) / 2 + \gamma/2 \right] = \frac{1}{\nu \pi T} \left[\psi \left(\frac{1}{2} \right) - \psi \left(\frac{1}{2} - \frac{\nu}{4\pi T} \right) \right],$$

with the first term

$$\frac{1}{8T^2} + \nu \frac{1}{2\pi^3 T^3} \frac{7}{8} \zeta(3) + \frac{\nu^2}{4 \cdot 96 T^4} + \dots \ . \tag{3.86}$$

For $\nu > 0$, the integration contour is closed in the lower half-plane, and the same result is obtained with ν replaced by $-\nu$. Thus one finds

$$L(\nu, \mathbf{0}) - L(0, \mathbf{0}) = \mathcal{N}(0)\left[\psi\left(\frac{1}{2}\right) - \psi\left(\frac{1}{2} + \frac{|\nu|}{4\pi T}\right)\right]$$

$$\approx -\mathcal{N}(0)\left[\frac{\pi}{8T}|\nu| - \nu^2 \frac{1}{2\pi^2 T^2}\frac{7}{8}\zeta(3) + \dots\right]. \quad (3.87)$$

The $\mathbf{k}$-dependence at $\nu = 0$ is obtained by expanding directly

$$L(0, \mathbf{k}) = \frac{T}{V}\sum_{\omega, \mathbf{p}} \frac{1}{i\omega - \xi(\mathbf{p} + \mathbf{k})}\frac{1}{-i\omega - \xi(\mathbf{p})} - \frac{1}{g}$$

$$= T\sum_{n=0}^{\infty}\frac{1}{V}\sum_{\omega, \mathbf{p}} \frac{1}{[i\omega - \xi(\mathbf{p})]^{n+1}}\left(\frac{\mathbf{p}\mathbf{k}}{m} + \frac{\mathbf{k}^2}{2m}\right)^n \frac{1}{-i\omega - \xi(\mathbf{p})} - \frac{1}{g}. \quad (3.88)$$

The sum over $\mathbf{p}$ may be split into radial and angular integrals [compare (3.62)]:

$$\frac{1}{V}\sum_{\mathbf{p}}\int \frac{d^3 p}{(2\pi)^3} \approx \mathcal{N}(0)\int d\xi \int \frac{d\hat{\mathbf{p}}}{4\pi}, \quad (3.89)$$

where the second integral runs over all unit vectors $\hat{\mathbf{p}}$ as follows:

$$\int \frac{d\hat{\mathbf{p}}}{4\pi} = \int_{-\pi}^{\pi} \frac{d\phi}{2\pi}\int_{-1}^{1}\frac{d\cos\theta}{2}. \quad (3.90)$$

Here θ and ϕ are the spherical angles of the momenta $\hat{\mathbf{p}}$.

The momentum integrals in (3.88) receive only contributions from a thin shell around the Fermi sphere, where $|\mathbf{p}| \approx p_F$, and there are only small corrections of the order $\mathcal{O}(T_D/T_F) \approx 10^{-3}$. Introducing now the *Fermi velocity* $v_F \equiv p_F/m$, for convenience, and performing the ξ-integrals in the form

$$\int d\xi \frac{1}{(i\omega - \xi)^{n+1}}\frac{1}{-i\omega - \xi} = (-i\,\mathrm{sgn}\,\omega)^n \frac{\pi}{2^n|\omega|^{n+1}}, \quad (3.91)$$

we find

$$L(0, \mathbf{k}) \approx 2\mathcal{N}(0)\mathrm{Re}\sum_{n=0}^{\infty}(-i)^n \frac{\pi}{2^n|\omega|^{n+1}}\int \frac{d\hat{\mathbf{p}}}{4\pi}\left(v_F \hat{\mathbf{p}}\mathbf{k} + \frac{\mathbf{k}^2}{2m}\right)^n - \frac{1}{g}. \quad (3.92)$$

For $\mathbf{k} = 0$, we recover the logarithmically divergent sum

$$L(0, \mathbf{0}) = \mathcal{N}(0)\sum_{\omega}\frac{\pi}{|\omega|} - \frac{1}{g}. \quad (3.93)$$

This is just another representation of the energy integral (3.65). It can therefore be made finite by the same cutoff and subtraction procedure as before.

The higher powers can be summed via formula (3.78) with the result

$$L(0, \mathbf{k}) = L(0, \mathbf{k}) + 2\mathcal{N}(0)\mathrm{Re}\sum_{n=1}^{\infty}\frac{(-i)^n}{2^n \pi^n T^n}(1 - 2^{-(n+1)})\zeta(n+1)\int \frac{d\hat{\mathbf{p}}}{4\pi}\left(v_F\hat{\mathbf{p}}\mathbf{k} + \frac{\mathbf{k}^2}{2m}\right)^n$$

$$= L(0, \mathbf{0}) + \mathcal{N}(0)\,\mathrm{Re}\int \frac{d\hat{\mathbf{p}}}{4\pi}\left[\psi\left(\frac{1}{2}\right) - \psi\left(\frac{1}{2} - i\left(v_F\hat{\mathbf{p}}\mathbf{k} + \frac{\mathbf{k}^2}{2m}\right)\frac{1}{4\pi T}\right)\right]. \quad (3.94)$$

Comparing this with Eq. (3.87), we see that the full $\mathbf{k}$- and ν-dependence is obtained by adding $|\nu|/4\pi T$ to the arguments of the second Digamma function. This can also be checked by a direct calculation. In the long-wavelength limit in which $kv_F/T \ll 1$, one has also

$$\frac{k^2/2m}{T} \approx \frac{k}{p_F}\frac{kv_F}{T} \ll \frac{kv_F}{T}, \tag{3.95}$$

and one may truncate the sum after the quadratic term as follows:

$$L(0,\mathbf{k}) = L(0,\mathbf{0}) + \sum \Lambda_{kl}(0)k_k k_l, \tag{3.96}$$

where

$$\Lambda_{kl}(0) = -2\mathcal{N}(0)\frac{1}{4\pi^2 T^2}\frac{7}{8}\zeta(3)v_F^2 \int \frac{d\hat{\mathbf{p}}}{4\pi}\hat{p}_k\hat{p}_l. \tag{3.97}$$

The angular integration yields

$$\int \frac{d\hat{\mathbf{p}}}{4\pi}\hat{p}_k\hat{p}_l = \frac{1}{3}\delta_{kl}, \tag{3.98}$$

so the lowest terms in the expansion of $L(\nu,\mathbf{k})$ are, for $kv_F \ll T$ and $\nu \ll T$,

$$L(\nu,\mathbf{k}) \approx L(0,\mathbf{0}) - \mathcal{N}(0)\left[\frac{\pi}{8T}|\nu| + \frac{1}{6\pi^2 T^2}\frac{7}{8}\zeta(3)v_F^2 k^2\right]. \tag{3.99}$$

The second term in (3.97) may also be conveniently calculated in x-space. For large $x \gg 1/p_F$, the Green function behaves like

$$G_0(\mathbf{x},\omega) \approx -\frac{m}{2\pi|\mathbf{x}|}\exp\left[ip_F|\mathbf{x}|\,\mathrm{sgn}\,\omega - \frac{|\omega|}{v_F}|\mathbf{x}|\right], \tag{3.100}$$

so that the second spatial derivatives of $\Delta(x)$ contribute to the expansion (3.52) a term

$$\int dx\left[\frac{1}{2}\int d^3x'\,T\sum_{\omega_n}G_0(\mathbf{x}-\mathbf{x}',\omega_n)G_0(\mathbf{x}-\mathbf{x}',-\omega_n)(x-x')_i(x-x')_j\right]\Delta^*(x)\nabla_i\nabla_j\Delta(x). \tag{3.101}$$

The expression in parentheses becomes

$$\frac{1}{2}\int d^3z\,T\sum_{\omega_n}\left(\frac{m}{2\pi|z|}\right)^2\exp\left(-2\frac{|\omega_n|}{v_F}|z|\right)z_i z_j$$

$$= \frac{1}{24}\delta_{ij}T\int d^3z\frac{1}{\sinh 2\pi|z|T/v_F} = \delta_{ij}\frac{7\zeta(3)}{48}\mathcal{N}(0)\frac{v_F^2}{\pi^2 T^2}, \tag{3.102}$$

which makes (3.101) agree with (3.96).

In many formulas to come it is useful to introduce the characteristic length parameter

$$\xi_0 \equiv \sqrt{\frac{7\zeta(3)}{48}} \frac{v_F}{\pi T_c}. \tag{3.103}$$

In proper physical units, the right-hand side carries a factor $\hbar/k_B$. Inserting $\zeta(3) \approx 1.202057$ from (3.81), this becomes

$$\xi_0 \approx 0.4187 \times \frac{v_F}{\pi T_c}. \tag{3.104}$$

Using $T_F \equiv \mu \equiv p_F^2/2m$, the right-hand side of (3.103) can also be written as

$$\xi_0 = \sqrt{\frac{7\zeta(3)}{48}} \frac{2 T_F p_F^{-1}}{\pi T_c} \approx 0.25 \times \frac{T_F}{T_c} p_F^{-1}. \tag{3.105}$$

In most superconductors, T_c is of the order of one degree Kelvin, about $1/1000$ of the Fermi temperature T_F. The length parameter ξ_0 is of the order of 1000 Å.

The low-frequency and long-wavelength result (3.99) corresponds, in the collective action (3.56), to a term[5]

$$\mathcal{A}_2[\Delta^*, \Delta] \approx -i\mathcal{N}(0) T \sum_{\nu \ll T, \mathbf{k}} \Delta^*(\nu, \mathbf{k}) \left\{ \left(1 - \frac{T}{T_c}\right) - \xi_0^2 \mathbf{k}^2 - \frac{\pi}{8T} |\nu| \right\} \Delta(\nu, \mathbf{k}). \tag{3.106}$$

This implies that for $T \leq T_c$, the field Δ has a correlation function

$$\overline{\Delta^*(\nu_n, \mathbf{k}) \Delta(\nu_m, \mathbf{k}')} = -(2\pi)^3 \delta^{(3)}(\mathbf{k} - \mathbf{k}') \frac{1}{T} \delta_{nm} \frac{1}{\mathcal{N}(0)} \left[-\frac{\pi}{8T} |\nu_n| + \left(1 - \frac{T}{T_c}\right) - \xi_0^2 \mathbf{k}^2 \right]^{-1}. \tag{3.107}$$

The spectrum of collective excitations can be extracted from this expression by continuing the energy back to real values from the upper half of the complex plane:

$$k_0 = -i\frac{8}{\pi}(T - T_c) - i\frac{8T}{\pi} \xi_0^2 \mathbf{k}^2. \tag{3.108}$$

These excitations are purely dissipative.

If the system is close enough to the critical temperature, all interaction terms except $\mathcal{A}_4[\Delta^*, \Delta]$ become irrelevant because of their high dimensions. And in $\mathcal{A}_4$ only the momentum-independent contributions are of interest, again because they have the lowest dimension.

---

[5]Note that only the Matsubara frequency $\nu_0 = 0$ satisfies the condition $\nu \ll T$. The neighborhood of $\nu_0 = 0$ with the linear behavior $|\nu|$ becomes visible only after analytic continuation of (3.107) to the retarded Green function which amounts to replacing $|\nu_n| \to -ik_0$.

Its calculation is standard, applying the procedure in Eq. (3.101) to the higher terms in the expansion (3.49):

$$
\mathcal{A}_4[\Delta^*\Delta] = \frac{i}{4}\mathrm{Tr}\,\mathrm{Tr}_{\mathrm{spin}}\left[\left(\frac{i}{i\partial_t - \xi(-i\boldsymbol{\nabla})}\Delta\delta\right)\left(\frac{i}{i\partial_t + \xi(i\boldsymbol{\nabla})}\delta\right)\Delta^*\delta\right]^2
$$

$$
= -\frac{i}{2}\int dx_1 dx_2 dx_3 dx_4 G_0(x_1 x_2)\tilde{G}_0(x_2 x_3)G_0(x_3 x_4)\tilde{G}_0(x_4 x_1)\Delta^*(x_1)\Delta(x_2)\Delta^*(x_3)\Delta(x_4)
$$

$$
\approx -\frac{1}{2}\int dx\,|\Delta(x)|^4 \int d^3 x_2 d^3 x_3 d^3 x_4
$$

$$
\times\ T\sum_{\omega_n}[G_0(\mathbf{x}-\mathbf{x}_2,\omega_n)G_0(\mathbf{x}_3-\mathbf{x}_2,-\omega_n)G_0(\mathbf{x}_3-\mathbf{x}_4,\omega_n)G_0(\mathbf{x}-\mathbf{x}_4,-\omega_n)]
$$

$$
\equiv -\frac{\beta}{2}\int dx|\Delta(x)|^4. \tag{3.109}
$$

The coefficient β is computed as follows:

$$
\beta = T\sum_{\omega_n \mathbf{p}}\frac{1}{(\omega_n^2 + \xi_{\mathbf{p}}^2)^2} \approx \mathcal{N}(0)T\sum_{\omega_n}\int d\xi \frac{1}{(\omega_n^2 + \xi^2)^2} = \mathcal{N}(0)\frac{\pi}{2}T\sum_{\omega_n}\frac{1}{|\omega_n|^3}
$$

$$
= \mathcal{N}(0)\frac{7\zeta(3)}{8(\pi T_c)^2} = 6\mathcal{N}(0)\frac{\xi_0^2}{v_F^2} = 9\frac{\rho}{m^2}\frac{\xi_0^2}{v_F^4} \approx 10^{-3}\frac{p_F^3}{T_F T_c^2}. \tag{3.110}
$$

In proper physical units, the right-hand side carries a factor $1/\hbar^2$. The time-independent part of this action at the classical level has been derived long time ago by Gorkov on the basis of Green function techniques [11, 12]. His technical manipulations are exactly the same as presented here. The difference lies only in the theoretical foundation [15, 16, 17, 19] and the ensuing prescriptions on how to improve the approximations. Our action of (3.8) is the *exact* translation of the fundamental theory into pair fields. These fields can be turned into quantum fields in the standard fashion, by going from functional formalism to the operator language. The result is a perturbation theory of Δ-fields with (3.107) as a free propagator and $\mathcal{A}_n, n > 2$ treated as perturbations. The higher terms $\mathcal{A}_6, \mathcal{A}_8, \ldots$ are very weak residual interactions as far as long distance questions are concerned. In fact, for the calculation of the critical indices, $\mathcal{A}_2$ and $\mathcal{A}_4$ contain the whole relevant information about the system.

3.2.1 Inclusion of Electromagnetic Fields into the Pair Field Theory

The original action $\mathcal{A}$ in (3.3) can be made invariant under general spacetime-dependent gauge transformations

$$
\psi(\mathbf{x}, T) \to \exp[-i\Lambda(\mathbf{x}, t)]\psi(\mathbf{x}, t). \tag{3.111}
$$

Such transformations can be absorbed into an electromagnetic vector potential

$$
A = (\varphi, \mathbf{A}) \tag{3.112}
$$

by letting it transform via the addition of a pure derivative

$$\varphi \ \to\ \varphi - \frac{1}{e}\partial_t \Lambda, \qquad A_i \ \to\ A_i + \frac{c}{e}\nabla_i \Lambda. \tag{3.113}$$

Then the action with the well-known minimal replacement

$$\mathcal{A}[\psi^*, \psi]\Big|_{i\partial_t \to i\partial_t + \frac{1}{e}\varphi, -\nabla_i \to -i\nabla_i + \frac{c}{e}A_i} \tag{3.114}$$

is invariant under (3.111) and (3.113). Adding to this the action of the electromagnetic field itself in the Coulomb gauge, $\nabla\mathbf{A} = 0$, we arrive at the complete action of the superconductor:

$$\mathcal{A}_{\text{sc}} = \mathcal{A}[\psi^*, \psi]\Big|_{i\partial_t \to i\partial_t + \frac{1}{e}\varphi, -\nabla_i \to -i\nabla_i + \frac{c}{e}A_i} + \frac{1}{8\pi}\int dx \left(-\varphi\nabla^2\varphi + \frac{1}{c^2}A^2 + A\nabla^2 A\right). \tag{3.115}$$

Since the final pair action (3.8) describes the same system as the initial action (1.45), it certainly has to possess the same invariance after inclusion of electromagnetism. From the constraint equation (3.4) we see

$$\Delta(x, x') \to \exp\{-i[\Lambda(x) + \Lambda(x')]\}\,\Delta(x, x'). \tag{3.116}$$

For the local pair field appearing in (3.42) this gives

$$\Delta(x) \to \exp[-2i\Lambda(x)]\,\Delta(x). \tag{3.117}$$

Near the critial temperature, we approximate the electron pair action in (6.75) by a sum of the quadratic action $\mathcal{A}_2$ in Eq. (3.106) and a fourth-order term $\mathcal{A}_4$ of Eq. (3.109). The first is made gauge invariant by the minimal substitution

$$i\partial_t \to i\partial_t + 2e\varphi, \qquad -i\nabla_i \to -i\nabla_i + 2\frac{e}{c}A_i,$$
$$k_0 \to k_0 + 2e\varphi, \qquad k_i \to k_i + 2\frac{e}{c}A_i. \tag{3.118}$$

This leads to the full time-dependent Lagrangian close to the critical point:

$$\begin{aligned}
\mathcal{L} \ =\ & \frac{\mathcal{N}(0)\pi}{8T}\Delta^*(x)(-\partial_t + 2ie\varphi)\Delta(x) + \mathcal{N}(0)\left(1 - \frac{T}{T_c}\right)\Delta^*\Delta \\
& - \mathcal{N}(0)\xi_0^2\left(\nabla_i - 2i\frac{e}{c}A_i\right)\Delta^*(x)\left(\nabla_i + 2i\frac{e}{c}A_i\right)\Delta(x) \\
& - 3\mathcal{N}(0)\frac{\xi_0^2}{v_F^2}|\Delta(x)|^4 + \frac{1}{8\pi}\left(-\varphi\nabla^2\varphi + \frac{1}{c^2}\dot{A}^2 + A\nabla^2 A\right).
\end{aligned} \tag{3.119}$$

The discussion of this Lagrangian is standard. At the classical level there are, above T_c, doubly charged pair states that possess a chemical potential

$$\mu_{\text{pair}} = L(0) = \mathcal{N}(0)\left(1 - \frac{T}{T_c}\right) = \frac{3}{2}\frac{\rho}{m^2}\frac{1}{p_F^2}\left(1 - \frac{T}{T_c}\right) < 0; \qquad T > T_c. \tag{3.120}$$

In proper physical units, the right-hand side carries a factor $1/\hbar^3$.

Below T_c the chemical potential becomes positive causing an instability which settles, due to the stabilizing fourth-order term, at a nonzero field value (the "gap")

$$\Delta_0(T) = \sqrt{\frac{\mu_{\text{pair}}}{\beta}} = \sqrt{\frac{8}{7\zeta(3)}\pi T_c \left(1 - \frac{T}{T_c}\right)^{1/2}}. \tag{3.121}$$

Inserting $\zeta(3) \approx 1.202057$, this is approximately

$$\Delta_0(T) \approx 3.063 \times T_c \left(1 - \frac{T}{T_c}\right)^{1/2}. \tag{3.122}$$

The new vacuum obviously breaks gauge invariance spontaneously: the field Δ will now fluctuate in size with a chemical potential

$$\mu_{\text{pair}} = -2\mathcal{N}(0)\left(1 - \frac{T}{T_c}\right) < 0; T < T_c. \tag{3.123}$$

For static spatially constant pair fields, the energy density in the Landau expansion up to $|\Delta|^4$ is given by the minimum of

$$f = \mu_{\text{pair}}|\Delta|^2 + \frac{\beta}{2}|\Delta|^4. \tag{3.124}$$

The minimum lies at the gap (3.154) where it has the value

$$f^{\text{min}} = \frac{\mu_{\text{pair}}}{2}|\Delta|^2 = -\frac{\mu_{\text{pair}}^2}{2\beta} = -\frac{\rho}{m^2}\left(1 - \frac{T}{T_c}\right)\frac{\hbar^2}{2\xi_0^2} \times \frac{1}{4}. \tag{3.125}$$

Its negative value is the so-called *condensation energy density* $f_c = -f^{\text{min}}$.

Due to the gradient terms in (3.119), spatial changes of the absolute size of the field Δ can take place over a length scale

$$\xi_c(T) \equiv \sqrt{\frac{\text{coefficient of } |\boldsymbol{\nabla}\Delta|^2}{|\mu_{\text{pair}}|}} = \xi_0 \left(1 - \frac{T}{T_c}\right)^{-1/2}, \tag{3.126}$$

called the temperature-dependent coherence length [11, 12]. The azimuthal fluctuations experience a different fate in the absence of electromagnetism; they have a vanishing chemical potential due to the invariance of $\mathcal{L}$ under phase rotations. As an electromagnetic field is turned on, the new center of oscillations (3.121) is seen in (3.119) to generate a mass term $1/8\pi\mu_A^2 A^2$ for the photon. The vector potential acquires a mass

$$\mu_A^2 = 8\pi \text{ coefficient of } A^2 \text{ in } |\boldsymbol{\nabla}\Delta|^2 \text{ -term} = 8\omega\frac{4e^2}{c^2}\mathcal{N}(0)\xi_0^2\Delta_0^2. \tag{3.127}$$

This mass limits the penetration of the magnetic fields into a superconductor. The penetration depth is defined as [11, 12]:

$$\lambda(T) \equiv \mu_A^{-1} = \sqrt{\frac{3}{\pi \mathcal{N}(0)} \frac{c}{4 e v_F}} \left(1 - \frac{T}{T_c}\right)^{-1/2}$$

$$= \sqrt{\frac{3\pi}{8}} \sqrt{\frac{c}{v_F \alpha}} p_F^{-1} \left(1 - \frac{T}{T_c}\right)^{-1/2}. \tag{3.128}$$

Here we have introduced the fine-structure constant

$$\alpha = \frac{e^2}{\hbar c} \approx \frac{1}{137}. \tag{3.129}$$

The ratio

$$\kappa(T) \equiv \frac{\lambda(T)}{\xi(T)} = \sqrt{\frac{9\pi^3}{14\zeta(3)}} \sqrt{\frac{c}{v_F \alpha} \frac{T_c}{T_F}} \approx 4.1 \times \sqrt{\frac{c}{v_F \alpha} \frac{T_c}{T_F}} \tag{3.130}$$

is the Ginzburg-Landau parameter that decides whether it is energetically preferable for the superconductor to have flux lines invading it or not. For $\kappa > 1/\sqrt{2}$ they do invade, and the superconductor is said to be of *type II*, for $\kappa < 1/\sqrt{2}$ they don't, and the superconductor is of *type I*.

3.3   Far below the Critical Temperature

We have seen in the last section that for T smaller than T_c the chemical potential of the pair field becomes positive, causing oscillations around a new minimum which is the gap value Δ_0 given by (3.121). That formula was based on the expansion (3.134) of the pair action and can be valid only as long as $\Delta \ll T_c$, i.e., $T \approx T_c$. If T drops far below T_c, one must account for Δ_0 non-perturbatively by inserting it as an open parameter into G_Δ of the collective action (3.8) and by extremizing $\mathcal{A}[\Delta^*, \Delta]$. If the extremum lies at $\Delta_0(x, x')$, we insert

$$\Delta(x, x') = \Delta_0(x - x') + \Delta'(x, x') \tag{3.131}$$

into the collective action (3.8)

$$\mathcal{A}[\Delta^*, \Delta] = \pm \frac{i}{2} \text{Tr} \log \left[i G_{\Delta_0 + \Delta'}^{-1}(x, x') \right] + \frac{1}{2} \int dx dx' |\Delta_0 + \Delta'(x, x')|^2 \frac{1}{V(x, x')}, \tag{3.132}$$

and expand everything in powers of $\Delta'(x, x')$. The Green function of the pair field is expanded around

$$G_{\Delta_0}(x, x') = i \begin{pmatrix} [i\partial_t - \xi(-i\boldsymbol{\nabla})]\delta & -\Delta_0 \\ -\Delta_0^\dagger & \mp i[\partial_t - \xi(i\boldsymbol{\nabla})]\delta \end{pmatrix}^{-1} (x, x') \tag{3.133}$$

as follows:

$$\mathbf{G}_\Delta = \mathbf{G}_{\Delta_0} - i\mathbf{G}_{\Delta_0}\begin{pmatrix} 0 & \Delta' \\ \Delta'^* & 0 \end{pmatrix}\mathbf{G}_{\Delta_0} - \mathbf{G}_{\Delta_0}\begin{pmatrix} 0 & \Delta' \\ \Delta'^* & 0 \end{pmatrix}\mathbf{G}_{\Delta_0}\begin{pmatrix} 0 & \Delta' \\ \Delta'^* & 0 \end{pmatrix}\mathbf{G}_{\Delta_0} + \dots .$$

(3.134)

This replaces the expansion (3.134).

3.3.1 The Gap

In the underlying theory of fields ψ^*, ψ, the matrix $\mathbf{G}_{\Delta_0}$ collects the bare one-particle Green functions:

$$\mathbf{G}_{\Delta_0}(x, x') = \begin{pmatrix} \overbrace{\psi(x)\psi^\dagger(x')} & \overbrace{\psi(x)\psi(x')} \\ \overbrace{\psi^\dagger(x)\psi^\dagger(x')} & \overbrace{\psi^\dagger(x)\psi(x')} \end{pmatrix}.$$

(3.135)

The off-diagonal (also called *anomalous*) Green functions are nonvanishing. This signalizes that for $T < T_c$, a condensate is present in the vacuum. The presence of Δ_0 causes a linear dependence of the action (3.132) on $\Delta'(x, x')$:

$$\mathcal{A}_1[\Delta'^*, \Delta'] = \pm\mathrm{Tr}\left[\mathbf{G}_{\Delta_0}\begin{pmatrix} 0 & \Delta' \\ \Delta'^* & 0 \end{pmatrix}\right]$$
$$+ \frac{1}{2}\int dx dx'\left[\Delta_0^*(x - x')\Delta'(x, x')\frac{1}{V(x, x')} + \text{c.c.}\right].$$

(3.136)

The gap function may now be determined optimally by minimizing the action with respect to $\delta\Delta'$ at $\Delta' = 0$ which amounts to the elimination of $\mathcal{A}_1[\Delta'^*, \Delta']$. Taking the functional derivative of (3.136) gives the *gap, equation*

$$\Delta_0(x - x') = \pm V(x - x')\,\mathrm{tr}_{2\times2}\left[\mathbf{G}_{\Delta_0}(x, x')\frac{\tau^-}{2}\right],$$

(3.137)

where $\tau^-/2$ is the matrix $\begin{pmatrix} 0 & 0 \\ 1 & 0 \end{pmatrix}$ in the 2×2 dimensional matrix space of (3.9). If the potential is instantaneous, the gap has a factor $\delta(t - t')$, i.e.,

$$\Delta_0(x - x') \equiv \delta(t - t') \times \Delta_0(\mathbf{x} - \mathbf{x}'),$$

(3.138)

and the Fourier transform of the spatial part satisfies

$$\Delta_0(\mathbf{p}) = \pm\frac{T}{V}\sum_{\omega,\mathbf{p}'} V(\mathbf{p} - \mathbf{p}')\,\mathrm{tr}_{2\times2}\left[\mathbf{G}_{\Delta_0}(\omega, \mathbf{p}')\frac{\tau^-}{2}\right].$$

(3.139)

Inverting (3.133) renders the propagator:

$$\mathbf{G}_{\Delta_0}(\tau, x) =$$

(3.140)

$$\mp\frac{T}{V}\sum_{\omega,\mathbf{p}} \exp\left[-i(\omega\tau - \mathbf{p}\mathbf{x})\right]\frac{1}{\omega^2 + \xi^2(\mathbf{p}) \mp |\Delta_0(\mathbf{p})|^2}\begin{pmatrix} \mp[i\omega + \xi(\mathbf{p})] & \Delta_0(\mathbf{p}) \\ \Delta_0^*(\mathbf{p}) & [i\omega - \xi(\mathbf{p})] \end{pmatrix},$$

so that the gap equation (3.139) takes the explicit form

$$\Delta_0(\mathbf{p}) = -\frac{T}{V}\sum_{\omega,\mathbf{p}'} V(\mathbf{p}-\mathbf{p}')\frac{\Delta_0(\mathbf{p}')}{\omega^2+\xi_{\mathbf{p}'}^2 \mp |\Delta_0(\mathbf{p}')|^2}. \tag{3.141}$$

Performing the frequency sum yields

$$\Delta_0(\mathbf{p}) = -\frac{1}{V}\sum_{\mathbf{p}'} V(\mathbf{p}-\mathbf{p}')\frac{\Delta_0(\mathbf{p}')}{2E_{\mathbf{p}'}}\tanh^{\mp 1}\frac{E_{\mathbf{p}'}}{2T}, \tag{3.142}$$

where

$$E_{\mathbf{p}} = \sqrt{\xi_{\mathbf{p}}^2 \mp |\Delta_0(\mathbf{p})|^2}. \tag{3.143}$$

For the case of the superconductor with an attractive local potential

$$V(x-x') = -g\delta^{(3)}(\mathbf{x}-\mathbf{x}')\delta(t-t') \tag{3.144}$$

this becomes

$$\Delta_0 = g\frac{T}{V}\sum_{\omega,\mathbf{p}} \frac{\Delta_0}{\omega^2+\xi_{\mathbf{p}}^2+|\Delta_0|^2} = \left[g\frac{1}{V}\sum_{\mathbf{p}}\frac{1}{2E_{\mathbf{p}}}\tanh\frac{E_{\mathbf{p}}}{2T}\right]\Delta_0. \tag{3.145}$$

There is a nonzero gap if

$$\frac{1}{V}\sum_{\mathbf{p}}\frac{1}{2E_{\mathbf{p}}}\tanh\frac{E_{\mathbf{p}}}{2T} = \frac{1}{g}. \tag{3.146}$$

Let $T = T_c$ denote the critical temperature at which the gap vanishes. At that temperature, the gap vanishes and $E_{\mathbf{p}} = \xi_{\mathbf{p}}$, so that Eq. (3.145) determines the same T_c as the previous Eq. (3.72) which were derived for $T \approx T_c$ in a different fashion. The result (3.145) holds for any temperature.

The full temperature dependence of the gap cannot be obtained in closed form from (3.146). For $T \approx T_c$ one may expand directly (3.145) in powers of Δ_0:

$$1 = g\frac{T}{V}\sum_{\omega,\mathbf{p}}\left\{\frac{1}{\omega^2+\xi_{\mathbf{p}}^2} - \Delta_0^2\frac{1}{\left[\omega^2+\xi_{\mathbf{p}}^2\right]^2} + \ldots\right\} \tag{3.147}$$

The first sum on the right-hand side yields the same integral as in (3.65), and we obtain

$$\begin{aligned}
1 &= g\mathcal{N}(0)\left[\log\frac{\omega_D}{T}2\frac{e^\gamma}{\pi} - \Delta_0^2\frac{7\zeta(3)}{8\pi^2 T^2} + \ldots\right] \\
&= 1 + \mathcal{N}(0)\left[\left(1-\frac{T}{T_c}\right) - \Delta_0^2\frac{7\zeta(3)}{8\pi^2 T^2} + \ldots\right]. \tag{3.148}
\end{aligned}$$

From this we find

$$\Delta_0^2(T) \approx \frac{8}{7\zeta(3)} \pi^2 T_c^2 \left(1 - \frac{T}{T_c}\right), \tag{3.149}$$

in agreement with (3.121). For very small temperatures, on the other hand, Eq. (3.145) can be written as

$$
\begin{aligned}
1 &= g\mathcal{N}(0) \int_0^{\omega_D} \frac{d\xi}{\sqrt{\xi^2 + \Delta_0^2}} \left[1 - 2\exp(-\sqrt{\xi^2 + \Delta_0^2}/T) - \dots\right] \\
&= g\mathcal{N}(0) \left[\log \frac{2\omega_D}{\Delta_0} - 2K_0(\Delta_0/T)\right] + \dots .
\end{aligned} \tag{3.150}
$$

For small T, the function $2K_0(\Delta_0/T)$ vanishes exponentially fast like

$$2K_0\left(\frac{\Delta_0}{T}\right) \to \frac{1}{\Delta_0} \sqrt{2\pi T \Delta_0} e^{-\Delta_0/T}. \tag{3.151}$$

At $T = 0$ we find the gap

$$\Delta_0(0) = 2\omega_D e^{-1/g\mathcal{N}(0)}. \tag{3.152}$$

Using Eq. (3.72), this is related to the critical temperature T_c by

$$\Delta_0(0) = \pi e^{-\gamma} T_c \approx 1.76 \times T_c. \tag{3.153}$$

This value is approached by the the temperature-dependent gap $\Delta_0(T)$ exponentially fast for $T \to 0$, since from (3.150)

$$\log \frac{\Delta_0(T)}{\Delta_0(0)} \approx \frac{\Delta_0(T)}{\Delta_0(0)} - 1 \approx -\frac{1}{\Delta_0(0)} \sqrt{2\pi T \Delta_0(0)} e^{-\Delta_0(0)/T}. \tag{3.154}$$

For arbitrary T, the calculation of (3.268) is conveniently done via the expansion into Matsubara frequencies

$$\frac{1}{2E} \tanh \frac{E}{2T} = \frac{1}{2E} T \sum_{\omega_n} \left(\frac{1}{i\omega_n + E} - \frac{1}{i\omega_n - E}\right) = T \sum_{\omega_n} \frac{1}{\omega_n^2 + \xi^2 + \Delta_0^2}. \tag{3.155}$$

This can be integrated over ξ and we find

$$\log \frac{T}{T_c} = 2\pi T \sum_{\omega_n > 0} \left(\frac{1}{\sqrt{\omega_n^2 + \Delta_0^2}} - \frac{1}{\omega_n}\right). \tag{3.156}$$

Here it is convenient to introduce the auxiliary dimensionless quantity

$$\delta = \frac{\Delta_0}{\pi T}, \tag{3.157}$$

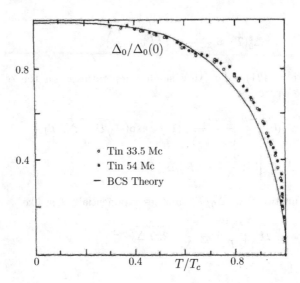

FIGURE 3.7 Energy gap of a superconductor as a function of temperature. The points are from ultrasonic attenuation data at two different frequencies measured by Morse and Bohm [20].

and a reduced version of the Matsubara frequencies:

$$x_n \equiv (2n + 1)/\delta. \tag{3.158}$$

Then the gap equation (3.156) takes the form

$$\log \frac{T}{T_c} = \frac{2}{\delta} \sum_{n=0}^{\infty} \left(\frac{1}{\sqrt{x_n^2 + 1}} - \frac{1}{x_n} \right). \tag{3.159}$$

The temperature dependence of Δ_0 is plotted in Fig. 3.7. The behavior in the vicinity of the critical temperature T_c can be extracted from Eq. (3.159) by expanding the sum under the assumption of small δ and large x_n. The leading term gives

$$\log \frac{T}{T_c} \approx -\frac{2}{\delta} \sum_{n=0}^{\infty} \frac{1}{2x_n^3} = -\delta^2 \sum_{n=0}^{\infty} \frac{1}{(2n+1)^2} = -\delta^2 \frac{7}{8} \zeta(3) \tag{3.160}$$

so that

$$\delta^2 \approx \frac{8}{7\zeta(3)} \left(1 - \frac{T}{T_c} \right) \tag{3.161}$$

and

$$\frac{\Delta_0}{T_c} = \pi \delta = \pi \sqrt{\frac{8}{7\zeta(3)}} \left(1 - \frac{T}{T_c} \right)^{1/2} \approx 3.063 \times \left(1 - \frac{T}{T_c} \right)^{1/2}, \tag{3.162}$$

as before in (3.122).

3.3.2   The Free Pair Field

The quadratic part of the action (3.132) in the pair fields Δ' reads

$$
\mathcal{A}_2[\Delta'^*, \Delta'] = \pm \frac{i}{4} \text{Tr} \left[\mathbf{G}_{\Delta_0} \begin{pmatrix} 0 & \Delta' \\ \Delta'^* & 0 \end{pmatrix} \mathbf{G}_{\Delta_0} \begin{pmatrix} 0 & \Delta' \\ \Delta'^* & 0 \end{pmatrix} \right]
$$
$$
+ \frac{1}{2} \int dx dx' |\Delta(x, x')|^2 \frac{1}{V(x, x')}, \tag{3.163}
$$

with an equation of motion

$$
\begin{pmatrix} \Delta'(x, x') \\ \Delta'^*(x, x') \end{pmatrix} = \mp \frac{i}{2} V(x, x') \text{tr}_{2 \times 2} \left[\mathbf{G}_{\Delta_0} \begin{pmatrix} 0 & \Delta' \\ \Delta'^* & 0 \end{pmatrix} \mathbf{G}_{\Delta_0} \begin{pmatrix} 0 & \Delta' \\ \Delta'^* & 0 \end{pmatrix} \frac{\tau^{\pm}}{2} \right] (x, x'), \tag{3.164}
$$

rather than (3.21). Inserting the momentum space representation (3.140) of $\mathbf{G}_{\Delta_0}$, this renders the two equations

$$
\Delta'(P|q) = -\frac{T}{V} \sum_{p'} V(P - P') \left[l_{11}(P'|q) \Delta'(P'|q) + l_{12}(P'|q) \Delta'^*(P'|q) \right], \tag{3.165}
$$

$$
\Delta'^*(P|q) = -\frac{T}{V} \sum_{p'} V(P - P') \left[l_{11}(P'|q) \Delta'^*(P'|q) + l_{12}(P'|q) \Delta'(P'|q) \right], \tag{3.166}
$$

where (with $P_0 \equiv i\omega$)

$$
l_{11}(P|q) = \frac{\omega^2 - \nu^2/4 + \xi_{\mathbf{q}/2 + \mathbf{P}} \, \xi_{\mathbf{q}/2 - \mathbf{P}}}{\left[(\omega + \nu/2)^2 + E_{\mathbf{q}/2 + \mathbf{P}}^2 \right] \left[(\omega - \nu/2)^2 + E_{\mathbf{q}/2 - s\mathbf{P}}^2 \right]}, \tag{3.167}
$$

$$
l_{12}(P|q) = \pm \frac{\Delta_0^2 (\mathbf{q}/2 + \mathbf{P})}{\left[(\omega + \nu/2)^2 + E_{\mathbf{q}/2 + \mathbf{P}}^2 \right] \left[(\omega - \nu/2)^2 + E_{\mathbf{q}/2 - s\mathbf{P}}^2 \right]}. \tag{3.168}
$$

Thus for $T \ll T_c$ the simple bound-state problem (3.31) takes quite a different form due to the presence of the off-diagonal terms in the propagator (3.140).

Note that the parenthesis on the right-hand side of Eqs. (3.165) and (3.166) contain precisely the Bethe-Salpeter wave function of the bound state (compare (3.24), (3.26) in the gapless case)

$$
\psi(P|q) \equiv \pm \frac{i}{2} \text{tr}_{2 \times 2} \left[\mathbf{G}_{\Delta_0} \left(\frac{q}{2} + P \right) \begin{pmatrix} 0 & \Delta'(P|q) \\ \Delta'^*(P|q) & 0 \end{pmatrix} \mathbf{G}_{\Delta_0} \left(P - \frac{q}{2} \right) \frac{\tau^*}{2} \right]
$$
$$
= l_{11}(P|q) \Delta'(P|q) l_{12}(P'|q) \Delta'^*(P|q). \tag{3.169}
$$

Not much is known on the general properties of the solutions of equations (3.166). Even for the simple case of a $\delta^{(4)}(x - x')$ function potential, only the long wavelength spectrum has been studied. There is, however, one important solution which always occurs for $T < T_c$ due to symmetry considerations: The original action (1.45) is symmetric under phase transformations

$$
\psi \to e^{-i\alpha} \psi, \tag{3.170}
$$

guaranteeing the conservation of the particle number. If the pair fields oscillate around a nonzero value $\Delta_0(x - x')$, this symmetry is spontaneously broken (since the complex c-number does not take part in such a phase transformation). As a consequence, there must now be an excitation of the system related to the infinitesimal symmetry transformation. This is known as the *Goldstone theorem*. If the whole system is transformed at once, we are dealing with momentum $\mathbf{q} = 0$. The symmetry ensures that this state has also a vanishing energy $q_0 = 0$. Indeed, suppose the gap equation would has a non-trivial solution $\Delta_0(P) \equiv 0$. Then we can easily see that a solution of the bound-state equations (3.165) and (3.166) at $q = 0$ would be

$$\Delta'(P|q = 0) \equiv i\Delta_0(P). \tag{3.171}$$

Take

$$l_{11}(P|q = 0) = \frac{\omega^2 + \xi_{\mathbf{P}}^2}{\omega^2 + E_{\mathbf{P}}^2}, \tag{3.172}$$

and insert (3.171) into (3.166). The associated gap is

$$
\begin{aligned}
\Delta_0(P) &= \frac{T}{V}\sum_{P'} V(P - P')\left\{ \frac{1}{(\omega'^2 + E_{\mathbf{P}'}^2)^2}\left[\omega^2 + \xi_{\mathbf{P}'}^2 \mp |\Delta_0(P')|^2\right] \right\} \\
&= -\frac{T}{V}\sum_{P'} V(P - P')\frac{\Delta_0(P')}{\omega'^2 + E_{\mathbf{P}'}^2},
\end{aligned} \tag{3.173}
$$

i.e., the bound-state equation at $q = 0$ reduces to the gap equation. Moreover, due to (3.169), the expression

$$\psi_0(P|q = 0) \equiv \frac{1}{\omega^2 + E_{\mathbf{P}}^2}\Delta_0(P) \tag{3.174}$$

is the Bethe-Salpeter wave function of the bound state with $q = 0$. If the potential is instantaneous, it is possible to calculate the *equal-time amplitude* $\psi_0(\mathbf{x} - \mathbf{x}', \tau) \equiv \psi(\mathbf{x}, \tau; \mathbf{x}'\tau)$. Doing the sum over ω in (3.173) we find

$$
\begin{aligned}
\psi_0(\mathbf{x} - \mathbf{x}', \tau) &= \int \frac{d^3P}{(2\pi)^3}e^{i\mathbf{P}(\mathbf{x}-\mathbf{x}')}T\sum_{\omega}\psi_0(\mathbf{P}|q = 0) \\
&= \int \frac{d^3P}{(2\pi)^3}e^{i\mathbf{P}(\mathbf{x}-\mathbf{x}')}\tanh^{\mp 1}\frac{E_{\mathbf{P}}}{2T}\frac{\Delta_0(\mathbf{P})}{2E_{\mathbf{P}}}.
\end{aligned} \tag{3.175}
$$

Note that the time-dependence of this amplitude happens to be trivial since the bound state has no energy. The $q = 0$-bound state described by the wave function $\psi_0(\mathbf{x} - \mathbf{x}', \tau) = \psi_0(\mathbf{x} - \mathbf{x}', 0)$ is called a *Cooper pair*.

In configuration space, (3.173) amounts to a Schrödinger type of equation:

$$-2E_{(-i\boldsymbol{\nabla})}\psi_0(\mathbf{x}) = V(\mathbf{x})\psi(\mathbf{x}). \tag{3.176}$$

This may be interpreted as the $q = 0$ -bound state of two quasi-particles whose energies are

$$E_{\mathbf{P}} = \sqrt{\xi^2(\mathbf{P}) \mp |\Delta_0(\mathbf{P})|^2}. \tag{3.177}$$

Note that Eq. (3.176) is non-linear since $\Delta_0(\mathbf{P})$ in $E_{\mathbf{P}}$ is a functional of $\psi_0(x)$. In order to establish contact with the standard discussion of pairing effects via canonical transformations (see Ref. [11]), a few comments may be useful. Let us restrict the discussion to instantaneous potentials. From equation (3.140) one sees that the propagator $\mathbf{G}_\Delta$ can be diagonalized by means of an ω-independent Bogoliubov transformation

$$B_{\mathbf{p}} = \begin{pmatrix} u_{\mathbf{p}}^* & \mp v_{\mathbf{p}}^* \\ -v_{\mathbf{p}} & u_{\mathbf{p}} \end{pmatrix}, \tag{3.178}$$

where

$$|u_{\mathbf{p}}|^2 = \frac{1}{2}\left[1 + \frac{\xi_{\mathbf{p}}}{E_{\mathbf{p}}}\right], \quad |v_{\mathbf{p}}|^2 = \mp\frac{1}{2}\left[1 - \frac{\xi_{\mathbf{p}}}{E_{\mathbf{p}}}\right], \quad 2u_{\mathbf{p}}v_{\mathbf{p}}^* = \frac{\Delta_0(\mathbf{p})}{E_{\mathbf{p}}}. \tag{3.179}$$

Since

$$|u_{\mathbf{p}}|^2 \mp |v_{\mathbf{p}}|^2 = 1, \tag{3.180}$$

one finds

$$B_{\mathbf{p}}^{-1} = \begin{pmatrix} u_{\mathbf{p}} & \pm v_{\mathbf{p}}^* \\ v_{\mathbf{p}} & u_{\mathbf{p}}^* \end{pmatrix} = \left\{ \begin{array}{ll} \sigma_3 B_{\mathbf{p}}^\dagger \sigma_3 & \text{for bosons} \\ B_{\mathbf{p}}^\dagger & \text{for fermions} \end{array} \right\}. \tag{3.181}$$

Thus $B_{\mathbf{p}}$ is a unitary spin rotation in the Fermi case, and a non-unitary element of the non-compact group $SU(1,1)$ [21]. After the transformation, the propagator is diagonal:

$$\begin{aligned} \mathbf{G}_{\Delta_0}^d(\omega, \mathbf{p}) &= B_{\mathbf{p}} \mathbf{G}_{\Delta_0}(\omega, \mathbf{p}) B_{\mathbf{p}}^\dagger \\ &= -\begin{pmatrix} (i\omega - E_{\mathbf{p}})^{-1} & \\ & \pm(i\omega + E_{\mathbf{p}})^{-1} \end{pmatrix}. \end{aligned} \tag{3.182}$$

The poles in the complex ω-plane may be interpreted as quasi-particles of energy

$$E_{\mathbf{p}} = \sqrt{\xi_{\mathbf{p}}^2 \mp |\Delta_0(\mathbf{p})|^2}. \tag{3.183}$$

In fact, we can introduce new creation and annihilation operators

$$\begin{pmatrix} \alpha_{\mathbf{p}}(\tau) \\ \beta_{-\mathbf{p}}^\dagger(\tau) \end{pmatrix} = B_{\mathbf{p}} \begin{pmatrix} a_{\mathbf{p}}(\tau) \\ a_{-\mathbf{p}}^\dagger(\tau) \end{pmatrix}. \tag{3.184}$$

Their propagators would be

$$
\mathbf{G}^d_{\Delta_0}(\tau - \tau', \mathbf{p}) \equiv \begin{pmatrix} \overline{\alpha_{\mathbf{p}}(\tau)\alpha^\dagger_{\mathbf{p}}(\tau')} & \overline{\alpha_{\mathbf{p}}(\tau)\beta_{-\mathbf{p}}(\tau')} \\ \overline{\beta^\dagger_{-\mathbf{p}}(\tau)\alpha^\dagger_{\mathbf{p}}(\tau')} & \overline{\beta^\dagger_{-\mathbf{p}}(\tau)\beta_{-\mathbf{p}}(\tau')} \end{pmatrix}
$$

$$
= T \sum_\omega e^{-i\omega(\tau - \tau')} \mathbf{G}^d_{\Delta_0}(\omega, \mathbf{p}) . \tag{3.185}
$$

At equal "times", where $\tau' = \tau + \epsilon$, the frequency sums may be performed with the result

$$
\sum_\omega \mathbf{G}^d_{\Delta_0}(\omega, \mathbf{p}) = \begin{pmatrix} \pm n^{b,f}_{\mathbf{p}} & 0 \\ 0 & \pm 1 + n^{b,f}_{\mathbf{p}} \end{pmatrix} , \tag{3.186}
$$

where $n^{b,f}_{\mathbf{p}}$ are the usual Bose and Fermi occupation factors for the quasi-particle energy (3.177):

$$
n^{b,f}_{\mathbf{p}} = \frac{1}{e^{E_{\mathbf{p}}/T} \mp 1} . \tag{3.187}
$$

The corresponding frequency sum for the original propagator becomes

$$
T \sum_\omega \mathbf{G}_{\Delta_0}(\omega, \mathbf{p}) = T \sum_\omega B^{-1}_{\mathbf{p}} \mathbf{G}^d_{\Delta_0}(\omega, \mathbf{p}) B^{-1\dagger}_{\mathbf{p}} \tag{3.188}
$$

$$
= \begin{pmatrix} \pm|v_{\mathbf{p}}|^2 \tanh^{\mp 1}\dfrac{E_{\mathbf{p}}}{2T} \pm n^{b,f}_{\mathbf{p}} & u_{\mathbf{p}}v^*_{\mathbf{p}}\tanh^{\mp 1}\dfrac{E_{\mathbf{p}}}{2T} \\ u^*_{\mathbf{p}}v_{\mathbf{p}}\tanh^{\pm 1}\dfrac{E_{\mathbf{p}}}{2T} & \pm|u_{\mathbf{p}}|^2\tanh^{\mp 1}\dfrac{E_{\mathbf{p}}}{2T} - n^{b,f}_{\mathbf{p}} \end{pmatrix} . \tag{3.189}
$$

The off-diagonal elements of $\mathbf{G}_{\Delta_0}$ describe, according to Eq. (3.135), the anomalous vacuum expectation values

$$
\langle \psi(\mathbf{x}, \tau)\psi(\mathbf{x}, \tau) \rangle = \int \frac{d^3p}{(2\pi)^3} e^{i\mathbf{p}\mathbf{x}} u_{\mathbf{p}}v^*_{\mathbf{p}} \tanh^{\mp 1}\frac{E_{\mathbf{p}}}{2T}
$$

$$
= \int \frac{d^3p}{(2\pi)^3} e^{i\mathbf{p}\mathbf{x}} \frac{\Delta_0(\mathbf{p})}{2E_{\mathbf{p}}} \tanh^{\mp 1}\frac{E_{\mathbf{p}}}{2T} .
$$

According to Eq. (3.175), this coincides with the Schrödinger type of wave function of the bound state $\langle \psi(\mathbf{x}, \tau)\psi(\mathbf{x}, \tau)|B(q) \rangle$ at $q = 0$.

After this general discussion let us now return to the superconductor. The quadratic part (3.163) of the action (3.132) in the pair fields Δ' reads, with the local interaction [generalizing (3.50) to $T < T_c$],

$$
\mathcal{A}_2[\Delta'^*, \Delta'] = -\frac{i}{2}\mathrm{Tr}\left[\mathbf{G}_{\Delta_0}\begin{pmatrix} 0 & \Delta' \\ \Delta'^* & 0 \end{pmatrix}\mathbf{G}_{\Delta_0}\begin{pmatrix} 0 & \Delta' \\ \Delta'^* & 0 \end{pmatrix}\right] - \frac{1}{g}\int dx|\Delta'(x)|^2 . \tag{3.190}
$$

This action can be written in momentum space in a form that generalizes (3.56):

$$
\mathcal{A}_2[\Delta'^*, \Delta'] = \frac{1}{2}\frac{T}{V}\sum_k [\Delta'^*(k)L_{11}(k)\Delta'(k) + \Delta'(-k)L_{22}(k)\Delta'(-k)
$$

$$
+ \Delta'^*(k)L_{12}(k)\Delta'^*(-k) + \Delta'(-k)L_{21}(k)\Delta'(k)] . \tag{3.191}
$$

The Lagrangian matrix elements $L_{ij}(k)$ are obtained by inserting the Fermi form of the propagator (3.140) into (3.190) [compare (3.166), (3.168)]. Setting $\nu = ik_0$ one has:

$$
\mathcal{A}_2[\Delta'^*, \Delta'] = -\frac{1}{2}\frac{T}{V}\sum_{\omega,\mathbf{p}} \frac{1}{\left(\omega + \frac{\nu}{2}\right)^2 + E^2_{\mathbf{p}+\frac{\mathbf{k}}{2}}} \frac{1}{\left(\omega - \frac{\nu}{2}\right)^2 + E^2_{\mathbf{p}-\frac{\mathbf{k}}{2}}}
$$
$$
\times \mathrm{Tr}\left[\left(\begin{array}{cc} i\left(\omega + \frac{\nu}{2}\right) + \xi_{\mathbf{p}+\frac{\mathbf{k}}{2}} & \Delta_0 \\ \Delta_0^* & i\left(\omega + \frac{\nu}{2}\right) - \xi_{\mathbf{p}+\frac{\mathbf{k}}{2}} \end{array}\right)\left(\begin{array}{cc} 0 & \Delta'(k) \\ \Delta'^*(k) & 0 \end{array}\right)\right.
$$
$$
\left.\times \left(\begin{array}{cc} i\left(\omega - \frac{\nu}{2}\right) + \xi_{\mathbf{p}-\frac{\mathbf{k}}{2}} & \Delta_0 \\ \Delta_0^* & i\left(\omega - \frac{\nu}{2}\right) - \xi_{\mathbf{p}-\frac{\mathbf{k}}{2}} \end{array}\right)\left(\begin{array}{cc} 0 & \Delta'(-k) \\ \Delta'^*(k) & 0 \end{array}\right)\right]
$$
$$
- \frac{1}{g}\sum_k \Delta'^*(k)\Delta'(k). \tag{3.192}
$$

This is equal to

$$
\mathcal{A}_2[\Delta'^*, \Delta'] = \frac{1}{2}\frac{T}{V}\sum_{\omega_n,\mathbf{p}}\left\{\left[\left(\omega + \frac{\nu}{2}\right)^2 + E^2_{\mathbf{p}+\frac{\mathbf{k}}{2}}\right]\left[\left(\omega - \frac{\nu}{2}\right)^2 + E^2_{\mathbf{p}-\frac{\mathbf{k}}{2}}\right]\right\}^{-1}
$$
$$
\times \left\{\left[\omega^2 - \frac{\nu^2}{4} + \xi_{\mathbf{p}+\frac{\mathbf{k}}{2}}\,\xi_{\mathbf{p}-\frac{\mathbf{k}}{2}}\right][\Delta'^*(k)\Delta'(k) + \Delta'(-k)\Delta'^*(-k)]\right.
$$
$$
\left.- |\Delta_0|^2\,[\Delta'^*(k)\Delta'^*(-k) + \Delta'(k)\Delta'(-k)]\right\} - \frac{1}{g}\sum_k \Delta'^*(k)\Delta'(k). \tag{3.193}
$$

From this, we read off the coefficients in (3.191):

$$
\begin{aligned}
L_{11}(k) &= L_{22}(k) = \frac{T}{V}\sum_{\omega,\mathbf{p}} l_{11}(p|k) \\
&= \int \frac{d^3p}{(2\pi)^3}T\sum_{\omega_n} \frac{\omega_n^2 - \nu^2/4 + \xi_+\xi_-}{\left[\left(\omega_n + \frac{\nu}{2}\right)^2 + E_+^2\right]\left[\left(\omega_n - \frac{\nu}{2}\right)^2 + E_-^2\right]} - \frac{1}{g}, \tag{3.194}
\end{aligned}
$$

and

$$
\begin{aligned}
L_{12}(k) &= [L_{21}(k)]^* = \frac{T}{V}\sum_{\omega,\mathbf{p}} l_{12}(p|k) \\
&= -\int \frac{d^3p}{(2\pi)^3}|\Delta_0|^2 T\sum_{\omega_n} \frac{1}{\left[\left(\omega_n + \frac{\nu}{2}\right)^2 + E_+^2\right]\left[\left(\omega_n - \frac{\nu}{2}\right)^2 + E_-^2\right]}. \tag{3.195}
\end{aligned}
$$

Here $\xi_\pm \equiv \xi_{\mathbf{p}\pm\frac{\mathbf{k}}{2}}$ and $E_\pm \equiv E_{\mathbf{p}\pm\frac{\mathbf{k}}{2}}$, so that with $\mathbf{v} \equiv \mathbf{p}/m$ we get

$$
\left\{\begin{array}{c} E_+ \\ E_- \end{array}\right\} = \sqrt{\left\{\begin{array}{c} \xi_+^2 \\ \xi_-^2 \end{array}\right\} + |\Delta_0|^2} \approx E \pm \frac{1}{2}\mathbf{v}\mathbf{k}\frac{\xi}{E} + \frac{1}{8}(\mathbf{v}\mathbf{k})^2\frac{|\Delta_0|^2}{E^3} + \dots, \tag{3.196}
$$

$$\left\{ \begin{array}{c} \xi_+ \\ \xi_- \end{array} \right\} \equiv \frac{(\mathbf{p} \pm \mathbf{k}/2)^2}{2m} = \frac{\mathbf{p}^2}{2m} \pm \frac{1}{2}\mathbf{vk} + \frac{\mathbf{k}^2}{8m} \approx \xi \pm \frac{1}{2}\mathbf{vk} + \dots , \qquad (3.197)$$

with ξ and $E = \sqrt{\xi^2 + |\Delta_0|^2}$. The integral over d^3p can be split into a size and a directional integral, according to (3.89), and we can approximate $\mathbf{v} \equiv {}_F/m \approx v_F\hat{\mathbf{p}}$. We now rearrange the terms in the sum in such a way that we obtain combinations of single sums of the type

$$T \sum_{\omega_n} \frac{1}{i\omega_n - E_+}, \qquad (3.198)$$

which lead to the Fermi distribution function

$$T \sum_{\omega_n} \frac{1}{i\omega_n - E} = n_E^f \equiv \frac{1}{e^{E/T} + 1} \qquad (3.199)$$

with the property

$$n_E^f = 1 - n_{-E}^f. \qquad (3.200)$$

In contrast to (1.104) and (3.34), we label here the occupation numbers with the energies as subscripts. If we introduce the notation $\omega_\pm \equiv \omega \pm \nu/2$, the first term in the sum (3.195) for $L_{12}(k)$ can be decomposed as follows:

$$\frac{1}{(\omega_+^2 + E_+^2)(\omega_-^2 + E_-^2)}$$
$$= \frac{1}{4E_+E_-} \left(\frac{1}{i\omega_+ + E_+} - \frac{1}{i\omega_+ - E_+} \right) \left(\frac{1}{i\omega_- + E_-} - \frac{1}{i\omega_- - E_-} \right)$$
$$= \frac{1}{4E_+E_-} \left\{ -\frac{1}{E_+ + E_- - i\nu} \left(\frac{1}{i\omega_+ - E_+} - \frac{1}{i\omega_- - E_-} \right) \right.$$
$$+ \frac{1}{E_+ + E_- + i\nu} \left(\frac{1}{i\omega_+ + E_+} - \frac{1}{i\omega_- + E_-} \right)$$
$$- \frac{1}{E_+ - E_- + i\nu} \left(\frac{1}{i\omega_+ + E_+} - \frac{1}{i\omega_- + E_-} \right)$$
$$+ \left. \frac{1}{E_+ - E_- - i\nu} \left(\frac{1}{i\omega_+ - E_+} - \frac{1}{i\omega_- - E_-} \right) \right\}. \qquad (3.201)$$

We now use the summation formula (1.104) and the fact that the frequency shifts ν in $\omega_\pm$ do not appear in the final result. They amount to a mere discrete translation in the infinite sum (3.199). Collecting the different terms we find

$$L_{12}(k) = [L_{21}(k)]^* = -\int \frac{d^3p}{(2\pi)^3} |\Delta_0|^2 \frac{1}{2E_-E_+} \qquad (3.202)$$
$$\times \left\{ \frac{E_+ + E_-}{(E_+ + E_-)^2 + \nu^2} \left(1 - n_{E_+}^f - n_{E_-}^f \right) + \frac{E_+ - E_-}{(E_+ - E_-)^2 + \nu^2} \left(n_{E_+}^f - n_{E_-}^f \right) \right\}.$$

In Eq. (3.194) for $L_{11}(k)$, we decompose

$$\frac{\omega_n^2 - \nu^2/4 + \xi_+\xi_-}{(\omega_+^2 + E_+^2)(\omega_-^2 + E_-^2)} = \frac{1}{2}\left\{\frac{1}{\omega_+^2 + E_+^2} + \frac{1}{\omega_-^2 + E_-^2}\right.$$
$$\left. - (E_+^2 + E_-^2 + \nu^2 - 2\xi_+\xi_-)\frac{1}{(\omega_+^2 + E_+^2)(\omega_-^2 + E_-^2)}\right\}. \quad (3.203)$$

When summing the first two terms, we use the formula

$$T\sum_\omega \frac{1}{\omega^2 + E^2} = \frac{1}{2E}(n_{-E}^f - n_E^f) = \frac{1}{2E}\tanh\frac{E}{2T}. \quad (3.204)$$

In the last term, the right-hand factor is easily evaluated as before: Replacing the factor $E_-^2 + E_+^2 + \nu^2$, once by $(E_- + E_+)^2 + \nu^2 - 2E_-E_+$ and once by $(E_- - E_+)^2 + \nu^2 + 2E_-E_+$, we obtain immediately

$$L_{11}(k) = L_{22} = \int \frac{d^3p}{(2\pi)^3}\left\{\frac{E_+E_- + \xi_+\xi_-}{2E_+E_-}\frac{E_+ + E_-}{(E_+ + E_-)^2 + \nu^2}\left(1 - n_{E_+}^f - n_{E_-}^f\right)\right.$$
$$\left. - \frac{E_+E_- - \xi_+\xi_-}{2E_+E_-}\frac{E_+ - E_-}{(E_+ - E_-)^2 + \nu^2}\left(n_{E_+}^f - n_{E_-}^f\right)\right\} - \frac{1}{g}. \quad (3.205)$$

Let us study in more detail the static case and consider only the long-wavelength limit of small $\mathbf{k}$. Hence, we shall take $\nu = 0$ and study only the lowest orders in $\mathbf{k}$. At $\mathbf{k} = 0$ we find from (3.202)

$$L_{12}(0) = -\mathcal{N}(0)|\Delta_0|^2 \int_{-\infty}^{\infty} d\xi \left\{\frac{1}{4E^3}\left(1 - 2n_E^f\right) + \frac{1}{2E^2}n_E^{f\prime}\right\}. \quad (3.206)$$

Inserting $E = \sqrt{\xi^2 + \Delta_0^2}$, this can be rewritten as

$$L_{12}(0) = -\frac{1}{2}\mathcal{N}(0)\phi(\Delta_0). \quad (3.207)$$

Here we have introduced the so-called *Yoshida function*

$$\phi(\Delta_0) \equiv \Delta_0^2 \int_0^{\infty} d\xi \left\{\frac{1}{E^3}\left(1 - 2n_E^f\right) + \frac{2}{E^2}n_E^{f\prime}\right\}. \quad (3.208)$$

Now we observe that

$$\partial_\xi\left(\frac{\xi}{\Delta_0^2 E}n_E^f\right) = \frac{1}{E^3}n_E^f + \left(\frac{1}{\Delta_0^2} - \frac{1}{E^2}\right)n_E^{f\prime}, \quad (3.209)$$

so that we can bring (3.208) to the form

$$\phi(\Delta_0) \equiv |\Delta_0|^2 \int_0^{\infty} d\xi \left\{\frac{1}{E^3} + \frac{2}{\Delta_0^2}n_E^{f\prime} - 2\partial_\xi\left(\frac{\xi}{\Delta_0^2 E}n_E^f\right)\right\}. \quad (3.210)$$

The surface term vanishes, and the first integral in Eq. (3.210) can be done to arrive at the more convenient expression

$$\phi(\Delta_0) = 1 + 2 \int_0^\infty d\xi \, n^{f\prime}{}_E = 1 - \frac{1}{2T} \int_0^\infty d\xi \frac{1}{\cosh^2(E/2T)}. \tag{3.211}$$

We now turn to Eq. (3.205) which reads at $\nu = 0$, $\mathbf{k} = 0$:

$$L_{11}(0) = \mathcal{N}(0) \frac{1}{2} \int_0^\infty d\xi \left\{ \frac{E^2 + \xi^2}{E^3} \left(1 - 2n_E^f\right) - 2\frac{\Delta_0^2}{E^2} n^{f\prime}{}_E - \frac{1}{g} \right\}. \tag{3.212}$$

Here we observe that due to the gap equation (3.141), $L_{11}(k)$ can also be expressed in terms of the Yoshida function $\phi(\Delta_0)$ as

$$L_{11}(0) = -\frac{1}{2} \mathcal{N}(0) \phi(\Delta_0). \tag{3.213}$$

For $T \approx 0$, this function approaches zero exponentially fast. The full temperature behavior is best calculated by using the Matsubara sum (3.204) to write

$$\begin{aligned}
\phi(\Delta_0) &= 2T \sum_{\omega_n} \int d\xi \frac{\Delta_0^2}{(\omega_n^2 + E^2)^2} = -2\Delta_0^2 T \sum_{\omega_n} \frac{\partial}{\partial \omega_n^2} \int d\xi \frac{1}{\omega_n^2 + \xi^2 + \Delta_0^2} \\
&= -2\Delta_0^2 T \sum_{\omega_n} \frac{\partial}{\partial \omega_n^2} \frac{\pi}{\sqrt{\omega_n^2 + \Delta_0^2}} = 2T\pi \sum_{\omega_n > 0} \frac{1}{\sqrt{\omega_n^2 + \Delta_0^2}^3}.
\end{aligned} \tag{3.214}$$

Using again the variables δ and x_n from (3.157) and (3.158), this becomes

$$\phi(\Delta_0) = \frac{2}{\delta} \sum_{n=0}^\infty \frac{1}{\sqrt{x_n^2 + 1}^3}. \tag{3.215}$$

For $T \to T_c$ and small δ we have

$$\phi(\Delta_0) \approx 2\delta^2 \sum_{n=0}^\infty \frac{1}{(2n+1)^3} = 2\delta^2 \frac{7\zeta(3)}{8} \approx 2\left(1 - \frac{T}{T_c}\right). \tag{3.216}$$

In the limit $T \to 0$, the sum turns into an integral. Using the formula

$$\int_0^\infty dx \frac{x^{\mu-1}}{(x^2+1)^\nu} = \frac{1}{2} B(\mu/2, \nu - \mu/2) \tag{3.217}$$

with $B(x,y) = \Gamma(x)\Gamma(y)/\Gamma(x+y)$, we see that

$$\phi(\Delta_0) \underset{T=0}{=} 1. \tag{3.218}$$

Note that we can also write $L_{11}(0)$ as

$$L_{11}(0) = -\frac{3}{4M^2 v_F^2} \rho_s \tag{3.219}$$

with

$$\rho_s \equiv \rho\,\phi(\Delta_0). \tag{3.220}$$

The function ρ_s is the superfluid mass density. To justify this, let us calculate the bending energies of the collective field $\Delta(x)$. For this, we expand $L_{11}(\varepsilon, \mathbf{k})$ and $L_{12}(\varepsilon, \mathbf{k})$ at $\nu = 0$ into powers of the momentum $\mathbf{k}$ up to $\mathbf{k}^2$. Let us denote the zero-frequency parts of $L_{11}(k)$ and $L_{12}(k)$ by $L_{11}(\mathbf{k})$ and $L_{11}(\mathbf{k})$, respectively, with the explicit form

$$L_{11}(\mathbf{k}) = \frac{T}{V}\sum_{\omega_n,\mathbf{p}} \frac{\omega^2 + \xi_+\xi_-}{(\omega^2 + E_+^2)(\omega^2 + E_-^2)} - \frac{1}{g}, \tag{3.221}$$

$$L_{12}(\mathbf{k}) = -\frac{T}{V}\sum_{\omega_n,\mathbf{p}} \frac{\Delta_0^2}{(\omega^2 + E_+^2)(\omega^2 + E_-^2)}. \tag{3.222}$$

Inserting the expansions

$$\xi_+\xi_- = \xi^2 - \frac{1}{4}(\mathbf{vk})^2 + \dots,$$

$$\left\{\begin{array}{c} E_+^2 \\ E_-^2 \end{array}\right\} = E^2 \pm \xi\mathbf{vk} + \frac{1}{2}(\mathbf{vk})^2 + \dots, \tag{3.223}$$

we have

$$L_{11}(\mathbf{k}) - L_{12}(\mathbf{k}) \approx \int \frac{d^3p}{(2\pi)^3} T\sum_\omega \frac{\omega^2 + \Delta_0^2 + \xi^2 - \frac{1}{4}(\mathbf{vk})^2}{(\omega^2 + E^2)^2\left[1 + \frac{1}{2}(\mathbf{vk})^2\frac{\omega^2 - \xi^2 + \Delta_0^2}{(\omega^2 + E^2)^2}\right]} - \frac{1}{g} + \dots$$

$$= \int \frac{d^3p}{(2\pi)^3}\left\{\left(T\sum_\omega \frac{1}{\omega^2 + E^2} - \frac{1}{g}\right)\right.$$

$$\left. + T\sum_{\omega_n}\left[\frac{1}{4}\frac{1}{(\omega^2 + E^2)^2} - \frac{\omega^2 + \Delta_0^2}{(\omega^2 + E^2)^3}\right](\mathbf{vk})^2\right\} + \dots. \tag{3.224}$$

Due to the gap equation, the first parentheses vanish so that we are left with

$$L_{11}(\mathbf{k}) - L_{12}(\mathbf{k}) \approx \mathcal{N}(0)\int \frac{d\hat{\mathbf{p}}}{4\pi}\left(\frac{\mathbf{pk}}{m}\right)^2$$

$$\times \int_{-\infty}^\infty d\xi\left[\frac{1}{4}\frac{1}{(\omega^2 + \xi + \Delta_0^2)^2} - \frac{\omega^2 + \Delta_0^2}{(\omega^2 + \xi^2 + \Delta_0^2)^3}\right]. \tag{3.225}$$

Similarly, we obtain

$$L_{12}(\mathbf{k}) \approx -\mathcal{N}(0)\int \frac{d\hat{\mathbf{p}}}{4\pi}\int_{-\infty}^\infty d\xi\left\{\frac{\Delta_0^2}{(\omega^2 + \xi^2 + \Delta_0^2)^2}\right.$$

$$\left. + (\mathbf{vk})^2\left[\frac{1}{2}\frac{1}{(\omega^2 + \xi^2 + \Delta_0^2)^3} - \frac{\omega^2 + \Delta_0^2}{(\omega^2 + \xi^2 + \Delta_0^2)^4}\right]\right\}. \tag{3.226}$$

Using the integrals

$$\int_{-\infty}^{\infty} d\xi \frac{1}{\left(\omega^2 + \xi^2 + \Delta_0^2\right)^{2,3,4}} = \left(\frac{1}{2}, \frac{3}{8}, \frac{5}{16}\right) \frac{\pi}{\sqrt{\omega^2 + \Delta_0^2}^{3,5,7}}, \tag{3.227}$$

we find

$$L_{11}(\mathbf{k}) - L_{12}(\mathbf{k}) \approx -\frac{\mathcal{N}(0)}{4} \frac{(\mathbf{vk})^2}{\Delta_0^2} \int \frac{d\hat{\mathbf{p}}}{4\pi} \phi(\Delta_0), \tag{3.228}$$

$$L_{12}(\mathbf{k}) \approx -\frac{\mathcal{N}(0)}{2} \phi(\Delta_0) + \frac{\mathcal{N}(0)}{8} (\mathbf{vk})^2 \int \frac{d\hat{\mathbf{p}}}{4\pi} \bar{\phi}(\Delta_0), \tag{3.229}$$

where $\phi(\Delta_0)$ is again the Yoshida function (3.215), while $\bar{\phi}(\Delta_0)$ is a further gap function:

$$\bar{\phi}(\Delta_0) = 2\Delta_0^4 \pi T \sum_{\omega_n > 0} \frac{1}{\sqrt{\omega_n^2 + \Delta_0^2}^5} \equiv \frac{2}{\delta} \sum_{n=0}^{\infty} \frac{1}{\sqrt{x_n^2 + 1}^5}. \tag{3.230}$$

For $T \approx T_c$, this behaves like

$$\bar{\phi}(\Delta_0) \approx 2\delta^4 \sum_{n=0}^{\infty} \frac{1}{(2n+1)^5} = \delta^4 \frac{31\zeta(5)}{16}, \tag{3.231}$$

and thus, by (3.161), the temperature behavior is

$$\bar{\phi}(\Delta_0) \approx 21.8144 \times \left(1 - \frac{T}{T_c}\right)^2. \tag{3.232}$$

In the limit $T \to 0$, on the other hand, the sum turns into an integral whose value is, by formula (3.217),

$$\bar{\phi}(\Delta_0^2) \underset{T=0}{=} \frac{2}{3}. \tag{3.233}$$

Altogether, we find for the energy density the gradient terms

$$f_{\text{grad}}(x) = \frac{1}{4m^2} \left[\rho_{ij}^{11} \nabla_i \Delta^*(x) \nabla_j \Delta(x) / \Delta_0^2 + \operatorname{Re} \rho_{ij}^{12} \nabla_i \Delta^*(x) \nabla_j \Delta^*(x) / \Delta_0^2 \right]. \tag{3.234}$$

Here we have dropped the primes on the field gradients, since the additional constant Δ_0 does not matter. The first coefficient is given by

$$\rho_{ij}^{11} = \frac{3\rho}{2} \int \frac{d\hat{\mathbf{p}}}{4\pi} \hat{p}_i \hat{p}_j \left[\phi(\Delta_0) - \frac{1}{2} \bar{\phi}(\Delta_0) \right], \tag{3.235}$$

the second by

$$\rho_{ij}^{12} = -\frac{3\rho}{2} \int \frac{d\hat{\mathbf{p}}}{4\pi} \hat{p}_i \hat{p}_j \frac{1}{2} \bar{\phi}(\Delta_0), \tag{3.236}$$

where we have expressed $\mathcal{N}(0)$ in terms of the mass density of electrons ρ via the relation (3.62). Performing the angular integral gives

$$\rho_{ij}^{11} = \frac{1}{2}\rho\left[\phi(\Delta_0^2) - \frac{1}{2}\bar{\phi}(\Delta_0)\right]\delta_{ij},\tag{3.237}$$

$$\rho_{ij}^{12} = -\frac{1}{4}\rho\bar{\phi}(\Delta_0)\delta_{ij}.\tag{3.238}$$

Decomposing the collective field $\Delta(x)$ into size $|\Delta(x)|$ and phase $\varphi(x)$,

$$\Delta(x) = |\Delta(x)|e^{i\varphi(x)},\tag{3.239}$$

the energy density reads

$$f_{\text{grad}}(x) = \frac{1}{4m^2}\left\{(\rho^{11} - \rho^{12})(\nabla\phi)^2 + (\rho^{11} + \rho^{12})(\nabla|\Delta(x)|)^2/\Delta_0^2\right\}.\tag{3.240}$$

Introducing, in addition, the notation

$$\bar{\rho}_s \equiv \rho\bar{\phi}(\Delta)\tag{3.241}$$

and adding to the energy density the earlier $\mathbf{k} = 0$ result, we find the total quadratic fluctuation energy density

$$f_{\text{grad}}(x) = \rho_s(\nabla\varphi)^2 + (\rho_s - \bar{\rho}_s)(\nabla|\Delta(x)|)^2/\Delta_0^2 + 6\rho_s(\nabla|\Delta(x)|)^2)/v_F^2.\tag{3.242}$$

The behavior of ρ_s and $\bar{\rho}_s$ for all $T \le T_c$ is shown in Fig. 3.8.

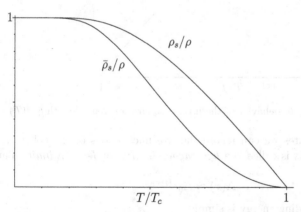

FIGURE 3.8 Temperature behavior of the superfluid density ρ_s/ρ (Yoshida function) and the gap function $\bar{\rho}_s/\rho$.

The phase fluctuations are of infinite range, the size fluctuations have a finite range characterized by the temperature-dependent coherence length

$$\xi(T) = \sqrt{\frac{v_F^2}{6\Delta^2}\frac{\rho_s - \bar{\rho}_s}{\rho_s}}.\tag{3.243}$$

For T close to T_c, the second ratio tends towards unity, while Δ^2 goes to zero according to Eq. (3.121). Thus we recover the previous result (3.105) for the coherence length:

$$\xi(T) \approx \xi_0 \left(1 - \frac{T}{T_c}\right),\tag{3.244}$$

with

$$\xi_0 = \sqrt{\frac{7\zeta(3)}{48}} \frac{v_F}{\pi T_c} \approx 0.419 \times \frac{v_F}{\pi T_c}.\tag{3.245}$$

For $T \to 0$, $\xi(T)$ tends exponentially fast against

$$\xi(0) = \frac{e^\gamma}{3} \frac{v_F}{\pi T} \approx 0.591 \times \frac{v_F}{\pi T} \approx 1.4179 \times \xi_0.\tag{3.246}$$

The behavior of $\xi_0^2/\xi^2(T)$ is displayed in Fig. 3.9.

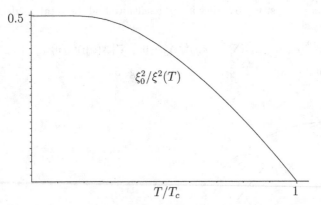

FIGURE 3.9 Temperature behavior of the inverse square coherence length $\xi^{-2}(T)$.

At low temperatures we can ignore the size fluctuations of the collective field parameter $\Delta(x)$. This is called the *hydrodynamic limit* or *London limit*. Thus we approximate

$$\Delta(x) \approx \Delta_0 e^{i\phi(x)}.\tag{3.247}$$

In this limit, the bending energy is simply

$$f_{\mathrm{grad}}(x) = \frac{1}{4m^2}\rho_s[\partial_i\phi(x)]^2.\tag{3.248}$$

By studying the behavior of this expression under Galilei transformations we identify the superfluid velocity of the condensate

$$\mathbf{v}_s = \frac{1}{2m}\boldsymbol{\nabla}\phi.\tag{3.249}$$

In terms of this, the energy density takes the form

$$f_{\text{grad}}(x) = \frac{1}{2}\rho_s \mathbf{v}_s^2. \tag{3.250}$$

This shows that ρ_s is the *superfluid density* of the condensate.

For temperatures close to zero, the sum over Matsubara frequencies $T\sum_{\omega}$ may also be performed as an integral $\int d\omega/2\pi$, and the result is [recall (3.221), (3.222)]

$$L_{11}(k) = L_{22}(k) = \frac{T}{V}\sum_{\omega,\mathbf{p}} l_{11}(p|k) - \frac{1}{g} = \frac{1}{V}\sum_{\mathbf{p}} \frac{E_+ E_- + \xi_+ \xi_-}{2E_+ E_-} \frac{E_+ + E_-}{(E_+ + E_-)^2 + \nu^2} - \frac{1}{g}, \tag{3.251}$$

$$L_{12}(k) = L_{21}(k) = \frac{T}{V}\sum_{\omega,\mathbf{p}} l_{12}(p|k) = -|\Delta_0|^2 \frac{1}{V}\sum_{\mathbf{p}} \frac{1}{2E_+ E_-} \frac{E_+ + E_-}{(E_+ + E_-)^2 + \nu^2}. \tag{3.252}$$

We now express $1/g$ in terms of the gap equation (3.268) at $T = 0$,

$$\frac{1}{g} = \sum_{\mathbf{p}} \frac{1}{2E(\mathbf{p})}, \tag{3.253}$$

so that the last term in $L_{11}(k)$ provides us with a subtraction of the sum.

The energies of fundamental excitations are obtained by diagonalizing the action $\mathcal{A}_2[\Delta'^*, \Delta']$, and by searching for zero eigenvalues of the matrix $L(k)$ via

$$L_{11}(k)L_{22}(k) - L_{12}^2(k) = 0. \tag{3.254}$$

Since $L_{11}(k) = L_{22}(k)$, this amounts to the two equations

$$L_{11}(k) = L_{12}(k), \tag{3.255}$$

and

$$L_{11}(k) = -L_{12}(k). \tag{3.256}$$

These equations can be solved for small k. Expanding them to fourth order in ν and k [22], and using once more the gap equation (3.253) at $T = 0$, one finds

$$L_{11}(k) = -\frac{M^2 v_F}{4\pi^2}\left(1 + \frac{\nu^2}{3\Delta_0^2} + \frac{v_F^2 k^2}{9\Delta_0^2} - \frac{v_F^2 \nu^2 k^2}{30\Delta_0^4} - \frac{\nu^4}{20\Delta_0^4} - \frac{v_F^4 k^4}{100\Delta_0^4} + \ldots\right), \tag{3.257}$$

$$L_{12}(k) = -\frac{M^2 v_F}{4\pi^2}\left(1 - \frac{\nu^2}{6\Delta_0^2} - \frac{v_F^2 k^2}{18\Delta_0^2} + \frac{v_F^2 \nu^2 k^2}{45\Delta_0^4} + \frac{\nu^4}{30\Delta_0^4} + \frac{v_F^4 k^4}{150\Delta_0^4} + \ldots\right). \tag{3.258}$$

We have ignored terms such as $k^4/M^2\Delta^2$ compared to $v_F^2 k^4/\Delta^4$ since the Fermi energy is much larger than the gap in a superconductor, i.e., $Mv_F^2/2 \ll \Delta$. With these expansions, the Eq. (3.255) has the solution with small $k_0 = -i\nu$:

$$k_0 = \pm c|\mathbf{k}|(1 - \gamma \mathbf{k}^2), \quad c \equiv \frac{v_F}{\sqrt{3}}, \quad \gamma = \frac{v_F^2}{45\Delta_0^2}. \tag{3.259}$$

The Eq. (3.256) can be solved directly for small $\mathbf{k}$ and $k_0 = -i\nu$. Using (3.252) and (3.253) one can write $-L_{11}(k) - L_{12}(k) = 0$ as[6]

$$\frac{1}{V}\sum_{\mathbf{p}}\left[\frac{1}{2E} + \frac{(\Delta_0^2 - EE' - \xi\xi')(E+E')}{2EE'[(E+E')^2 + \nu^2]}\right] = 0. \tag{3.260}$$

For small $\mathbf{k}$, this leads to the energies [22]

$$k_0^{(n)} = 2\Delta_0 + \Delta_0\left(\frac{v_F\mathbf{k}}{2\Delta_0}\right)^2 z_n, \tag{3.261}$$

with z_n being the solutions of the integral equation

$$\int_{-1}^{1}dx\int_{-\infty}^{\infty}dy\frac{x^2 - z}{x^2 + y^2 - z} = 0. \tag{3.262}$$

Setting $e^t = \left(\sqrt{1-z} + 1\right)/\left(\sqrt{1-z} - 1\right)$ this is equivalent to

$$\sinh t + t = 0, \tag{3.263}$$

which has infinitely many solutions t_n, starting with

$$t_1 = 2.251 + i4.212, \tag{3.264}$$

and tending asymptotically to

$$t_n \approx \log[\pi(4n-1)] + i\left(2\pi n - \frac{\pi}{2}\right). \tag{3.265}$$

The excitation energies are

$$k_0^{(n)} = 2\Delta_0 - \frac{v_F^2}{4\Delta_0}\mathbf{k}^2\frac{1}{\sinh^2 t_n/2}. \tag{3.266}$$

Of these, only the first one at $k_0^{(1)} \approx 2\Delta_0 + (.24 - .30i)v_F^2/4\Delta_0^2\mathbf{k}^2$ lies on the second sheet and may have observable consequences, while the others are hiding under lower and lower sheets of the two-particle branch cut from $2\Delta_0$ to ∞. The cut is logarithmic due to the dimensionality of the surface of the Fermi sea at $T = 0$.

The basic strength of the Hubbard-Stratonovich transformation is that the two ways of eliminating the four-particle interaction via formula (3.1) and (2.4) yield both a complete description of the system, once in terms of a bilocal pair field $\Delta(x, x')$, and once in terms of the local scalar field $\varphi(x)$. In practice, however, this is an important weakness. In either description the effects of the other can be recovered by summing an infinity of diagrams formed with the other collective

---

[6]For $T \neq 0$, each result appears with a factor $\frac{1}{2}\left(\tanh\frac{E}{2T} + \tanh\frac{E'}{2T}\right)$, to which we must add once more the entire expression with E' replaced by $-E'$.

quantum field. Such an infinite set of diagrams can unfortunately never be calculated and summed. The only way out of this dilemma is to go to classical collective fields $\Delta(x, x')$ and $\varphi(x)$.

The way how to do this has been shown for simple systems in the textbook [23], and for large-order effects in general quantum field theory in the textbook [24]. Here we shall present the method for the example of the superconducting local four-particle interaction (3.41).

3.4 From BCS to Strong-Coupling Superconductivity

The above calculations were valid only for weak coupling since they were based on the assumption that, in momentum space, only a small layer of electrons in the neighborhood of the Fermi sphere is subject to the phonon-induced attraction. This was implied by the approximation (3.61) which was used to simplify the evaluation of the gap function in (3.59). This approximation explained the phenomena observed in all old-fashioned superconductors known until 1986 (recall Fig. 3.1).

As described in the beginning of this chapter, this was the year that superconductivity was discovered in a completely new class of material by Johannes Georg Bednorz and Karl Alexander Müller [5]. A first attempt to explain this phenomenon was to free the BCS-theory from the weak-coupling assumption, by allowing the phonon attraction to extend beyond the thin shell around the Fermi surface in momentum space. For strong coupling g, the attraction can lead to the formation of tightly bound bosons rather than loosely bound Cooper pairs. These can undergo Bose-Einstein condensation, and all calculations have to be reconsidered. Initially, this was done in Refs. [25, 47]. More recently the same mechanism has come under renewed investigation with the hope that it might explain experimental data of high-T_c superconductivity in underdoped cuprate samples [26, 53]. These materials show an anomalous behavior in the normal phase well above the superconductive transition T_c. In particular, they exhibit an anomalous temperature dependence in resistivity, specific heat, spin susceptibility, and similar properties. Angle Resolved Photoemission Spectroscopy (ARPES) indicates the existence of a *pseudogap* in the single-particle excitation spectrum [30, 56]. This manifests itself in a significant suppression of low-frequency spectral weight. This suppression is in contrast to the complete suppression in the presence of an ordinary gap.

Thus we study what happens if we increase the coupling strength g. We shall be able to explain a few of the features of high-T_c superconductors, although many of them will remain unexplained. In particular one cannot explain the fact that the physical system undergoes merely a crossover from the pseudogap-regime to the normal phase, rather than performing a proper phase transition of the model.

In an exactly solvable field theoretic model, a strong attraction leads to two phase transitions: first a second order one in which tightly bound bosonic bound states are formed, and a second one of the Berezinskii-Kosterlitz-Thouless type [31], in which these undergo a Bose-Einstein condensation [34]. This has been discussed in Chapter 23 of the textbook Ref. [35].

There is a possible connection between the presence of a pseudogap and the existence of an *anomalous* normal state above T_c in underdoped cuprates, as was pointed out experimentally in [36, 37]. ARPES experiments have shown that the superconducting gap below T_c in both underdoped and optimally doped materials have the same magnitude and wave vector dependence as a pseudogap well above T_c. Further experimental facts display the pseudogap behavior above T_c:

1. Experiments on YBCO [38, 39] observe a significant suppression of in-plane conductivity $\sigma_{ab}(\omega)$ at frequencies below 500 cm^{-1}. They begin at temperatures much above T_c. The temperature T^*, where the suppression starts, increases with decreasing doping. This is confirmed by recent experiments [37, 40] on underdoped samples which show clearly the increase of resistivity for decreasing temperature if T drops below a certain value.

2. Specific heat experiments [41] also clearly display a pseudogap behavior much above T_c.

3. Nuclear Magnetic Resonance (NMR) and some neutron scattering observations [42, 43] show that, below a temperature T^* which lies much higher than T_c, magnetic response starts decreasing. Within the model to be studied the connection of pseudogap and loss of magnetic response was studied theoretically in [44].

4. Experiments reviewed in Ref. [26] on optical conductivity [45, 46] and on tunneling exhibit the opening of a pseudogap in underdoped and optimally doped cuprates.

The model to be presented here was investigated in various ways in a number of papers [47, 48]. We shall describe now the superconductive phase at small but finite temperature and the pseudogaped normal phase where mean-field methods should also be reliable [44, 50]. Indeed, the paramagnetic susceptibility was studied in the anomalous phase in Ref. [44], and the experiments exhibit the pseudogap behavior.

Some years ago, analytic results were obtained within the same model for the entire crossover region at $T = 0$ in a three-dimensional system [51]. For the two-dimensional system, similar results were obtained earlier at $T = 0$ in [52].

In this chapter we shall begin by reproducing the results of Refs. [51, 52] for $\Delta(0)$. Then we extend the results to find the temperature behavior of gap and pseudogap as well as of thermodynamic functions.

3.5 Strong-Coupling Calculation of the Pair Field

We work with the same action as before in (3.39), but make two important changes with respect to the earlier weak-coupling calculations. First, the gap equation (3.268) is evaluated without the approximation of Eq. (3.65), which led to the simple results (3.147) and (3.150). Second, the chemical potential is no longer close to its zero-temperature limit ε_F, but has a pronounced temperature behavior. To find it, we have to evaluate an extra equation obtained from the derivative of the Euclidean action with respect to the chemical potential. This yields the equation for the particle densiy:

$$\rho = \frac{N}{V} = \sum_{\mathbf{p}} \rho_{\mathbf{p}} = \frac{1}{V} \sum_{\mathbf{p}} \left(1 - \frac{\xi_{\mathbf{p}}}{E_{\mathbf{p}}} \tanh \frac{E_{\mathbf{p}}}{2T} \right). \tag{3.267}$$

Taking the gap equation (3.146) and converting the momentum sum into an integral over ξ and an integral over the directions $\hat{\mathbf{p}}$, as in (3.60), we arrive at the three-dimensional gap equation

$$\frac{1}{g} = \frac{1}{V} \sum_{\mathbf{p}} \frac{1}{2E_{\mathbf{p}}} \tanh \frac{E_{\mathbf{p}}}{2T} = \kappa_3 \int_{-\mu}^{\infty} d\xi \frac{\sqrt{\xi^2 + \mu}}{2\sqrt{\xi^2 + \Delta^2}} \tanh \frac{\sqrt{\xi^2 + \Delta^2}}{2T}, \tag{3.268}$$

where the constant is $\kappa_3 = m^{3/2}/\sqrt{2}\pi^2$.

Instead of the coupling constant one can parametrize the attractive δ-function attraction by the renormalized coupling constant

$$\frac{1}{g_R} = \frac{1}{g} - \frac{1}{V} \sum_{\mathbf{p}} \frac{1}{\epsilon_{\mathbf{p}}}. \tag{3.269}$$

and express the renormalized coupling constant in terms of the exprimentally measurable s-wave scattering length a:

$$\frac{1}{g_R} = -\frac{m}{4\pi\hbar^3} \frac{1}{a}. \tag{3.270}$$

The factor denominator 4π instead of 2π accounts for the fact that two equal masses have a reduced mass $m/2$, and the negative sign is there since an attractive positive g refers to a negative s-wave phase shift. The explanation of all this is in the textbook [35] in Chapter 9 Eqs. (9.264)–(9.266).

Thus we can write the gap equation (3.268) also as:

$$-\frac{m}{4\pi\hbar^3} \frac{1}{a} = \frac{1}{g_R} = \frac{1}{V} \sum_{\mathbf{p}} \left(\frac{1}{2E_{\mathbf{p}}} \tanh \frac{E_{\mathbf{p}}}{2T} - \frac{1}{2\epsilon_{\mathbf{p}}} \right). \tag{3.271}$$

In two-dimensions, the density of states is constant, and the gap equation becomes

$$\frac{1}{g} = \kappa_2 \int_{-\mu}^{\infty} d\xi \frac{1}{2\sqrt{\xi^2 + \Delta^2}} \tanh \frac{\sqrt{\xi^2 + \Delta^2}}{2T}, \tag{3.272}$$

with a constant $\kappa_2 = m/2\pi$. The particle number density in Eq. (3.267) can be integrated with the result

$$\rho = \frac{m}{2\pi} \left\{ \sqrt{\mu^2 + \Delta^2} + \mu + 2T \log \left[1 + \exp \left(-\frac{\sqrt{\mu^2 + \Delta^2}}{T} \right) \right] \right\}, \tag{3.273}$$

the right-hand side being a function $\rho(\mu, T, \Delta)$.

The δ-function potential produces a divergence similar to that in (3.65), which requires regularization. Since the momentum sum is running over the entire momentum space rather than merely over the vicinity of the Fermi sphere, the divergence is the same as in the calculation of the scattering amplitude for the δ-function potential. It can therefore be removed by a subtraction, based on going from g to the renormalized coupling g_R via Eq. (3.269).

For the temperature of pair dissociation we obtain the estimate:

$$T_{\text{dissoc}} \simeq \varepsilon_B / \log(\varepsilon_B/\varepsilon_F)^{3/2}. \tag{3.274}$$

This shows that, at strong couplings, T^* is indeed related to pair formation [50] which lies above the temperature of phase coherence [26, 47].

The gap in the spectrum of single-particle excitations has a special feature at the point where the chemical potential changes its sign [25, 57, 58]. The sign change occurs at the minimum of the Bogoliubov quasiparticle energy $E_{\mathbf{k}}$ where this energy defines the gap energy in the quasiparticle spectrum:

$$E_{\text{gap}} = \min[\xi_{\mathbf{k}}^2 + \Delta^2]^{1/2}. \tag{3.275}$$

Thus, for a positive chemical potential, the gap energy is given directly by the gap function Δ, whereas for a negative chemical potential, it is larger:

$$E_{\text{gap}} = \begin{cases} \Delta & \text{for} \quad \mu > 0, \\ (\mu^2 + \Delta^2)^{1/2} & \text{for} \quad \mu < 0. \end{cases} \tag{3.276}$$

In three dimensions at $T = 0$, Equations (3.267) and (3.268) were solved analytically in the entire crossover region in [51] to obtain Δ and μ as functions of the reduced chemical potential

$$\hat{\mu} \equiv \mu/\Delta. \tag{3.277}$$

This will be referred to as *crossover parameter*. The results for Δ and μ are:

$$\frac{\Delta}{\varepsilon_F} = \frac{1}{[\hat{\mu}I_1(\hat{\mu}) + I_2(\hat{\mu})]^{2/3}}, \tag{3.278}$$

$$\frac{\mu}{\varepsilon_F} = \hat{\mu}\frac{\Delta}{\varepsilon_F} = \frac{\hat{\mu}}{[\hat{\mu}I_1(\hat{\mu}) + I_2(\hat{\mu})]^{2/3}}, \tag{3.279}$$

where $\hat{\mu}$ is related to the s-wave scattering length by

$$\frac{1}{k_F a} = -\frac{4}{\pi}\frac{\hat{\mu}I_2(\hat{\mu}) - I_1(\hat{\mu})}{[\hat{\mu}I_1(\hat{\mu}) + I_2(\hat{\mu})]^{1/3}}. \tag{3.280}$$

Here we have introduced the functions

$$I_1(z) \equiv \int_0^\infty dx \frac{x^2}{[(x^2 - z)^2 + 1]^{3/2}}$$

$$= (1+z^2)^{1/4} E(\pi/2, y) - \frac{1}{4z_1^2(1+z^2)^{1/4}} F(\pi/2, z), \qquad (3.281)$$

$$I_2(z) \equiv \frac{1}{2} \int_0^\infty dx \frac{1}{[(x^2-z)^2+1]^{1/2}}$$

$$= \frac{1}{2(1+z^2)^{1/4}} F(\pi/2, y), \qquad (3.282)$$

where

$$x^2 \equiv \frac{k^2}{2M} \frac{1}{\Delta}, \qquad y^2 \equiv \frac{z_1^2}{(1+z^2)^{1/2}}, \qquad z_1 \equiv \frac{1}{2}(\sqrt{1+z^2}+z), \qquad (3.283)$$

and $E(\pi/2, x)$ and $F(\pi/2, x)$ are the standard elliptic integrals.

The calculations for $D = 2$ are similar. They were obtained in Ref. [52] and need not be presented here. The results are

$$\frac{\Delta}{\varepsilon_F} = \frac{2}{\hat{\mu} + \sqrt{1+\hat{\mu}^2}}, \qquad (3.284)$$

from (3.278), and

$$\frac{\mu}{\varepsilon_F} = \frac{\mu}{\Delta} \frac{\Delta}{\varepsilon_F} = \frac{2\hat{\mu}}{\hat{\mu} + \sqrt{1+\hat{\mu}^2}}, \qquad (3.285)$$

from (3.278). The first equation relates the gap Δ to μ by

$$\left(\frac{\Delta}{\varepsilon_F}\right)^2 + \left(\frac{\mu}{\varepsilon_F}\right)^2 = 2, \qquad (3.286)$$

which shows that the gap disappears at the critical value $\mu = \mu_c \equiv \sqrt{2}\varepsilon_F$. Combining this with Eq. (3.273), we find that at μ_c, the particle number density has the critical value

$$\rho = \rho_c \equiv \frac{m}{\pi}\mu_c. \qquad (3.287)$$

If the number density exceeds this critical value, the gap disappears, and the superconductor becomes a normal metal. In Fig. 3.10, we plot the three- and two-dimensional quantities Δ and μ as a function of the ratio $\hat{\mu} = \mu/\delta$.

Let us also calculate the pair binding energy ε_B from the bound-state equation

$$-\frac{1}{g} = \frac{1}{V} \sum_{\mathbf{k}} \frac{1}{\mathbf{k}^2/m + \varepsilon_B} = \frac{m}{2\pi} \int_{-\hat{\mu}}^\infty dz \frac{1}{2z + \varepsilon_B/\Delta + 2\hat{\mu}}. \qquad (3.288)$$

After performing the elementary integrals, we find

$$\frac{\varepsilon_B}{\Delta} = \sqrt{1+\hat{\mu}^2} - \hat{\mu}. \qquad (3.289)$$

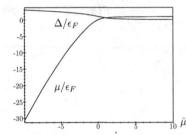
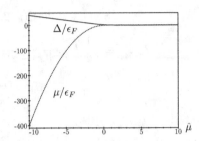

FIGURE 3.10 Gap function Δ and chemical potential μ at zero temperature as functions of the crossover parameter $\hat{\mu}$ in $D = 3$ (left-hand plot) and $D = 2$ (right-hand plot).

By combining (3.289) with (3.284) we find the dependence of the ratio ϵ_0/ϵ_F on the crossover parameter $\hat{\mu}$:

$$\frac{\varepsilon_B}{\varepsilon_F} = 2\frac{\sqrt{1 + \hat{\mu}^2} - \hat{\mu}}{\sqrt{1 + \hat{\mu}^2} + \hat{\mu}}. \qquad (3.290)$$

These relations can easily be extended to non-zero temperature. In the ensuing analysis of the gap or pseudogap function at fixed coupling strength we no longer consider the carrier density as fixed, but rather assume the system to be in contact with a reservoir of a fixed chemical potential $\mu = \mu(1/k_F a_s)$. This will be most convenient for deriving simple analytic results for the finite-temperature behavior of the system.[7]

The idea of the present discussion is to produce, at strong couplings and zero temperature, a large mean-field gap $\Delta e^{i\gamma(\mathbf{x})}$ with a rather rigid phase factor $\gamma(\mathbf{x}) \equiv \gamma$. As the temperature is raised, the phase fluctuations increase, depending on the stiffness. At a certain temperature to be identified with T_c, the stiffness becomes so small that the phase fluctuations become decoherent. The gap is still there, but the fluctuations are so violent that order is destroyed. This is the pseudogap regime. As the temperature is raised further, it reaches some value T^* where the mean-field gap is completely destroyed.

Let us now turn to the region near zero temperature, where we can derive exact results for the gap. From (3.278) we extract the asymptotic behavior in the three-dimensional case for $\hat{\mu} > 1$. In this region one can assume density of states to be roughly constant, since the integrand of (3.268) is peaked in the narrow region near $\xi = 0$. The small-T behavior is

$$\Delta(T) = \Delta(0) - \Delta(0)\sqrt{\frac{\pi}{2}}\sqrt{\frac{T}{\Delta(0)}} e^{-\Delta(0)/T} \left[1 + \mathrm{erf}\left(\left[\frac{\sqrt{\hat{\mu}^2 + 1} - 1}{T/\Delta(0)}\right]^{1/2}\right)\right], \qquad (3.291)$$

---

[7]In Ref. [44], the temperature dependence of the chemical potential was calculated numerically within a "fixed carrier density model", where it turned out to be very weak in comparison with the strong dependence on the coupling strength.

where $\text{erf}(x)$ is the error function. Since the density of states is nearly constant in this limit, the same equation holds in two-dimensions — apart from a modified gap $\Delta(0)$ given by (3.284).

In the weak-coupling limit, $\hat{\mu} = \mu/\Delta(0)$ tends to infinity, and the expression above approaches exponentially fast the well-known BCS-result:

$$\Delta(T) = \Delta(0) - [2\pi\Delta(0)T]^{1/2}e^{-\Delta(0)/T}. \tag{3.292}$$

For strong couplings with $\hat{\mu} < -1$, where the three-dimensional momentum integrals are no longer peaked on a thin shell around the Fermi surface, the gap is given by

$$\Delta(T) = \Delta(0) - \frac{8}{\sqrt{\pi}}\sqrt{-\hat{\mu}}\left(\frac{\Delta(0)}{T}\right)^{3/2}e^{-\sqrt{\hat{\mu}^2+1}\,\Delta(0)/T}. \tag{3.293}$$

Near $T = 0$, the gap $\Delta(T)$ tends exponentially fast to $\Delta(0)$.

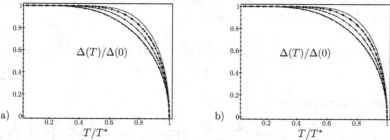

FIGURE 3.11 Temperature dependence of the gap function in three (a) and two (b) dimensions. The solid line corresponds to the crossover parameter $\hat{\mu} = 10$ (which lies in the BCS regime), crosses represent $\hat{\mu} = 0$ (i.e., the intermediate regime), and lines with boxes and circles represent $\hat{\mu} = -2$ and $\hat{\mu} = -5$, respectively, and the dashed line represents $\hat{\mu} = -10$ (i.e., the strong-coupling regime).

In two dimensions, the behavior is similar:

$$\Delta(T) = \Delta(0) - \frac{\Delta(0)}{2}E_1\left(\sqrt{\hat{\mu}^2 + 1}\,\Delta(0)/T\right), \tag{3.294}$$

where E_1 is the exponential integral $E_1(z) = \int_z^\infty e^{-t}/t\, dt$. For very strong couplings, this becomes:

$$\Delta(T) = \Delta(0) - \frac{\Delta(0)}{2}\frac{T}{\Delta(0)}\frac{1}{\sqrt{\hat{\mu}^2+1}}e^{-\sqrt{\hat{\mu}^2+1}\,\Delta(0)/T}. \tag{3.295}$$

For a plot of the temperature dependence of the gaps and the associated transition temperatures T^*, see Figs. 3.11 and Figs. 3.12.

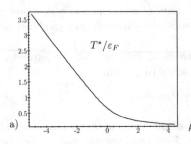 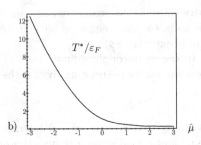

FIGURE 3.12 Dependence of T^* on the crossover parameter in three (a) and two (b) dimensions.

Let us also calculate the grand-canonical free energy F_G near $T = 0$ which we shall denote by $\Omega = F_G$, as often done in thermodynamics

$$\Omega = \sum_\mathbf{k} \left\{ \frac{\Delta^2}{2\sqrt{\xi_\mathbf{k}^2 + \Delta^2}} \tanh \frac{\sqrt{\xi_\mathbf{k}^2 + \Delta^2}}{2T} - 2T \log \left[2 \cosh \frac{\sqrt{\xi_\mathbf{k}^2 + \Delta^2}}{2T} \right] + \xi_\mathbf{k} \right\}. \quad (3.296)$$

Here and in the following formulas, $\Delta(0)$ will be abbreviated as Δ. In three dimensions, Eq. (3.296) turns into the integral

$$\frac{\Omega}{V} = \kappa_3 \int_{-\mu}^\infty d\xi \sqrt{\xi + \mu} \left[\frac{\Delta^2}{2\sqrt{\xi^2 + \Delta^2}} \tanh \frac{\sqrt{\xi^2 + \Delta^2}}{2T} - 2T \log \left(2 \cosh \frac{\sqrt{\xi^2 + \Delta^2}}{2T} \right) + \xi \right].$$

$$(3.297)$$

In two dimensions, the integral is

$$\frac{\Omega}{V} = \kappa_2 \int_{-\mu}^\infty d\xi \left[\frac{\Delta^2}{2\sqrt{\xi^2 + \Delta^2}} \tanh \frac{\sqrt{\xi^2 + \Delta^2}}{2T} - 2T \log \left(2 \cosh \frac{\sqrt{\xi^2 + \Delta^2}}{2T} \right) + \xi \right]. \quad (3.298)$$

The two expressions are regularized by subtracting their normal-state values $\Omega_n = \Omega(\Delta = 0)$.

For weak couplings, the thermodynamic potential (3.297) has the temperature dependence

$$\frac{\Omega_s}{V} \equiv \frac{\Omega - \Omega_n}{V} = \kappa_3 \sqrt{\mu} \left[-\frac{\Delta^2}{4} + \frac{1}{2}\mu|\mu| - \frac{1}{2}\mu\sqrt{\mu^2 + \Delta^2} \right]. \quad (3.299)$$

In the BCS-limit ($\hat{\mu} \to \infty$) this reduces to the well-known result

$$\frac{\Omega_s}{V} = -\kappa_3 \sqrt{\mu}\Delta^2/2. \quad (3.300)$$

In two dimensions, we find a result valid for any coupling strength:

$$\frac{\Omega_s}{V} = \kappa_2 \left[-\frac{\Delta^2}{4} + \frac{1}{2}\mu|\mu| - \frac{1}{2}\mu\sqrt{\mu^2 + \Delta^2} \right], \quad (3.301)$$

with the same BCS-limit as in (3.300):

$$\frac{\Omega_s}{V} = -\kappa_2 \Delta^2 / 2. \tag{3.302}$$

In both three- and two-dimensional cases, the low-temperature corrections to the BCS-limit are $\pi T^2 / 3$.

In the opposite limit of strong couplings, we find in three dimensions

$$\frac{\Omega}{V} = -\frac{\pi}{64} \kappa_3 \Delta^{5/2} (\hat{\mu}^2)^{-3}. \tag{3.303}$$

The gap $\Delta(0)$ has, by Eq. (3.278), the strong-coupling limit

$$\Delta(0) \underset{\hat{\mu} \to -\infty}{\approx} \varepsilon_F [16/3\pi]^{3/2} |\hat{\mu}|^{1/3}, \tag{3.304}$$

yielding the large-$\hat{\mu}$ behavior

$$\frac{\Omega}{V} \sim -\kappa_3 \varepsilon_F^{5/2} \frac{\pi}{64} \left(\frac{16}{3\pi} \right)^{15/4} |\hat{\mu}|^{-2/3}. \tag{3.305}$$

In two dimensions, we substitute the gap function Δ of Eq. (3.284) into the thermodynamic potential (3.301) and obtain, for strong couplings where $\mu < 0$, the result

$$\frac{\Omega}{V} \equiv 0. \tag{3.306}$$

Let us now turn to the entropy. In three dimensions near $T = 0$ it is given in the weak-coupling limit $\hat{\mu} = \mu/\Delta \to \infty$ by

$$\frac{S}{V} = \kappa_3 \sqrt{\hat{\mu}} \sqrt{\frac{8\pi\Delta^4}{T}} e^{-\Delta/T}, \tag{3.307}$$

and in two dimensions by

$$\frac{S}{V} = \kappa_2 \sqrt{\frac{8\pi\Delta^3}{T}} e^{-\Delta/T}. \tag{3.308}$$

In the strong-coupling limit where $\hat{\mu} = \mu/\Delta \ll -1$, the results are in three dimensions

$$\frac{S}{V} = \kappa_3 \frac{\sqrt{\pi}}{4} T^{1/2} \Delta \sqrt{\hat{\mu}^2} e^{-\hat{\mu}\Delta/T}, \tag{3.309}$$

and in two dimensions:

$$\frac{S}{V} = -2\kappa_2 \Delta \sqrt{\hat{\mu}^2} e^{-\hat{\mu}\Delta/T}. \tag{3.310}$$

From the entropy, we easily derive the heat capacity at a constant volume c_V. In three dimensions and near $T = 0$, it becomes near $T = 0$:

$$c_V = \kappa_3 \sqrt{2\pi\Delta^4}\sqrt{\hat{\mu}}\frac{2\Delta}{T^{3/2}}e^{-\Delta/T},$$ (3.311)

and in two dimensions:

$$c_V = \kappa_3\frac{\sqrt{\pi}}{4}T^{-1/2}\Delta^2\hat{\mu}^2 e^{-\hat{\mu}\Delta/T}.$$ (3.312)

The strong-coupling behavior is in three dimensions

$$c_V = \kappa_3\frac{\sqrt{\pi}}{4}T^{-1/2}\Delta^2\hat{\mu}^2 e^{-\sqrt{\hat{\mu}^2}\Delta/T},$$ (3.313)

and in two dimensions

$$c_V = 2\kappa_2\frac{\Delta^2}{T}\hat{\mu}^2 e^{-\sqrt{\hat{\mu}^2}\Delta/T}.$$ (3.314)

3.6 From BCS Superconductivity Near T_c to the Onset of Pseudogap Behavior

We now turn to the region near T^*, for which we derive the asymptotic behavior of the ratios $\Delta(T)/T^*$ and $\Delta(T)/\Delta(0)$, as well as other thermodynamic quantities. In doing so, we shall consider $\Delta(T)/T$ as a small parameter of the problem. In our calculations near T^* it is convenient to use the reduced chemical potential $\tilde{\mu} \equiv \mu/2T^*$ rather than $\hat{\mu} = \mu/\Delta$ as a crossover parameter (which also tends to ∞ in the weak-, and to $-\infty$ in the strong-coupling limit). In three dimensions, we find for weak couplings

$$\left[\frac{\Delta(T)}{2T^*}\right]^2 = \frac{\left(1-\frac{T}{T^*}\right)\left(1+\tanh\frac{\tilde{\mu}}{2}\right)}{\frac{1}{4}\left[\frac{1}{\tilde{\mu}/2} - \frac{1}{(\tilde{\mu}/2)^2}\tanh\frac{\tilde{\mu}}{2}\right] + \left(\frac{2}{\pi}\right)^2\left(1+\frac{2}{\pi}\arctan\frac{\tilde{\mu}}{\pi}\right)}.$$ (3.315)

In the limit $\tilde{\mu} \to \infty$ this tends to the BCS-result

$$\frac{\Delta(T)}{T_c} \simeq 3\sqrt{1 - \frac{T}{T_c}}.$$ (3.316)

In the opposite limit of strong couplings, both T^* and $\Delta(0)$ approach infinity. The ratio $\Delta(T)/T$ near T^* goes to zero exponentially as a function of the crossover parameter $\tilde{\mu}$:

$$\left[\frac{\Delta(T)}{2T^*}\right]^2 = \frac{16}{\sqrt{2\pi}}\left(1 - \frac{T}{T^*}\right)\left(-\frac{\tilde{\mu}}{2}\right)^{3/2}e^{\tilde{\mu}}.$$ (3.317)

In two-dimensions the same near-T^* formula (3.315) is applicable. In the weak-coupling limit, it reproduces once more the BCS-result (3.316). In the strong-coupling limit, we find once more that the ratio $\Delta(T)/T$ tends to zero exponentially and behaves now on the crossover parameter $\tilde{\mu}$ like

$$\left[\frac{\Delta(T)}{2T^*}\right]^2 = 2\left[\frac{1}{4} - \left(\frac{2}{\pi}\right)^2\right]^{-1}\left(\frac{\tilde{\mu}}{2}\right)\left(1 - \frac{T}{T^*}\right)e^{\tilde{\mu}}. \tag{3.318}$$

Let us calculate the dependence of T^* on the crossover parameter $\tilde{\mu}$ in the strong-coupling limit. In three dimensions, we obtain from Eq. (3.268) the relation

$$\frac{T^*}{\varepsilon_F} = \left(\frac{1}{3}\right)^{2/3}e^{-2\tilde{\mu}/3}. \tag{3.319}$$

This is solved for T^* (up to a logarithm) by

$$T^* \simeq -\frac{2}{3}\mu\log^{-1}\left(-\mu/\varepsilon_F\right) \tag{3.320}$$

As a function of the crossover parameter $\hat{\mu}$, we obtain

$$\frac{T^*}{\varepsilon_F} \simeq \frac{1}{2}\left(\frac{16}{3\pi}\right)^{2/3}|\hat{\mu}|^{4/3}\log^{-1}\left(\sqrt{16/\pi}|\hat{\mu}|\right). \tag{3.321}$$

In two dimensions we find from (3.272)

$$\frac{T^*}{\varepsilon_F} = \frac{1}{2}e^{-\tilde{\mu}}, \tag{3.322}$$

and thus

$$T^* \simeq -\mu\log^{-1}\left(-\mu/\varepsilon_F\right). \tag{3.323}$$

As a function of $\hat{\mu}$, this implies

$$\frac{T^*}{\varepsilon_F} = 2\hat{\mu}^2\log^{-1}\left(2\sqrt{2}|\hat{\mu}|\right). \tag{3.324}$$

Let us also derive the dependence of the ratio $\Delta(0)/T^*$ on the crossover parameter in the strong-coupling region, which in three dimensions reads

$$\frac{\Delta(0)}{T^*} = \frac{4}{\sqrt{\pi}}\left(-\tilde{\mu}\right)^{1/4}e^{\tilde{\mu}/2}, \tag{3.325}$$

and in two dimensions

$$\frac{\Delta(0)}{T^*} = 2\sqrt{2}\left(-\tilde{\mu}\right)^{1/2}e^{\tilde{\mu}/2}. \tag{3.326}$$

In the weak-coupling regime, the results are in both three and two dimensions

$$\frac{\Delta(0)}{T^*} = \frac{\pi}{e^\gamma} \left[1 - \frac{\Delta^2(0)}{4\mu^2} \right]^{-1/2} = \frac{\pi}{e^\gamma} \left(1 - \frac{1}{4\hat{\mu}^2} \right)^{-1/2} \simeq \frac{\pi}{e^\gamma} \left(1 + \frac{1}{8\hat{\mu}^2} \right). \quad (3.327)$$

In the weak-coupling regime of three- and two dimensional systems, the temperature T^* is the following function of $\hat{\mu}$:

$$\frac{T^*}{\epsilon_F} \simeq \frac{e^\gamma}{\pi} \left(\frac{1}{\hat{\mu}} - \frac{3}{8\hat{\mu}^3} \right). \quad (3.328)$$

Using this, we can also calculate the asymptotic behavior of the ratio $\Delta(T)/\Delta(0)$ near T^*. In three dimensions, the strong-coupling limit is

$$\left[\frac{\Delta(T)}{\Delta(0)} \right]^2 = \frac{\sqrt{\pi}}{2} \left(-\frac{\tilde{\mu}}{2} \right) \left(1 - \frac{T}{T^*} \right), \quad (3.329)$$

and in two dimensions:

$$\left[\frac{\Delta(T)}{\Delta(0)} \right]^2 = \frac{1}{8} \left(\frac{4}{\pi^2} - \frac{1}{4} \right)^{-1} \left(1 - \frac{T}{T^*} \right). \quad (3.330)$$

At weak couplings, both the three- and two-dimensional gap functions behave like

$$\left[\frac{\Delta(T)}{\Delta(0)} \right]^2 = \frac{4\pi^2}{e^{2\gamma}} \frac{\left(1 - \frac{T}{T^*} \right) \left[1 + \tanh \frac{\tilde{\mu}}{2} \right]}{\frac{1}{4} \left[\frac{1}{\tilde{\mu}/2} - \frac{1}{(\tilde{\mu}/2)^2} \tanh \frac{\tilde{\mu}}{2} \right] + \left(\frac{2}{\pi} \right)^2 \left(1 + \frac{2}{\pi} \arctan \frac{\tilde{\mu}}{\pi} \right)}. \quad (3.331)$$

In order to calculate the thermodynamic potential near T^*, we expand the general expression (3.296) in powers of $\Delta(T)/\Delta(0)$ and, keeping only terms of the lowest order, we get

$$\frac{\Omega_s}{V} \simeq -\frac{(T^* - T)\Delta^2}{4T^*} \int \frac{d^D\mathbf{p}}{(2\pi)^D} \cosh^{-2} \frac{\xi}{2T^*}$$

$$-\frac{\Delta^4}{8} \int \frac{d^D\mathbf{p}}{(2\pi)^D} \frac{1}{\xi^2} \left(\frac{1}{2T^*} \cosh^{-2} \frac{\xi}{2T^*} - \frac{1}{\xi} \tanh \frac{\xi}{2T^*} \right), \quad (3.332)$$

where D is the space dimension. Recall again that we consider here the temperature behavior of the system at a fixed chemical potential $\mu(T, 1/k_F a_s) \approx \mu(0, 1/k_F a_s)$, and regularize Ω by subtracting $\Omega_n = \Omega(\Delta = 0)$. Then we obtain in three dimensions at weak-couplings near T^* the thermodynamic potential:

$$\frac{\Omega_s}{V} = -\kappa_3 \sqrt{\tilde{\mu}} \sqrt{T^*} \left\{ \frac{(T^* - T)\Delta^2}{2T^*} \left[1 + \tanh \frac{\tilde{\mu}}{2} \right] + \quad (3.333) \right.$$

$$\left. + \frac{\Delta^4}{4} \frac{1}{(2T^*)^2} \left[\frac{1}{4} \left(\frac{1}{\tilde{\mu}/2} - \frac{1}{(\tilde{\mu}/2)^2} \tanh \frac{\tilde{\mu}}{2} \right) + \left(\frac{2}{\pi} \right)^2 \left(1 + \frac{2}{\pi} \arctan \frac{\tilde{\mu}}{\pi} \right) \right] \right\}.$$

In the BCS-limit, this reduces to the well-known formula:

$$\frac{\Omega_s}{V} = -\kappa_3 \sqrt{\tilde{\mu}} \sqrt{T^*} \Delta^2 \left(1 - \frac{T}{T_c} - \frac{1}{2\pi^2} \frac{\Delta^2}{T_c^2} \right). \tag{3.334}$$

In the strong-coupling limit we have

$$\frac{\Omega_s}{V} = -\kappa_3 \left\{ \frac{\pi}{64} \Delta^4 (2T^*)^{-3/2} \left(-\frac{\tilde{\mu}}{2} \right)^{-3/2} + \left(1 - \frac{T}{T^*} \right) \Delta^2 \frac{\sqrt{\pi}}{2} \sqrt{T^*} e^{\tilde{\mu}} \right\}. \tag{3.335}$$

Using the asymptotic estimates derived above for the strong-coupling limit, and the fact that in this limit

$$\frac{\mu}{T^*} \simeq -\log\left(-\frac{\mu}{\varepsilon_F} \right) \simeq -\text{const} \times \log(|\hat{\mu}|), \tag{3.336}$$

we find near T^* the difference between the thermodynamic potential of the gapless and pseudogaped normal states:

$$\frac{\Omega_s}{V} \simeq -\text{const} \left(1 - \frac{T}{T^*} \right)^2 |\hat{\mu}|^{-3/2}. \tag{3.337}$$

In two dimensions near $T = T^*$, the thermodynamic potential of the gas of bound pairs is given by the formula

$$\frac{\Omega_s}{V} = -\kappa_2 \left\{ \frac{(T^* - T)\Delta^2}{2T^*} \left[1 + \tanh \frac{\tilde{\mu}}{2} \right] + \right. \tag{3.338}$$

$$\left. + \frac{\Delta^4}{4} \frac{1}{(2T^*)^2} \left[\frac{1}{4} \left(\frac{1}{\tilde{\mu}/2} - \frac{1}{(\tilde{\mu}/2)^2} \tanh \frac{\tilde{\mu}}{2} \right) + \left(\frac{2}{\pi} \right)^2 \left(1 + \frac{2}{\pi} \arctan \frac{\tilde{\mu}}{\pi} \right) \right] \right\},$$

which holds in the crossover region, where it can be approximated by

$$\frac{\Omega_s}{V} = -\kappa_2 \Delta^2 \left(1 - \frac{T}{T^*} - \frac{1}{2\pi^2} \frac{\Delta^2}{T^{*2}} \right). \tag{3.339}$$

In the strong-coupling limit it reads

$$\frac{\Omega_s}{V} = -\kappa_2 \left\{ \left(1 - \frac{T}{T^*} \right) \Delta^2 e^{\tilde{\mu}} + \frac{\Delta^4}{4} \frac{1}{(2T^*)^2} \left[\left(\frac{1}{4} - \frac{4}{\pi^2} \right) \frac{1}{\tilde{\mu}/2} \right] \right\}. \tag{3.340}$$

Using the earlier-derived asymptotic behavior, plus the limiting equation (3.336) which also holds in two dimensions, we derive the $\hat{\mu}$-behavior of the thermodynamic potential

$$\frac{\Omega_s}{V} \simeq -\text{const} \times \left(1 - \frac{T}{T^*} \right)^2 \log |\hat{\mu}|. \tag{3.341}$$

The entropy behaves in three dimensions in the weak-coupling regime near T^* like

$$\frac{S_s}{V} \equiv \frac{S - S_n}{V} = -\kappa_3 \sqrt{\tilde{\mu}} \sqrt{T^*} \frac{\Delta^2}{2T^*} \left[1 + \tanh \left(\frac{\tilde{\mu}}{2} \right) \right], \tag{3.342}$$

with the BCS-limit

$$\frac{S_s}{V} = -\kappa_3 \sqrt{\mu} \frac{\Delta^2}{T_c}. \tag{3.343}$$

The strong-coupling limit in three dimensions yields

$$\frac{S_s}{V} = -\kappa_3 \frac{\sqrt{\pi}}{2} \frac{\Delta^2}{\sqrt{T^*}} e^{\tilde{\mu}}. \tag{3.344}$$

Inserting in the above asymptotic formulas the quantities Δ, μ, and T^*, we find

$$\frac{S_s}{V} \simeq -\text{const} \times \left(1 - \frac{T}{T^*}\right) |\hat{\mu}|^{-5/3}. \tag{3.345}$$

In two dimensions, the entropy is given in the entire crossover region by

$$\frac{S_s}{V} = -\kappa_2 \frac{\Delta^2}{2T^*} \left[1 + \tanh \frac{\tilde{\mu}}{2}\right]. \tag{3.346}$$

In the BCS-limit, the behavior is

$$\frac{S_s}{V} = -\kappa_2 \frac{\Delta^2}{T^*}, \tag{3.347}$$

and in the strong-coupling limit:

$$\frac{S_s}{V} = -\kappa_2 \frac{\Delta^2}{T^*} e^{\tilde{\mu}}. \tag{3.348}$$

Using corresponding asymptotic formulas for Δ, μ and T^* in two dimensions, this depends on $\hat{\mu}$ as

$$\frac{S_s}{V} = -\text{const} \times \left(1 - \frac{T}{T^*}\right) \hat{\mu}^{-2}. \tag{3.349}$$

Let us now derive the specific heat. From the derivative with respect to the temperature we find, in three dimensions and at weak-couplings near T^*:

$$\frac{C_s}{V} = 2T\kappa_3 \sqrt{\tilde{\mu}} \sqrt{T^*} \frac{\left(1 + \tanh \frac{\tilde{\mu}}{2}\right)^2}{\frac{1}{4} \left[\frac{1}{\tilde{\mu}/2} - \frac{1}{(\tilde{\mu}/2)^2} \tanh \frac{\tilde{\mu}}{2}\right] + \left(\frac{2}{\pi}\right)^2 \left(1 + \frac{2}{\pi} \arctan \frac{\tilde{\mu}}{\pi}\right)}, \tag{3.350}$$

which has the well-known BCS-limit

$$\frac{C_s}{V} \simeq \kappa_3 \sqrt{\tilde{\mu}} \sqrt{T^*} \pi^2 T_c. \tag{3.351}$$

In the strong-coupling limit, the result is

$$\frac{C_s}{V} = \kappa_3 16\sqrt{2} T^{*3/2} \left(-\frac{\tilde{\mu}}{2}\right)^{3/2} e^{2\tilde{\mu}}. \tag{3.352}$$

Inserting earlier- derived asymptotic formulas we see that C_s tends to zero in the strong-coupling limit as

$$\frac{C_s}{V} \sim \text{const} \times |\hat{\mu}|^{-2}. \tag{3.353}$$

In two dimensions, the result for the entire crossover region is

$$\frac{C_s}{V} = 2T^* \kappa_2 \frac{\left(1 + \tanh \dfrac{\tilde{\mu}}{2}\right)^2}{\dfrac{1}{4}\left[\dfrac{1}{\tilde{\mu}/2} - \dfrac{1}{(\tilde{\mu}/2)^2} \tanh \dfrac{\tilde{\mu}}{2}\right] + \left(\dfrac{2}{\pi}\right)^2 \left(1 + \dfrac{2}{\pi} \arctan \dfrac{\tilde{\mu}}{\pi}\right)}. \tag{3.354}$$

In the BCS-limit, this becomes

$$\frac{C_s}{V} \simeq \kappa_2 \pi^2 T^*, \tag{3.355}$$

and in the strong-coupling limit

$$\frac{C_s}{V} = 4\kappa_2 \sqrt{\tilde{\mu}}\sqrt{T^*} \left(\frac{1}{4} - \frac{4}{\pi^2}\right)^{-1} e^{2\tilde{\mu}}. \tag{3.356}$$

As a function of $\hat{\mu}$, the result is

$$\frac{C_s}{V} \sim \text{const} \times \hat{\mu}^{-2}. \tag{3.357}$$

From the above calculation near T^* we see that both quantities S_s and C_s tend rapidly to zero with growing coupling strength in the pseudogaped regime (like a power of the crossover parameter $\hat{\mu}$ or with an exponential dependence on $\hat{\mu}$).

Note that, in the strong-coupling regime, the modified gap function $\sqrt{\mu^2 + \Delta^2} = \sqrt{\hat{\mu}^2 + 1}\,\Delta$ of Eq. (3.276) enters the expressions for thermodynamic quantities below T^* in the same way as the ordinary gap Δ in the BCS-limits (3.309), (3.310), (3.313), and (3.314).

3.7 Phase Fluctuations in Two Dimensions and Kosterlitz-Thouless Transition

In the previous sections we have calculated the properties of the theory in the mean-field approximation. One of the results is the bending energy of the complex gap function $\Delta(\mathbf{x}) = |\Delta(\mathbf{x})|e^{i\theta(\mathbf{x})}$. Most important are the stiffness fluctuations of the phase angle $\theta(\mathbf{x})$. These determine the superfluid density ρ_s. As the temperature increases, the stiffness decreases just like in an ordinary solid. In two dimensions, the defects in the phase angle field are observable as vortices and antivortices. These attract each other with logarithmic potentials, thus behaving like a two-dimensional gas of electric charges. When the softening of the stiffness proceeds, one reaches a temperature T_{BKT}, where the vortex pairs separate. This so-called *pair unbinding transition* was discovered by Berezinskii and developed further by Kosterlitz and Thouless, and named after them [54]. At the transition, the stiffness collapses in the same way as in the melting process of a crystal, where the elastic constants collapse. We shall now present a calculation of the stiffness, first in the two-dimensional system. There the phase fluctuations are quite violent so that the Mermin-Wagner-Hohenberg-Coleman theorem [55] forbids the existence of a strict long-range order. It rather leads to a power behavior of correlation functions for all temperatures below $T_{\rm BKT}$.

The crossover of the Kosterlitz-Thouless transition from weak to strong coupling was first considered in the 1990s by [48, 49]. It was also studied by means of an XY-model, whose stiffness of phase fluctuations was derived from a fixed nonvanishing modulus of the order parameter Δ [44]. Writing the spacetime-dependent order parameter as $\Delta(x)e^{i\theta(x)}$, the partition function may be expressed as a functional integral

$$Z(\mu, T) = \int \Delta \mathcal{D}\Delta \mathcal{D}\theta \, e^{-\mathcal{A}^{\rm e}_{\Delta}}. \tag{3.358}$$

The exponent contains a Euclidean collective action $\mathcal{A}^{\rm e}_{\Delta} = -i\mathcal{A}[\Delta^*, \Delta]$:

$$\mathcal{A}^{\rm e}_{\Delta} = \frac{1}{g} \int_0^{\beta} d\tau \int dx \, \Delta^2(x) - {\rm Tr} \log{(G^{\rm e}_{\Delta})^{-1}} + {\rm Tr} \log{(G^{\rm e}_{\Delta_0})^{-1}}. \tag{3.359}$$

This can be extracted from (3.8) by a Wick rotation of the time t to imaginary values $-i\tau$. The right-hand side contains the inverse Euclidean Green functions of the fermions in the collective pair field. These can be read off the Wick-rotated version of Eq. (3.47), which reads for fermions

$$(G^{\rm e}_{\Delta})^{-1} = \begin{pmatrix} [-\partial_{\tau} - \xi(-i\boldsymbol{\nabla})]\delta(x-x'') & \Delta(x, x'') \\ \Delta^*(x, x'') & [i\partial_{\tau} + \xi(i\boldsymbol{\nabla})]\delta(x-x'') \end{pmatrix}. \tag{3.360}$$

This may also be written in a matrix notation as

$$(G^{\rm e}_{\Delta})^{-1} = -\hat{I}\partial_{\tau} + \tau_3 \left(\frac{\boldsymbol{\nabla}^2}{2M} + \mu \right) + \tau_1 \Delta(\tau, \mathbf{x}), \tag{3.361}$$

where $\hat{I} = \tau_0$, τ_1, τ_3 are Pauli matrices.

We want to study the action as a functional of the phase fluctuations of $\Delta(x)$. For this we can in principle proceed as in Eq. (3.132). In the present context, however, we shall be interested mainly in the bending energy of the phase fluctuations. Then it is more convenient to absorb the phase factor of $\Delta = e^{i\theta(x)}\Delta_0$ into the external fermions before integrating them out in the functional integral. Then the inverse Green function has the matrix form

$$(G^{\rm e}_{\Delta})^{-1} \equiv (G^{\rm e}_{\Delta_0})^{-1} - \Sigma = -\hat{I}\partial_{\tau} + \tau_3 \left(\frac{\boldsymbol{\nabla}^2}{2M} + \mu \right) + \tau_1 \Delta_0 - \Sigma, \tag{3.362}$$

where

$$\Sigma \equiv \tau_3 \Sigma_3 + \tau_0 \Sigma_0 \equiv \tau_3 \left[\frac{i\partial_{\tau}\theta}{2} + \frac{(\boldsymbol{\nabla}\theta)^2}{8M} \right] - \hat{I}\left[\frac{i\boldsymbol{\nabla}^2\theta}{4M} + \frac{i\boldsymbol{\nabla}\theta(\tau, \mathbf{x})\boldsymbol{\nabla}}{2M} \right]. \tag{3.363}$$

By expanding the collective action (3.359) in powers of Σ, we derive the gradient expansion

$$\mathcal{A}_\Delta^e = \mathcal{A}_\Delta^{e,\text{pot}} + \mathcal{A}_\Delta^{e,\text{grad}}, \tag{3.364}$$

where

$$\mathcal{A}_\Delta^{e,\text{grad}} = \text{Tr} \sum_{n=1}^{\infty} \frac{1}{n}(G_{\Delta_0}^e \Sigma)^n. \tag{3.365}$$

From now on we shall neglect the subscript of Δ_0 as being superfluous because we work only at the extremal value of Δ, and study only phase modulations. The first term in the expansion (3.365) is

$$\mathcal{A}_\Delta^{e(1)} = \int_0^\beta d\tau \int d^D x \sum_{n=-\infty}^{\infty} \int d^D k \,\text{tr}[G_\Delta^e(i\omega_n, \mathbf{k})\tau_3] \left[\frac{i\partial_\tau\theta}{2} + \frac{(\boldsymbol{\nabla}\theta)^2}{8m}\right], \tag{3.366}$$

with

$$G_\Delta^e(i\omega_n, \mathbf{k}) = -\frac{i\omega_n \hat{I} + \tau_3 \xi_{\mathbf{k}} - \tau_1 \Delta}{\omega_n^2 + \xi_{\mathbf{k}}^2 + \Delta^2}. \tag{3.367}$$

After summing over the Matsubara frequencies and integrating over $\mathbf{k}$, we obtain

$$\mathcal{A}_\Delta^{e(1)} = \int_0^\beta d\tau \int d^D x \, \rho(\mu, T, \Delta) \left[\frac{i\partial_\tau\theta}{2} + \frac{(\boldsymbol{\nabla}\theta)^2}{8m}\right], \tag{3.368}$$

with $\rho(\mu, T, \Delta)$ given by (3.267).

For the second expansion term we obtain two contributions. The first is

$$\mathcal{A}_\Delta^{e(2,1)} = -\frac{1}{2}\int_0^\beta d\tau \int d^D x \, K(\mu, T, \Delta) \left[\frac{i\partial_\tau\theta}{2} + \frac{(\boldsymbol{\nabla}\theta)^2}{8M}\right]^2, \tag{3.369}$$

where $K(\mu, T, \Delta)$ is the integral

$$K(\mu, T, \Delta) = \frac{M}{8\pi}\left(1 + \frac{\mu}{\sqrt{\mu^2 + \Delta^2}} \tanh\frac{\sqrt{\mu^2 + \Delta^2}}{2T}\right). \tag{3.370}$$

The second term is

$$\mathcal{A}_\Delta^{(2,2)} = -\int_0^\beta d\tau \int d^D x \frac{1}{32\pi^2 M^2}\int d^D k \frac{k^2}{\cosh^2[\sqrt{\xi_{\mathbf{k}}^2 + \Delta^2}/2T]}(\boldsymbol{\nabla}\theta)^2. \tag{3.371}$$

Combining (3.368), (3.369), and (3.371), we obtain

$$\begin{aligned}
\mathcal{A}_\Delta^{e\,\text{grad}} = \frac{1}{2}\int_0^\beta d\tau \int d^D x \,[&\rho(\mu, T, \Delta)i\partial_\tau\theta \\
&+ J(\mu, T, \Delta(\mu, T))(\boldsymbol{\nabla}\theta)^2 + K(\mu, T, \Delta(\mu, T))(\partial_\tau\theta)^2], \tag{3.372}
\end{aligned}$$

where $J(\mu, T, \Delta)$ is the stiffness coefficient

$$J(\mu, T, \Delta) = \frac{1}{4M}\rho(\mu, T, \Delta) - \frac{T}{4\pi}\int_{-\mu/2T}^{\infty} dx \frac{x + \mu/2T}{\cosh^2\sqrt{x^2 + \Delta^2/4T^2}}. \qquad (3.373)$$

At the temperature T^* where the modulus of Δ vanishes, also the stiffness disappears. The gradient energy corresponds to an XY-model with a Hamiltonian [54, 33]:

$$H = \frac{J}{2}\int d\mathbf{x}[\boldsymbol{\nabla}\theta(\mathbf{x})]^2. \qquad (3.374)$$

The only difference with reprect to the standard XY-model lies in the dependence of the stiffness constant J on the temperature. Here this is determined from the solutions of gap- and number-equations (3.268) and (3.267). Clearly, in this model the Berezinskii-Kosterlitz-Thouless transition always takes place below T^*. In the XY-model with vortices of a high fugacity, the temperature of the phase transition is determined by a simple formula [33, 59]:

$$T_{\mathrm{BKT}} = \frac{\pi}{2}J. \qquad (3.375)$$

The transition point is found from the divergence of the average square distance of the vortex-antivortex pair. The two attract each other by a Coulomb potential $v(r) = 2\pi J \log(r/r_0)$, and the average square distance is

$$\langle r^2 \rangle \propto \int_{r_0}^{\infty} d^2\mathbf{x}\, r^2 e^{-(2\pi J/T)\log(r/r_0)} \propto \frac{1}{4 - 2\pi J/T}. \qquad (3.376)$$

This diverges indeed at the temperature (3.375). In our case, T_{BKT} is determined self-consistently from the equation

$$T_{\mathrm{BKT}} = \frac{\pi}{2}J(\mu, T_{\mathrm{BKT}}, \Delta(\mu, T_{\mathrm{BKT}})). \qquad (3.377)$$

Using Eqs. (3.373) and (3.375), we can easily see that T_{BKT} tends to zero when the pair attraction vanishes. In general, the behavior of T_{BKT} for strong and weak couplings is found by the following considerations. We observe that the particle number n does not vary appreciably if the temperature lies in the interval $0 < T < T^*$, so that weak-coupling estimates for T_{BKT} derived within the model under the assumption of a temperature-independent chemical potential practically coincide with those derived from a fixed fermion density. Further it is immediately realized that in the weak-coupling limit, $\Delta(T_{\mathrm{BKT}}, \mu)/T_{\mathrm{BKT}}$ is a small parameter. At zero coupling, the stiffness $J(\mu, T_{\mathrm{BKT}}, \Delta(\mu, T_{\mathrm{BKT}}))$ vanishes identically, so that an estimate of J at weak couplings requires calculating a lowest-order correction to the second term of Eq. (3.373) which is a term proportional to $\Delta(T_{\mathrm{BKT}}, \mu)/T_{\mathrm{BKT}}$. Thus the weak-coupling approximation expression to the stiffness reads:

$$J(T) \simeq \frac{7\zeta(3)}{16\pi^3}\varepsilon_F \frac{\Delta^2(T)}{T^{*2}}. \qquad (3.378)$$

Inserting here the BCS value (3.316), and equating J with the critical stiffness (3.375), we obtain the weak-coupling equation for T_{BKT}:

$$T_{BKT} \simeq \frac{\varepsilon_F}{4}\left(1 - \frac{T_{BKT}}{T^*}\right), \qquad (3.379)$$

where $\varepsilon_F = (\pi/M)\rho$ is the Fermi energy of free fermions in two dimensions.

It is useful to introduce the reduced dimensionless temperatures $\tilde{T}_{KT} \equiv T_{KT}/\varepsilon_F$ and $\tilde{T}^* = T^*/\varepsilon_F$, which are both small in the weak-coupling limit. Then we rewrite Eq. (3.379) as

$$\tilde{T}_{KT} \simeq \frac{1}{4}\frac{1}{1 + 1/4\tilde{T}^*}. \qquad (3.380)$$

For small $\tilde{T}^*$, we may expand

$$\tilde{T}_{KT} \approx \tilde{T}^* - 4\tilde{T}^{*2}. \qquad (3.381)$$

This equation shows explicitly how T_{BKT} merges with T^* for decreasing coupling strength.

For weak coupling, T_{BKT} behaves like

$$T_{BKT} \approx \frac{e^\gamma}{\pi}\Delta(0). \qquad (3.382)$$

The merging of the two temperatures in the weak-coupling regime is displayed in Fig. 3.13.

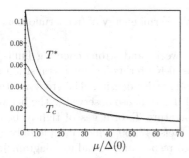 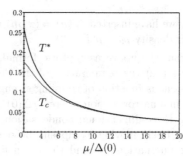

FIGURE 3.13 Dependence of the pair-formation temperature T^* on the chemical potential. Dashed lines represent the pair condensation temperature T_c. The left figure is for $D = 2$ dimensions, where $T_c = T_{BKT}$, the right for $D = 3$.

Consider now the opposite limit of strong couplings. There we see, using Eqs. (3.377), (3.267), (3.268), and (3.373) for T_{BKT}, $\rho(T, \mu)$, and $\Delta(T, \mu)$, that T_{BKT} tends to a constant value. We further observe that in the strong-coupling limit, $\Delta(T_{BKT})$ is always situated close to the zero-temperature value of $\Delta(T_{BKT}, \mu) \approx \Delta(T = 0, \mu)$. Taking this into account, we derive an estimate for the second term in (3.373), thus obtaining the strong-coupling equation for T_{BKT}:

$$T_{BKT} \simeq \frac{\pi}{8}\left\{\frac{\rho}{M} - \frac{T_{BKT}}{\pi}\exp\left[-\frac{\sqrt{\mu^2 + \Delta^2(T_{BKT}, \mu)}}{T_{BKT}}\right]\right\}. \qquad (3.383)$$

With the approximation $\Delta(T_{\mathrm{BKT}}, \mu) \approx \Delta(T = 0, \mu)$, we find that the first term in the exponent tends, in the strong-coupling limit, to a constant $\Delta^2(T_{\mathrm{BKT}}, \mu)/2\mu T_{\mathrm{BKT}} \to -4$. The first term in brackets tends to $-\infty$. Hence Eq. (3.383) has the limiting form

$$T_{\mathrm{BKT}} \simeq \frac{\pi}{8} \frac{\rho}{m} \left\{ 1 - \frac{1}{8} \exp\left[\frac{2\mu}{\varepsilon_F} - 4 \right] \right\}. \tag{3.384}$$

Thus for increasing coupling strength, the phase-decoherence temperature T_{BKT} tends rapidly towards a constant:

$$T_{\mathrm{BKT}} \simeq \frac{\pi}{8} \frac{\rho}{m}. \tag{3.385}$$

In this limit we know, from Eq. (3.267), that the difference in the carrier density at zero temperature, $\rho(T = 0)$, becomes equal to $\rho(T = T_{KT})$, so that our limiting result coincides with that obtained in the "fixed carrier density model":

$$T_{\mathrm{BKT}} = \frac{\varepsilon_F(n_0)}{8} = \frac{\pi}{8m} \epsilon_0, \tag{3.386}$$

where we have inserted $\varepsilon_F(n) = (\pi/m)\rho$ for the Fermi energy of free fermions at the carrier density $\rho_0 = \rho(T = 0)$.

From the above asymptotic formulas for weak- and strong-coupling limits we observe that the temperature of the Berezinskii-Kosterlitz-Thouless transition is a monotonous function of coupling strength and carrier density. The crossover takes place in a narrow region where $\mu/\Delta(0) \in (-1, 1)$. It is also observed in the behavior of the three-dimensional condensation temperature T_c of a gas of tightly bound, almost freely moving, composite bosons. In Refs. [47, 50] which include only quadratic fluctuations around the mean field (corresponding to ladder diagrams), T_c was shown to tend to a constant free Bose gas value $T_c = [\rho/2\zeta(3/2)]^{2/3}\pi/M$, with no dependence on the internal structure of the boson.

Here we find a similar result in two dimensions, where T_{KT} tends to a constant, depending only on the mass $2M$ and the density $\rho/2$ of the pairs. No dependence on the coupling strength remains. The only difference with respect to the three-dimensional case is that here the transition temperature $T_c = T_{KT}$ is linear in the carrier density n, while growing like $\rho^{2/3}$ in three dimensions. Our limiting result (3.386) agrees with Refs. [48] and [44]. There exists a corresponding equation for the temperature T^* in the strong-coupling limit $\varepsilon_0 \gg \varepsilon_F$:

$$T^* \simeq \frac{\varepsilon_0}{2} \frac{1}{\log \varepsilon_0/\varepsilon_F}. \tag{3.387}$$

3.8 Phase Fluctuations in Three Dimensions

In this section we discuss, in a completely analogous way, the fluctuations in three dimensions. For small temperatures, where $\Delta(T)$ is close to $\Delta(0)$, we obtain from (3.373):

$$J_{3D}(\mu, T, \Delta) = \frac{1}{4m}\rho(\mu, T, \Delta) - \frac{\sqrt{2M}}{16\pi^2}\frac{1}{T}\int_{-\mu}^{\infty}d\xi\,\frac{(\xi + \mu)^{3/2}}{\cosh^2(\sqrt{\xi^2 + \Delta^2}/2T)}, \qquad (3.388)$$

governing the phase fluctuations via an effective XY-model

$$H = \frac{J_{3D}}{2}\int d^3\mathbf{x}[\boldsymbol{\nabla}\theta(\mathbf{x})]^2. \qquad (3.389)$$

The temperature of the phase transition in this model can reasonably be estimated using mean-field methods for the lattice 3D XY-model [33]:

$$T_{3D}^{MF} \simeq 3J_{3D}a. \qquad (3.390)$$

Then the lattice spacing of the theory [33] is $a = 1/n_b^{1/3}$, where n_b denotes the number of vortex-antivortex pairs.

In the weak-coupling limit, the stiffness coefficient becomes, approximating T_c by T^*,

$$J_{3D} \approx \frac{7}{48\pi^4}\zeta(3)\frac{p_F^3}{M}\frac{\Delta^2}{T^{*2}}. \qquad (3.391)$$

This is precisely the coefficient of the gradient term in the Ginzburg-Landau expansion. In the weak-coupling limit, the two temperatures merge according to the formula

$$\tilde{T}_c = \tilde{T}^* - \alpha\tilde{T}^{*5/2}, \qquad (3.392)$$

which contains a larger power of $\tilde{T}^*$ in the second term, as well as a smaller prefactor $\alpha = (2\pi^2)^{2/3}/2 \approx 3.65$. Formula (3.392) is compared with the two-dimensional expression in (3.381). The merging behavior of the two T^*-curves is displayed in Fig. 3.13.

In the strong-coupling limit of the theory, there exist tightly bound composite bosons, and the phase stiffness tends asymptotically to

$$J = \frac{\rho}{4m} - \frac{3\sqrt{2\pi m}}{16\pi^2}T^{3/2}e^{-\sqrt{\mu^2 + \Delta^2}/T}. \qquad (3.393)$$

For small T this goes rapidly to

$$J_{BE} = \frac{\rho}{4m}. \qquad (3.394)$$

An estimate for the critical temperature, obtained via the mean-field treatment of the 3D XY-model on the lattice, reads in this limit:

$$T_c = \frac{3}{2m}\left[\left(\frac{\rho}{2}\right)^{2/3} - \frac{1}{\rho^{1/3}}\frac{1}{2^{7/6}\pi^{3/2}}T_c^{3/2}m^{3/2}e^{-\sqrt{\mu^2 + \Delta^2}/T_c}\right]. \qquad (3.395)$$

This quickly tends from below to the value:

$$T_c^{\text{3D XY}} = \frac{3\rho^{2/3}}{2^{5/3}m} = \varepsilon_F \frac{3}{(6\pi^2)^{2/3}} \simeq 0.2\varepsilon_F. \tag{3.396}$$

The result is very close to the temperature of the condensation of bosons of mass $2M$ and density $\rho/2$, which was obtained after including the effect of Gaussian fluctuations on the mean-field equation for the particle number [47, 50] (as discussed above), yielding

$$T_c^{\text{Bosons}} = [\rho/2\zeta(3/2)]^{2/3}\pi/m = 0.218\varepsilon_F. \tag{3.397}$$

Let us remark that the separation of T^* and T_c has an analogy in ferroelectrics and magnets. These also contain two separate characteristic temperatures, for example in the latter case — the Stoner- and the Curie-temperature. It also can be studied more precisely in a simple field theoretic model in $2+\varepsilon$ dimensions with an $O(n)$ symmetry for large n. In such a model, the existence of two small parameters ε and $1/n$ has permitted us to *prove* the existence of two transitions, and to exhibit clearly their different physical origins [34].

3.9   Collective Classical Fields

The introduction of a fluctuating pair field $\Delta_{\alpha\beta}$ via the Hubbard-Stratonovich transformations (1.79) and (1.80), together with the identity (3.42), is an exact procedure. It allows to re-express the interaction in the partitian function in the form (3.42) which contains the fundamental field only quadratically. However, since our calculations of the physical properties will eventually be merely *approximate*, the exactness of the transformation is not a virtue, but turns out to be a handicap. A better approximation is more useful than an exact expression in the wrong environment. We have seen in Chapter 2 that there exists another possibility of eliminating the fourth-order interaction term with the help of a completely different real density field φ. That procedure led also to an exact reformulation of the theory in terms of a fluctuating collective quantum field. The question as to which of the two formulations is better depends on the phenomena which one wants to study. The phenomena emerging in one formulation from a low-order approximation may require, in the other formulation, the summation of infinitely many diagrams. Usually, this is a hard task, so we need a procedure where the collective effects in each possible channel emerge from a low-order calculation. The way out is found by giving up the attempt of rewriting the functional integral of the theory in terms of a fluctuating *collective quantum field*. Instead, we must resort to a non-fluctuating *collective classical field*. Such a theory was developed in the context of quantum mechanics under the name of *Variational Perturbation Theory* (VPT) [23]. This theory has been shown to produce exponentially fast converging results which contain the effects of all possible collective phenomena. They are based on the introduction and subsequent optimization of a variety of collective classical fields. The theory has been extended to QFT in the textbook [24].

Following the classical collective field approach we modify the classical action in a trivial way, rather than applying the functional integral identity (3.42) in the partition function (3.1). Initially, the classical action consists of a sum

$$\mathcal{A} = \mathcal{A}_0 + \mathcal{A}_{\text{int}}, \qquad (3.398)$$

with a free term

$$\mathcal{A}_0 = \int d^3x \, dt \, \psi_\alpha^\dagger(\mathbf{x}, t)[i\partial_t - \xi(-i\boldsymbol{\nabla})]\psi_\alpha(\mathbf{x}, t), \qquad (3.399)$$

and an interaction term

$$\mathcal{A}_{\text{int}} = \frac{g}{2} \sum_{\alpha,\beta} \int d^3x \, dt \, \psi_\alpha^*(\mathbf{x}, t)\psi_\beta^*(\mathbf{x}, t)\psi_\beta(\mathbf{x}, t)\psi_\alpha(\mathbf{x}, t). \qquad (3.400)$$

To this we add and subtract a dummy term which has the form of a simple spacetime-dependent *mass term*:

$$\mathcal{A}_\mathcal{M} = -\frac{1}{2} \int \int d^3x \, dt \, f^\dagger(\mathbf{x}, t)\mathcal{M}(\mathbf{x}, t)f(\mathbf{x}, t). \qquad (3.401)$$

It contains in a mass matrix

$$\mathcal{M}_{\alpha\beta} \equiv \begin{pmatrix} \Sigma_{\alpha\beta} & \Delta_{\alpha\beta} \\ \Delta_{\alpha\beta}^* & \pm\Sigma_{\alpha\beta} \end{pmatrix}, \qquad (3.402)$$

which depends an as yet undetermined off-diagonal trial field $\Delta_{\alpha\beta}(x)$ and a diagonal field $\Sigma_{\alpha\beta}(x)$. Explicitly, the mass term (3.401) reads

$$\mathcal{A}_\mathcal{M} = -\int d^3x \, dt \left[\pm\psi_\alpha^*\Sigma_{\alpha\beta}\psi_\beta + \frac{1}{2}\left(\pm\psi_\alpha^*\Sigma_{\alpha\beta}\psi_\beta + \psi_\beta\Delta_{\beta\alpha}^*\psi_\alpha + \psi_\alpha^*\Delta_{\alpha\beta}\psi_\beta^* \right) \right]. \qquad (3.403)$$

After adding and subtracting $\mathcal{A}_\mathcal{M}$, we reorganize the action (3.398): We change the free part $\mathcal{A}_0$ to $\mathcal{A}_0^{\text{new}} \equiv \mathcal{A}_0 + \mathcal{A}_\mathcal{M}$, and the interaction to the new subtracted interaction $\mathcal{A}_{\text{int}}^{\text{new}} \equiv \mathcal{A}_{\text{int}} - \mathcal{A}_\mathcal{M}$. In terms of the four-components fields $f(x)$ introduced in Eq. (3.46), the new free action may be written in a 4×4-matrix form analogous to (3.47):

$$\mathcal{A}_0^{\text{new}} = \int d^4x \, f^*(x)A_{\Sigma,\Delta}^{\text{new}}f(x), \qquad (3.404)$$

where $A_{\Sigma,\Delta}^{\text{new}}$ is the same functional matrix as before in (3.78), except that it contains, in addition to the pair field $\Delta_{\alpha\beta}$, the density field $\Sigma_{\alpha\beta}$ in the form

$$A_{\Sigma,\Delta}^{\text{new}} = \begin{pmatrix} [i\partial_t - \xi(-i\boldsymbol{\nabla})]\,\delta_{\alpha\beta} - \Sigma_{\alpha\beta}(x) & -\Delta_{\alpha\beta}(x) \\ -\Delta_{\alpha\beta}^*(x) & \mp[i\partial_t + \xi(i\boldsymbol{\nabla})]\,\delta_{\alpha\beta} \mp \Sigma_{\alpha\beta}(x) \end{pmatrix} \delta(x - x'), \qquad (3.405)$$

with $\Sigma_{\alpha\beta}$ and $\Delta_{\alpha\beta}$ being matrices in spin space.

An important difference with respect to the earlier treatment is that $\Sigma_{\alpha\beta}$ and $\Delta_{\alpha\beta}$ are now *nonfluctuating classical fields*. They will be determined at the end of the calculation by an optimization process. The new interaction $\mathcal{A}_{\text{int}}^{\text{new}} \equiv \mathcal{A}_{\text{int}} - \mathcal{A}_{\mathcal{M}}$ reads explicitly

$$\mathcal{A}_{\text{int}}^{\text{new}} = \frac{g}{2}\sum_{\alpha,\beta}\int d^3x dt\, \psi_\alpha^*(\mathbf{x},t)\psi_\beta^*(\mathbf{x},t)\psi_\beta(\mathbf{x},t)\psi_\alpha(\mathbf{x},t)$$
$$+ \int d^3x dt\,\left[\pm\psi_\alpha^*\Sigma_{\alpha\beta}\psi_\beta + \frac{1}{2}\left(\psi_\beta\Delta_{\beta\alpha}^*\psi_\alpha + \psi_\alpha^*\Delta_{\alpha\beta}\psi_\beta^*\right)\right]. \qquad (3.406)$$

We now calculate the partition function of the action $\mathcal{A} = \mathcal{A}_0^{\text{new}} + \mathcal{A}_{\text{int}}^{\text{new}}$ in perturbation theory. To zeroth order, the quadratic terms produce the collective action

$$\mathcal{A}_0[\Delta,\Sigma] = \pm\frac{i}{2}\text{Tr}\log\left[i\mathbf{G}_{\Delta,\Sigma}^{-1}(x,x')\right], \qquad (3.407)$$

where the 4×4-matrix $\mathbf{G}_{\Delta,\Sigma}$ denotes the propagator which satisfies the functional equation

$$\begin{pmatrix} [i\partial_t - \xi(-i\boldsymbol{\nabla})]\delta_{\alpha\beta} - \Sigma_{\alpha\beta}(x) & -\Delta_{\alpha\beta}(x) \\ -\Delta_{\alpha\beta}^*(x) & \mp[i\partial_t + \xi(i\boldsymbol{\nabla})]\delta_{\alpha\beta} \mp \Sigma_{\alpha\beta}(x) \end{pmatrix}\mathbf{G}_{\Delta,\Sigma}(x,x') = i\delta(x-x'). \qquad (3.408)$$

To first order in perturbation theory we calculate the expectation value of the interaction (3.42) using Wick's theorem. First we have

$$\begin{aligned}\langle\psi_\alpha^*(\mathbf{x},t)\psi_\beta^*(\mathbf{x},t)\psi_\beta(\mathbf{x},t)\psi_\alpha(\mathbf{x},t)\rangle &= \langle\psi_\alpha^*(\mathbf{x},t)\psi_\alpha(\mathbf{x},t)\rangle\langle\psi_\beta^*(\mathbf{x},t)\psi_\beta(\mathbf{x},t)\rangle \\ &\pm \langle\psi_\alpha^*(\mathbf{x},t)\psi_\beta(\mathbf{x},t)\rangle\langle\psi_\beta^*(\mathbf{x},t)\psi_\alpha(\mathbf{x},t)\rangle \\ &+ \langle\psi_\alpha^*(\mathbf{x},t)\psi_\beta^*(\mathbf{x},t)\rangle\langle\psi_\beta(\mathbf{x},t)\psi_\alpha(\mathbf{x},t)\rangle. \end{aligned} \qquad (3.409)$$

We now introduce the expectation values[8]

$$G_{\alpha\beta}^{\Delta*}(\mathbf{x},t) \equiv \langle\psi_\alpha^*(\mathbf{x},t)\psi_\beta^*(\mathbf{x},t)\rangle, \qquad G_{\alpha\beta}^{\Delta}(\mathbf{x},t) \equiv \langle\psi_\beta(\mathbf{x},t)\psi_\alpha(\mathbf{x},t)\rangle, \qquad (3.410)$$
$$G_{\alpha\beta}^{\Sigma}(\mathbf{x},t) \equiv \langle\psi_\alpha^*(\mathbf{x},t)\psi_\beta(\mathbf{x},t)\rangle, \qquad G_{\alpha\beta}^{\Sigma*}(\mathbf{x},t) \equiv \langle\psi_\beta^*(\mathbf{x},t)\psi_\alpha(\mathbf{x},t)\rangle. \qquad (3.411)$$

Then we can rewrite the interaction as

$$\begin{aligned}\langle\mathcal{A}_{\text{int}}\rangle &= (1/2g)\int d^3x dt\,(\tilde{\Sigma}_{\alpha\alpha}\tilde{\Sigma}_{\beta\beta} \pm \tilde{\Sigma}_{\alpha\beta}\tilde{\Sigma}_{\beta\alpha} + \tilde{\Delta}_{\beta\alpha}^*\tilde{\Delta}_{\beta\alpha} \\ &\quad - (1/2g)\int d^3x dt\,(\pm2\tilde{\Sigma}_{\alpha\beta}\Sigma_{\alpha\beta} + \tilde{\Delta}_{\beta\alpha}\Delta_{\beta\alpha}^* + \tilde{\Delta}_{\alpha\beta}^*\Delta_{\beta\alpha}), \end{aligned} \qquad (3.412)$$

$$\begin{aligned}\langle\mathcal{A}_{\text{int}}\rangle &= (g/2)\int d^3x dt\,(G_{\alpha\alpha}^{\Sigma}G_{\beta\beta}^{\Sigma} \pm G_{\alpha\beta}^{\Sigma}G_{\beta\alpha}^{\Sigma} + \tilde{G}_{\beta\alpha}^{\Delta*}\tilde{G}_{\beta\alpha}^{\Delta} \\ &\quad - (1/2)\int d^3x dt\,(\pm2G_{\alpha\beta}^{\Sigma}\Sigma_{\alpha\beta} + G_{\beta\alpha}^{\Delta}\Delta_{\beta\alpha}^* + G_{\alpha\beta}^{\Delta*}\Delta_{\beta\alpha}). \end{aligned} \qquad (3.413)$$

---

[8]As in Eq. (3.44), the hermitian adjoint $\Delta_{\uparrow\downarrow}^*(x)$ comprises transposition in the spin indices, i.e., $\Delta_{\uparrow\downarrow}^*(x) = [\Delta_{\downarrow\uparrow}(x)]^*$.

The total action is then

$$\mathcal{A}_1[\Delta, \Sigma] = \mathcal{A}_0[\Delta, \Sigma] + (g/2) \int d^3x dt \, (G^\Sigma_{\alpha\alpha} G^\Sigma_{\beta\beta} \pm G^\Sigma_{\alpha\beta} G^\Sigma_{\beta\alpha} + \tilde{G}^{\Delta*}_{\beta\alpha} \tilde{G}^\Delta_{\beta\alpha}$$
$$- (1/2) \int d^3x dt \, (\pm 2G^\Sigma_{\alpha\beta} \Sigma_{\alpha\beta} + G^\Delta_{\beta\alpha} \Delta^*_{\beta\alpha} + G^{\Delta*}_{\alpha\beta} \Delta_{\beta\alpha}). \quad (3.414)$$

Now we observe that

$$\frac{\delta}{\delta \Delta^*_{\alpha\beta}} \mathcal{A}_0[\Delta, \Sigma] = \frac{1}{2} G^\Delta_{\alpha\beta}, \qquad \frac{\delta}{\delta \Sigma_{\alpha\beta}} \mathcal{A}_0[\Delta, \Sigma] = G^\Sigma_{\alpha\beta}. \quad (3.415)$$

This shows that the first-order collective action

$$\mathcal{A}_1[\Delta, \Sigma] = \mathcal{A}_0[\Delta, \Sigma] + \langle \mathcal{A}_{\text{int}} \rangle \quad (3.416)$$

is automatically extremal in the variational parameters $\Sigma_{\alpha\beta}$ and $\Delta_{\alpha\beta}$, and that their extremal values are

$$\Sigma_{\alpha\beta} = g G^\Sigma_{\alpha\beta}, \quad \Delta_{\alpha\beta} = g G^\Delta_{\alpha\beta}. \quad (3.417)$$

Moreover, if we insert the extremal solutions (3.417) into the first-order collective action $\mathcal{A}_1[\Delta, \Sigma]$, it becomes

$$\mathcal{A}_1[\Delta, \Sigma] = \pm \frac{i}{2} \text{Tr} \log \left[i\mathbf{G}^{-1}_{\Delta,\Sigma}(x, x') \right] - \frac{g}{2} \int d^3x dt \, (-G^\Sigma_{\alpha\alpha} G^\Sigma_{\beta\beta} \pm G^\Sigma_{\alpha\beta} G^\Sigma_{\beta\alpha} + \tilde{G}^{\Delta*}_{\beta\alpha} \tilde{G}^\Delta_{\beta\alpha}),$$
$$(3.418)$$

or

$$\mathcal{A}_1[\Delta, \Sigma] = \pm \frac{i}{2} \text{Tr} \log \left[i\mathbf{G}^{-1}_{\Delta,\Sigma}(x, x') \right] - \frac{2}{g} \int d^3x dt \, (-\Sigma_{\alpha\alpha} \Sigma_{\beta\beta} \pm \Sigma_{\alpha\beta} \Sigma_{\beta\alpha} + \Delta^*_{\beta\alpha} \Delta_{\beta\alpha}).$$
$$(3.419)$$

3.9.1 Superconducting Electrons

We now focus attention upon electrons of spin 1/2 where the interaction (3.420) is simply[9]

$$\mathcal{A}_{\text{int}} = g \int d^3x dt \, \psi^*_\uparrow(\mathbf{x}, t) \psi^*_\downarrow(\mathbf{x}, t) \psi_\downarrow(\mathbf{x}, t) \psi_\uparrow(\mathbf{x}, t), \quad (3.420)$$

and the Wick contractions (3.409) are

$$\langle \psi^*_\uparrow(\mathbf{x}, t) \psi^*_\downarrow(\mathbf{x}, t) \psi_\downarrow(\mathbf{x}, t) \psi_\uparrow(\mathbf{x}, t) \rangle = \langle \psi^*_\uparrow(\mathbf{x}, t) \psi_\uparrow(\mathbf{x}, t) \rangle \langle \psi^*_\downarrow(\mathbf{x}, t) \psi_\downarrow(\mathbf{x}, t) \rangle$$
$$\pm \langle \psi^*_\uparrow(\mathbf{x}, t) \psi_\downarrow(\mathbf{x}, t) \rangle \langle \psi^*_\downarrow(\mathbf{x}, t) \psi_\uparrow(\mathbf{x}, t) \rangle$$
$$+ \langle \psi^*_\uparrow(\mathbf{x}, t) \psi^*_\downarrow(\mathbf{x}, t) \rangle \langle \psi_\downarrow(\mathbf{x}, t) \psi_\uparrow(\mathbf{x}, t) \rangle. \quad (3.421)$$

---

[9]Note that $g > 0$ is the attractive case.

In the absence of a magnetic field we expect that

$$G_{\downarrow\uparrow}^{\Delta}(\mathbf{x},t) \equiv G^{\Delta}(\mathbf{x},t), \qquad\qquad G_{\uparrow\downarrow}^{\Delta*}(\mathbf{x},t) \equiv G^{\Delta*}(\mathbf{x},t), \qquad (3.422)$$
$$G_{\uparrow\uparrow}^{\Sigma}(\mathbf{x},t) \equiv G^{\Sigma}(\mathbf{x},t) = G_{\downarrow\downarrow}^{\Sigma}(\mathbf{x},t), \qquad G_{\uparrow\downarrow}^{\Sigma}(\mathbf{x},t) \equiv G_{\downarrow\uparrow}^{\Sigma}(\mathbf{x},t) \equiv 0. \quad (3.423)$$

Then the interaction (3.413) becomes

$$\langle \mathcal{A}_{\rm int} \rangle = \frac{1}{g}(\tilde{\Sigma}^2 + |\tilde{\Delta}|^2) - (2G^{\Sigma}\Sigma + G^{\Delta*}\Delta + G^{\Delta}\Delta^*). \qquad (3.424)$$

Now the action at the extremum (3.419) reads

$$\mathcal{A}_1[\Delta, \Sigma] = \mathcal{A}_0[\Delta, \Sigma] - \frac{1}{g}(\Sigma^2 + |\Delta|^2). \qquad (3.425)$$

If the interactions are strong and the attraction is not confined to a narrow layer around the Fermi sphere, the present quantum field theory needs subtractions of the same type as encountered before. We must add a mass counterterm to the interaction (3.406)

$$\mathcal{A}^{\rm div} = -gG_{\Delta}^{\rm div}\frac{1}{2}\int d^3x dt \left(\psi_\alpha \psi_\alpha + \psi_\alpha^* \psi_\alpha^* \right), \qquad (3.426)$$

with a divergent integral $G_{\Delta}^{\rm div}$. Then all equations become finite if we replace the inverse coupling constant g by the renormalized one g_R.

Note that if we assume that Σ vanishes identically, the extremum of the one-loop action $\mathcal{A}_1[\Delta, \Sigma]$ gives the same result as the one obtained from the mean-field collective quantum field action (3.8), which reads for the present δ-function attraction

$$\mathcal{A}_1[\Delta, 0] = \mathcal{A}_0[\Delta, 0] - \frac{1}{g}|\Delta|^2. \qquad (3.427)$$

On the other hand, if we extremize the action (3.425) at $\Delta = 0$, we find the extremum from the expression

$$\mathcal{A}_1[0, \Sigma] = \mathcal{A}_0[0, \Sigma] - \frac{1}{g}\Sigma^2. \qquad (3.428)$$

The essential difference between the collective *quantum* field theory and the collective *classical* field theory is only manifest at higher orders. In the collective quantum field theory based on the Hubbard-Stratonovich transformation where a functional integral remains over the fluctuating pair field $\Delta_{\alpha\beta}(x)$, there are higher-order diagrams to be calculated with the help of the propagators of the collective field. These are extremely complicated quantities, and this makes all higher diagrams formed with them practically impossible to integrate. Moreover, and most importantly, they contain infinities which *cannot* be removed by counter terms.

There exists so far no technique that would allow a renormalization of the field theory based on the fluctuating collective field $\Delta_{\alpha\beta}(x)$.

In contrast to that, the higher-order diagrams in the present theory can *all* be calculated with ordinary free-particle propagators G^Δ and G^Σ of Eqs. (3.410) and (3.411), using the interaction (3.420). Even that can become tedious for higher orders in g. But *all* encountered infinities can be compensated by divergent counter terms. These have all the *same form* as the terms which are already present in the original action (3.398). They are either diagonal or off-diagonal mass terms which are quadratic in the original fields, or their gradients, or they are interaction terms.

Moreover, there is a simple rule to find the higher terms of the theory [33]. One calculates the diagrams with only the four-particle interaction, and collects the contributions to order g^n in a term $\mathcal{A}_n[\Delta, \Sigma]$. Then one replaces $\mathcal{A}_n[\Delta, \Sigma]$ by $\mathcal{A}_n[\Delta - \varepsilon g \Delta, \Sigma - \varepsilon g \Sigma]$ and re-expands all results up to the order g^n, forming an expression $\sum_{n=0}^{N} \varepsilon^n \bar{\mathcal{A}}_n[\Delta, \Sigma]$. Finally one sets ε equal to $1/g$.[10] If the result of these operations up to order N is denoted by $\sum_{n=0}^{N} \tilde{\mathcal{A}}_n[\Delta, \Sigma]$, we arrive at the final action by an expression like (3.425):

$$\mathcal{A}_1[\Delta, \Sigma] = \mathcal{A}_0[\Delta, \Sigma] + \sum_{n=1}^{N} \tilde{\mathcal{A}}_n[\Delta, \Sigma] - \frac{1}{g}(\Sigma^2 + |\Delta|^2). \qquad (3.429)$$

Note that this action must be merely extremized, as any action in a classical treatment. There are no more quantum fluctuations in the *classical collective fields* Δ, Σ. At the extremum, the action (3.429) is directly the grand-canonical potential.

3.10 Strong-Coupling Limit of Pair Formation

Our goal is to understand the phenomena arising in a Fermi liquid at low temperature in an external magnetic field. In order to set up a theory at strong couplings, we shall work in a four-dimensional "world crystal" discussed in the textbook [83]. The forth dimension represents the inverse temperature of the system. We shall treat the electrons and holes with the help of relativistic fields. After the calculation we can go back to the non-relativistic limit. In that limit, the Klein-Gordon wave function reduces to the Schrödinger field multiplied with a phase factor $e^{i\frac{mc^2 t}{\hbar}}$, just as in Schrödinger's original derivation of his time-displacement operator $H_S = \mathbf{k}^2/2m$:

$$e^{i\frac{\mathbf{k}\mathbf{x}}{\hbar}} e^{-i\frac{\sqrt{\mathbf{k}^2 + m^2 c^2} ct}{\hbar}} e^{i\frac{mc^2 t}{\hbar}} \approx e^{i\frac{\mathbf{k}\mathbf{x}}{\hbar}} e^{-i\frac{\mathbf{k}^2}{2m\hbar}t} = e^{i\frac{\mathbf{k}\mathbf{x}}{\hbar}} e^{-iH_S\frac{t}{\hbar}}. \qquad (3.430)$$

The attraction between any two particles can be tuned as a function of an external magnetic field. Ultimately it can be made so strong that the coupling constant reaches the unitary limit of infinite s-wave scattering length by means of a so-called *Feshbach resonance*. This phenomenon is discussed in detail in Subsection 9.2.8 of the textbook [35] and more recently in the review [84]. At that point, the Cooper

---

[10] The alert reader will recognize here the so-called *square-root trick* of Chapter 5 in the textbook Ref. [23].

pairs which form in the weak-coupling limit at low temperature and make the system a BCS superconductor, become so strongly bound that they behave like elementary bosonic particles. These form a Bose-Einstein condensate (BEC). At low temperature, the condensate behaves like a superconductor in which vortices can form behaving like bosonic quasi-particles. We study what happens with this condensate in the neighborhood of the unitarity limit.

By setting up strong-coupling equations for the fermions moving in a Bose-Einstein condensate we find that in $2+\epsilon$ dimensions they couple to the gas of boson pairs encircling them, thus forming new fermionic quasi-particles. These can bind, in their own right, to bosonic pairs which condense at low enough temperature and form a new type of condensate. That condensation happens at a much higher temperature than the first condensation process, so that it may be at the origin of high-T_c superconductivity.

The problem of understanding the behavior of a Fermi gas as a function of temperature has been investigated with reasonable success in a review paper by Randeria and Taylor [27], which is discussed in great detail in [28]. The results are summarized in Figure 3 of their paper which we reprint in Fig. 3.14. The abcissa shows the inverse of the s-wave scattering length which is infinite at $1/k_F a = 0$. Near the origin, the figure has a weak-coupling regime where pairs form, and the Fermi liquid is dominated by the physics of these Cooper pairs. The right-hand part of the figure is denoted as "Normal Bose Liquid".

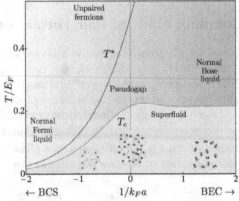

FIGURE 3.14 Qualitative phase diagram of the BCS-BEC crossover as a function of temperature T/ϵ_F and coupling $1/k_F a$, where k_F is the Fermi momentum and a the scattering length. The picture shows schematically the evolution from the BCS limit with large Cooper pairs to the BEC limit with tightly bound molecules. Unitarity ($1/k_F a = 0$) corresponds to strongly interacting pairs with size comparable to k_F^{-1}. The pair-formation crossover scale T^* diverges away from the transition temperature T_c, below which a condensate exists and the system is superfluid, as the attraction increases. Reproduced from Figure 3 of Randeria and Taylor in Ref. [27].

In Ref. [85] it has been argued that their Randeria-Taylor-results should be improved as follows. The region to the right of the vertical dashed line is still dominated by strongly interacting fermions. In addressing the many-body problem at finite temperature we can incorporate the relevant s-wave scattering physics via a "zero-range" contact potential in the Hamiltonian for spinor wave functions, $\psi_\sigma = \psi_\downarrow(x)$ for $\sigma = -1$ and $\psi_\sigma = \psi_\uparrow(x)$ for $\sigma = +1$, and start with

$$\beta\mathcal{H} = \int_0^\beta d\tau \int d\mathbf{x} \left[\mathcal{H}_0 - g\psi_\uparrow^\dagger(x)\psi_\downarrow^\dagger(x)\psi_\downarrow(x)\psi_\uparrow(x)\right], \qquad (3.431)$$

where $x = (\mathbf{x}, \tau)$, and $\beta = 1/T$ is the inverse temperature. The kinetic energy of fermions with mass m and chemical potential μ is collected in

$$\mathcal{H}_0 = \sum_\sigma \psi_\sigma^\dagger(x)[\partial_\tau - \nabla^2/2m - \mu]\psi_\sigma(x). \qquad (3.432)$$

It will eventually be necessary to include, in addition to the two-body interaction, also three-body interactions [61, 62].

Let an attractive interaction be parametrized by a positive coupling constant $g(\Lambda)$, where $\Lambda = \pi\ell_0^{-1}$ represents the inverse range of the four-fermion interaction (3.431). The distance ℓ_0 is usually much smaller than the lattice spacing ℓ of the mean separation between two atoms ($\ell_0 \ll \ell$). The bare coupling constant determines the gap size by the gap equation (3.268) or its renormalized version (3.271):

$$\frac{1}{g(\Lambda)} = \frac{1}{V} \sum_{|\mathbf{k}|<\Lambda} \frac{1}{2E_\mathbf{k}} \tanh \frac{E_\mathbf{k}}{2T}, \qquad \frac{1}{g_R} = \frac{1}{V} \sum_{|\mathbf{k}|<\Lambda} \left(\frac{1}{2E_\mathbf{k}} \tanh \frac{E_\mathbf{k}}{2T} - \frac{1}{\epsilon_\mathbf{k}}\right), \quad (3.433)$$

where $E_\mathbf{k}$ denotes the quasipaticle energy, and the symbol $\sum_{\omega_n,|\mathbf{k}|<\Lambda}$ contains the phase-space integral plus the sum over the *Matsubara frequencies* $\omega_n = 2\pi T n$ for $n = 0, \pm1, \pm2, \cdots$. Let us go to the regime of large enough temperature where we the system is in the nomal phase and the gap vanishes. Then we are dealing with free fermions which will participate in the strong-coupling expansion, and the unrenormalized coupling constant fulfills the equation

$$\frac{1}{g(\Lambda)} = \frac{1}{V} \sum_{|\mathbf{k}|<\Lambda} \frac{1}{2\epsilon_\mathbf{k}} \tanh \frac{\epsilon_\mathbf{k}}{2T}. \qquad (3.434)$$

If we now allow for scattering between the fermions with an s-wave scattering length a, that satisfies the equation [compare with Eq. (3.271)]:

$$\frac{m}{4\pi a} = -\frac{1}{g(\Lambda)} + \frac{1}{V} \sum_{|\mathbf{k}|<\Lambda} \frac{1}{2\epsilon_\mathbf{k}} \tanh \frac{\epsilon_\mathbf{k}}{2T}. \qquad (3.435)$$

In $d = 3$ dimension, this can be written as

$$\frac{4\pi a}{mg(\Lambda)} = \frac{k_F a}{4\pi\hbar^3 b} \mathcal{S}_3(T) - 1, \qquad (3.436)$$

where

$$S_3(T) \equiv \int_0^1 dt \tanh\left[\frac{\epsilon_F}{T} \frac{\pi^2 t^2}{8b^2}\right] \tag{3.437}$$

is the reduced temperature-dependent phase-space sum in (3.433). Although we shall work eventually with a continuum model, we may adopt a "world-crystal" lattice language and define an effective lattice spacing ℓ by setting the Fermi wave number equal to

$$k_F \approx \frac{(3\pi^2)^{1/3}}{\ell}. \tag{3.438}$$

The so-called half-filling electron density is simply $n \approx 1/\ell^3 \approx k_F^3/(3\pi^2)$. Moreover, we have introduced the dimensionless length parameter $b = \ell_0 k_F/2 = 2^{-1}(3\pi^2)^{1/3}\ell_0/\ell \ll 1$ as the relative width of the electron density around the surface of their Fermi sphere (its "thickness"). The definition of b ensures the normalization $S_3(0) = 1$. For low temperatures and small a, Eq. (3.433) reduces properly to its well-known BCS expressions [72, 74]. In the opposite limit of strong-coupling where $1/k_F a = 0$, we arrive at the so-called unitary fermion gas.

In the weak-coupling region, one is confronted with spontaneous symmetry breaking and Cooper-pair formation in the Bardeen-Cooper-Schrieffer (BCS) model of superconductivity. There the standard *collective quantum field treatment* is to introduce $\mathcal{C}(x) = \psi_\downarrow(x)\psi_\uparrow(x) = \Delta(x)e^{i\theta(x)}$ and to express Eq. (3.431) in terms of quadratic fermion fields. After integrating out the fermion fields one obtains the collective quantum field action [72, 74]:

$$\mathcal{A}[\Delta^*, \Delta] = \frac{1}{2}\int_0^\beta d\tau \int d\mathbf{x} \left\{-\frac{i}{2}\text{Tr}\ln\left(i\mathbf{G}_\Delta^{-1}\right) + \frac{|\Delta(x)|^2}{g}\right\}, \tag{3.439}$$

where $\mathbf{G}_\Delta^{-1}$ is the inverse operator of quadratic fermion fields, which is a functional of $\Delta(x)$ and its spacetime derivatives. After performing a regularization of the coupling strength via the experimental s-wave scattering length a in (3.433), we adopt the approximation of a uniform static saddle point $\mathcal{C}(x) \approx \langle \mathcal{C}(0)\rangle = \Delta$ that satisfies the saddle-point condition $\delta\mathcal{A}/\delta\Delta = 0$, as well as the fermion number $N = -\delta\mathcal{A}/\delta\Delta$. Then the renormalized gap and number equations are obtained from (3.268) and (3.267) [72, 74, 75]:

$$\frac{m}{4\pi a} = \frac{1}{V}\sum_{\mathbf{k}}\left[\frac{1}{2\epsilon_{\mathbf{k}}} - \frac{\tanh(\beta E_{\mathbf{k}}/2)}{2E_{\mathbf{k}}}\right], \quad n = \frac{1}{V}\sum_{\mathbf{k}}\left[1 - \frac{\xi_{\mathbf{k}}}{E_{\mathbf{k}}}\tanh\left(\frac{\beta E_{\mathbf{k}}}{2}\right)\right]. \tag{3.440}$$

Here $E_{\mathbf{k}} = \sqrt{\xi_{\mathbf{k}}^2 + \Delta^2}$ with $\xi_{\mathbf{k}} = \epsilon_{\mathbf{k}} - \mu$. In the weak-coupling limit where $1/k_F a \to -\infty$ and the temperature is a small $T/\mu \ll 1$, the chemical potential is close to the Fermi energy $\mu \approx \epsilon_F = \hbar^2 k_F^2/2m = \hbar^2(3\pi^2 n)^{2/3}/2m$ and one obtains the BCS results [75]: There the critical temperature is $T_c = 1.1\epsilon_F \exp(-\pi/2k_F|a|)$, and the energy gap is $\Delta_0 \equiv \Delta(T = T_c) = 1.76T_c$. The Cooper-pair size ξ_{pair} is much larger than the lattice spacing $\xi_{\text{pair}} \sim k_F^{-1}\exp(\pi/2k_F|a|) \gg k_F^{-1}$. This implies that the Cooper pair $C(x) = \psi_\downarrow(x)\psi_\uparrow(x)$ is a loosely bound pair of spin-up electron and spin-down electron. In the domain of the infrared-stable fixed point, the Cooper-pair size ξ_{pair} sets the physical scale for scaling laws. The critical temperature T_c is scaled by the gap value Δ_0 as $\Delta(T)/\Delta_0 = 1.74\,(1 - T/T_c)^{1/2}$.

As the four-fermion attractive coupling g or $1/k_F a$ increases, it is expected that the Cooper pairs become tightly bound states of bosons. They form a superfluid Bose liquid, provided the temperature T is less than the crossover temperature T^* of Cooper-pair formation ($T < T^*$). Otherwise, the Cooper pairs dissociate into two fermions and form a normal Fermi liquid of unpaired fermions ($T > T^*$). It was qualitatively shown that the crossover temperature T^* of Cooper-pair formations diverges away from the transition temperature T_c as the four-fermion attraction increases. They approach each other in the weak-coupling $1/k_F a \to -\infty$ regime of BCS [74].

The crossover from the weak-coupling BCS pairing to strong-coupling BEC of tightly-bound pairs, as a function of the attractive interaction (3.431), has long been of interest to theoretical physicists. For that it is important to study the pair-formation crossover temperature T^*, and the transition temperature T_c, as well as the phase diagram of T/ϵ_F versus $1/k_F a$, in particular the infrared (IR) scaling domain $1/k_F a \to -\infty$ for the BCS-limit and the ultra-violate (UV) scaling domain $1/k_F a \to 0^\pm$ in the unitarity limit.

Inspired by strong-coupling quantum field theories [76, 77], we calculate the two-point Green functions of the composite boson and fermion fields. We use a strong-coupling expansion to diagonalize the Hamiltonian into a bilinear form of the composite fields. This produces composite-particle spectra in the strong-coupling phase. We find that the Fermi liquid of composite fermions coexists with the Bose liquid of composite bosons in the pseudogap region ($T_c < T < T^*$) as well as in the BEC region ($T < T_c$). The lattice representation of the Hamiltonian (3.431), for one-electron per cubic lattice site (half filling), reads

$$\beta\mathcal{H} = \beta\sum_{i,\sigma=\uparrow,\downarrow}(\ell^d)\psi_\sigma^\dagger(i)\Big[-\nabla^2/2m\ell^2 - \mu\Big]\psi_\sigma(i) - g\beta\sum_i(\ell^d)\psi_\uparrow^\dagger(i)\psi_\downarrow^\dagger(i)\psi_\downarrow(i)\psi_\uparrow(i). \quad (3.441)$$

Here each fermion field is defined at a lattice site "i" as $\psi_\uparrow(i) = \psi_\uparrow(x)$ or $\psi_\downarrow(i) = \psi_\downarrow(x)$, the parameter d is the spatial dimension, and the index i runs over all lattice sites. The fermion field ψ_σ has a length dimension $[\ell^{-d/2}]$, and the four-fermion coupling g has a dimension $[\ell^{d-1}]$. The Laplace operator ∇^2 is defined in d spacetime dimensions as

$$\nabla^2\psi_\sigma(i) \equiv \sum_{\hat{\ell}}\Big[\psi_\sigma(i+\hat{\ell}) + \psi_\sigma(i-\hat{\ell})\Big] - 2\psi_\sigma(i) \Rightarrow 2\Big[\sum_{\hat{\ell}}\cos(k\hat{\ell}) - 1\Big]\psi_\sigma(k)$$

$$\approx -k^2\ell^2\psi_\sigma(k). \quad (3.442)$$

The vectors $\hat{\ell}$ for $l = 1,\ldots,d$ indicate the orientated lattice space vectors to the nearest neighbors, and $\psi_\sigma(k)$ are the Fourier components of $\psi_\sigma(i)$ in momentum-space. In the last line we assume that $k^2\ell^2 \ll 1$. The chemical potential μ controls the density of free fermions with a dispersion $\epsilon_{\mathbf{k}} = \mathbf{k}^2/2m$.

Following the framework of chiral gauge field theories [77], we calculate the expansion in the strong-coupling limit. Here we relabel $\beta \ell^d \to \beta$ and $2m\ell^2 \to 2m$, so the lattice spacing ℓ is effectively set equal to unity; this rescales $\psi_\sigma(i) \to (\beta g)^{1/4}\psi_\sigma(i)$ and $\psi_\sigma^\dagger(i) \to (\beta g)^{1/4}\psi_\sigma^\dagger(i)$. The Hamiltonian (3.441) can therefore be written as $\beta \mathcal{H} = \sum_i [h\mathcal{H}_0(i) + \mathcal{H}_{\text{int}}(i)]$, where the hopping parameter is $h \equiv \beta/(\beta g)^{1/2}$, and

$$\mathcal{H}_0(i) = \sum_{\sigma=\uparrow,\downarrow} \mathcal{H}_0^\sigma(i) \equiv \sum_{\sigma=\uparrow,\downarrow} \psi_\sigma^\dagger(i)(-\nabla^2/2m - \mu)\psi_\sigma(i), \qquad (3.443)$$

$$\mathcal{H}_{\text{int}}(i) \equiv -\psi_\uparrow^\dagger(i)\psi_\downarrow^\dagger(i)\psi_\downarrow(i)\psi_\uparrow(i). \qquad (3.444)$$

The partition function is given by

$$\mathcal{Z} = \Pi_{i,\sigma} \int d\psi_\sigma(i)d\psi_\sigma^\dagger(i) \exp(-\beta\mathcal{H}), \qquad (3.445)$$

$$\langle \cdots \rangle = \mathcal{Z}^{-1}\Pi_{i,\sigma} \int d\psi_\sigma(i)d\psi_\sigma^\dagger(i)(\cdots)\exp(-\beta\mathcal{H}). \qquad (3.446)$$

Fermion fields $\psi_\uparrow$ and $\psi_\downarrow$ are one-component Grassman variables, $\psi_\sigma(i)\psi_{\sigma'}(j) = -\psi_{\sigma'}(j)\psi_\sigma(i)$ and

$$\int d\psi_\sigma(i)\psi_{\sigma'}(j) = \delta_{\sigma,\sigma'}\delta_{ij}, \quad \int d\psi_\sigma^\dagger(i)\psi_{\sigma'}^\dagger(j) = \delta_{\sigma,\sigma'}\delta_{ij}, \qquad (3.447)$$

and all others vanish.

In the strong-coupling limit $h \to 0$ for $g \to \infty$ and finite T, the kinetic terms (3.443) are neglected, and the partition function (3.445) becomes the one-site integral at the spatial point "i"

$$\Pi_i \int_{i\downarrow} \int_{i\uparrow} \exp\left(\psi_\uparrow^\dagger(i)\psi_\downarrow^\dagger(i)\psi_\downarrow(i)\psi_\uparrow(i)\right) = -\Pi_i \int_{i\downarrow} \psi_\downarrow(i)^\dagger\psi_\downarrow(i) = (1)^\mathcal{N}, \qquad (3.448)$$

where $\mathcal{N}$ is the total number of lattice sites, $\int_{i\uparrow} \equiv \int[d\psi_\uparrow^\dagger(i)d\psi_\uparrow(i)]$ and $\int_{i\downarrow} \equiv \int[d\psi_\downarrow^\dagger(i)d\psi_\downarrow(i)]$. The strong-coupling expansion can now be performed in powers of the hopping parameter h, so that it is a *hopping expansion*.

3.11   Composite Bosons

In the strong-coupling phase, we first consider a composite bosonic pair field $\mathcal{C}(x) = \psi_\downarrow(x)\psi_\uparrow(x)$. We want to study its two-point function

$$G(x) = \langle \psi_\downarrow(0)\psi_\uparrow((0), \psi_\uparrow^\dagger(x)\psi_\downarrow^\dagger(x)\rangle = \langle \mathcal{C}(0), \mathcal{C}^\dagger(x)\rangle. \qquad (3.449)$$

Here the fermion fields are not re-scaled by $(\beta g)^{1/4}$, and x stands for the point at the nearest lattice site labeled by "i". The leading strong-coupling approximation to (3.449) is

$$G(x) = \frac{\delta^{(d)}(x)}{\beta g}. \qquad (3.450)$$

The first correction is obtained by using the one-site partition function $Z(i)$ and the integral

$$\langle \psi_\uparrow \psi_\downarrow \rangle \equiv \frac{1}{Z(i)} \int_{i\downarrow} \int_{i\uparrow} \psi_\uparrow(i) \psi_\downarrow(i) e^{-h\mathcal{H}_0(i) - \beta\mathcal{H}_{\text{int}}(i)},$$

$$= h^2 \sum_{\hat{\ell}}^{\text{ave}} \psi_\uparrow(i;\hat{\ell}) \sum_{\hat{\ell}'}^{\text{ave}} \psi_\downarrow(i;\hat{\ell}') \approx h^2 \sum_{\hat{\ell}}^{\text{ave}} \psi_\uparrow(i;\hat{\ell}) \psi_\downarrow(i;\hat{\ell}), \qquad (3.451)$$

where the non-trivial result needs $\psi_\uparrow^\dagger(i)$ and $\psi_\downarrow^\dagger(i)$ fields in the hopping expansion of $e^{-h\mathcal{H}_0(i)}$, and $\sum_{\hat{\ell}}^{\text{ave}} \psi_\sigma(i;\hat{\ell}) \equiv \sum_{\hat{\ell}} \left[\psi_\sigma(i+\hat{\ell}) + \psi_\sigma(i-\hat{\ell}) \right]$. When integrating over fields $\psi_{\uparrow,\downarrow}(i)$ at the site "i" in the expansion of (3.449), the first corrected (3.449) reads:

$$G(x) = \frac{\delta^{(d)}(x)}{\beta g} + \frac{1}{\beta g} \left(\frac{\beta}{2m} \right)^2 \sum_{\hat{\ell}} \left[G^{\text{nb}}(x+\hat{\ell}) + G^{\text{nb}}(x-\hat{\ell}) \right]. \qquad (3.452)$$

Here $\delta^{(d)}(x)$ is a spatial δ-function and $G^{\text{nb}}(x \pm \hat{\ell})$ is the Green function (3.449) without integration over the fields ψ_σ at the neighbor site x. Note that the nontrivial contributions come only from kinetic hopping terms $\propto (h/2m)^2 = (1/\beta g)(\beta/2m)^2$. The chemical potential term $\mu\psi_\sigma^\dagger(i)\psi_\sigma(i)$ in the Hamiltonian $\mathcal{H}_0$ (3.443) does not contribute to the hopping.

Replacing $G^{\text{nb}}(x \pm \hat{\ell})$ by $G(x \pm \hat{\ell})$ converts Eq. (3.452) into a recursion relation for $G(x)$, which actually takes into account all high-hopping corrections in the strong-coupling expansion. Going to momentum space we obtain

$$G(q) = \frac{1}{\beta g} + \frac{2}{\beta g} \left(\frac{\beta}{2m} \right)^2 G(q) \sum_{\hat{\ell}} \cos(q\hat{\ell}). \qquad (3.453)$$

This equation is solved by

$$G(q) = \frac{(2m/\beta\ell)^2}{4\ell^{-2} \sum_{\hat{\ell}} \sin^2(q\hat{\ell}/2) + M_B^2}, \qquad (3.454)$$

where we have returned to the original lattice spacing ℓ by replacing back $\beta \to \beta\ell^3$ and $2m \to 2m\ell^2$. This implies that in the strong-coupling effective Hamiltonian, the two-fermion field $\mathcal{C} = \psi_\downarrow \psi_\uparrow$ possesses a massive composite boson mode, i.e., a bosonic bound state with a propagator

$$gG(q) = \frac{gR_B^2/2M_B}{(q^2/2M_B) + M_B/2} = \frac{gR_B^2}{q^2 + M_B^2}, \qquad (q\ell \ll 1). \qquad (3.455)$$

This has a pole at the mass M_B with a residue gR_B^2:

$$M_B^2 = \left[g(2m)^2(\ell/\beta) - 2d \right] \ell^{-2} > 0, \qquad R_B^2 = (2m/\beta\ell)^2. \qquad (3.456)$$

The condition $M_B^2 = 0$ determines a critical curve that satisfies in $d = 3$ dimensions

$$T - \frac{1}{g}\frac{3}{2m^2\ell} = 0. \tag{3.457}$$

The strong-coupling effective Hamiltonian of the composite boson field $\mathcal{C}$ associated with the propagator (3.455) can be written as

$$\beta\mathcal{H}_{\text{eff}}^B = \beta\sum_i (\ell^3) Z_B^{-1}\mathcal{C}^\dagger(i)\Big[-\nabla^2/2M_B\ell^2 - \mu_B\Big]\mathcal{C}(i), \tag{3.458}$$

where $\mu_B = -M_B/2$ is the chemical potential and $Z_B = gR_B^2/2M_B$ the wave-function renormalization constant. As long as Z_B is finite, we renormalize the elementary fermion field and the composite boson field as

$$\psi \to (gR_B^2)^{-1/4}\psi, \quad \text{and} \quad \mathcal{C} \to (2M_B)^{1/2}\mathcal{C}. \tag{3.459}$$

Now the composite boson field $\mathcal{C}$ behaves like a quasi particle in Eq. (3.458). Contrary to a loosely-packed bound state of two electrons in a Cooper pair formed at a small s-wave scattering length $(k_Fa)^{-1} \ll 0$ in the weak-coupling region, this is a tightly-packed bound pair, i.e., the proper bound state of a Feshbach resonance for $(k_Fa)^{-1} \gg 0$ in the strong-coupling region.

At weak coupling, the bound states are composed of two constituent fermions $\psi_\downarrow(k_1)$ and $\psi_\uparrow(k_2)$ around the Fermi surface, $k_1 \approx k_2 \approx k_F$ and $k_2 - k_1 = q \ll k_F$. The form factor or wave-function renormalization $Z_B \propto gT^2$ (3.455) relates to the bound-state size ξ_{boson}. As $gT^2 \to 0$, Z_B decreases and $\mathcal{C}(x)$ decribes a loosely-bound Cooper pair. The vanishing wave function renormalization constant indicates the fact that the bosonic bound state pole dissolves into two fermionic constituent cuts [78]. At this dissociation scale, i.e., at the crossover temperature T^*, the phase transition takes place which leads to a normal Fermi liquid of unpaired fermions. Limited by the validity of strong-coupling expansion, we have not been able to quantitatively obtain the dissociation scale T^* as it results from the inverse scattering length $1/a$. At the unitary point $1/k_Fa = 0$, we can estimate the crossover temperature as $T^* \approx \epsilon_B/\log(\epsilon_B/\epsilon_F)^{3/2}$ [74, 27], and the binding energy ϵ_B from (3.290) as

$$\frac{\epsilon_B}{\epsilon_F} = 2\frac{\sqrt{1+\hat{\mu}^2}-\hat{\mu}}{\sqrt{1+\hat{\mu}^2}-\hat{\mu}}. \tag{3.460}$$

Inserting here the crossover parameter $\hat{\mu} = \mu_B/M_B = -1/2$, obtained from taking M_B as the mass gap at the unitary point $1/k_Fa=0$, we find $\epsilon_B/\epsilon_F = 5.24$ and $T^*/\epsilon_F = 4.86$.

Note that the mass term $M_B^2\mathcal{C}\mathcal{C}^\dagger(\mathbf{x})$ changes its sign from $M_B^2 > 0$ to $M_B^2 < 0$ and the pole M_B in (3.455) becomes imaginary, implying the second-order phase transition from the symmetric phase to the condensed phase [77]. The vanishing boson mass $M_B^2 = 0$ gives rise to the critical curve (3.457), which can also be written as

$$\frac{T_c}{\epsilon_F} = \frac{T_c^u(T_c)}{\epsilon_F}\left[1 - \frac{4\pi\hbar^3 b}{\mathcal{S}_3(T_c)}\frac{1}{ak_F}\right], \tag{3.461}$$

where $T_c^u = T_c^u(T_c = T_c^u)$ is the temperature at the unitarity point $1/k_F a = 0$. The prefactor is a solution of the following equation:

$$\frac{T_c^u(T_c)}{\epsilon_F} = (3\pi^2)^{-1/3} \frac{3\mathcal{S}_3(T_c)}{(4\pi)^2 b}. \tag{3.462}$$

Indeed, by inserting Eqs. (3.436) and (3.438) into Eq. (3.461), we obtain

$$\begin{aligned}
\frac{T_c}{\epsilon_F} &= \frac{m}{4\pi a} \frac{4\pi a}{g(\Lambda)} \frac{3}{2m^2 \ell \epsilon_F} = \frac{m}{4\pi a} \left[\frac{k_F a}{4\pi \hbar^3 b} \mathcal{S}_3(T) - 1 \right] \frac{3}{2m^2 \ell \epsilon_F} \\
&= \frac{3 k_F \mathcal{S}_3(T_c)}{2m(4\pi)^2 \hbar^3 b \, \ell \epsilon_F} \left[1 - \frac{4\pi \hbar^3 b}{\mathcal{S}_3(T_c)} \frac{1}{k_F a} \right],
\end{aligned} \tag{3.463}$$

Going here the unitary point $1/k_F a = 0$ and returning to physical dimension, we verify (3.462).

We now calculate the phase diagram numerically for the Fermi layer thickness parameters $b = 0.02, 0.03$ corresponding to the ratios $\ell_0/\ell = 0.013, 0.02$. We find $T_c^u/\epsilon_F \approx 0.31, 0.2$ and plot the result of (3.461) in Fig. 3.15. In contrast to the practically horizontal phase boundary estimated by Randeria and Taylor in their Fig. 3 of Ref. [27] (reprinted in Fig. 3.14), we obtain a decreasing critical temperature T_c as a function of $1/k_F a \geq 0$. At an "infinite" coupling strength we find $T_c = 0$ where $1/k_F a \to \mathcal{S}_3(0)/4\pi \hbar^3 b = 1/4\pi \hbar^3 b$. The limit $T_c \to 0$ at $g_c \to \infty$ is taken while keeping the product $T_c g_c$ constant, so that the hopping parameter $h \propto 1/(Tg)^{1/2}$ is a small number. This means that the composite pairs are very massive at this "infinite" coupling point, and their kinetic energies are negligible. Here we have a quantum phase transition. Viewing the four-fermion interaction as an attractive potential, this "infinite"-coupling point indicates the most tightly bound state located at the lowest energy level of the potential, with a scattering length a that is of the order of $-2\pi \ell_0$. If the attraction comes from a δ-function, the length parameters a and b vanish, while $1/k_F a \to \infty$, recovering the nearly horizontal critical line presented in Fig. 3.14. Note that Eq. (3.461) is inapplicable in the weak-coupling regime of BCS where $1/k_F a \ll 0$. It holds only in the strong-coupling regime with $1/k_F a \gg 0$. There we find a superfluid phase with tightly bound composite bosons which have undergone a Bose-Einstein condensation (BEC). This means that the pair field $\mathcal{C}(x) \equiv |\Delta(x)| e^{i\theta(x)}$ has developed a nonzero vacuum expectation value in the same way as it happened in the BCS weak-coupling regime.

Analogously, we consider the composite field of an electron and a hole, i.e., the plasmon field $\mathcal{P}(x) = \psi_\downarrow^\dagger(x) \psi_\uparrow(x)$. The same calculations are applied for the two-point Green function $G_\mathcal{P}(x) = \langle \mathcal{P}(0), \mathcal{P}^\dagger(x) \rangle$. In the lowest nontrivial order of a strong-coupling expansion, we obtain the same result as in (3.455) and (3.456), indicating a tightly bound state of a plasmon field. Its Hamiltonian has the same form as in Eq. (3.458), only that $\mathcal{C}(i)$ has been replaced by $\mathcal{P}(i)$. This is not a surprise since the pair field $\mathcal{C}(x)$ and the plasmon $\mathcal{P}(x)$ field appear in the same way in the strongly interacting Hamiltonian (3.441). However, the pair field $\mathcal{C}(x)$ is a charged field

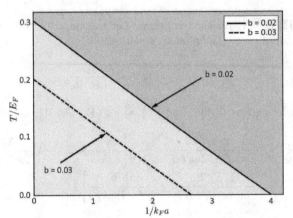

FIGURE 3.15 Qualitative phase diagram in the unitarity limit. Solving Eq. (3.438), the transition temperature T_c/ϵ_F is plotted as a function of $1/k_F a \geq 0$ for the selected parameters $b = 0.02, 0.03$. The "infinite"-coupling points are shown to lie at one of the corresponding zeros $(1/k_F a)_{T_c=0} = 4.0, 2.7$. These are points of quantum phase transition. Above the critical line is a normal liquid consisting of massive composite bosons and fermions. Below the critical line lies a superfluid phase with a new type of BEC, that involves composite massive fermions.

whereas the plasmon field $\mathcal{P}(x)$ is neutral, so that they can be different up to a relative phase $\theta(x)$. We select the relative phase such that $\langle|\mathcal{P}(x)|\rangle = \langle|\mathcal{C}(x)|\rangle = \Delta(x)$.

In the weak-coupling regime $1/k_F a \to -\infty$ we go into the super-fluid phase of BCS, where the ground state is parametrized by the minimum of the vacuum expectation value. Here the Cooper field $\mathcal{C}(x)$ has the expectation value $\langle\mathcal{C}(x)\rangle \equiv \Delta \neq 0$, rather than forming a condensate of the plasmon field $\mathcal{P}(x)$. This makes the difference between the charged Cooper pair and the neutral plasmon pair of an electron and a hole, which does not contribute to low-temperature superconductivity in the BCS-limit.

In the strong-coupling limit, we also consider the two-point Green function of the Cooper field $\mathcal{C}(x)$ with a plasmon field $\mathcal{P}(x)$:

$$G_{\mathcal{M}}(x) \;=\; \langle \mathcal{P}(0), \mathcal{C}^\dagger(x)\rangle = \langle \psi_\downarrow^\dagger(0)\psi_\uparrow(0)\psi_\uparrow^\dagger(x)\psi_\downarrow^\dagger(x)\rangle. \tag{3.464}$$

From it we shall see their correlations and mixing. The same calculations up to the lowest nontrivial order of the strong-coupling expansion lead to

$$G_{\mathcal{M}}(x) \;=\; \frac{1}{\beta g}\left(\frac{\beta}{2m}\right)^2 \sum_{\hat{\ell}}\left[G_{\mathcal{M}}(x + \hat{\ell}) + G_{\mathcal{M}}(x - \hat{\ell})\right]. \tag{3.465}$$

Going to Fourier space, this is solved to be identically vanishing, as one field is charged while the other is neutral.

3.12 Composite Fermions

To exhibit the presence of composite fermions in the strong-coupling effective Hamiltonian $\mathcal{H}$ of Eqs. (3.443) and (3.444), we use the Cooper field $\mathcal{C}(x) = \psi_\downarrow(x)\psi_\uparrow(x)$ and calculate the two-point Green functions:

$$S_{LL}(x) \equiv \langle \psi_\uparrow(0), \psi_\uparrow^\dagger(x) \rangle, \tag{3.466}$$

$$S_{ML}(x) \equiv \langle \psi_\uparrow(0), \mathcal{C}^\dagger(x)\psi_\downarrow(x) \rangle = \langle \psi_\uparrow(0), [\psi_\uparrow^\dagger(x)\psi_\downarrow^\dagger(x)]\psi_\downarrow(x) \rangle, \tag{3.467}$$

$$S_{ML}^\dagger(x) \equiv \langle \psi_\downarrow^\dagger(0)\mathcal{C}(0), \psi_\uparrow^\dagger(x) \rangle = \langle \psi_\downarrow^\dagger(0)[\psi_\downarrow(0)\psi_\uparrow(0)], \psi_\uparrow^\dagger(x) \rangle, \tag{3.468}$$

$$S_{MM}(x) \equiv \langle \psi_\downarrow^\dagger(0)\mathcal{C}(0), \mathcal{C}^\dagger(x)\psi_\downarrow(x) \rangle = \langle \psi_\downarrow^\dagger(0)[\psi_\downarrow(0)\psi_\uparrow(0)], [\psi_\uparrow^\dagger(x)\psi_\downarrow^\dagger(x)]\psi_\downarrow(x) \rangle. \tag{3.469}$$

Here the fermion fields are not re-scaled by $(\beta g)^{1/4}$. Using the following fermionic integrals:

$$\langle \psi_\uparrow^\dagger \rangle \equiv \frac{1}{Z(i)} \int_{i\downarrow} \int_{i\uparrow} \psi_\uparrow^\dagger(i)e^{-h\mathcal{H}_0(i)-\beta\mathcal{H}_{\text{int}}(i)} \approx h^3 \sum_{\hat{\ell}}^{\text{ave}} \psi_\uparrow^\dagger(i;\hat{\ell})\psi_\downarrow(i;\hat{\ell})\psi_\uparrow^\dagger(i;\hat{\ell}), \tag{3.470}$$

$$\langle \psi_\uparrow^\dagger\psi_\downarrow\psi_\downarrow^\dagger \rangle \equiv \frac{1}{Z(i)} \int_{i\downarrow} \int_{i\uparrow} \psi_\uparrow^\dagger(i)\psi_\downarrow(i)\psi_\downarrow^\dagger(i)e^{-h\mathcal{H}_0(i)-\mathcal{H}_{\text{int}}(i)} = h \sum_{\hat{\ell}}^{\text{ave}} \psi_\uparrow^\dagger(x;\hat{\ell}), \tag{3.471}$$

$$\langle \psi_\uparrow\psi_\uparrow^\dagger\psi_\downarrow\psi_\downarrow^\dagger \rangle \equiv \frac{1}{Z(i)} \int_{i\downarrow} \int_{i\uparrow} \psi_\uparrow(i)\psi_\uparrow^\dagger(i)\psi_\downarrow(i)\psi_\downarrow^\dagger(i)e^{-h\mathcal{H}_0(i)-\mathcal{H}_{\text{int}}(i)} = 1, \tag{3.472}$$

we obtain, by analogy with (3.449)–(3.452), for the composite-fermion Green functions (3.466)–(3.469) the recursion relations

$$S_{LL}(x) = \frac{1}{\beta g}\left(\frac{\beta}{2m}\right)^3 \sum_{\hat{\ell}}[S_{ML}(x+\hat{\ell}) + S_{ML}(x-\hat{\ell})], \tag{3.473}$$

$$S_{ML}(x) = \frac{\delta^{(d)}(x)}{\beta g} + \frac{1}{\beta g}\left(\frac{\beta}{2m}\right)\sum_{\hat{\ell}}[S_{LL}(x+\hat{\ell}) + S_{LL}(x-\hat{\ell})], \tag{3.474}$$

$$S_{MM}(x) = \frac{1}{\beta g}\left(\frac{\beta}{2m}\right)\sum_{\hat{\ell}}[S_{ML}^\dagger(x+\hat{\ell}) + S_{ML}^\dagger(x-\hat{\ell})]. \tag{3.475}$$

Transforming these three two-point functions to momentum space $S_X(p) = \sum_x e^{-ipx}S_X(x)$ where X stands for the symbol-pairs LL, ML, MM, respectively, we obtain the recursion relations

$$S_{LL}(p) = \frac{1}{\beta g}\left(\frac{\beta}{2m}\right)^3 \left[2\sum_{\hat{\ell}}\cos(p\hat{\ell})\right]S_{ML}(p), \tag{3.476}$$

$$S_{ML}(p) = \frac{1}{\beta g} + \frac{1}{\beta g}\left(\frac{\beta}{2m}\right)\left[2\sum_{\hat{\ell}}\cos(p\hat{\ell})\right]S_{LL}(p), \tag{3.477}$$

$$S_{MM}(p) = \frac{1}{\beta g}\left(\frac{\beta}{2m}\right)\left[2\sum_{\hat{\ell}}\cos(p\hat{\ell})\right]S_{ML}^\dagger(p). \tag{3.478}$$

They are solved by

$$S_{ML}(p) = \frac{(1/\beta g)}{1 - (1/\beta g)^2 (\beta/2m)^4 \left[2\sum_{\hat{\ell}} \cos(p\hat{\ell})\right]^2}. \qquad (3.479)$$

From this $S_{LL}(p)$ is found via (3.476) and $S_{MM}(p)$ via (3.478). By analogy with Eqs. (3.454) and (3.455) for the composite boson, we find

$$\begin{aligned}
S(p) &= R_B^{-1} S_{LL}(p) + 2R_B^{-2} S_{ML}(p) + R_B^{-3} S_{MM}(p) \\
&= \frac{2}{4\ell^{-2}\sum_{\hat{\ell}} \sin^2(p\hat{\ell}/2) + M_F^2} \Rightarrow \frac{2}{p^2 + M_F^2}, \quad (p\ell \ll 1), \qquad (3.480)
\end{aligned}$$

where R_B and $M_F^2 = M_B^2$ are given by (3.456) in the lowest order calculation. We have returned to the original lattice spacing ℓ by re-substituting $\beta \to \beta\ell^3$ and $2m \to 2m\ell^2$. Equation (3.480) represents a composite fermion consisting of the elementary fermion $\psi_\uparrow$ and the three-fermion state $\mathcal{C}(x)\psi_\downarrow^\dagger(x)$,

$$\Psi_\uparrow(x) = R_B^{-1/2}\psi_\uparrow(x) + R_B^{-3/2}\mathcal{C}(x)\psi_\downarrow^\dagger(x) \Rightarrow g^{1/4}\psi_\uparrow(x) + g^{3/4}\mathcal{C}(x)\psi_\downarrow^\dagger(x). \qquad (3.481)$$

The three-fermion state $\mathcal{C}(x)\psi_\downarrow^\dagger(x)$ is made of a hole $\psi_\downarrow^\dagger(x)$ that is "dressed" by a cloud of Cooper pairs. The associated two-point Green function satisfies

$$\begin{aligned}
\langle \Psi_\uparrow(0), \Psi_\uparrow^\dagger(x) \rangle &= \langle \psi_\uparrow(0), \psi_\uparrow^\dagger(x) \rangle + \langle \psi_\uparrow(0), \mathcal{C}^\dagger(x)\psi_\downarrow(x) \rangle \\
&+ \langle \mathcal{C}(0)\psi_\downarrow^\dagger(0), \psi_\uparrow^\dagger(x) \rangle + \langle \mathcal{C}(0)\psi_\downarrow^\dagger(0), \mathcal{C}^\dagger(x)\psi_\downarrow(x) \rangle, \qquad (3.482)
\end{aligned}$$

and its momentum transform is given by Eq. (3.480). A similar result holds for the spin-down composite fermion field $\Psi_\downarrow(x) = R_B^{-1/2}\psi_\downarrow(x) + R_B^{-3/2}\mathcal{C}(x)\psi_\uparrow^\dagger(x)$.

Defining the quantity $gS(p)$ as the propagator of the composite fermion $S_{\text{Fermion}}(p)$, the composite fermion can be described by the strong-coupling effective Hamiltonian:

$$\beta\mathcal{H}_{\text{eff}}^F = \beta \sum_{i,\sigma=\uparrow,\downarrow} (\ell^3) Z_F^{-1} \Psi_\sigma^\dagger(i) \left[-\nabla^2/2M_F\ell^2 - \mu_F \right] \Psi_\sigma(i). \qquad (3.483)$$

Its chemical potential is $\mu_F = -M_F/2$ and its wave function renormalization constant is $Z_F = g/M_F$. Following the renormalization (3.459) of elementary fermion fields, we renormalize the composite fermion field $\Psi_{\uparrow,\downarrow} \Rightarrow (Z_F)^{-1/2}\Psi_{\uparrow,\downarrow}$, which behaves like a quasi-particle in Eq. (3.483), analogously to the composite boson (3.458). The negatively charged (e) three-fermion state is a combination of a twice negatively charged $(2e)$ Cooper field $\mathcal{C}(x) = \psi_\downarrow(x)\psi_\uparrow$ of two electrons with a once positively charged hole in $\psi_\downarrow(x)$. Similarly, positively charged $(-e)$ composite fermion fields $\Psi_\uparrow^\dagger(x)$ or $\Psi_\downarrow^\dagger(x)$ are composed of two-hole states $\mathcal{C}^\dagger\psi_\uparrow^\dagger\psi_\downarrow^\dagger$ combined with a single hole state $\psi_\uparrow^\dagger$ or $\psi_\downarrow^\dagger$. Suppose that two constituent electrons $\psi_\downarrow(k_1)$ and $\psi_\uparrow(k_2)$ of the Cooper pair field are combined with a constituent hole $\psi_\uparrow(k_3)$ lying close to the Fermi surface $k_1 \approx k_2 \approx k_3 \approx k_F$, then the Cooper bound state $q = k_2 - k_1 \ll k_F$

and the three-fermion bound state $p = k_1 - k_2 + k_3 \approx k_3 \approx k_F$ lie also around the Fermi surface. As a result, the composite fermion bound states $\Psi_{\uparrow,\downarrow}$ live around the Fermi surface as well.

Due to the Cooper field $\mathcal{C}(x)$ and the plasmon field $\mathcal{P}(x)$ appearing in an equivalent way in the interacting Hamiltonian (3.441), the same results (3.480)–(3.483) are obtained for the case of a plasmon field $\mathcal{P}(x) = \psi_\downarrow^\dagger(x)\psi_\uparrow(x)$ combined with another electron or hole. The associated composite fermion field is given by

$$\Psi_\uparrow^P(x) = R_B^{-1/2}\psi_\uparrow(x) + R_B^{-3/2}\mathcal{P}(x)\psi_\downarrow(x) \Rightarrow g^{1/4}\psi_\uparrow(x) + g^{3/4}\mathcal{P}(x)\psi_\downarrow(x), \quad (3.484)$$

whose two-point Green function reads:

$$\begin{aligned}
\langle \Psi_\uparrow^P(0), \Psi_\uparrow^{P\dagger}(x)\rangle &= \langle \psi_\uparrow(0), \psi_\uparrow^\dagger(x)\rangle + \langle \psi_\uparrow(0), \mathcal{P}^\dagger \psi_\downarrow^\dagger(x)\rangle \\
&+ \langle \mathcal{P}\psi_\downarrow(0), \psi_\uparrow^\dagger(x)\rangle + \langle \mathcal{P}\psi_\downarrow(0), \mathcal{P}^\dagger\psi_\uparrow^\dagger(x)\rangle. \quad (3.485)
\end{aligned}$$

For the spin-down field we find similarly $\Psi_\downarrow^P(x) = R_B^{-1/2}\psi_\downarrow(x) + R_B^{-3/2}\mathcal{P}(x)\psi_\uparrow(x)$. The composite fermions can be represented in the strong-coupling effective Hamiltonian of Eq. (3.483) with $\Psi_\sigma(i) \to \Psi_\sigma^P(i)$, following the renormalization (3.459) of elementary fermion fields, and a renormalization $\Psi_{\uparrow,\downarrow}^P \Rightarrow (Z_F)^{-1/2}\Psi_{\uparrow,\downarrow}^P$. The charged three-fermion states $\mathcal{P}\psi_{\uparrow,\downarrow}$ or $\mathcal{P}^\dagger\psi_{\uparrow,\downarrow}^\dagger$ are composed of one electron or one hole combined with a neutral plasmon field $\mathcal{P}(x) = \psi_\downarrow^\dagger(x)\psi_\uparrow(x)$ or $\mathcal{P}^\dagger(x) = \psi_\uparrow^\dagger(x)\psi_\downarrow(x)$ of an electron and a hole. The composite fermion fields $\Psi_{\uparrow,\downarrow}^P(x)$ are composed of a three-fermion state $\mathcal{P}\psi_{\uparrow,\downarrow}$ that consists of a plasmon in combination with a further elementary fermion $\psi_\uparrow$ or $\psi_\downarrow$.

The same thing is true for its charge-conjugate state. Suppose that a constituent electron $\psi_\downarrow(k_1)$ and a hole $\psi_\downarrow^\dagger(k_2)$ are combined with another constituent electron $\psi_\uparrow(k_3)$, and suppose that all momenta lie around the Fermi surface $k_1 \approx k_2 \approx k_3 \approx k_F$. Let the plasmon have the momentum $q = k_2 - k_1 \ll k_F$, and the composite fermion bound state have momenta $p = k_1 - k_2 + k_3 \approx k_3 \approx k_F$ near the Fermi surface. We can consider a three-fermion state $\mathcal{C}(x)\psi_\downarrow^\dagger(x)$ in Eq. (3.481). It can be written as

$$\mathcal{C}(x)\psi_\downarrow^\dagger(x) = \psi_\downarrow(x)\psi_\uparrow(x)\psi_\downarrow^\dagger(x) = -\psi_\downarrow^\dagger(x)\psi_\uparrow(x)\psi_\downarrow(x) = -\mathcal{P}(x)\psi_\downarrow(x). \quad (3.486)$$

This implies that the three-fermion state $\mathcal{C}(x)\psi_\downarrow^\dagger(x)$ is the same as the three-fermion state $\mathcal{P}(x)\psi_\downarrow(x)$ up to a definite phase factor $e^{i\pi}$. As a result, the composite fermion field $\Psi_\sigma(x)$ in (3.481) is the same as the composite fermion $\Psi_\sigma^P(x)$ in (3.484), up to a definite phase factor.

3.13 Conclusion and Remarks

In the weak-coupling limit, as the running energy scale becomes smaller corresponding to an increase of the lattice spacing ℓ, the limit $1/k_F a \to -\infty$ produces an IR-stable fixed point. Its scaling domain is described by an effective Hamiltonian of

BCS physics. The temperature region $T \sim T_c$ is characterized by the energy scale $\Delta_0 = \Delta(T_c)$. In elementary particle physics, this is analogous to the IR-stable fixed point and scaling domain of an effective Lagrangian of the Standard Model (SM). The physics at this electroweak scale is recapitulated in Ref. [86, 71].

As the running energy scale becomes larger, which happens when the lattice spacing ℓ becomes smaller, the coupling g becomes stronger and $1/k_F a$ becomes finite, crossing over to the region where the Cooper pairs are getting tightly bound. Their size becomes smaller and smaller and the width of the Feshbach resonance becomes sharper and sharper. At negative $1/k_F a \to 0$, the fermion system approaches the unitarity limit, where the Feshbach resonance turns into a tightly bound composite boson which behaves like an elementary scalar particle.

In Ref. [87], it is shown that at zero temperature and for $d > 2$, the unitarity limit of negative $1/k_F a \to 0^-$ and positive $1/k_F a \to 0^+$ represents an UV-stable fixed point of large coupling. The couplings $g > g_{UV}$ and $g < g_{UV}$ approach g_{UV}, as the running energy scale becomes larger (which happens when the lattice spacing ℓ becomes smaller). In the scaling domain of this UV-stable fixed point in the unitarity limit $1/k_F a \to 0^\pm$, where $T^* > T \to T_c$, an effective Hamiltonian of composite bosons and fermions is realized with characteristic scales $M_{B,F}(T)$. These considerations apply to the dimension $d = 2 + \epsilon > 0$ case. In elementary particle physics, this is analogous to the fixed point in the UV-regime with its scaling domain, where an effective Lagrangian of composite particles is realized with the characteristic scale probably in TeV range. The effective Lagrangian preservers SM chiral gauge symmetries and composite particles are made of SM elementary fermions [89].

We have shown here that the effective Hamiltonians (3.458) and (3.483) exist for composite bosons and fermions if $1/k_F a \geq 0$ at different values of temperature T. In the first regime $T \in (T^*, T_c)$, one finds a mixed liquid of composite bosons and fermions with the pseudogap $M_{F,B}(T)$. It is expected to dissolve to a normal unpaired Fermi gas at the crossover temperature T^*. These composite quasi particles are either charged or neutral. They behave like superfluids up to a relatively high crossover temperature T^*. In a second regime $T < T_c$, the superfluid phase of composite bosons undergoes BEC and one finds in the ground state the coexistence of BEC and semi-degenerate fermions $\Psi_\uparrow(x)$ and $\Psi_\downarrow(x)$. The latter couple to the BEC background to form massive quasi-particles of fermion type, moreover they form tightly bound states $\Psi_\uparrow \Psi_\downarrow$ or $\Psi_\uparrow^\dagger \Psi_\downarrow$, which are new bosonic quasi-particles producing a new condensate of the Bose-Einstein type. In both cases, whenever the Coulomb repulsion between electrons can be compensated by "phonons" in an analogous way to either composite bosons via a Feshbach resonance, or new bosonic quasi-particles via a composite-fermion pair state, this would result in superconductivity and superfluity at high temperature $T_c \propto \mathcal{O}(\epsilon_F)$. The scale of that is the result of a large coherent mass gap $M_{F,B}(T)$, being much larger than the BCS gap. The coherent supercurrents consist of composite fermions and bosons. These features, which we have discussed for $1/k_F a \geq 0$, are expected to be also true in $1/k_F a \ll 0$, only with a much smaller scale $M_{F,B}(T)$. Due to the presence of composite fermions in addition to composite bosons, we expect a further suppression of the low-energy spectral weight for single-particle excitations and for the material following a harder equation of state. Its observable consequences include a further T-dependent suppression of heat capacity and gap-like dispersion in the density-of-states and spin susceptibility. Moreover, we discuss the quantum critical point and speculate upon the phase of complex quasi-particles involved. It is known that the limit $1/k_F a \ll 0$ produces an IR-stable fixed point, and its scaling domain is described by an effective Hamiltonian of BCS physics with the gap scale $\Delta_0 = \Delta(T_c)$ in $T \sim T_c \lesssim T^*$. This is analogous to the IR-stable fixed point and scaling domain of an effective Lagrangian of the Standard Model (SM), which contains up to the electroweak scale all relevant fields of elementary particle physics [71, 86].

The unitarity limit $1/k_F a \to 0^{\pm}$ representing a scale invariant point [88] was formulated in a renormalization group framework [87], implying an UV-stable fixed point of large coupling. The couplings $g > g_{\mathrm{UV}}$ and $g < g_{\mathrm{UV}}$ approach g_{UV}, as the running energy scale becomes larger. In the scaling domain of this UV-stabe fixed point $1/k_F a \to 0^{\pm}$ and $T \to T_c^u$, an effective Hamiltonian of composite bosons and fermions is realized with a characteristic scale

$$M_{B,F}(T) = \left| \frac{T - T_c^u}{T_c^u} \right|^{\nu/2} \frac{(2d)^{1/2}}{(3\pi^2)^{1/d}} k_F, \quad T \gtrsim T_c^u, \tag{3.487}$$

where $\nu = 1$ is the critical exponent derived from the β-function which determines the scaling laws. Equation (3.487) shows that the relevant cutoff are the Fermi momentum k_F and the physical correlation length $\xi \propto M_{B,F}^{-1}$. The last characterizes the size of composite particles via their form factor $Z_{B,F} \propto M_{B,F}^{-1}$ (3.458) and (3.483). This domain should be better explored experimentally. The analogy was discussed in elementary particle physics with anticipations of the UV-scaling domain at TeV scales and an effective Lagrangian of composite particles made by SM elementary fermions including those of Majorana type [89].

Appendix 3A Auxiliary Strong-Coupling Calculations

Here we present the following one-site functional integrals over fermion fields that are useful for obtaining the recursion relations of two-point Green functions of composite boson and fermion fields. The integral of one field $\psi_\uparrow(i)$ at the point "i" is defined as,

$$\langle \psi_\uparrow^\dagger(x) \rangle \equiv \frac{1}{Z(i)} \int_{i\downarrow} \int_{i\uparrow} \psi_\uparrow^\dagger(i) e^{-h\mathcal{H}_0(i) - \beta\mathcal{H}_{\mathrm{int}}(i)}, \tag{3A.1}$$

where $\int_{i\downarrow} \equiv \int [d\psi_\downarrow^\dagger(i) d\psi_\downarrow(i)]$ and $\int_{i\uparrow} \equiv \int [d\psi_\uparrow^\dagger(i) d\psi_\uparrow(i)]$. To have a non-vanishing integral $\int_{i\uparrow}$, it needs a $\psi_\uparrow(x)$-field in the expansion of $e^{-h\mathcal{H}_0^\uparrow(x)}$,

$$
\begin{aligned}
\langle \psi_\uparrow(i) \rangle &= \frac{h}{Z(i)} \sum_{\ell}^{\mathrm{ave}} \psi_\uparrow^\dagger(i;\hat{\ell}) \int_{i\downarrow} e^{-h\mathcal{H}_0^\downarrow(i)} \int_{i\uparrow} \psi_\uparrow(i) \psi_\uparrow^\dagger(i) e^{-\mathcal{H}_{\mathrm{int}}(i)}, \\
&= -\frac{h}{Z(i)} \sum_{\ell}^{\mathrm{ave}} \psi_\uparrow^\dagger(i;\hat{\ell}) \int_{i\downarrow} e^{-h\mathcal{H}_0^\downarrow(i)}
\end{aligned}
\tag{3A.2}
$$

where $\sum_{\hat{\ell}}^{\text{ave}} \psi_\uparrow^\dagger(i;\hat{\ell}) = \sum_{\hat{\ell}} \left[\psi_\uparrow^\dagger(i,\hat{\ell}) + \psi_\uparrow^\dagger(i,-\hat{\ell}) \right]$ and

$$\int_{i\uparrow} \psi_\uparrow(i)\psi_\uparrow^\dagger(i)e^{-\mathcal{H}_{\text{int}}(i)} = \frac{1}{\psi_\downarrow^\dagger(i)\psi_\downarrow(i)} = -1, \tag{3A.3}$$

using Eq. (3.448). The integral $\int_{i\downarrow}$ needs to have $\psi_\downarrow(i)$ and $\psi_\downarrow^\dagger(i)$ fields in expansion of $e^{-h\mathcal{H}_0^\downarrow(i)}$,

$$\langle\psi_\uparrow(i)\rangle = \frac{h^3}{Z(i)} \sum_{\hat{\ell}}^{\text{ave}} \psi_\uparrow^\dagger(i;\hat{\ell}) \sum_{\hat{\ell}'}^{\text{ave}} \psi_\downarrow(i;\hat{\ell}') \sum_{\hat{\ell}''}^{\text{ave}} \psi_\downarrow^\dagger(i;\hat{\ell}'') \int_{i\downarrow} \psi_\downarrow(i)\psi_\downarrow^\dagger(i),$$

$$\approx h^3 \sum_{\hat{\ell}}^{\text{ave}} \psi_\uparrow^\dagger(i;\hat{\ell})\psi_\downarrow(i;\hat{\ell})\psi_\downarrow^\dagger(i;\hat{\ell}), \tag{3A.4}$$

where the three fields $\psi_\uparrow^\dagger$, $\psi_\downarrow$ and $\psi_\downarrow^\dagger$ are approximately at the same point $i + \hat{\ell}$ for the lowest non-trivial contribution.

The integral of two fields $\psi_\uparrow(i)\psi_\downarrow(i)$ at "i" is defined as

$$\langle\psi_\uparrow(i)\psi_\downarrow(i)\rangle \equiv \frac{1}{Z(i)} \int_{i\downarrow} \int_{i\uparrow} \psi_\uparrow(i)\psi_\downarrow(i)e^{-h\mathcal{H}_0(i)-\beta\mathcal{H}_{\text{int}}(i)}. \tag{3A.5}$$

To have a non-trivial result, it needs $\psi_\uparrow^\dagger(i)$ and $\psi_\downarrow^\dagger(i)$ fields in the expansion of $e^{-h\mathcal{H}_0(i)}$,

$$\langle\psi_\uparrow(i)\psi_\downarrow(i)\rangle = h^2 \sum_{\hat{\ell}}^{\text{ave}} \psi_\uparrow(i;\hat{\ell}) \sum_{\hat{\ell}'}^{\text{ave}} \psi_\downarrow(i;\hat{\ell}') \approx h^2 \sum_{\hat{\ell}}^{\text{ave}} [\psi_\uparrow(i;\hat{\ell})\psi_\downarrow(i;\hat{\ell})]. \tag{3A.6}$$

The integral of three fields at the site "i" is defined as

$$\langle\psi_\uparrow^\dagger(i)\psi_\downarrow(i)\psi_\downarrow^\dagger(i)\rangle \equiv \frac{1}{Z(i)} \int_{i\downarrow} \int_{i\uparrow} \psi_\uparrow^\dagger(i)\psi_\downarrow(i)\psi_\downarrow^\dagger(i)e^{-h\mathcal{H}_0(i)-\mathcal{H}_{\text{int}}(i)}$$

$$= \frac{h}{Z(i)} \sum_{\hat{\ell}}^{\text{ave}} \psi_\uparrow^\dagger(x;\hat{\ell}) \int_{i\downarrow} \int_{i\uparrow} \psi_\uparrow(i)\psi_\uparrow^\dagger(i)\psi_\downarrow(i)\psi_\downarrow^\dagger(i)e^{-h\mathcal{H}_0^\downarrow(i)-\mathcal{H}_{\text{int}}(i)}$$

$$= h \sum_{\hat{\ell}}^{\text{ave}} \psi_\uparrow^\dagger(x;\hat{\ell}), \tag{3A.7}$$

where a $\psi_\uparrow(i)$ field comes from the expansion of $e^{-h\mathcal{H}_0^\uparrow(i)}$.

The integral of the four fermion fields at site "i" reads:

$$\langle\psi_\uparrow(i)\psi_\uparrow^\dagger(i)\psi_\downarrow(i)\psi_\downarrow^\dagger(i)\rangle \equiv \frac{1}{Z(i)} \int_{i\downarrow} \int_{i\uparrow} \psi_\uparrow(i)\psi_\uparrow^\dagger(i)\psi_\downarrow(i)\psi_\downarrow^\dagger(i)e^{-h\mathcal{H}_0(i)-\mathcal{H}_{\text{int}}(i)} = 1. \tag{3A.8}$$

Appendix 3B   Propagator of the Bilocal Pair Field

Consider the Bethe-Salpeter equation (3.23) with a potential λV instead of V

$$\Gamma = -i\lambda V G_0 G_0 \Gamma. \tag{3B.1}$$

Take this as an eigenvalue problem in λ at fixed energy-momentum $q = (q^0, \mathbf{q}) = (E, \mathbf{q})$ of the bound states. Let $\Gamma_n(P|q)$ be all solutions, with eigenvalues $\lambda_n(q)$. Then the convenient normalization of Γ_n is:

$$-i \int \frac{d^4 P}{(2\pi)^4} \Gamma_n^\dagger (P|q) G_0 \left(\frac{q}{2} + P\right) G_0 \left(\frac{q}{2} - P\right) \Gamma_{n'}(P|q) = \delta_{nn'}. \tag{3B.2}$$

If all solutions are known, there is a corresponding completeness relation (the sum may comprise an integral over a continuous part of the spectrum)

$$-i \sum_n G_0 \left(\frac{q}{2} + P\right) G_0 \left(\frac{q}{2} - P\right) \Gamma_n(P|q)\Gamma_n^\dagger(P'|q) = (2\pi)^4 \delta^{(4)}(P - P'). \tag{3B.3}$$

This completeness relation makes the object given in (3.37) the correct propagator of Δ. In order to see this, write the free Δ action $\mathcal{A}_2[\Delta^\dagger \Delta]$ as

$$\mathcal{A}_2 = \frac{1}{2}\Delta^\dagger \left(\frac{1}{\lambda V} + iG_0 \times G_0\right) \Delta, \tag{3B.4}$$

where we have used λV instead of V. The propagator of Δ would have to satisfy

$$\left(\frac{1}{\lambda V} + iG_0 \times G_0\right) \overrightarrow{\Delta \Delta^\dagger} = i. \tag{3B.5}$$

Performing this calculation on (3.29), one has indeed for Γ_n and λ_n, by virtue of (3B.1), the equation

$$\left(\frac{1}{\lambda V} + iG_0 \times G_0\right) \times \left\{-i\lambda \sum_n \frac{\Gamma_n \Gamma_n^\dagger}{\lambda - \lambda_n(q)}\right\}$$

$$= -i\lambda \sum_n \frac{\frac{1}{\lambda V}\Gamma_n \Gamma_n^\dagger + iG_0 \times G_0 \Gamma_n \Gamma_n^\dagger}{\lambda - \lambda_n(q)}$$

$$= i\lambda \sum_n \frac{-\frac{\lambda_n(q)}{\lambda} + 1}{\lambda - \lambda_n(q)} (-iG_0 \times G_0 \Gamma_n \Gamma_n^\dagger)$$

$$= i\left(-i\sum_i G_0 \times G_0 \Gamma_n \Gamma_n^\dagger\right) = i. \tag{3B.6}$$

Note that the expansion of the propagator in powers of λ, namely

$$\overrightarrow{\Delta \Delta^\dagger} = i \sum_k \left(\sum_n \left(\frac{\lambda}{\lambda_n(q)}\right)^k \Gamma_n \Gamma_n^\dagger\right), \tag{3B.7}$$

corresponds to the graphical sum over one, two, three, etc. exchanges of the potential λV. For $n = 1$ this is immediately obvious due to (3B.1):

$$i \sum_n \frac{\lambda}{\lambda_n(q)} \Gamma_n \Gamma_n^\dagger = \sum \frac{\lambda}{\lambda_n(q)} \lambda_n(q) V G_0 \times G_0 \Gamma_n \Gamma_n^\dagger = i\lambda V. \tag{3B.8}$$

For $n = 2$ one can rewrite, using the orthogonality relation,

$$i \sum_n \left(\frac{\lambda}{\lambda_n(q)} \right)^2 \Gamma_n \Gamma_n^\dagger = \sum_{nn'} \frac{\lambda}{\lambda_n(q)} \Gamma_n \Gamma_n^\dagger G_0 \times G_0 \Gamma_{n'} \Gamma_{n'}^\dagger \frac{\lambda}{\lambda_{n'}(q)} = \lambda V G_0 \times G_0 \lambda V, \tag{3B.9}$$

which displays the exchange of two λV terms with particles propagating in between. The same procedure applies at any order in λ. Thus the propagator has the expansion

$$\overset{\frown}{\Delta \Delta^\dagger} = i\lambda V - i\lambda V G_0 \times G_0 i\lambda V + \dots. \tag{3B.10}$$

If the potential is instantaneous, the intermediate $\int dP_0/2\pi$ can be performed replacing

$$G_0 \times G_0 \rightarrow \frac{i}{E - E_0(\mathbf{P}|q)}, \tag{3B.11}$$

where

$$E_0(\mathbf{P}|q) = \xi \left(\frac{\mathbf{q}}{2} + \mathbf{P} \right) + \xi \left(\frac{\mathbf{q}}{2} - \mathbf{P} \right)$$

is the free particle energy which may be considered as the eigenvalue of an operator H_0. In this case the expansion (3B.10) reads

$$\overset{\frown}{\Delta \Delta^\dagger} = i \left(\lambda V + \lambda V \frac{1}{E - H_0} \lambda V + \dots \right) = i\lambda V \frac{E - H_0}{E - H_0 - \lambda V}. \tag{3B.12}$$

We see it related to the resolvent of the complete Hamiltonian as

$$\overset{\frown}{\Delta \Delta^\dagger} = i\lambda V (R\lambda V + 1), \tag{3B.13}$$

where

$$R \equiv \frac{1}{E - H_0 - \lambda V} = \sum_n \frac{\psi_n \psi_n^\dagger}{E - E_n}, \tag{3B.14}$$

with ψ_n being the Schrödinger amplitudes in standard normalization. We can now easily determine the normalization factor N in the connection between Γ_n and the Schrödinger amplitude ψ_n. In the instantaneous case, Eq. (3B.2) gives

$$\int \frac{d^3 P}{(2\pi)^3} \Gamma_n^\dagger(\mathbf{P}|q) \frac{1}{E - H_0} \Gamma_{n'}(\mathbf{P}|q) = \delta_{nn'}. \tag{3B.15}$$

Inserting ψ from (3.29) renders the orthogonality relation

$$\frac{1}{N^2} \int \frac{d^3 P}{(2\pi)} \psi_n^\dagger (E - H_0) \psi_{n'}(\mathbf{P}|q) = \delta_{nn'}. \tag{3B.16}$$

But since

$$(E - H_0)\psi = \lambda V \psi, \tag{3B.17}$$

the orthogonality relation reads also

$$\frac{1}{N^2} \int \frac{d^3 P}{(2\pi)^3} \psi_n^\dagger (\mathbf{P}|q) \lambda V \psi_{n'} (\mathbf{P}|q) = \delta_{nn'}. \tag{3B.18}$$

For wave functions ψ_n in standard normalization, the integral expresses the differential

$$\lambda \frac{dE}{d\lambda}.$$

For a typical calculation of a resolvent, the reader is referred to Schwinger's treatment of the Coulomb problem [60]. His result may directly be used for a propagator of electron hole pairs bound to excitons.

Appendix 3C   Fluctuations Around the Composite Field

Here we show that the quantum mechanical fluctuations around the *classical* equations of motion

$$\Delta(x,y) = V(x - y)\psi(x)\psi(y) \tag{3C.1}$$

are quite simple to calculate. This will be compared with the collective plasmon field in equation

$$\varphi(x) = \int dy V(x,y)\psi^\dagger(y)\psi(y). \tag{3C.2}$$

For this let us compare the Green functions of $\Delta(x,y)$ [or $\varphi(x)$] or with those of the composite operators on the right-hand side of Eqs. (3C.1) or (3C.2). The Green functions of Δ [or φ] are generated by adding to the original actions (3.3) [or (2.4)], respectivly, the external currents $1/2 \int dx dy (\Delta(y,x)I^\dagger(x,y) + \text{c.c.})$ [or $\int dx \, \varphi(x)I(x)$] to the final actions (3.8) [or (2.12)], respectively, and by forming functional derivatives $\delta/\delta K$ [or $\delta/\delta I$]. The Green functions of the composite operators $\psi(x)\psi(y)$ [or $\psi^\dagger(x) \, \psi(x)$], on the other hand, are obtained by adding to the original actions (3.3) [or (2.4)] the source terms

$$\frac{1}{2} \int dx \int dy \, [V(x - y)\psi(x)\psi(y)K^\dagger(x,y) + \text{c.c.} \tag{3C.3}$$

[or

$$\int dx \left(\int dy V(x,y)\psi^\dagger(y)\psi(y) \right) I(x) \,],$$ (3C.4)

and by forming functional derivatives $\delta/\delta K$ [or $\delta/\delta I$]. To do this most simply we observe that the source K can be included in the final actions by replacing in (3.8)

$$\Delta(x,y) \to \Delta'(x,y) = \Delta(x,y) - K(x,y),$$ (3C.5)

[or in (2.12)

$$\varphi(x) \to \varphi'(x) = \varphi(x) - \int dx' I(x') V(x',x) \,].$$ (3C.6)

If one now shifts the functional integrations to the new translated variables and drops the irrelevant superscript "prime", the combined action can be rewritten as

$$A[\Delta^*,\Delta] = \pm \frac{i}{2}\mathrm{Tr}\log\left(i\mathbf{G}_\Delta^{-1}\right) + \frac{1}{2}\int dxdx' |\Delta(x,x')|^2 \frac{1}{V(x,x')}$$
$$+ \frac{i}{2}\int dxdx'\, j^\dagger(x)\mathbf{G}_\Delta(x,x')j(x')\frac{1}{V(x,x')} \qquad (3C.7)$$
$$+ \frac{1}{2}\int dxdx' \left\{ \Delta(y,x)K^\dagger(x,y) + \text{h.c.} \right\} + \frac{1}{2}\int dxdx' |K(x,x')|^2 V(x,x'),$$

[or

$$A[\varphi] = \pm i\mathrm{Tr}\log(iG_\varphi^{-1}) + \frac{1}{2}\int dxdx'\,\varphi(x)V^{-1}(x,x')\varphi(x') + i\int dxdx'\,\eta^\dagger(x)G_\varphi(x,x')\eta(x)$$
$$+ \int dx\,\varphi(x)I(x) + \frac{1}{2}\int dxdx'\,I(x)V(x,x')I(x') \,].$$ (3C.8)

In this form the actions display clearly the fact that derivatives with respect to the sources K or I coincide exactly, except for all possible insertions of the direct interaction V. For example, the propagators of the plasmon field $\varphi(x)$ and of the composite operator $\int dy V(x,y)\psi^\dagger(y)\psi(y)$ are related by

$$\overline{\varphi(x)\varphi(x')} = -\frac{\delta^{(2)}Z}{\delta I(x)\delta I(x')} = V^{-1}(x,x') - \frac{\delta^{(2)}Z}{\delta K(x)\delta K(x')}$$ (3C.9)
$$= V^{-1}(x,x') + \langle 0|\hat{T}\left(\int dy V(x,y)\psi^\dagger(y)\psi(\varphi) \right)\left(\int dy' V(x'y')\psi^\dagger(y')\psi^\dagger(y')\psi(y') \right)|0\rangle,$$

where $\hat{T}$ is the time-ordering operator. Similarly, one finds for the pair fields:

$$\overline{\Delta(x,x')[\Delta(y,y')]}^\dagger = \delta(x-y)\delta(x'-y')iV(x-x')$$
$$+ \langle 0|\hat{T}\left(V(x',x)\psi(x')\psi(x) \right)\left(V(y',y)\psi^\dagger(y)\psi^\dagger(y') \right)|0\rangle.$$ (3C.10)

Note that the latter relation is manifestly displayed in the representation (3B.10) of the propagator Δ. Since

$$\overline{\Delta\Delta^\dagger} = iVG^{(4)}V,$$

one has from (3C.10)

$$\langle 0|\hat{T}[V(\psi\psi)(\psi^\dagger\psi^\dagger V)]|0\rangle = V\,G^{(4)}V, \tag{3C.11}$$

which is certainly true, since $G^{(4)}$ is the full four-point Green function. In the equal-time situation in the presence of an instantaneous potential, $G^{(4)}$ is replaced by the resolvent R.

Notes and References

[1] J. Bardeen, L.N. Cooper, and J.R. Schrieffer, Phys. Rev. **108**, 1175 (1957).
See also the little textbook from the Russian school:
N.N. Bogoliubov, E.A. Tolkachev, and D.V. Shirkov, *A New Method in the Theory of Superconductivity*, Consultants Bureau, New York, 1959.

[2] N.N. Bogoliubov, Sov. Phys. JETP **7**, 41 (1957).

[3] D. Jerome, A. Mazaud, M. Ribault, and K. Bechgaard, Journal de Physique Lettres **41**, 4 (1980);
See also the textbook
A.G. Lebed (ed.), *The Physics of Organic Superconductors and Conductors*, Springer Series in Material Science, **110**, (2008).

[4] See the historic paper on the web under http://www.superconductors.org/history.htm.

[5] J.G. Bednorz and K.A. Müller, Z. Physik, B **64** 189 (1986).

[6] From the MS-thesis of
P.J. Ray, Niels Bohr Institute, Copenhagen, Denmark, November 2015, DOI:10.6084/m9.figshare.2075680.v2.

[7] See the www addresses:
http://superconductors.org/type2.htm;
http://www.users.qwest.net/~csconductor/Experiment_Guide/History of Superconductivity.htm.

[8] See the article in Nature:
http://www.nature.com/nature/journal/v525/n7567/abs/nature14964.html.

[9] For the introduction and use of such bilocal fields in particle physics see
H. Kleinert, Erice Lectures 1976 on Particle Physics (ed. by A. Zichichi);
Phys. Letters B **62**, 429 (1976); B **59**, 163 (1975).

[10] H. Kleinert, Fortschr. Phys. **26**, 565 (1978).

[11] See for example:
A.A. Abrikosov, L.P. Gorkov, and I.E. Dzyaloshinski, *Methods of Quantum Field Theory in Statistical Physics*, Dover, New York (1975);
L.P. Kadanoff, and G. Baym, *Quantum Statistical Mechanics*, Benjamin, New York (1962);
A.L. Fetter and J.D. Walecka, *Quantum Theory of Many-Particle Systems*, McGraw-Hill, New York (1971).

[12] See the last of Ref. [11] or
D. Saint-James, G. Sarma, and E.J. Thomas, *Type II Superconductivity*, Pergamon Press, New York (1969).

[13] See Eq. (9.266) in the textbook:
H. Kleinert, *Particles and Quantum Fields*, World Scientific, Singapore, 2016.

[14] M. Randeria, W. Zwerger, and M.W. Zwierlein, In W. Zwerger, editor, The BCS-BEC Crossover and the Unitary Fermi Gas, Lecture Notes in Physics. Springer, 2012.

[15] J. Hubbard, Phys. Rev. Letters **3**, 77 (1959);
B. Mühlschlegel, J. Math. Phys. **3**, 522 (1962);
J. Langer, Phys. Rev. **134**, A 553 (1964);
T.M. Rice, Phys. Rev. **140**, A 1889 (1965); J. Math. Phys. **8**, 1581 (1967);
A.V. Svidzinskij, Teor. Mat. Fiz. **9**, 273 (1971);
D. Sherrington, J. Phys. **C4**, 401 (1971).

[16] F.W. Wiegel, Phys. Reports **C16** 57 (1975);
V.N. Popov, *Kontinual'nye Integraly v Kvantovoj Teorii Polja i Statisticheskoj Fizike*, Atomizdat, Moscow (1976); Engl. translation by J. Niederle and L. Hlavatý in *Functional Integrals in Quantum Field Theory and Statistical Physics* (Mathematical Physics and Applied Mathematics).

[17] The first authors to employ such identities were
P.T. Mathews and A. Salam, Nuovo Cimento **12**, 563 (1954); **2**, 120 (1955).

[18] An excellent review on this equation is given by
N. Nakanishi, Progr. Theor. Phys. Suppl. **43**, 1 (1969).

[19] H.E. Stanley, *Phase Transitions and Critical Phenomena*, Clarendon Press, Oxford, 1971;
F.J. Wegner, in *Phase Transitions and Critical Phenomena*, ed. by C. Domb and M.S. Green, Academic Press 1976, p. 7;
E. Brézin, J.C. Le Guillou, and J. Zinn-Justin, *ibid.*, p. 125.
See also
L. Kadanoff, Rev. Mod. Phys. **49**, 267 (1977).

[20] R.W. Morse and H.V. Bohm, Phys. Rev. **108**, 1094 (1957).

[21] J.G. Valatin and D. Butler, Nuovo Cimento X, 37 (1958);
R.W. Richardson, J. Math. Phys. **9**, 1327 (1968).

[22] V.A. Adrianov and V.N. Popov, Theor. Math. Fiz. **28**, 340 (1976).

[23] H. Kleinert, *Path Integrals in Quantum Mechanics, Statistics, Polymer Physics, and Financial Markets*, World Scientific, Singapore, 2009 (klnrt.de/b5).

[24] H. Kleinert and V. Schulte-Frohlinde, *Critical Properties of ϕ^4-Theories*, World Scientific, Singapore 2001, pp. 1–489 (`klnrt.de/b8`).

[25] A.J. Leggett, in *Modern Trends in the Theory of Condensed Matter*, edited by A. Pekalski and J. Przystawa, Lecture Notes in Physics, Vol. 115 (Springer-Verlag, Berlin, 1980), p. 13.

[26] M. Randeria. *Precursor Pairing Correlations and Pseudogaps*, in G. Iadonisi, J. R. Schrieffer, and M. L. Chiafalo, editors, Proceedings of the International School of Physics Enrico Fermi, Course CXXXVI on High Temperature Superconductors, pages 53-75. IOS Press, 1998 (cond-mat/9710223).

[27] M. Randeria and E. Taylor, Annual Review of Condensed Matter Physics, **5**, 209 (2014).

[28] M. Randeria, W. Zwerger and M. Zwierlein, In W. Zwerger, editor, The BCS-BEC Crossover and the Unitary Fermi Gas, Lecture Notes in Physics. Springer, 2012.

[29] J.R. Engelbrecht, A. Nazarenko, M. Randeria, and E. Dagotto, Phys. Rev. B **57**, 13406 (1997) (cond-mat/9705166).

[30] H. Ding *et al.*, Nature **382**, 51 (1996).

[31] J.V. José, *40 Years of Berezinskii-Kosterlitz-Thouless Theory*, World Scientific, Singapore, 2013. The last two authors have received the Nobel prize in 2016, together with
F.D.M. Haldane [32]. See also the textbooks cited in Ref. [33].

[32] F.D.M. Haldane, Phys. Rev. B **15**, 2477 (1977), Phys. Rev. Letters **61**, 1988 (2015).

[33] H. Kleinert, *Gauge Fields in Condensed Matter*, Vol. 1: Superflow and Vortex Lines (Disorder Fields, Phase Transitions); Vol. 2: Stresses and Defects (Differential Geometry, Crystal Melting), World Scientific, Singapore, 1989.
D.J. Thouless, *Topological Quantum Numbers in Nonrelativistic Physics*. World Scientific, Singapore, 1998.

[34] H. Kleinert and E. Babaev, Phys. Lett. B **438**, 311 (1998.)

[35] H. Kleinert, *Particles and Quantum Fields*, World Scientific, Singapore, 2016.

[36] J.M. Harris *et al.*, Phys. Rev. Lett. **79**, 143 (1997).

[37] T. Ito *et al.*, Phys. Rev. Lett. **70**, 3995 (1993);

[38] L.D. Rotter *et al.*, Phys. Rev. Lett. **67**, 2741 (1991).

[39] J. Orenstein *et al.*, Phys. Rev. B **42**, 6342 (1990).

[40] B. Bucher *et al.*, Phys. Rev. Lett. **70**, 2012 (1993).

[41] J. Loram *et al.*, Phys. Rev. Lett. **71**, 1740 (1993) and Physica C **235-240**, 134 (1994).

[42] Y.J. Uemura, in *Proceedings of the Workshop in Polarons and Bipolarons in High-T_c Superconductors and Related Materials*, Cambridge, 1994, ed. by E. Salje *et al.* (Cambridge Univ. Press, 1995), pp. 453-460.

[43] Y.J. Uemura, in *Proceedings of the CCAST Symposium on High-T_c Superconductivity and the C_{60} Family*, Beijing, China, 1994, ed. by S. Feng and H.C. Ren (Gordon and Breach, New York, 1995), pp. 113.

[44] V.P. Gusynin, V.M. Loktev, and S.G. Sharapov, JETP Lett. **65** 182 (1997); JETP **88**, 685 (1999) (cond-mat/9709034);
V.M. Loktev, S.G. Sharapov, Cond. Matter Physics (Lviv) **11**, 131 (1997) (cond-mat/9706285);
V.M. Loktev, R.M. Quick, and S.G. Sharapov, Physica C **314**, 233 (1999) (cond-mat/9804026).

[45] A. Puchkov, D. Basov, and T. Timusk, J. Phys. Condensed Matter **8**, 10049 (1996).

[46] C.C. Homes *et al.*, Phys. Rev. Lett. **71**, 1645 (1993).

[47] P. Nozières and S. Schmitt-Rink, J. Low. Temp. Phys. **59**, 195 (1985).

[48] M. Drechsler and W. Zwerger, Ann. Phys. (Germany) **1**, 15 (1992).

[49] V.J. Emery and S.A. Kivelson, Nature **374**, 434 (1995).

[50] J.R. Engelbrecht, M. Randeria, and C.A.R. Sá de Melo, Phys. Rev. B **55**, 15153 (1997).

[51] M. Marini, F. Pistolesi, and G.C. Strinati, Eur. Phys. J. **1**, 151 (1998) (cond-mat/9703160).

[52] M. Randeria, J.-M. Duan, and L. Shieh, Phys. Rev. B **41**, 327 (1990).

[53] Y. J. Uemura *et al.*, Phys. Rev. Lett. **66**, 2665 (1991); Nature (London) **352**, 605 (1991).

[54] V.L. Berezinskii, Zh. Eksp. Teor. Fiz. **59**, 907 (1971);
J.M. Kosterlitz and D.J. Thouless. J. Phys. C **6**, 1181 (1973).

[55] N.D. Mermin and H. Wagner, Phys. Rev. Lett. **17**, 1113 (1966);
P.C. Hohenberg, Phys. Rev. **158**, 383 (1967);
S. Coleman, Comm. Math. Phys. **31**, 259 (1973).

[56] D.S. Marshall *et al.*, Phys. Rev. Lett. **76**, 4841 (1996).

[57] M. Randeria, in *Bose-Einstein Condensation*, edited by A. Griffin, D.W. Snoke and S. Stringari, New York, Cambridge University Press, 1995, p. 355.

[58] D.M. Eagles, Phys. Rev. **186**, 456 (1969).

[59] P. Minnhagen, Rev. Mod. Phys. **59**, 1001 (1987).

[60] J. Schwinger, J. Math. Phys. **5**, 1606 (1964).

[61] R.J. Fletcher, R. Lopes, J. Man, N. Navon, R.P. Smith, M.W. Zwierlein, Z. Hadzibabic, Science **355**, 377 (2017).

[62] A. Gammal, T. Frederico, L. Tomio, and P. Chomaz. J. Phys. B (At. Mol. Opt. Phys.) **33**, 4053 (2000).

[63] Y. Nambu and G. Jona-Lasinio, Phys. Rev. **122**, 345 (1961).

[64] M. Tinkham, *Superconductivity*, Dover, London, 2012.

[65] F. Englert and R. Brout, Phys. Rev. Lett. **13**, 321 (1964).

[66] P.W. Higgs, Phys. Lett. **12**, 132 (1964); Phys. Rev. Lett. **13**, 508 (1964); Phys. Rev. **145** 1156 (1966).

[67] G.S. Guralnik, C.R. Hagen, and T.W.B. Kibble, Phys. Rev. Lett. 13 (1964) 585.

[68] T.W.B. Kibble, Phys. Rev. **155**, 1554 (1967).

[69] S. Coleman and E. Weinberg, Phys. Rev. D **7**, 1888 (1973).

[70] E. Weinberg, *Radiative Corrections as the Origin of Spontaneous Symmetry Breaking*, Harvard University, Cambridge, Massachusetts, 1973.

[71] See also the application to the Standard model of electroweak interactions by M. Fiolhais and H. Kleinert, Physics Letters A **377**, 2195 (2013) (http://klnrt.de/402).
S.-S. Xue, Phys. Lett. B **737**, 172 (2014).

[72] H. Kleinert, Fortschr. Physik 26, 565 (1978) (http://klnrt.de/55).

[73] See Eq, (9.264) to (9.266) of the textbook [35].

[74] C.A.R. Sá de Melo, M. Randeria, and J.R. Engelbrecht, Phys. Rev. Lett. **71**, 3202 (1993).

[75] J.R. Engelbrecht, M. Randeria, and C.A.R. Sá de Melo, Phys. Rev. B 55, 15153, 1997.

[76] C.M. Bender, F. Cooper, G.S. Guralnik, D.H. Sharp, Phys. Rev. D **19**, 1865 (1979);
C.M. Bender, F. Cooper, G.S. Guralnik, P. Roskies, D.H. Sharp, Phys. Rev. Lett. **43**, 537 (1979); Phys. Rev. D **23**, 2999 (1981);
E. Eichten and J. Preskill, Nucl. Phys. B **268**, 176 (1986);
M. Creutz and C. Rebbi, M. Tytgat, and S.-S. Xue, Phys. Lett. B **402**, 341 (1997).

[77] S.-S. Xue, Phys. Lett. B **381**, 277 (1996); Nucl. Phys. B **486**, 282 (1997); Phys. Rev. D **64**, 094504 (2001)

[78] This is reminiscent of the discussion whether helium nuclei are elementary or composite particles, based on the vanishing form-factors and binding energy of composite particles. See the articles by
S. Weinberg, Phys. Rev. **130**, 776 (1963), **131**, 440 (1963), **133**, B232 (1963), **137**, B672 (1965).

[79] Y. Nishida and D.T. Son, Phys. Rev. Lett. **97**, 050403 (2006).

[80] Y. Nishida and D.T. Son, Phys. Rev. A **75**, 063617 (2007).

[81] M. Randeria, J.M. Duan, and L.Y. Shieh, Phys. Rev. Lett. **62**, 981 (1989).

[82] C.A.R. Sá de Melo, M. Randeria, and J. R. Engelbrecht, Phys. Rev. Lett. **71**, 3202 (1993).

[83] H. Kleinert, *Multivalued Fields in Condensed Matter, Electromagnetism, and Gravitation*, World Scientific, Singapore 2009 (http://klnrt.de/b11).

[84] W. Zwerger, Proceedings of the International School of Physics "Enrico Fermi" - Course 191 "Quantum Matter at Ultralow Temperatures" edited by M. Inguscio, W. Ketterle, S. Stringari. and G. Roati, IOS Press, Amsterdam, and SIF Bologna, 2016, pp. 63-142 (https://arxiv.org/abs/1608.00457).

[85] S.-S. Xue and H. Kleinert, *Composite Fermions and their Pair States in a Strongly-Coupled Fermi Liquid.* (http://klnrt.de/415) (arXiv:1708.04023).

[86] S.-S. Xue, Phys. Lett. B **721**, 347 (2013), *ibid* **727**, 308 (2013); **744**, 88 (2015). JHEP **11**, 027 (2016) (arXiv:1605.01266).

[87] see for example, the papers by
P. Nikolić and S. Sachdev, Phys. Rev. A **75**, 033608 (2007) (arXiv:cond-mat/0609106).

[88] T.-L. Ho. Phys. Rev. Lett. **92**, 090402 (2004).

[89] S.-S. Xue, Phys. Lett. B **737**, 172 (2014) and JHEP **05**, 146 (2017) (arXiv:1601.06845).

[90] J.R. Engelbrecht, M. Randeria, and C.A.R. Sá de Melo, Phys. Rev. B, **55** 15153 (1997).

[91] C.A.R. Sá de Melo, M. Randeria, and J.R. Engelbrecht, Phys. Rev. Lett., **71**, 3202 (1993).

[92] Y. Nishida and D. T. Son, in W. Zwerger (ed.) *The BCS-BEC Crossover and the Unitary Fermi Gas*, Lecture Notes in Physics, Springer (2011).

[93] M.M. Forbes, S. Gandolfi, and A. Gezerlis, Phys. Rev. Lett. **106**, 235303 (2011).

[94] A. Bulgac, J.E. Drut, and P. Magierski, Phys. Rev. A **78**, 023625 (2008).

4

Superfluid ^{3}He

The explanation of the phenomenon of superconductivity by Bardeen, Cooper, and Schrieffer in 1957 [1], and a little later by Bogoliubov and his school [2], prompted a search for similar phenomena in other Fermi systems, such as fermionic nuclei [3] and, in particular, liquid ^{3}He [4]. While nuclear forces did, in principle, allow a direct application of the BCS formalism [5], it was soon realized [6] that in ^{3}He the strong repulsive core of the interatomic potential would not permit the formation of s-wave Cooper pairs as in superconductors. Thus, if anything similar to Cooper pairing should occur, it had to be in a nonzero angular momentum.

4.1 Interatomic Potential

If we take a look at the shape of the potential shown in Fig. 4.1, we see that the hard core starts at a radius of about $r \approx 2.5\,\text{Å}$. At $r \approx 3\,\text{Å}$ there is a minimum of roughly $-10\,\text{K}$. Beyond this, the potential approaches zero with the van der Waals behavior r^{-6}. It is obvious that the hard core prevents the formation of s-wave bound states since the wave function must vanish at zero relative distance. There is, however, the possibility of bound states in nonzero angular momentum states. Let

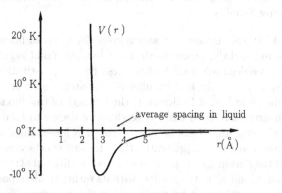

FIGURE 4.1 Interatomic potential between ^{3}He atoms as a function of the distance r.

us estimate its probable magnitude. As in superconductors, only the fermions close to the surface of the Fermi sphere in momentum space are capable of substantial interactions. They move with a momentum $p \approx p_F \approx 8 \times 10^{-19}$ g cm/sec. For angular momenta $l = 0, \hbar, 2\hbar, 3\hbar$, the impact parameter, i.e., the distance at which the particles pass one another, is of the order of $l/p_F \approx 0$, 1.25Å, 2.5 Å, 3.75 Å, ... With the repulsive core rising at $r \leq 2.5$ Å, it was estimated that the lowest partial wave having a chance of showing a bound state would be the d-wave. In fact, the first quantitative analyses indeed suggested that d-wave pairs do form a superfluid condensate, and a first extension of the entire BCS formalism was undertaken for this case [8].

The situation is, however, more complicated than in superconductors. There are strong many-body effects which have been neglected in these first considerations. The strong-coupling effects lead to a screening of the fundamental interatomic potential, so that the partial wave estimates had to be modified. Moreover, the hard core together with the Pauli exclusion principle generate strong spin-spin correlations. As a consequence, there is a pronounced resonance in the dynamic susceptibility (see Fig. 4.2), which is usually referred to as a *paramagnon excitation*. The exchange

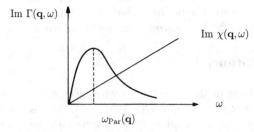

FIGURE 4.2 Imaginary part of the susceptibility caused by repeated exchange of spin fluctuations, as a function of energy ω. There is a pronounced peak whose sharpness increases with decreasing $\mathbf{q}$. Thus, for small $\mathbf{q}$, there are long-lived excitations in the system which are called *paramagnons*. The straight line shows the imaginary part of the susceptibility for a free Fermi system.

of these particle-like states between two atoms gives rise to an additional attraction between parallel spins, and this enhances the bound states of odd angular momenta.

It would clearly be desirable to calculate these effects quantitatively from first principles, i.e., from an n-body Hamiltonian of ^{3}He-atoms with the fundamental interaction $V(r)$ shown in Fig. 4.1. However, the strength of this interaction makes the calculation an extremely hard task. Therefore we decide to take the evidence from experiment showing that Cooper pairs form at a lower angular momentum than expected, namely at $l = 1$. Apparently, the screening effects weaken somewhat the hard core, and the paramagnons provide sufficient additional attraction between parallel spins to cause binding in the p-wave with its rather small impact parameter. By statistics, an $l = 1$ -state must be symmetric in the spin wave function, so that its total spin is necessarily $S = 1$ (spin triplet) [9, 10, 11].

Given the difficulties in calculating which orbital wave is the leading one in binding the Cooper pairs, also the estimates for the transition temperature T_c were initially quite inaccurate. Early estimates started as high as $T_c \approx 0.1$ K. They were lowered successively when a phase transition was not yet seen.

Experimentally, the transition was discovered in 1972 [7] at 2.7 mK, when cooling liquid ^{3}He down along the melting curve. A second transition was found at 2.1 mK. Since then, the superfluid phases of ^{3}He have attracted increasing experimental and theoretical attention. On the one hand, these pose the practical challenge of achieving and maintaining ultra-low temperatures. On the other hand, the observed phenomena show macroscopic system in anisotropic quantum statea passessing rather interesting collection excitations. Many surprising properties have been found and are probably waiting for their discovery. They form a beautiful field of applications with many theoretical methods still being developed for different branches of physics.

At the microscopic level, they are based on converting the fundamental action involving ^{3}He atoms to an alternative, equivalent form in which collective excitations can be studied most directly. This was done by generalizing the treatment of superconductors to superfluid ^{3}He.

4.2 Phase Diagram

The reason why measurements were first performed along the melting curve lies in the simplicity of the cooling technique and of the temperature control via the so-called *Pomeranchuk effect*. It is useful to keep in mind how temperatures in the milli-Kelvin range can be reached and maintained: First, the system is pre-cooled to roughly 77 K by working inside a Dewar container filled with liquid nitrogen. Embedded in this is another container filled with liquid ^{4}He which maintains, at atmospheric pressure, a temperature of 4 K. Enclosed in this lies a dilution refrigerator. This exploits the fact that liquid ^{3}He, when brought into contact with ^{4}He, forms a well-defined interface. Across it, diffusion takes place in the same way as in the evaporation process across a water surface. This lowers the temperature. The process can be made cyclic just like in an ordinary evaporation refrigerator. Temperatures of a few mK were easily reached. In the beginning, the dilution cooling was used only down to around 100 mK. From there on, the Pomeranchuk effect was exploited. This is based on the observation that according to the *Clausius-Clapeyron equation*,

$$\frac{dP}{dT} = \frac{S_{\text{liquid}} - S_{\text{solid}}}{V_{\text{liquid}} - V_{\text{solid}}}, \tag{4.1}$$

the temperature is lowered by increasing the pressure, since the entropy of the liquid becomes smaller than that of the solid in spite of its larger volume. Thus, in order to cool the system, one just has to compress it.

If one wants to measure the phase diagram away from the melting curve, adiabatic demagnetization may be used in addition to the Pomeranchuk effect. The best magnetic materials for this purpose are either CMN (cereous magnesium nitrate) or copper. In the first material, the magnetic moments of the electrons are demagnetized, in the second, the demagnetization is done on the nuclei. There for copper, temperatures of a few mK can be maintained for several days.

With such techniques, the phase diagram has been measured for very low pressures (see Fig. 4.3). The two phases originally discovered along the melting curve are

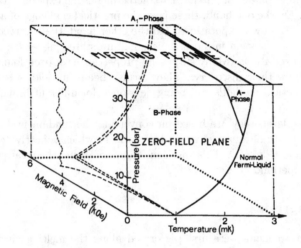

FIGURE 4.3 Phase diagram of ^{3}He plotted against temperature, pressure, and magnetic field H. At $H = 0$ there are two phases, A and B. For increasing magnetic fields, a widening phase A$_1$ develops. At pressures above the melting plane, the liquid solidifies.

called A and B. For large magnetic fields, there is another phase, called A$_1$, which forms between the A-phase and the normal liquid. In order to improve visibility in Fig. 4.3, we have exaggerated the corresponding temperature interval.

Many properties of these three phases have meanwhile been investigated experimentally, and they are all in best agreement with the theoretical description via p-wave spin triplet Cooper pairs.

While it was hard to predict the precise value of the transition temperature, the finally observed value $T_c = 2.7$ mK is in perfect scale with respect to that of low-temperature superconductors (see Table 4.1).

TABLE 4.1 A factor of roughly 1000 separates the characteristic length scales of superconductors and ^{3}He.

	T_F	mass	T_c
Superconductor	1000 K	1 m_{electron}	2.7 K
^{3}He	1 K	1000 m_{electron}	2.7 mK

4.3 Preparation of Functional Integral

4.3.1 Action of the System

It will be convenient to consider the system as a grand canonical ensemble in which the particle number can fluctuate while its average remains fixed (recall Subsection 1.1.7). Then, instead of the Hamiltonian H, the time evolution is driven by the grand-canonical energy $H_G = H - \mu N$ where H is the Hamiltonian, and N counts the particle number:

$$N = \int d^3x \, \psi^*(\mathbf{x}, t)\psi(\mathbf{x}, t). \tag{4.2}$$

The Lagrangian multiplyer μ is the chemical potential, as usual. Then the total field action reads for free fields

$$\mathcal{A}_0 = \int d^4x \, \psi^*(x) i\hbar \partial_t \psi(x) - \int dt \, (H_0 - \mu N), \tag{4.3}$$

where H_0 is the free Hamiltonian, so that

$$H_0 - \mu N = \int d^3x \, \psi^\dagger(x) \left(-\hbar^2 \frac{\nabla^2}{2m} - \mu \right) \psi(x), \tag{4.4}$$

The interaction contains the pairing term

$$\mathcal{A}_{\text{int}} \equiv -\frac{1}{2} \int d^4x d^4x' \, \psi^*(x)\psi^*(x')V(x', x)\psi(x')\psi(x). \tag{4.5}$$

As in the previous Chapter 3, we shall use a four-vector notation for spacetime. The italic symbol x indicates space and time with the four components $x^\mu \equiv (x^0, \mathbf{x})$. We shall also write $d^4x \equiv dt d^3\mathbf{x} = dt dx^1 dx^2 dx^3$. The potential $V(x, x')$ may be approximated by an instantaneous and time-independent function of the distance between $\mathbf{x}$ and $\mathbf{x}'$:

$$V(x', x) = \delta(t' - t)V(\mathbf{x}' - \mathbf{x}). \tag{4.6}$$

The dominant part of $V(\mathbf{x}' - \mathbf{x})$ consists in the van der Waals molecular potential thet was displayed in Fig. 4.1.

4.3.2 Dipole Interaction

In contrast to electrons in a superconductor, the ^{3}He-atoms are electrically neutral, so that there are no Coulomb forces at atomic distances. There is, however, a weak nuclear magnetic moment $\gamma \approx 2.04 \times 10^4$ (gauss sec)$^{-1}$ causing an additional small spin-spin dipole interaction

$$H_{\text{dd}} = \gamma^2 \int d^3x' d^3x \, v_{ab}(|\mathbf{x}' - \mathbf{x}|) \, \psi^*(\mathbf{x}', t)\frac{\sigma_a}{2}\psi(\mathbf{x}', t)\psi^*(\mathbf{x}, t)\frac{\sigma_b}{2}\psi(\mathbf{x}, t), \tag{4.7}$$

with the dipole potential

$$v_{ab}(|\mathbf{x}' - \mathbf{x}|) = -\partial_a \partial_b \frac{1}{r} = \left[\delta_{ab} - 3\frac{(x' - x)_a (x - x)_b}{|\mathbf{x}' - \mathbf{x}|^2} \right] \frac{1}{r^3}. \tag{4.8}$$

Due to its smallness, this interaction is negligible in the normal Fermi liquid. In the highly sensitive superfluid phase, however, it has interesting consequences, causing a variety of domain structures. There the Hamiltonian (4.4) with the dipole interaction H_{dd} is sufficient to explain quantitatively most of the properties of the normal and superfluid ^{3}He. As stated in the beginning of this chapter, the condensate of Cooper pairs is a very sensitive system. We shall see that many of its interesting phenomena are a direct manifestation of the very small dipole coupling (even though this is merely a *hyperfine* interaction).

4.3.3 Euclidean Action

As shown in the previous Chapter 3, the thermodynamic action governing the statistical mechanics of the fluid is obtained by analytic continuation to imaginary time τ:

$$-i\mathcal{A} \to \mathcal{A}^T = \int_0^{\hbar\beta} d\tau d^3x \, \psi^*(x)\hbar\partial_\tau\psi(x) + \int_0^{\hbar\beta} d\tau \, (H - \mu N + H_{dd}), \qquad (4.9)$$

with $\psi(x)$ in this Euclidean expression standing for $\psi(\mathbf{x}, \tau)$, and $\beta \equiv 1/k_BT$ as usual. In the partition function, the path integral extends over all fields $\psi(x) = \psi(\mathbf{x}, \tau)$ which are antiperiodic under the replacement $\tau \to \tau + \hbar\beta$:

$$\psi(\mathbf{x}, \tau) = -\psi(\mathbf{x}, \tau + \hbar\beta). \qquad (4.10)$$

Faced with the action (4.9) it appears, at first sight, quite hopeless to attempt any perturbative treatment. First of all, the potential $V(r)$ has a strongly repulsive core. Moreover, from the experimental density we can estimate the average distance between the atoms in the liquid to be about 3.5 Å, where the potential is still of considerable strength. The salvation from this difficulty is provided by Landau's observation that many features of this strongly interacting Fermi liquid will obey the same laws as in a free Fermi system:

1. The specific heat behaves like $C_V \sim T$.

2. The susceptibility behaves like $\chi \sim$ const.

3. The compressibility behaves, for small T but in the normal liquid, like $\kappa \sim$ const.

In fact, all free Fermi liquid laws for these quantities are valid, provided we replace the atomic mass $m_{^3\mathrm{He}}$ by an effective mass m_{eff} which is a few times larger than the true mass $m_{^3\mathrm{He}}$. The factor ranges from 3 to 6, depending on whether one works close to zero or close to the melting pressure ($\sim$ 35 bar). Apart from that, there is a simple multiplicative renormalization by a factor which can be attributed to molecular field effects, similar to what happens in Weiss' theory of ferromagnetism.

Landau's interpretation of this phenomenon is the following: By restricting one's attention to low-energy and momentum properties of a system, the strong-interaction problems simplify considerably. The rapid fluctuations cause an almost instantaneous re-adjustment of the particle distribution. For this reason, if slow and long-wavelength disturbances are applied to the system, several ^{3}He atoms that are in their mutual range of interaction will respond simultaneously as a cluster, called *quasiparticle*, with an effective mass larger than the atomic mass. The residual interaction between these quasiparticles is very smooth and weak since any potential hole, which could appear as a result of a small displacement in the liquid, is immediately filled up and screened away by a rapid redistribution of the atoms. It is this screening effect, mentioned in the introduction, which makes quantitative calculations managable at least at the level of quasiparticles. Apparently, the fast fluctuations generate a new effective action of approximately the same form as (4.3), except that $\psi(x)$ has to be read as a quasiparticle field, m as the effective mass m_{eff} which is a few times larger than the true mass $m_{^3\text{He}}$.

The potential $V(x - x')$ in (4.5) is the residual effective potential between the quasiparticles. The energy range of integrations in the Fourier decomposition of the fields is, however, limited to some cutoff frequency ω_{cutoff} beyond which the effective action becomes invalid. Using the path integral formulation of the partition function we shall quite easily be able to rewrite the fundamental expression $\mathcal{A}$ in terms of quasiparticle fields, at least in principle.

4.3.4 From Particles to Quasiparticles

It was argued that fluctuations cause a significant screening of the potential. The screened lumps of particles move almost freely but with a larger effective mass. In order to formulate this situation we first need a precise distinction between fast and slow fluctuations. For this we expand the field in a Fourier series

$$\psi(\mathbf{x}, t) = \frac{1}{\sqrt{(t_b - t_a)V}} \sum_{\omega_n, \mathbf{k}} e^{-i\omega_n t + i\mathbf{k}\mathbf{x}}, \tag{4.11}$$

where V is the spatial volume of the system and ω_n are the Matsubara frequencies

$$\omega_n \equiv \frac{2\pi (n + 1/2)}{t_b - t_a}, \tag{4.12}$$

which enforce the anti-periodic boundary condition (4.10). Apparently, there are natural energy and momentum scales ω_s and k_s, so that a separation of the field into slow and long-wavelength and fast and short-wavelength makes sense:

$$\psi(\mathbf{x}, t) \equiv \psi_s(\mathbf{x}, t) + \psi_h(\mathbf{x}, t) \tag{4.13}$$

$$= \frac{1}{\sqrt{(t_b - t_a)V}} \left(\sum_{\substack{|\omega_n| < \omega_s \\ |\mathbf{k}| < k_s}} e^{-i\omega_n t + i\mathbf{k}x} \psi(\omega_n, \mathbf{k}) + \sum_{\substack{|\omega_n| \geq \omega_s \\ |\mathbf{k}| > k_s}} e^{-i\omega_n t + i\mathbf{k}x} \psi(\omega_n, \mathbf{k}) \right).$$

This can be used to simplify the path integral. The two terms are referred to as *soft* and *hard components* of the field $\psi(\mathbf{x}, t)$. When written in energy momentum space, the functional integral measure may be separated accordingly:

$$\int \mathcal{D}\psi \mathcal{D}\psi^* = \int \mathcal{D}\psi_s \mathcal{D}\psi_s^* \int \mathcal{D}\psi_h \mathcal{D}\psi_h^*$$

$$\equiv \prod_{|\omega|<\omega_s, |\mathbf{k}|<k_s} \frac{d\psi(\omega_n, \mathbf{k}) d\psi^*(\omega_n, \mathbf{k})}{2\pi i} \prod_{|\omega|\geq\omega_s, |\mathbf{k}|\geq k_s} \frac{d\psi(\omega_n, \mathbf{k}) d\psi^*(\omega_n, \mathbf{k}))}{2\pi i}. \qquad (4.14)$$

If we now perform the path integral over the hard components we remain with a partition function

$$Z = \int \mathcal{D}\psi_s \mathcal{D}\psi_s^* e^{i\mathcal{A}_s[\psi_s^*, \psi_s]}, \qquad (4.15)$$

where $\mathcal{A}_s[\psi_s^*, \psi_s]$ is a functional of only the soft components. The point of Landau's argument is now that due to the high quality of the free Fermi gas laws there seems to exist an optimal choice for ω_s and k_s so that the action looks like the action of the initial ^{3}He particles, except that the new fields $\psi_s(\mathbf{x}, t)$ have a larger effective mass m^* and that the interactions are much weaker than in the original fundamental form (4.3).

Certainly, the actual calculation of the path integral over the fast components is extremely difficult due to the strength of the interactions. We shall therefore accept Landau's argument on phenomenological grounds and see its justification in the successful derivation of the physical properties of the liquid.

At first sight, the precise choice of ω_s and k_s seems to be a rather ad hoc matter and one might fear that all results derived from the partition function (4.14) depend strongly on which values are taken. It is gratifying to note, however, that this is not really true. Only the prediction as to the size of the transition temperature T_c varies strongly with ω_s, k_s. But in all final results ω_s, k_s can be eliminated in favor of the observable temperature T_c. In this way, any arbitrariness is removed. This is completely analogous to the independence of all physical amplitudes on the cutoff in renormalizable field theory.

For the phenomena of superfluidity, the optimal choice of ω_s, k_s will be so that ω_s is about ten times larger than the transition temperature T_c while k_s comprises approximately ten atomic distances (i.e., $k_s \approx 2\pi/10\text{Å}$). In this way quasiparticle fields are well enough localized in space and time to describe excitations with frequencies between zero and one MHz, corresponding roughly to $1\, k_B T_c$ in ^{3}He, and wavelengths of up to about 100 Å.

4.3.5 Approximate Quasiparticle Action

We are thus confronted with a simplified problem of calculating the partition function over soft field components ψ_s. For brevity, the subscripts will be dropped. The soft field *quanta* are precisely what Landau introduced as quasiparticles. Since we are not able to calculate $\mathcal{A}_s$ explicitly, we have to deduce its structure from experimental facts. As argued above, the action must account for the free-particle-like behavior of the specific heat and of the susceptibility with a characteristic transition temperature T_c modified by a simple renormalization factor.

TABLE 4.2 Pressure dependence of Landau parameters F_1, F_0, and F_0^S of ^{3}He, together with the molar volume v and the effective mass ratio m^*/m. The values of v, m^*/m and F_1 are from Greywall [12], whereas F_0, F_0^S are from Wheatley [13], except for corrections using more recent values of m^*/m. The values at $P = 34.39$ bar are from Wheatley [13].

$P(\text{bar})$	$v(\text{cm}^3)$	m^*/m	F_1	F_0	F_0^S
0	36.84	2.80	5.39	9.30	−0.695
3	33.95	3.16	6.49	15.99	−0.723
6	32.03	3.48	7.45	22.49	−0.733
9	30.71	3.77	8.32	29.00	−0.742
12	29.71	4.03	9.09	35.42	−0.747
15	28.89	4.28	9.85	41.73	−0.753
18	28.18	4.53	10.60	48.46	−0.757
21	27.55	4.78	11.34	55.20	−0.755
24	27.01	5.02	12.07	62.16	−0.756
27	26.56	5.26	12.79	69.43	−0.755
30	26.17	5.50	13.50	77.02	−0.754
33	25.75	5.74	14.21	84.79	−0.755
34.39	25.50	5.85	14.56	88.47	−0.753

Let us briefly take a look at the experimental situation: For a free Fermi gas with g spin states in a volume V, the total density of states per unit energy at the Fermi surface for the two spin-1/2 states is $2\mathcal{N}(0)$, and

$$\mathcal{N}(0) = \frac{m p_F}{2\pi^2\hbar^3} = \frac{3}{4\hbar^3}\frac{\rho}{p_F^2} \qquad (4.16)$$

is the density of states at the surface of the Fermi sea [recall Eq. (3.62)], where $\rho = MN/V$ is the particle density. As before in Eq. (3.64), the quantity

$$p_F = \left(3\pi^2\right)^{1/3}\left(\frac{N}{V}\right)^{1/3}\hbar \approx g \times 10^{-20}\text{g cm/sec} \qquad (4.17)$$

is the Fermi momentum of free spin-1/2 particles. The associated Fermi velocity $v_F \equiv p_F/m$ varies from $5.5 \cdot 10^3$ cm/sec at zero pressure to about $3 \cdot 10^3$ cm/sec at melting pressure (see Table 4.2 above).

From this it is easy to calculate the three quantities *specific heat*, *magnetic susceptibility*, and *compressibility*:[1]

$$C_V = N\frac{m p_F}{3\hbar^3}k_B T, \qquad (4.18)$$

---

[1]The standard derivation is omitted at this point since it will appear anyhow later. Note that we keep the physical constants $\hbar$ and k_B explicit in these formulas.

$$\chi_N = \frac{\gamma^2}{4}\frac{mp_F}{\pi^2}\frac{1}{\hbar}, \tag{4.19}$$

$$\kappa_T = \frac{m}{\rho^2}\frac{p_F}{\pi^2\hbar^3}. \tag{4.20}$$

The parameter $\gamma = 2\mu/\hbar$ is the gyromagnetic ratio of the ^{3}He nucleus, and μ is the nuclear magnetic moment.

Experimentally one finds the linear behavior of c_V below 20 mK to be enhanced by a factor 6 to 14 for pressures ranging from atmospheric to 35 bar (melting pressure). This enhancement may be attributed to a change in the effective mass from m to a larger value m^*. The effective mass is defined so that the energy of quasiparticles starts for small momentum as $\mathbf{p}^2/2m^* + \mathcal{O}(\mathbf{p}^4)$. Let $\Sigma(\omega, \mathbf{p})$ be the self-energy of the fermions after all interactions are taken into account. Then the fermion Green may be written as [generalizing (1.95)]

$$G(\mathbf{x}, t; \mathbf{x}', t') = -\int \frac{d\omega}{2\pi}\frac{d^3p}{(2\pi)^3}e^{\omega(t-t')+i\mathbf{p}(\mathbf{x}-\mathbf{x}')}\frac{1}{i\omega - \xi^*(\omega, \mathbf{p})}. \tag{4.21}$$

where

$$\xi^*(\omega, \mathbf{p}) \equiv \xi(\mathbf{p}) + \Sigma(\omega, \mathbf{p}). \tag{4.22}$$

Expanding this in powers of ω and $\mathbf{p}^2$, the chemical potential is shifted from μ to the renormalized value $\mu + \Sigma(0, \mathbf{0})$, and the effective mass is found from the equation

$$\frac{m^*}{m} = \frac{1 - i\partial_{i\omega}\xi^*(\omega, \mathbf{p})}{2m\,\partial\xi^*(\omega, \mathbf{p})/\partial\mathbf{p}^2}\bigg|_{\omega=0, \mathbf{p}^2=p_F^2} = \frac{1 - i\partial_{i\omega}\Sigma(\omega, \mathbf{p})}{1 + 2m\,\partial\Sigma(\omega, \mathbf{p})/\partial\mathbf{p}^2}\bigg|_{\omega=0, \mathbf{p}^2=p_F^2}. \tag{4.23}$$

It is customary to introduce the parameter F_1 so that this equation reads

$$\frac{m^*}{m} = 1 + \frac{F_1}{3}. \tag{4.24}$$

The precise values of F_1 can be seen in Table 4.2.

The spin susceptibility is found to be independent of temperature below 40 mK. If one inserts the effective mass m^* into the free-fermion formula (4.19) one finds a value about four times too small. This is attributed to molecular field effects. If the atomic magnetic moments are partially oriented, the magnetic field seen by an individual atom consists of the external field plus that of the other moments in the liquid. The enhancement factor is usually denoted as

$$\frac{1}{1 + F_0^S} \equiv \frac{1}{1 + Z_0/4}, \tag{4.25}$$

with $F_0^S \equiv Z_0/4$ being roughly -3 up to the melting pressure of ≈ 35 bar (see Table 4.2).

The compressibility κ_T, finally, is determined my measuring the velocity of sound. Inserting m^* into Eq. (4.20), we obtain it from

$$c = \frac{1}{\sqrt{\rho \kappa_T}} = \frac{v_F}{\sqrt{3}} \left(1 + \frac{F_1}{3}\right)^{1/2},$$ (4.26)

where

$$v_F = \frac{p_F}{m^*}$$ (4.27)

is the Fermi velocity for the effective mass m^*, which ranges from 5 to 3×10^3 cm/sec (see Table 4.2). Experimentally, formula (4.26) turns out to fail by a factor 3 to 10, a failure which is again attributed to molecular field effects upon the density field ρ. Thereby the sound velocity is multiplied by a correction factor $1/(1 + F_0)$, and becomes

$$c = \frac{v_F}{\sqrt{3}} \left[\left(1 + \frac{F_1}{3}\right)(1 + F_0)\right]^{1/2},$$ (4.28)

with F_0 ranging from 10 to 100 (Table 4.2). Thus the compressibiliy $\kappa_T = 1/c^2\rho$ is modified by a factor $1/(1 + F_0)$.

4.3.6 Effective Interaction

What an action can we set up to explain these low-frequency and small-momentum features of liquid ^{3}He observed in a wide range above the superfluid transition temperature? It appears simple to include the effective mass. All we have to do is choose a free-particle Hamiltonian

$$H_0 = \int d^3x \; \psi^*(x)\left(i\partial_t + \frac{\boldsymbol{\nabla}^2}{2m^*}\right)\psi(x),$$ (4.29)

where we have used natural units with $\hbar = 1$, $k_B = 1$. This naive approximation would indeed lead to the specific heat (4.18) with the mass m replaced by m^*, provided the number of quasiparticles is taken to be equal to the true particle number, so that also the Fermi momentum p_F. Recall that according to Eq. (4.17), this depends simply on the particle density N/V.

If one would set the system into motion by displacing all particle velocities

$$\mathbf{v} = \frac{\mathbf{P}}{m^*}$$ (4.30)

by a certain amount $\Delta\mathbf{v}$, the total momentum $\mathbf{P}$ of the system would change by

$$\Delta\mathbf{P} = \Delta\mathbf{v}\, Nm^*$$ (4.31)

rather than by the bare expression

$$\Delta\mathbf{P} = \Delta\mathbf{v}\, Nm.$$ (4.32)

This can only be corrected by introducing an additional interaction which, however, must not modify the previous calculation of the specific heat. Such interactions are well known in molecular field theories. We simply add to the free Hamiltonian a current interaction

$$H_{\text{curr}-\text{curr}} = \frac{1}{2\rho^*}\frac{F_1}{3}\int d^3x\,\psi^*(x)\frac{i}{2}\overset{\leftrightarrow}{\nabla}\psi(x)\psi^*(x)\frac{i}{2}\overset{\leftrightarrow}{\nabla}\psi(x), \qquad (4.33)$$

where $\overset{\leftrightarrow}{\nabla} \equiv \overset{\rightarrow}{\nabla} - \overset{\leftarrow}{\nabla}$ is the right-minus-left derivative, the constant F_1 denotes the coupling strength, and

$$\rho^* = \frac{m^*N}{V} \qquad (4.34)$$

is the mass density of quasi-particles. Then the kinematic properties of single quasi-particle states are automatically correct. Indeed, such a state has an energy

$$E = \frac{p^2}{2m^*}\left(1 + \frac{F_1}{3}\right), \qquad (4.35)$$

so that the velocity is

$$\mathbf{v} = \frac{\partial E}{\partial \mathbf{p}} = \frac{\mathbf{p}}{m^*}\left(1 + \frac{F_1}{3}\right), \qquad (4.36)$$

and the total momentum changes, upon a change $\Delta\mathbf{v}$ in the velocity, by

$$\Delta\mathbf{P} = \frac{\Delta\mathbf{v}Nm^*}{1 + F_1/3}, \qquad (4.37)$$

as it should, due to Eq. (4.32), if we use the relation (4.24).

The renormalization factors for susceptibility and compressibility have to be inferred in a similar manner.

It is nontrivial to see that the interaction (4.33) really leaves the specific heat in the form (4.18), only that m is replaced by m^*. When going from one Galilean frame of reference to another one that moves with velocity v, the energy changes by

$$\Delta H_v = -\int d^3x\,\psi^*(x)\frac{\overset{\leftrightarrow}{\nabla}}{2}\psi(x)\,\Delta\mathbf{v}. \qquad (4.38)$$

When turning on a magnetic field, the energy changes by

$$\Delta H_H = \int d^3x\,\psi^*(x)\frac{\sigma^a}{2}\psi(x)\,\gamma H_a, \qquad (4.39)$$

due to the interaction with the spin magnetic moments.

Finally, if a chemical potential is introduced by contact with a particle reservoir, the energy is modified by

$$\Delta H_\mu = -\int d^3x\,\psi^*(x)\psi(x)\,\mu. \qquad (4.40)$$

Thus, the current density

$$\mathbf{j} \equiv \frac{1}{2}\psi^* \overset{\leftrightarrow}{\nabla}\psi, \tag{4.41}$$

the spin density

$$s^a \equiv \psi^* \frac{\sigma^a}{2}\psi, \tag{4.42}$$

and the particle density

$$n \equiv \psi^*\psi, \tag{4.43}$$

all appear on the same footing.

We have seen that the quadratic current density coupling brings changes in the kinetic energy to the correct form

$$\frac{1}{m^*}\mathbf{p}d\mathbf{p} \to \frac{1}{m^*}\left(1 + \frac{F_1}{3}\right)\mathbf{p}d\mathbf{p} = \frac{\mathbf{p}}{m}d\mathbf{p}. \tag{4.44}$$

Thus we expect quadratic spin density and particle density couplings

$$H_{dd} = \frac{1}{2}\frac{F_0}{\rho^*}\int d^3x\, \psi^*(x)\psi(x)\psi^*(x)\psi(x), \tag{4.45}$$

$$H_{sd} = \frac{1}{2}\frac{F_1^S}{\rho^*}\int d^3x\, \psi^*(x)\frac{\sigma}{2}\psi(x)\psi^*(x)\frac{\sigma}{2}\psi(x), \tag{4.46}$$

to produce corresponding correction factors for changes in the magnetic and chemical energy density

$$\chi\, HdH \to \chi\left(1 + F_0\right)HdH, \tag{4.47}$$

$$\kappa\, \mu d\mu \to \kappa\left(1 + F_0^S\right)\mu d\mu. \tag{4.48}$$

These are needed to obtain agreement with experiment.

The above simple couplings are just the leading terms in the more complete multipole expansion

$$H_{int} = \frac{1}{2\rho^*}\sum_{l=0}^{\infty}\int d^3x \frac{F_l}{2l+1}\psi^*(x)\partial_{lm}\psi(x)\psi^*(x)\partial_{lm}\psi(x)$$

$$+ \frac{1}{2\rho^*}\sum_{l=0}^{\infty}\int d^3x \frac{F_l^S}{2l+1}\psi^*(x)\frac{\sigma^a}{2}\partial_{lm}\psi^*(x)\frac{\sigma^a}{\sqrt{2}}\partial_{lm}\psi(x). \tag{4.49}$$

The parameters F_l can depend also on the momentum transfer, i.e., the momentum of the composite field which is given by $-i^l$ times the spherical derivative of the angular momentum l with z-component m, to be denoted by $\psi^*\partial_{lm}\psi$. Such a dependence characterizes the form factor of the quasiparticles. The spherical derivative ∂_{lm} is a short notation for the product of l spatial derivatives which are combined to be traceless. In this way one projects out a definite angular momentum, for instance

$$\partial_{2m} \propto \partial_i\partial_j - \frac{1}{3}\delta_{ij}\partial^2, \tag{4.50}$$

$$\partial_{3m} \propto \partial_i\partial_j\partial_k - \frac{1}{5}\left(\delta_{ij}\partial^2\partial_k + 2 \text{ cyclic permutations}\right). \tag{4.51}$$

We shall choose the proportionality factor to comply with the spherical momentum definition in terms of spherical harmonics $Y_{lm}(\hat{\mathbf{k}})$:

$$k_{lm} = \sqrt{\frac{4\pi}{2l+1}} Y_{lm}(\hat{\mathbf{k}}) |\mathbf{k}|^l. \tag{4.52}$$

Since the labels m refer to a spherical basis, we must distinguish k_{lm} and $k_{lm}^* = (-1)^m k_{l-m}$. If one wants to form rotationally invariant objects, the labels must be contracted accordingly, for instance $\partial_{lm}^* \partial_{lm} = (-1)^m \partial_{lm} \partial_{l-m}$.

It will turn out that many phenomena depend only on the values of F_l at zero momentum transfer. Moreover, only the parameters which appear in Eq. (4.28) are easily accessible to experimental measurement.

4.3.7   Pairing Interaction

With the couplings introduced so far, the properties of the degenerate Fermi liquid can be explained within very simple approximations as long as the temperature is above the critical value T_c. As explained in the introduction, the superfluid properties below T_c require the formation of p-wave spin triplet Cooper pairs. This can only happen due to an additional attractive interaction which must consist of one screened version of the original potential V. Its accurate shape is unknown. This, however, turns out to be no handicap. The reason is the following: The attractive force is extremely weak. Therefore the Cooper pairs are only barely bound, as manifested by the fact that the critical temperature T_c is much smaller than the characteristic temperature unit of the system which is $T_F = p_F^2/2m$ (which is the Fermi energy of the system of the order of ≈ 1 K). This makes the radius of the bound-state wave functions much larger than $1/p_F \approx 1\,\text{Å}$. Its size will turn out to be a few hundred Å. For this reason, it does not matter what the detailed shape of $V(\mathbf{x}' - \mathbf{x})$ in Eq. (4.6) really is, and it can be chosen to be point-like with a range of a few Å. That has only a single bound state. It must only bind in a p-wave spin triplet state, we may directly write

$$H_{\text{pair}} = -\frac{3g}{4p_F^2} \int d^3x \, \psi^*(x) \frac{\sigma^a}{2} c^\dagger \overset{\leftrightarrow}{\nabla} \psi^*(x) \, \psi(x) c \frac{\sigma^a}{2} \overset{\leftrightarrow}{\nabla} \psi(x). \tag{4.53}$$

The matrix c is

$$c = i\sigma^2 = \begin{pmatrix} 0 & 1 \\ -1 & 0 \end{pmatrix}, \tag{4.54}$$

which ensures that $\psi c \sigma_a \psi$ transforms in the same way as $\psi^* \sigma^a \psi$, i.e., like a vector, due to the equivalence of the 2×2 rotation matrix U to its complex conjugate by $U^* = cUc^{-1}$.

4.4 Transformation from Fundamental to Collective Fields

While fundamental fields provide the theoretically most satisfactory way of *defining* the action of a theory, they are quite ineconomic as far as the description of low-energy and long-wavelength phenomena of systems like ^{3}He and superconductors is concerned. The reason is basically the following: Below the transition temperature T_c at which the superfluid phase arises, the binding between the fundamental particles in Cooper pairs results in an energy gap Δ of the single particle spectrum. This becomes

$$E(\mathbf{p}) = \sqrt{\xi^2(\mathbf{p}) + \Delta^2}. \tag{4.55}$$

For ^{3}He, the size of the gap is of the order of mK, while for most superconductors Δ lies in the K-regime. As a consequence, the propagator

$$\langle 0|T(\psi(x)\psi^*(y))|0\rangle \tag{4.56}$$

has no singularities in the energy plane below $E = \Delta$. A description of the rich set of physical phenomena with energies much smaller than Δ^2 such as zero-sound waves, spin waves etc. is quite complicated when employing the fundamental field $\psi(x)$. An infinite set of Feynman graphs is necessary even for a lowest order understanding of these phenomena. On the other hand, there are Green functions which directly display excitations of this type in the complex energy plane, for example those of the composite field operators

$$\langle 0|T(\psi^*(x)\psi(x)\psi^*(y)\psi(y))|0\rangle, \tag{4.57}$$

$$\langle 0|T\left(\psi^*(x)\frac{\sigma_a}{2}\psi(x)\psi^*(y)\frac{\sigma_b}{2}\psi(y)\right)|0\rangle. \tag{4.58}$$

Singularities which appear in such composite Green functions but not in (4.56) are called *collective excitations*. One may expect that the most economic description of the associated physical phenomena can be obtained by first transforming the full theory to the appropriate composite fields. Such transformations have, in fact, been studied long time ago in many-body theory at the quasiclassical level. For superconductors [14] and ^{3}He, the result is the so-called Ginzburg-Landau equation [15]. This equation has been extremely successful in explaining many low-energy properties of the system. The approximate methods leading to this equation have been described in general in Chapter 1. They have been applied to plasmons in Chapter 2 and to superconductors in Chapter 3. Following this method we add here to the sum of a free action (4.3) and the pair interaction (4.53):

$$\mathcal{A}_{\text{int}} = \int dt \, H_{\text{pair}} \tag{4.59}$$

---

[2]These will often be called "infrared" phenomena, for brevity.

a complete square involving an auxiliary collective field $A_{ai}(x)$:

$$\Delta \mathcal{A} = -\frac{1}{3g^2} \int d^4x \left| A_{ai}(x) - \frac{3g}{2p_F} \psi_i(x) \overset{\leftrightarrow}{\nabla}_i c \frac{\sigma_a}{2} \psi(x) \right|^2 . \tag{4.60}$$

This does not change the theory. Up to this point, $A_{ai}(x)$ is a nondynamical field, since its time derivative $\partial_t A_{ai}(x)$ does not appear in the action. Such a field can be eliminated from the action by solving the Euler-Lagrange equation $\delta \mathcal{A}/\delta A_{ai}(x) = 0$ which yields the relation

$$A_{ai}(x) = A_{ai}^{\psi\psi}(x) \equiv \frac{3g}{2p_F} \psi(x) i \overset{\leftrightarrow}{\nabla}_i c \frac{\sigma_a}{2} \psi(x). \tag{4.61}$$

At the classical level, A_{ai} coincides with the composite field of a pair of ^{3}He atoms in a p-wave spin triplet configuration. Since it serves to describe the collective phenomena it will, from now on, be called the *collective pair field* of liquid ^{3}He. Reinserting (4.61) into (4.60) gives $\Delta \mathcal{A} = 0$ so that, at the classical level, the addition of $\Delta \mathcal{A}$ really leaves the action unchanged.

As before in Chapter 3 for the case of superconductors, this remains true at the full quantum level. By analogy with that chapter, we consider the partition function of the theory

$$Z = \int \mathcal{D}\psi^* \mathcal{D}\psi \mathcal{D}A_a^* \mathcal{D}A_{ai} e^{i(\mathcal{A}_0 + \mathcal{A}_{\text{int}} + \Delta \mathcal{A})}. \tag{4.62}$$

The integral over the auxiliary field $\mathcal{D}A_{ai}$ is of the Gaussian type. It peaks for each spacetime point x when A_{ai} is equal to $A_{ai}^{\psi\psi}(x)$. Since the integral runs at each point from $-\infty$ to $+\infty$, the finite shift is irrelevant and the integral renders the same irrelevant constant for each x. The merit of choosing (4.60) for $\Delta \mathcal{A}$ lies in its eliminating the fourth-order term in the action in (4.62), so that the combination

$$\mathcal{A} = \mathcal{A}_0 + \mathcal{A}_{\text{int}} + \Delta \mathcal{A} \tag{4.63}$$

$$= \int d^4x \left\{ \psi^*(x) \left[i\partial_t - \xi(-i\boldsymbol{\nabla}) \right] \psi(x) + \left(A_{ai}^*(x) \psi i \tilde{\nabla}_i c \frac{\sigma_a}{2} \psi + \text{c.c.} \right) - \frac{1}{3g} A_{ai}^* A_{ai} \right\}$$

is quadratic in the fields $\psi(x)$. For this reason, the functional integral $\int \mathcal{D}\psi^* \mathcal{D}\psi$ can be performed in (4.62), and the result is a quantum field theory formulated entirely in terms of the pair field A_{ai}. In (4.63) we have gone over to a dimensionless right-minus-left derivative

$$\tilde{\boldsymbol{\nabla}} \equiv \frac{1}{2p_F} \overset{\leftrightarrow}{\boldsymbol{\nabla}}, \tag{4.64}$$

for convenience.

We now bring the path integral over Fermi fields to the standard form by rewriting the action as in Eq. (3.405), with the help of the four-component field which combines the $\psi(x)$ and $\psi^*(x)$ components into a single quadruplet

$$f = \begin{pmatrix} \psi \\ c\psi^* \end{pmatrix}. \tag{4.65}$$

Then (4.63) can be rewritten as

$$\mathcal{A} = \int d^4x \left[\frac{1}{2} f^*(x) \begin{pmatrix} i\partial_t - \xi(-i\boldsymbol{\nabla}) & i\tilde{\boldsymbol{\nabla}}_i \sigma_a A_{ai} \\ i\tilde{\boldsymbol{\nabla}}_i \sigma_a A_{ai}^* & i\partial_t + \xi(i\boldsymbol{\nabla}) \end{pmatrix} f(x) - \frac{1}{3g} A_{ai}^* A_{ai} \right]. \quad (4.66)$$

The derivatives $\tilde{\boldsymbol{\nabla}}_i$ are meant to act only on $f^*(x)$, $f(x)$ but not on the collective field $A_{ai}(x)$. Performing now the functional integral $\int \mathcal{D} f^* \mathcal{D} f$ with the help of the fermionic Gaussian functional formula (1.80), written as

$$\int \mathcal{D} f^* \mathcal{D} f e^{i\frac{1}{2} f^* M f} = e^{\frac{1}{2} \mathrm{Tr} \log M}, \quad (4.67)$$

we obtain

$$Z = \int \mathcal{D} A_{ai}^* \mathcal{D} A_{ai} e^{i\mathcal{A}_{\mathrm{coll}}[A^*,\, A]}. \quad (4.68)$$

The exponent contains the collective action

$$\mathcal{A}_{\mathrm{coll}}[A^*, A] = -\frac{i}{2} \mathrm{Tr} \log \begin{pmatrix} i\partial_t - \xi(-i\boldsymbol{\nabla}) & i\tilde{\boldsymbol{\nabla}}_i \sigma_a A_{ai} \\ i\tilde{\boldsymbol{\nabla}}_i \sigma_a A_{ai}^* & i\partial_t + \xi(i\boldsymbol{\nabla}_i) \end{pmatrix} - \frac{1}{3g} \int d^4x A_{ai}^*(x) A_{ai}(x). \quad (4.69)$$

The functional integral (4.68) over the fluctuating A_{ai}-field promotes this field from a collective classical field to a *collective quantum field* [17].

The Trace log part is treated as in the case of a superconductor [recall (3.9)–(3.19)]. It is rewritten as

$$-\frac{i}{2} \mathrm{Tr} \log \begin{pmatrix} i\partial_t - \xi(-i\boldsymbol{\nabla}) & 0 \\ 0 & i\partial_t + \xi(i\boldsymbol{\nabla}) \end{pmatrix} \quad (4.70)$$

$$-\frac{i}{2} \mathrm{Tr} \log \left[1 - i \begin{pmatrix} \dfrac{i}{i\partial_t - \xi(-i\boldsymbol{\nabla})} & 0 \\ 0 & \dfrac{i}{i\partial_t + \xi(i\boldsymbol{\nabla})} \end{pmatrix} \begin{pmatrix} 0 & i\tilde{\boldsymbol{\nabla}}_i \sigma_a A_{ai} \\ i\tilde{\boldsymbol{\nabla}}_i \sigma_a A_{ai}^* & 0 \end{pmatrix} \right],$$

and the first term can immediately be calculated, yielding the lowest contribution to $i\mathcal{A}_{\mathrm{coll}}[A^*, A]/\hbar$:

$$\mathrm{Tr} \log \left[(i\partial_t - \xi)\delta_{\alpha\beta} \right] = 2 \int \frac{d^3 p}{(2\pi)^3} \log \left[1 + e^{-\xi(\mathbf{p})/T} \right] \equiv -F_0/T, \quad (4.71)$$

where F_0 is the free energy of a free fermion system.

The second term can be expanded in powers of A_{ai} as follows:

$$i \sum_{n=1}^{\infty} \frac{(-i)^{2n}}{2n} \mathrm{Tr} \left[\frac{i}{i\partial_t - \xi(-i\boldsymbol{\nabla})} i\overset{\leftrightarrow}{\boldsymbol{\nabla}}_i \sigma_a A_{ai} \frac{i}{i\partial_t + \xi(i\boldsymbol{\nabla})} i\overset{\leftrightarrow}{\boldsymbol{\nabla}}_j \sigma_a A_{bj}^* \right]^n. \quad (4.72)$$

The lowest terms of this expansion correspond to the loop diagrams shown in Fig. 3.2. The free part of the collective action is given by the term in (4.72) with

$n = 1$, plus the last term in (4.69). By Fourer-decomposing the field in space and imaginary time

$$A_{ai}(x) = \frac{\cdot\; 1}{\sqrt{V/T}} \sum_k A_{ai}(k) e^{-ikx} \equiv \frac{1}{\sqrt{V/T}} \sum_{\omega_n, \mathbf{k}} A_{ai}(\omega_n, \mathbf{k}) e^{-i(\omega_n \tau - \mathbf{k}\mathbf{x})}, \qquad (4.73)$$

we find the action for static collective fields in which all $A_{ai}(\omega_n, \mathbf{k})$ with $\omega_n \neq 0$ are equal to zero, and only the $\omega_n = 0$ component $A_{ai}(0, \mathbf{k}) \equiv A_{ai}(\mathbf{k})$ is present,

$$\mathcal{A}_0[A^*, A] \approx \frac{\mathcal{N}(0)}{3} \int dt \sum_{\mathbf{k}} A_{ai}^*(\mathbf{k}) \left[\left(1 - \frac{T}{T_c} \right) \delta^{ij} - \frac{3}{5} \xi_0^2 \left(\mathbf{k}^2 \delta^{ij} + 2k^i k^i \right) \right] A_{aj}(\mathbf{k}). \quad (4.74)$$

Here

$$\xi_0 = \sqrt{\frac{7\zeta(3)}{48\pi^2}} \frac{v_F}{T_c} \approx 120\text{Å} \qquad (4.75)$$

is the basic coherence length of the superfluid[3], and T_c is the critical temperature. Its value is obtained by solving the gap equation [compare (3.65)–(3.71)]

$$\begin{aligned} 0 &= \int \frac{d^3p}{(2\pi)^3} \frac{1}{2\xi(\mathbf{p})} \tanh \frac{\xi(\mathbf{p})}{2T} - \frac{1}{g} \approx \mathcal{N}(0) \int_{-\omega_{\text{cutoff}}}^{\omega_{\text{cutoff}}} \frac{d\xi}{2\xi} \tanh \frac{\xi}{2T} - \frac{1}{g} \\ &= \mathcal{N}(0) \log \left(2 \frac{2^\gamma}{\pi} \frac{\omega_{\text{cutoff}}}{T} \right) - \frac{1}{g}, \end{aligned} \qquad (4.76)$$

where $\mathcal{N}(0)$ is the density of states at the surface of the Fermi sea in Eq. (3.62). The critical temperature is therefore determined by the equation [compare (3.72)]

$$T_c \equiv \omega_{\text{cutoff}} \, 2 \frac{e^\gamma}{\pi} e^{-1/g\mathcal{N}(0)}. \qquad (4.77)$$

Close to T_c, the right-hand side of Eq. (4.76) is approximately equal to $\mathcal{N}(0)\,(1 - T/T_c)$, and this leads to the first term of Eq. (4.74).

The lowest-order interaction in the collective action is fourth order in the $A_{ai}(x)$-fields. It becomes in the static case for long-wavelengths

$$\mathcal{A}_{\text{int}}[A^*, A] = -\int d^4x \left[\; \beta_1 A_{ai}^* A_{bj} A_{ai}^* A_{bj} + \beta_2 \left(A_{ai}^* A_{ai} \right)^2 \right. \qquad (4.78)$$

$$\left. + \beta_3 A_{ai}^* A_{aj} A_{bi}^* A_{bj} + \beta_4 A_{ai}^* A_{bi} A_{bj}^* A_{aj} + \beta_5 A_{ai}^* A_{bi} A_{aj}^* A_{bj} \right].$$

The coefficients β_i are found from the loop integral for $n = 2$ in the same way as in the case of the superconductor [recall (3.110)]:[4]

$$-2\beta_1 = \beta_2 = \beta_3 = \beta_4 = -\beta_5 = \frac{2}{5} \mathcal{N}(0) \frac{\xi_0^2}{v_F^2 \hbar^2} = \frac{3}{5} \frac{\rho}{m^2} \frac{\xi_0^2}{v_F^4 \hbar^2}. \qquad (4.79)$$

---

[3]The constant is $\zeta(3) \equiv \sum_{n=1}^{\infty} 1/n^3 \approx 1.202$. See Eq. (3.82).

[4]The coefficient β_2 is related to the coefficient β of the superconductor $\beta = 6\mathcal{N}(0)\xi_0^2/v_F^2\hbar^2$ in Eq. (3.110) by $\beta_2 = \beta/15$ (assuming that we take account of the different mass parameters).

The full interaction contains infinite powers of the collective field $A_{ai}(\mathbf{x})$. If one restricts the consideration to temperatures close to the critical point $(T \approx T_c)$, where μ_A is very small, the fields $A_{ai}(\mathbf{x})$ undergo large long-range fluctuations. As far as long-wavelength properties are concerned, higher and higher powers in $A_{ai}(\mathbf{x})$ become more and more irrelevant, due to the fact that the dimension of $A_{ai}(\mathbf{x})$ is 1/length. This type of discussion is standard in the renormalization group treatment of critical phenomena [18].

In x-space, the free part of the action can be written as

$$\mathcal{A}_0[A^*, A] = \int d^4x \left(\mu_A A_{ai}^* A_{ai} - \frac{K_1}{2} \partial_i A_{aj}^* \partial_i A_{aj} - \frac{K_2}{2} \partial_i A_{aj}^* \partial_j A_{ai} - \frac{K_3}{2} \partial_i A_{ai}^* \partial_j A_{aj} \right),$$

(4.80)

with the temperature dependence residing all in

$$\mu_A = \frac{1}{3} \mathcal{N}(0) \left(1 - \frac{T}{T_c} \right) = \frac{1}{2} \frac{\rho}{m^2} \frac{1}{v_F^2} \left(1 - \frac{T}{T_c} \right), \qquad (4.81)$$

which is related to the chemical potential in the superconductor calculation by a factor $1/3$. The stiffness constants satisfy

$$K_1 = \frac{2}{5} \mathcal{N}(0) \xi_0^2 = v_F^2 \hbar^2 \beta_2 = \frac{3}{5} \frac{\rho}{m^2} \frac{\xi_0^2}{v_F^2}, \qquad K_2 + K_3 = 2K_1. \qquad (4.82)$$

The static long-wavelength action consisting of the free part (4.80) and the interactions (4.78) is referred to as the "weak-coupling" Ginzburg-Landau action of ^{3}He. The functional integral (4.68) with this action defines a fluctuating field theory of the superfluid in the neighborhood of the critical temperature. From this, all universal critical properties can be calculated with great accuracy. The general procedure for doing this is amply described in the literature [18].

If fluctuation corrections are calculated, they do not change the general form of (4.78) and (4.80) for small and smooth fields $A_{ai}(\mathbf{x})$. Only the numerical values of the coefficients are modified, and will no longer satisfy the relations (4.79) and (4.82) obtained from the expansion (4.72). This is actually a consequence of the symmetry properties of the original action (4.56), which is invariant under separate rotations of spin, orbits, and phase of the fermion fields $\psi \to e^{i\alpha}\psi$. The collective action derived from the original action displays the same invariance. In the static long-wavelength limit with $T \approx T_c$, this leaves only the form (4.78) plus (4.80).

On the same symmetry grounds it is obvious that the dipole action (4.60) *cannot* be included by a mere change of the coefficients: The action contracts spatial with spin indices and is no longer invariant under separate spin and orbital rotations. It can be shown [19] that the collective form of the dipole action gives rise to an additional mass term for the A_{ab} field:

$$\mathcal{A}_d = g_d \int d^4x \left(A_{aa}^* A_{bb} + A_{ab}^* A_{ba} - \frac{2}{3} A_{ab}^* A_{ab} \right). \qquad (4.83)$$

The coupling of spatial and spin degrees results in the most interesting observable phenomena of the superfluid phase.

At the mean-field level, the integrand of the collective action yields the Ginzburg-Landau free energy to be used in the sequel:

$$
\begin{aligned}
f &= f_0 + f_{\text{int}} + f_{\text{d}} \\
&= -\mu_A A^*_{ai} A_{ai} + \frac{K_1}{2} \partial_i A^*_{aj} \partial_i A_{aj} + \frac{K_2}{2} \partial_i A^*_{aj} \partial_j A_{ai} + \frac{K_3}{2} \partial_i A^*_{ai} \partial_j A_{aj} \\
&\quad + \beta_1 A^*_{ai} A_{bj} A^*_{ai} A_{bj} + \beta_2 \left(A^*_{ai} A_{ai} \right)^2 \\
&\quad + \beta_3 A^*_{ai} A_{aj} A^*_{bi} A_{bj} + \beta_4 A^*_{ai} A_{bi} A^*_{bj} A_{aj} + \beta_5 A^*_{ai} A_{bi} A^*_{aj} A_{bj}. \\
&\quad - g_{\text{d}} \left(A^*_{aa} A_{bb} + A^*_{ab} A_{ba} - \frac{2}{3} A^*_{ab} A_{ab} \right).
\end{aligned}
\tag{4.84}
$$

The terms proportional to K_1, K_2, K_3 constitute the so-called gradient energy

$$
f_{\text{grad}} \equiv \frac{K_1}{2} \partial_i A^*_{aj} \partial_i A_{aj} + \frac{K_2}{2} \partial_i A^*_{aj} \partial_j A_{ai} + \frac{K_3}{2} \partial_i A^*_{ai} \partial_j A_{aj}.
\tag{4.85}
$$

4.5   General Properties of a Collective Action

The static action (4.78) with (4.80) describes the ^{3}He-liquid in terms of a complex 3×3 -matrix, i.e., an 18-component field called the *order field*. If the dipole interaction is left out, the action is invariant under global SU(2) × SU(2) × U(1) - transformations:

$$
\begin{aligned}
A_{ai} &\rightarrow R_{ab}(\boldsymbol{\varphi}^s) R_{ij}(\boldsymbol{\varphi}^o) e^{-2i\varphi} \\
&= \left(e^{-i\varphi^s \epsilon} \right)_{ab} \left(e^{-i2\varphi^o \epsilon} \right)_{ij} e^{-2i\varphi} A_{bj},
\end{aligned}
\tag{4.86}
$$

where $(\epsilon_a)_{bc} \equiv -i\epsilon_{abc}$ are the 3×3 matrix generators of the three-dimensional rotation group and the angular vectors $\boldsymbol{\varphi}^s$, $\boldsymbol{\varphi}^o$ denote the associated rotation parameters. Remembering the classical equality

$$
A_{ai}(x) = \frac{3g}{2p_F} \psi(x) i \overset{\leftrightarrow}{\nabla}_i c \frac{\sigma_a}{2} \psi(x)
\tag{4.87}
$$

we see that the first transformation corresponds to pure spin, the second to pure orbital rotations to the original field ψ. The last phase is associated with particle number conservation and is doubled because of the two fields occurring in (4.87).

Accordingly, there are three conserved Noether currents which are obtained by functional derivatives with respect to infinitesimal *x-dependent* symmetry transformations:
First there is the particle current density:

$$
j_i \equiv \frac{\delta \mathcal{A}}{\delta \partial_i \varphi}
\tag{4.88}
$$

$$
= i \left\{ K_1 A^*_{aj} \overset{\leftrightarrow}{\nabla}_i A_{aj} + K_2 \left(A^*_{aj} \overset{\leftrightarrow}{\nabla}_j A_{ai} - A^*_{ai} \overset{\leftrightarrow}{\nabla}_j A_{aj} \right) + K_3 \left(A^*_{ai} \overset{\leftrightarrow}{\nabla}_j A_{aj} - A^*_{ai} \overset{\leftrightarrow}{\nabla}_j A_{ai} \right) \right\}.
$$

This current density j_i coincides also with $1/m$ times the components T^{0i} of the energy momentum tensor. Indeed, under an infinitesimal Galilei transformation

$$\psi(x) \to e^{-im\mathbf{v}\cdot\mathbf{x}}\psi(x), \tag{4.89}$$

the collective field changes as follows:

$$A_{ai}(x) \to e^{-2im\mathbf{v}\cdot\mathbf{x}}A_{ai}(x), \tag{4.90}$$

so that the energy changes by

$$\delta E = m \int d^3x \, \mathbf{j}(x) \cdot \mathbf{v}. \tag{4.91}$$

Similarly, we derive, from an infinitesimal spin rotation, the conserved spin current density:

$$j_{ai}^{\text{spin}} \equiv \frac{\delta \mathcal{A}}{\delta \partial_i \varphi_a^s} = \epsilon_{abc}\left[K_1\left(A_{bj}^*\overset{\leftrightarrow}{\nabla}_iA_{cj} + A_{cj}^*\overset{\leftrightarrow}{\nabla}_iA_{bj}\right)\right. \tag{4.92}$$

$$\left. + K_2\left(A_{bj}^*\overset{\leftrightarrow}{\nabla}_jA_{cj} + A_{cj}^*\overset{\leftrightarrow}{\nabla}_jA_{bj}\right) + K_3\left(A_{bi}^*\overset{\leftrightarrow}{\nabla}_jA_{cj} + A_{cj}^*\overset{\leftrightarrow}{\nabla}_jA_{bi}\right)\right].$$

The orbital current density can be written as

$$m\,j_i^{\text{orb}} = \epsilon_{ijk}\left(x^jT^{0k} - x^kT^{0i}\right). \tag{4.93}$$

This is equal to

$$\mathbf{j}^{\text{orb}} \equiv \mathbf{x} \times \mathbf{j}, \tag{4.94}$$

since angular momentum density is the vector product of $\mathbf{x}$ with the momentum density $m\mathbf{j}$. Both orbital and spin currents have zero divergence:

$$\nabla \cdot \mathbf{j} = 0, \quad \nabla \cdot \mathbf{j}^{\text{orb}} = 0, \tag{4.95}$$

if the fields satisfy the Euler-Lagrange equation. The two currents follow from Noether's theorem and the invariance under *spatially independent* symmetry transformations.[5]

As a consequence of $\nabla \cdot \mathbf{j}^{\text{orb}} = 0$, the integral over (4.94), which is the total angular momentum

$$\mathbf{L} = \int d^3x \, \mathbf{x} \times \mathbf{j}^{\text{orb}}, \tag{4.96}$$

is a time-independent quantity.

Since the invariance of the collective action under (4.86) is a direct consequence of the original fundamental action being invariant under separate phase, spin, and orbital rotations defined as

$$\psi \to e^{-i\varphi}\psi, \tag{4.97}$$

$$\psi \to e^{-i\boldsymbol{\varphi}^s\cdot\boldsymbol{\sigma}}\psi, \tag{4.98}$$

$$\psi \to e^{-\boldsymbol{\varphi}^o\cdot(\mathbf{x}\times\nabla)}\psi, \tag{4.99}$$

---

[5]See Chapter 3 in the textbook [44] or Chapter 8 in the textbook [45].

the currents (4.88), (4.92), and (4.94) are simply the collective versions of the fundamental Noether currents following from the continuous symmetries under the transformations (4.97)–(4.99):[6]

$$j_i \equiv \frac{1}{2mi}\psi^* \overset{\leftrightarrow}{\boldsymbol{\nabla}}_i\psi,$$

$$j_{ai} \equiv \frac{1}{2mi}\psi^* \sigma_a \overset{\leftrightarrow}{\boldsymbol{\nabla}}_i\psi,$$

$$\mathbf{j}^{\text{orb}} \equiv \mathbf{x} \times \mathbf{j}. \tag{4.100}$$

Because of the invariance under (4.97)–(4.99) and the fourth-order form of the collective action, the theory at hand is what is called a $3 + 1$ dimensional $SU(2) \times SU(2) \times U(1)$ -symmetric linear σ-model. It is of the same type as the O(3)-symmetric Landau model of ferromagnetism.

When confronted with such a model, the discussion usually starts with the stability analysis of all possible vacuum states. One examines small oscillations of the field A_{ai} around its static ground state configurations. There the action $\mathcal{A}$ can be expressed in terms of the energy as

$$\mathcal{A} = -\int dt E = -\int dt d^3x \, f \tag{4.101}$$

A glance at Eq. (4.84) shows that small oscillations of A_{ai} around zero are stable as long as

$$\mu_A = \frac{\mathcal{N}(0)}{3}\left(1 - \frac{T}{T_c}\right) < 0, \qquad \text{i.e., } T > T_c. \tag{4.102}$$

As the temperature drops below the critical value T_c, the quadratic potential becomes unstable and the fourth-order term is needed to control the fluctuations. The field A_{ai} settles at some new minimum away from zero. Unfortunately, no full mathematical analysis is available on the minima for all possible configurations of the coefficients β_i. Among the many minima discussed in the literature [21], there are three which apparently have been found in the laboratory associated with the phases which were shown in Fig. 4.3. Each of these is non-unique due to a residual degeneracy and can be parametrized as follows:

A-phase

$$A_{ai}^0 = \Delta_A d_a \left(\phi^{(1)} + i\phi^{(2)}\right)_i. \tag{4.103}$$

Here $\mathbf{d}$, $\boldsymbol{\phi}^{(1)}$, $\boldsymbol{\phi}^{(2)}$ are arbitrary real unit vectors with $\boldsymbol{\phi}^{(1)} \perp \boldsymbol{\phi}^{(2)}$.

B-phase

$$A_{ai}^0 = \Delta_B R_{ai}(\hat{\mathbf{n}}, \theta)e^{i\varphi}. \tag{4.104}$$

---

[6] Recall Subsec. 3.5.3 in the textbook [44].

Here R_{ai} is an arbitrary rotation around an axis $\hat{\mathbf{n}}$ by an angle θ with φ being some phase angles.

A_1-phase

$$A_{ai} = \Delta_{A_1} \left(d^{(1)} + i d^{(2)} \right)_a \left(\phi^{(1)} + i \phi^{(2)} \right)_i. \tag{4.105}$$

Here $\mathbf{d}^{(1)}, \mathbf{d}^{(2)}$; $\boldsymbol{\phi}^{(1)}, \boldsymbol{\phi}^{(2)}$ are unit vectors with $\mathbf{d}^{(1)} \perp \mathbf{d}^{(2)}$ and $\boldsymbol{\phi}^{(1)} \perp \boldsymbol{\phi}^{(2)}$. The magnitudes of Δ are controlled by the free energy. In the three cases, this becomes:

$$
\begin{aligned}
f_A &= -2\mu_A \Delta_A^2 + (\beta_2 + \beta_4 + \beta_5)\, 4\Delta_A^4, \\
f_B &= -3\mu_A \Delta_B^2 + (\beta_1 + \beta_2)\, 9\Delta_B^4 + (\beta_3 + \beta_4 + \beta_5)\, 3\Delta_B^4, \\
f_{A_1} &= -4\mu_A \Delta_{A_1}^2 + 16\, (\beta_2 + \beta_4)\, \Delta_{A_1}^4.
\end{aligned}
\tag{4.106}
$$

For $\mu_A < 0$, the minima lie at the nonzero values

$$
\begin{aligned}
\Delta_A &= \sqrt{\frac{\mu_A}{4\beta_{245}}} = \pi T_c \sqrt{\frac{10}{7\zeta(3)}} \sqrt{1 - \frac{T}{T_c}}, \\
\Delta_B &= \sqrt{\frac{\mu_A}{6\beta_{12} + 2\beta_{345}}} = \pi T_c \sqrt{\frac{8}{7\zeta(3)}} \sqrt{1 - \frac{T}{T_c}}, \\
\Delta_{A_1} &= \sqrt{\frac{\mu_A}{8\beta_{24}}} = \pi T_c \sqrt{\frac{10}{7\zeta(3)}} \sqrt{1 - \frac{T}{T_c}},
\end{aligned}
\tag{4.107}
$$

where $\beta_{ij}, \beta_{ijk}, \ldots$ are short for $\beta_i + \beta_j,\ \beta_i + \beta_j + \beta_k,\ \ldots$. The minimal values are

$$
\begin{aligned}
f_A^{\min} &= \mu_A \Delta_A^2 = -\frac{\rho}{m^2}\left(1 - \frac{T}{T_c}\right)\frac{\hbar^2}{2\xi_0^2} \times \frac{5}{24}, \\
f_B^{\min} &= -\frac{3}{2}\mu_A \Delta_A^2 = -\frac{\rho}{m^2}\left(1 - \frac{T}{T_c}\right)\frac{\hbar^2}{2\xi_0^2} \times \frac{1}{4}, \\
f_{A_1}^{\min} &= -2\mu_A \Delta_{A_1}^2 = -\frac{\rho}{m^2}\left(1 - \frac{T}{T_c}\right)\frac{\hbar^2}{2\xi_0^2} \times \frac{5}{48},
\end{aligned}
\tag{4.108}
$$

respectively. Note that in the B-phase, the expression for the gap and for the energy (4.108) are the same as for the superconductor in Eqs. (3.154) and (3.125) (apart from the different mass values).

It is useful to introduce the various relevant β-values in the different phases,

1) B: $\beta_B^{-1} = \left(\beta_{12} + \frac{1}{3}\beta_{345}\right)^{-1} = \frac{6}{5}\beta_2^{-1}$

2) A: $\beta_A^{-1} = \left(\frac{2}{3}\beta_{245}\right)^{-1} = 1\,\beta_2^{-1}$

3) A_1: $\beta_{A_1}^{-1} = (2\beta_{24})^{-1} = \frac{1}{3}\,\beta_2^{-1}$

so that the gap is given by

$$\Delta = \sqrt{\frac{\mu_A}{6\beta}}, \tag{4.109}$$

and the minimal energy is the negative of the *condensation energy*

$$f^{\min} = -f_c = -\frac{\mu_A^2}{4\beta} = -\frac{5}{24}\frac{\rho}{m^2}\left(1 - \frac{T}{T_c}\right)\frac{\hbar^2}{2\xi_0^2} \times \frac{\beta_2}{\beta}. \tag{4.110}$$

The fluctuations around the new minima can be separated according to massive and massless ones. The massive ones all occur with a mass of the same order of magnitude as is found for the oscillations of Δ at the new minimum. This can be calculated as follows: Introducing

$$\Delta \to \Delta + \Delta', \tag{4.111}$$

we find for Δ'-oscillations

$$f = f^{\min} + \delta^2 f, \tag{4.112}$$

with a mass term twice the opposite of that in (4.106):

$$\delta^2 f = 4\mu_A \Delta'^2. \tag{4.113}$$

The massive oscillations in directions other than Δ have the same type of mass term except that it is accompanied by a numerical factor (determined by a Clebsch-Gordon coefficient). The massless oscillations arise for small displacements of the *direction* vectors $\mathbf{d}$ and $\boldsymbol{\phi}$ and the phase φ characterizing the minima. They are called Goldstone bosons.

Group-theoretically, the following considerations are useful. The action is invariant under the global transformations of the group $SU(2) \times SU(2) \times U(1)$. The infinitesimal transformations consist of those which change the directions of the minima and a subgroup leaving them invariant. The first ones coincide with the long-wavelength limit of Goldstone bosons oscillating around the new minimum. The mass of these oscillations is zero, since the action is invariant in the limit of infinite wavelength in which the small displacements become uniform rotations of $\mathbf{d}$, $\boldsymbol{\phi}$, φ. The subgroup of symmetry transformations which leave the directions at the minima invariant, but they mix the Goldstone modes with each other. These transformations describe the residual symmetry left for the physics of the Goldstone modes.

The collective field A_{ai} has 18 parameters while the above A_{ai} have only 6, 5, or 7 parameters in A, B, and A_1-phases, respectively. The above parametrizations of the vacuum, therefore, does not allow to describe *all* massive oscillations (only those of the size parameter Δ are included).

In field theoretic considerations a particular direction di^0 is usually chosen as a vacuum of the theory. The freedom of taking an arbitrary direction corresponds to an infinite degeneracy of the possible vacua. In ^{3}He physics such a uniform choice is usually not possible since, as we shall see, boundary effects do not permit the ground state to settle in a uniform direction of the A_{ai}^0 field. The "vacua" are nontrivial. In addition to boundaries, also external fields[7], currents[8], and topology may serve to

---

[7]For a general discussion and references see Ref. [17].

[8]Non-trivial helix-like textures in the presence of currents have first been found in Ref. [40].

stabilize different non-uniform field configurations. The latter fact establishes links with discussions of topologically interesting vacua in field theory.

As we have stressed repeatedly, we shall analyze the quantum liquid only with respect to those phenomena which take place at energies much smaller than the gap energy Δ. In this limit, all massive oscillations become unimportant (since their energy lies in the Δ regime). We can therefore assume Δ to be pinned down tightly at one of the degenerate extremal values (4.114) and allow only fluctuations of the *direction* of A_{ai}^0. This approximation, in which only the Goldstone modes are studied, is called the *hydrodynamic* or the *London limit* of the theory. In σ-models of field theory, this corresponds to letting the mass of the σ-particle (the σ-oscillations) go to infinity. This limit leaves only the pion as a dynamical field in what is called a nonlinear σ-model. In the following, we shall restrict our discussions to this hydrodynamic limit.

4.6 Comparison with O(3)-Symmetric Linear σ-Model

For comparison, we briefly recall the symmetry-breakdown in the simple $O(3)$-symmetric σ-model, also known as the classical Heisenberg model of ferromagnetism. There the free-energy density reads, for constant fields,

$$f = \frac{\mu^2}{2} \left(\pi_1^2 + \pi_2^2 + \pi_3^2 \right) + \frac{\lambda}{4} \left(\pi_1^2 + \pi_2^2 + \pi_3^2 \right)^2 . \tag{4.114}$$

For $\mu^2 < 0$, this has the following set of degenerate minima:

$$\pi_i^0 \equiv \Delta^0 d_i^0 \quad \text{with} \quad \Delta^0 = \sqrt{-\frac{\mu^2}{\lambda}}, \tag{4.115}$$

where d_i^0 is an arbitrary unit vector in three-space. The oscillations of $\pi_i \equiv \Delta d_i$ around π_i^0 consist of massive radial oscillations in Δ controlled by

$$f = -\frac{\mu^4}{4\lambda} + (-\mu^2)(\Delta - \Delta^0)^2 \tag{4.116}$$

and massless oscillations of d_i around the direction of d_i^0. If d_i^0 points along the $3-$axis, these oscillations can be parametrized as

$$d_i = \left(\frac{\pi_i'}{\Delta}, \sqrt{1 - \frac{\pi'^2}{\Delta^2}} \right) . \tag{4.117}$$

The energy depends only on Δ. Rotations leaving d_i^0 invariant transform the fluctuationg fields π_1' and π_2' among each other and correspond to the residual $O(2)$ symmetry after spontaneous symmetry breakdown of the original $O(3)$. The situation here is simpler than that for ^{3}He since the parametrization $\pi_i = \Delta d_i$ of the ground state can be used to cover the *entire* three-dimensional field space.

4.7   Hydrodynamic Properties Close to T_c

In the hydrodynamic limit the only degrees of freedom of the liquid consist in ground state configurations A^0_{ai} with slow spatial variations of the directional vectors. In the A-phase, in which

$$A^0_{ai} = \Delta_A d_a \phi_i$$
$$\phi_i \equiv \phi^{(1)}_i + i\phi^{(2)}_i, \qquad\qquad (4.118)$$

where $\phi^{(1)}_i$ and $\phi^{(2)}_i$ are orthogonal unit vectors, the magnitude Δ_A is pinned down at the potential minimum (4.108) with a value (4.107). The unit vectors d_a and $\phi^{(1)}_i$, $\phi^{(2)}_i$ vary in space. It is useful to visualize the physical meaning of these directions. Remembering the relation (4.87) expressing the collective field A_{ai} in terms of the pair of fundamental fields, the vector d_a indicates the direction along which the spin has the wave function $\frac{1}{\sqrt{2}} (\uparrow\downarrow + \downarrow\uparrow)$, along which the third spin component vanishes.[9] The plane in which the Cooper pair moves is given by the plane spanned by the unit vectors $\phi^{(1)}$ and $\phi^{(2)}$. It has become customary to introduce a vector

$$l \equiv \phi^{(1)} \times \phi^{(2)}, \qquad\qquad (4.119)$$

which denotes the direction of the intrinsic orbital angular momentum of the Cooper pairs in the condensate. For the completeness of the description, one has to specify, in addition, the azimuthal angle α of $\phi^{(1)}$ in the plane orthogonal to 1. This specification can be made unique, for example, by the following choice of parametrization:

$$\phi \equiv \phi^{(1)} + i\phi^{(2)} \qquad\qquad (4.120)$$
$$= e^{-i\alpha} \{ (-\sin\gamma, \ \cos\gamma, 0) + i(-\cos\beta\cos\gamma, \ -\cos\beta\sin\gamma, \sin\beta) \},$$
$$l \equiv (\sin\beta\cos\gamma, \ \sin\beta\sin\gamma, \cos\beta). \qquad\qquad (4.121)$$

Consider now the gradient energy density (4.85) in the hydrodynamic limit. Inserting the above parametrization of the order field, it becomes

$$f_{\mathrm{grad}} = \frac{1}{2}\Delta^2_A \Big\{ K_1|\nabla_i\phi_j|^2 + K_2\nabla_i \big[\phi^\dagger_j\nabla_j\phi_i - \phi^\dagger_i\nabla_j\phi_j\big] + K_{23}|\nabla_i\phi_i|^2$$
$$+ K_{23}|\phi\cdot\boldsymbol\nabla d_a|^2 + 2K_1(\nabla_i d_a)^2 \Big\}, \qquad\qquad (4.122)$$

with the notation $K_{12} \equiv K_2 + K_3$. The last term is a pure divergence and can be neglected in most discussions. Since the magnitude of all directional vectors is unity, the mass term in (4.80) and the fourth-order term (4.78) add up to the minimal values given in (4.108). Since Δ is tightly pinned down at that minimal value, any deviation of the energy from this minimum is completely determined by the derivative terms (4.122) of the Ginzburg-Landau expansion (4.84). These vanish

---

[9]To verify this, let $d_a = (0,0,1)_a$ so that $d_a(c\sigma^a)_{\alpha\beta} = -(\uparrow_\alpha\downarrow_\beta + \downarrow_\alpha\uparrow_\beta)$, where $\uparrow_\alpha$, $\downarrow_\beta$ are the spin-$\frac{1}{2}$ two-spinors with spin up and down, respectively.

for uniform field configurations and grow with increasing bending of the field lines. For this reason, the gradient-energy (4.122) is often referred to as *bending energy*.

The prefactor $\frac{1}{2}\Delta_A^2$ can be brought to a physically more transparent form: Using (4.76) and (4.107), we find in the weak-coupling limit:

$$
\begin{aligned}
\frac{1}{2}\Delta_A^2 &= \frac{1}{2}\frac{\mu_A}{4\beta_{245}} = \frac{1}{2}\frac{1}{3}\mathcal{N}(0)\frac{1}{\frac{8}{5}\mathcal{N}(0)\xi_0^2/v_F^2}\left(1-\frac{T}{T_c}\right) = \frac{5}{48}\frac{v_F^2}{\xi_0^2}\left(1-\frac{T}{T_c}\right) \\
&= \frac{\rho}{16m^2}\left(1-\frac{T}{T_c}\right)\frac{1}{K_1},
\end{aligned}
\tag{4.123}
$$

where ρ is the mass density of ^{3}He particles per unit volume. Now, if a collective excitation of wave vector $\mathbf{k}$ runs through the liquid, its energy density per particle is of the order $(\mathbf{k}^2/2m)(1-T/T_c)$. It grows with decreasing temperature due to the increasing condensation energy.

Instead of the complex vector $\boldsymbol{\phi}$, one may express the energy density in a somewhat more intuitive fashion by using the more physical vector $\mathbf{l}$ of (4.121). To this end we define a gradient vector called the *macroscopic superfluid velocity*:

$$
v_{si} = \frac{1}{2m}\phi^{(1)}\nabla_i\phi^{(2)} = \frac{i}{4m}\phi^\dagger\nabla_i\phi,
\tag{4.124}
$$

where the vectors $\phi^{(1)}$ and $\phi^{(2)}$ are those introduced in Eq. (4.103). Then the gradient part (4.85) of the free energy density takes the form (see Appendix 4A)

$$
\begin{aligned}
f_{\text{grad}} &= \frac{1}{2}\rho_s\mathbf{v}_s^2 - \frac{1}{2}\rho_0(\mathbf{l}\cdot\mathbf{v}_s)^2 + c\mathbf{v}_s\cdot(\nabla\times\mathbf{l}) - c_0(\mathbf{l}\cdot\mathbf{v}_s)[\mathbf{l}\cdot(\nabla\times\mathbf{l})] \\
&\quad + \frac{1}{2}K_s(\nabla\cdot\mathbf{l})^2 + \frac{1}{2}K_t[\mathbf{l}\cdot(\nabla\times\mathbf{l})]^2 + \frac{1}{2}K_b[\mathbf{l}\times(\nabla\times\mathbf{l})]^2 \\
&\quad + \frac{1}{2}K_1^d(\nabla_i d_a)^2 - \frac{1}{2}K_2^d(\mathbf{l}\cdot\nabla d_a)^2,
\end{aligned}
\tag{4.125}
$$

with the coefficients

$$
\begin{aligned}
\rho_s &= \Delta_A^2(K_1 + \tfrac{1}{2}K_{23})\,4m^2, & \rho_0 &= \Delta_A^2 K_{23}\,4m^2, \\
c &= \Delta_A^2 K_3 2m, & c_0 &= \rho_0/2m, \\
K_s &= K_t = \Delta_A^2 K_1, & K_b &= \Delta_A^2(K_1 + K_{23}), \\
K_1^d &= \rho_s/4m^2, & K_2^d &= \rho_0/4m^2.
\end{aligned}
\tag{4.126}
$$

In the weak-coupling limit (4.84) these expressions simplify to

$$
\begin{aligned}
\frac{\rho_s}{2} &= \rho_0 = 2mc_0 = 4mc \\
&= (2m)^2 2K_s = (2m)^2 2K_t = (2m)^2\frac{2}{3}K_b \\
&= (2m)^2\frac{1}{2}K_1^d = (2m^2)K_2^d.
\end{aligned}
\tag{4.127}
$$

The material constants $K_{s,b,t}$ parametrize the stiffness for the various fundamental deformations of the l-field which are illustrated in Fig. 4.4. They have been defined in the theory of liquid crystals to be discussed in Chapter 5 as *splay*, *bend* and *twist* deformations [22] (see Section 5.3 of Chapter 5). It is easily verified that in these three configurations the terms with K_s, K_b, and K_t give the dominant contributions. Indeed, if the spatial changes of the l field take place only in the z-direction, one can write the relevant part in gradient energy (4.125) as

$$f_{\text{grad stb}} \approx \frac{1}{2}K_s\left(\boldsymbol{\nabla}\cdot\mathbf{l}\right)^2 + \frac{1}{2}K_t\left[\mathbf{l}\cdot\left(\boldsymbol{\nabla}\times\mathbf{l}\right)\right]^2 + \frac{1}{2}K_b\left[\mathbf{l}\times\left(\boldsymbol{\nabla}\times\mathbf{l}\right)\right]^2 \qquad (4.128)$$

$$= \frac{1}{2}K_s\sin^2\beta\,\beta_z^2 + \frac{1}{2}K_t\sin^4\beta\,\gamma_z^2 + \frac{1}{2}K_b\left[\cos^2\beta\left(\sin^2\beta\,\gamma_z^2 + \beta_z^2\right)\right].$$

In the twist texture, l changes in the xy-plane from the x- to the $-x$-direction. Hence $\beta \equiv \pi/2$ is a constant, and only γ_z contributes to the gradient energy

$$f_{\text{grad t}} = \frac{1}{2}K_t\gamma_z^2. \qquad (4.129)$$

The other two textures are not that clearly separated: In both, $\gamma \equiv 0$ is constant, so that

$$f_{\text{grad sb}} = \frac{1}{2}K_s\sin^2\beta\,\beta_z^2 + \frac{1}{2}K_b\cos^2\beta\,\beta_z^2. \qquad (4.130)$$

In the splay case, l turns in the xz-plane from z to $-z$-direction. In the middle of the texture, i.e., in the place of largest β_z where angle β is $\pi/2$, the first term dominates. In the bend case, l turns in the xz-plane from x- to $-x$-direction. Thus, for the largest β_z where $\beta \approx \pi$, the second term dominates.

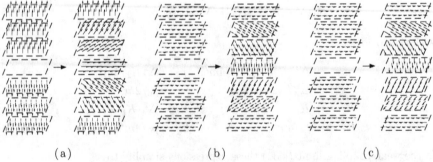

(a)   (b)   (c)

FIGURE 4.4 Three fundamental planar textures: splay (a), bend (b), and twist (c). The left-hand side of the figure shows field configurations with a singular plane where the fields reverse direction. Since the superfluid would have to be normal in this plane, it prefers the right-hand configuration in which the direction changes smoothly through a domain wall of finite size. The thickness ξ_d is determined by the competition of the dipole and bending energy.

The currents can now be calculated by inserting (4.118) into (4.87) and (4.88). For the mass current density we find

$$j_i = \rho_{s\,ij} v_{sj} + c_{ij} \left(\nabla \times \mathbf{l} \right)_j, \tag{4.131}$$

with the matrices

$$\rho_{s\,ij} \equiv \rho_s \delta_{ij} - \rho_0 l_i l_j,$$
$$c_{ij} = c \delta_{ij} - c_0 l_i l_j. \tag{4.132}$$

Note, that this result also follows directly from an infinitesimal Galilean transformation. If one multiplies A by $e^{2im\mathbf{v}\mathbf{x}}$, this leaves $\mathbf{l}$ invariant while changing the superfluid velocity (4.123) as follows

$$\mathbf{v}_s \to \mathbf{v}_s + \mathbf{v}. \tag{4.133}$$

This shows that $\mathbf{v}_s$ transforms indeed like any velocity (thus justifying its name). Using this transformation together with (4.91) on (4.125) yields again the current density (4.131). This current density describes the superflow of Cooper pairs in the rest frame of the normal liquid. The superfluid density is a tensor with a component longitudinal to $\mathbf{l}$, $\rho_s^{\parallel} = \rho_s - \rho_0$, and a transverse one, $\rho_{s\perp} = \rho_s$.

We now turn to the "orbital current". It describes the collective motion of the atoms *within* the Cooper pairs. It is similar to the current density $\nabla \times \mathbf{M}$ which appears in magnetostatics in the presence of magnetizable matter [23] in the Maxwell equation

$$\nabla \times \mathbf{B} = 4\pi \left(\mathbf{j} + \nabla \times \mathbf{M} \right). \tag{4.134}$$

The second current term describes the electronic current density flowing within the molecular orbits of the matter. In complete analogy to this, there is a local matter current associated with the rotation of ^{3}He atoms inside the Cooper pairs. This current contributes to the total superflow.

The spin current density can be derived similarly to the matter current density via the appropriate symmetry transformation which brings $A \to e^{-2\varphi^s \epsilon} A$ and $d_a \to d_a + \delta d_a$ with

$$\delta d_b = -2\varphi_a^s \epsilon_{abc} d_c. \tag{4.135}$$

Since the spin current density is defined by $j_{ai}^{\text{spin}} \equiv -\partial e / \partial_i \varphi_a^s$ we find directly, from the hydrodynamic energy (4.128):

$$j_{ai} = 2 \left(K_1^d \delta_{ij} - K_2^d l_i l_j \right) \epsilon_{abc} d_b \nabla_j d_c. \tag{4.136}$$

In order to keep as much analogy as possible with the superfluid velocity we may define a superspin velocity

$$v_{s\,ai} \equiv \frac{1}{2m} \epsilon_{abc} d_b \nabla_i d_c, \tag{4.137}$$

in terms of which the current density becomes

$$j_{ai} = 4m \left(K_1^d \delta_{ij}, -K_2^d l_i l_j \right) v_{saj} \tag{4.138}$$

where, again, there is a longitudinal term proportional to $K_1^d - K_2^d$ and a transverse one with a factor K_1^d.

Under a spin rotation (4.135), the velocity transforms according to

$$v_{sai} \to -2\varphi_a^s \epsilon_{abc} v_{sci} + \nabla_i \varphi^s / m. \tag{4.139}$$

The orbital angular momentum current density is obtained from (4.131) by forming the vector product with x.

The action is still incomplete since, until now, we have left out the dipole force. Inserting the parametrization (4.118) and (4.120) into (4.83), we find

$$f_{\rm d} = -2g_{\rm d} \Delta_A^2 \left[(\mathbf{d} \cdot \mathbf{l})^2 - \frac{1}{3} \right]. \tag{4.140}$$

Thus, the dipole force tends to align $\mathbf{d}$ and $\mathbf{l}$ parallel or antiparallel. This can physically be understood as follows: Let the atoms orbit around each other, say, in the xy-plane. If the spin points in the z-direction, the two nuclear moments have equal poles all the time adjacent to each other. In the $S_z = 0$ configuration they are, on the other hand, aligned so that opposite poles face each other for half the orbit. This corresponds to $\mathbf{d} \parallel \mathbf{l}$.

A comparison of the strength of the dipole energy with the main term (4.119) of the bending energy is possible if we write

$$f_{\rm d} = -\Delta_A^2 K_{23} \frac{1}{\xi_{\rm d}^2} (\mathbf{d} \cdot \mathbf{l})^2 + \text{const}. \tag{4.141}$$

Then, the dipole length

$$\xi_{\rm d} = \sqrt{K_{23}/2g_{\rm d}} \tag{4.142}$$

measures the length scale over which the direction of field lines has to vary appreciably in $f_{\rm d}$ of (4.141) in order to give the bending energy a comparable size with the dipole energy. The microscopic calculation yields $\xi_{\rm d} \approx 10^{-3}$ cm $(1 - T/T_c)$ which is two orders of magnitude larger than the coherence length ≈ 1000Å $(1 - T/T_c)$.

The small dipole energy (4.141) in the σ-model of ^{3}He plays a very similar role as the small PCAC-violation in σ-models of particle physics. Before $f_{\rm d}$ is turned on, all Goldstone modes are massless. With (4.141), the oscillations in which the relative angle between $\mathbf{d}$ and $\mathbf{l}$ vibrates produce a small mass. The experimental resonance frequency is $\Omega_A \approx 50$ kHz corresponding, energetically, to the temperature $T_A \approx 5 \times 10^{-7}$ K. It is, therefore, much smaller than the gap energy ($\approx m \cdot K$).

While $\mathbf{l}$ and $\mathbf{v}_s$-vectors have physically the most transparent meaning, they are dynamically not independent, since $v_{si} = \frac{1}{2m} \phi^{(1)} \nabla_i \phi^{(2)}$ involves derivatives of $\mathbf{l}$. In fact, the curl of $\mathbf{v}_s$ is related to the $\mathbf{l}$ field as follows

$$\nabla \times \mathbf{v}_s = \frac{1}{4m} \epsilon_{ijk} \, \mathbf{l} \cdot (\nabla_j \mathbf{l} \times \nabla_k \mathbf{l}). \tag{4.143}$$

For the proof, one forms the derivative of (4.124):

$$(\nabla \times \mathbf{v}_s)_i = \frac{1}{2m}\epsilon_{ijk}\nabla_j\left(\phi^{(1)}\nabla_k\phi^{(2)}\right)$$

$$= \frac{1}{2m}\epsilon_{ijk}\left(\nabla_j\phi^{(1)}\nabla_k\phi^{(2)}\right). \tag{4.144}$$

Since $\phi^{(1)}\nabla_k\phi^{(1)} = 0$ (due to $\phi^{(1)2} = 1$), $\nabla_k\phi^{(1,2)}$ has only a component along $\mathbf{l}$ and $\phi^{(2,1)}$. Thus

$$\nabla_j\phi^{(1)}\nabla_k\phi^{(2)} = \left(\mathbf{l}\cdot\nabla_j\phi^{(1)}\right)\left(\mathbf{l}\cdot\nabla_k\phi^{(2)}\right). \tag{4.145}$$

But $\mathbf{l}\cdot\nabla_j\phi^{(1,2)} = -\phi^{(1,2)}\nabla_j\mathbf{l}$ (due to $\phi^{(2)}\mathbf{l} = 0$) so that we can write

$$(\nabla \times \mathbf{v}_s)_i = \frac{1}{4m}\epsilon_{ijk}\left[\left(\phi^{(1)}\nabla_j\mathbf{l}\right)\left(\phi^{(2)}\cdot\nabla_k\mathbf{l}\right) - \left(\phi^{(1)}\nabla_k\mathbf{l}\right)\left(\phi^{(2)}\nabla_j\mathbf{l}\right)\right]. \tag{4.146}$$

From this, Eq. (4.142) follows directly since $\phi^{(1)}$, $\phi^{(2)}$, $\mathbf{l}$ are an orthonormal triplet. The relation (4.142) will be powerful in relating the flow vortices to the geometric properties of the container of the liquid. For, if one takes the scalar product of (4.142) with $\mathbf{l}$, we find [25]

$$2m\mathbf{l}\cdot(\nabla \times \mathbf{v}_s) = \frac{1}{2}\epsilon_{ijk}l_i\left[\mathbf{l}\cdot(\nabla_j\mathbf{l}\times\nabla_k\mathbf{l})\right] = K. \tag{4.147}$$

The right-hand side is the Gaussian curvature of a surface cutting normally through the $\mathbf{l}$ field. If there is a closed normal surface anywhere inside the liquid, the integral over k gives 2π times the Euler invariant characteristic E of a closed surface. This characteristic is

$$E = 2(1-m) \tag{4.148}$$

for a surface equivalent to a sphere with m handles (see Fig. 4.5). Performing the same integral over the left-hand side renders 2π times the number of singular vortex lines which have to enter the closed surface at some place. Indeed, consider a closed contour on top of the closed surface (see Fig. 4.6).

Let $\mathbf{t}$ be the tangent vector and $\mathbf{n} = \mathbf{l}\times\mathbf{t}$ be the normal vector of the contour inside the surface. Since $\phi^{(1)}$, $\phi^{(2)}$ lie in the tangent plane they can be spanned as follows:

$$\phi^{(1)} = \cos\theta\,\mathbf{n} + \sin\theta\,\mathbf{t},$$
$$\phi^{(2)} = -\sin\theta\,\mathbf{n} + \cos\theta\,\mathbf{t}. \tag{4.149}$$

As one proceeds a little way along the surface the tangential component of $\mathbf{v}_s$ is

$$2m\mathbf{v}_s\cdot\mathbf{t} = ds\phi^{(1)}\times\frac{d}{ds}\phi^{(2)}$$

$$= ds\left[\frac{d\theta}{ds} + \mathbf{t}\cdot\left(\mathbf{l}\times\frac{d}{ds}\mathbf{t}\right)\right]. \tag{4.150}$$

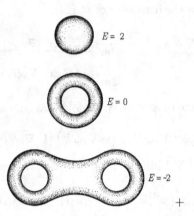

FIGURE 4.5 Sphere with one, two, or no handles, and their respective Euler characteristics $E = 0, = 2$, or 2.

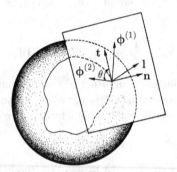

FIGURE 4.6 Local tangential coordinate system $\mathbf{n}$, $\mathbf{t}$, $\mathbf{i}$ for an arbitrary curve on the surface of a sphere.

The second term

$$\gamma = \mathbf{t} \cdot \left(1 \times \frac{d}{ds} \mathbf{t} \right) \tag{4.151}$$

is called the geodesic curvature since it describes the rate of change of $\mathbf{t}$ away from the $\mathbf{t}$ direction (it is zero on the equator of a sphere). If we now convert the integral over the left-hand side of (4.147) into a contour integral and increase the contour throughout the surface leaving out all singular points, the result is

$$\sum_i \oint ds \left[\frac{d\theta}{ds} + \gamma \right] \tag{4.152}$$

with the sum over all enclosed singularities. If the circles are infinitesimal, the surface can be considered as a plane and the integral over the geodesic curvature renders

$$\oint ds\,\gamma = 2\pi. \tag{4.153}$$

The integral over $d\theta/ds$, on the other hand, depends on the vertex strength N_i at the point i as

$$2\pi(N_i - 1). \tag{4.154}$$

For, if there is no vortex, the vectors $\phi^{(1)}$, $\phi^{(2)}$ stay fixed in space along the contour. Thus, the intrinsic coordinate θ of (4.149) changes by -2π. If, on the other hand, there is a vortex with $\phi^{(1)}, \phi^{(2)}$ rotating N_i times around 1 in the positive sense, when going around the contour, there will be an additional change of $2\pi \cdot N_i$. Thus the Euler characteristic determines the number of vortex lines passing through any closed surface normal to the l-field inside the liquid. This theorem will be useful for the discussion to follow.

In the B-phase, in which

$$A_{ai}^0 = \Delta_B R_{ai}\left(\theta\right)e^{i\varphi}, \tag{4.155}$$

the magnitude of Δ_B is pinned down at the potential minimum (4.108) with a value (4.107). Only the angles θ and φ are allowed to vary. Due to (4.80), the gradient energy becomes

$$f_{\text{grad}} = \frac{1}{2}\Delta_B^2\left[K_1\delta_{ij}\delta_{kl} + \frac{1}{2}K_{23}\left(\delta_{il}\delta_{jk} + \delta_{ik}\delta_{jl}\right)\right]\nabla_k\left(R_{ai}\left(\theta\right)e^{-i\varphi}\right)\nabla_l\left(R_{aj}(\theta)e^{i\varphi}\right). \tag{4.156}$$

The derivative factor can be rewritten as

$$\nabla_k\varphi\nabla_l\varphi + \nabla_k R_{ai}\nabla_l R_{aj} + \dots, \tag{4.157}$$

with mixed terms $\nabla R\nabla\varphi$ vanishing in the contraction with the tensor (4.152), for symmetry reasons. If we parametrize small oscillations in θ as

$$R_{ai}(\theta) = R_{aj}(\theta_0)R_{ji}(\theta), \tag{4.158}$$

the energy becomes

$$f_{\text{grad}} = \frac{1}{2}\Delta_B^2\left\{K_1\left[3(\boldsymbol{\nabla}\varphi)^2 + 2(\nabla_i\tilde{\theta}_j)^2\right]\right.$$
$$\left. + K_{23}\left[(\boldsymbol{\nabla}\varphi)^2 + (\nabla_i\tilde{\theta}_j)^2 - \frac{1}{2}(\boldsymbol{\nabla}\tilde{\theta})^2 - \frac{1}{2}(\nabla_i\tilde{\theta}_j\nabla_j\tilde{\theta}_i)\right]\right\}. \tag{4.159}$$

Using the result of Eq. (4.107) together with (4.123), we have

$$\frac{1}{2}\Delta_B^2 = \frac{8}{10}\frac{1}{2}\Delta_A^2 = \frac{8}{10}\frac{1}{16m^2}\rho\left(1 - \frac{T}{T_c}\right)\frac{1}{K_1}, \tag{4.160}$$

which can be used to bring the energy to the form

$$
\begin{aligned}
f_{\text{grad}} &= \frac{1}{4m^2} \frac{1}{2} \rho_s^B \left\{ \frac{3}{5} (\boldsymbol{\nabla}\varphi)^2 + \frac{2}{5} (\nabla_i \tilde{\theta}_j)^2 \right. \\
&+ \left. \frac{K_{23}}{2K_1} \left[\frac{2}{5} (\boldsymbol{\nabla}\varphi)^2 + \frac{2}{5} (\nabla_i \tilde{\theta}_j)^2 - \frac{1}{5} (\boldsymbol{\nabla}\tilde{\theta})^2 - \frac{1}{5} \nabla_i \tilde{\theta}_j \nabla_j \hat{\theta}_i \right] \right\}.
\end{aligned}
\tag{4.161}
$$

Here, we have introduced the quantity

$$
\rho_s^B \equiv 2\rho \left(1 - \frac{T}{T_c} \right),
\tag{4.162}
$$

which for $T \approx T_c$ is the superfluid density of the rmB-phase. The current density can be obtained either by inserting (4.158) into (4.88), or by performing $\varphi \to \varphi + 2\delta\varphi$ in (4.161):

$$
j_i = \rho_s^B \frac{1}{2m^2} \frac{1}{5} (3 + K_{23}/K_1) \nabla_i \varphi.
\tag{4.163}
$$

The spin current density may be obtained by inserting (4.158) into (4.92), from which we find

$$
\begin{aligned}
j_{ai} &= R_{aa'}(\boldsymbol{\theta}_0) \tilde{j}_{a'i}, \\
\tilde{j}_{ai} &= -\frac{1}{2m^2} \rho_s \left\{ \frac{2}{5} \left(1 + \frac{K_{23}}{2K_1} \right) \nabla_i \hat{\theta}_a - \frac{1}{5} \frac{K_{23}}{2K_1} \nabla_a \tilde{\theta}_i + \delta_{ia} \boldsymbol{\nabla} \cdot \tilde{\boldsymbol{\theta}} \right\}.
\end{aligned}
\tag{4.164}
$$

4.8 Bending the Superfluid ^{3}He-A

The experimental interest lies in the possibility of preparing many nontrivial field configurations by gaining control over the directions of l- and d-vectors. Their presence can be detected by magnetic and sonic resonances. The principal means of enforcing certain field directions are the following:

1. *External Magnetic Fields*
   These try to enforce d⊥H with a strength comparable to the dipole energy if $\mathbf{H} \approx 35$ Oe. The energy is proportional to $(\mathbf{d} \cdot \mathbf{H})^2$. The microscopic reason for this collective effect is clear. The field $\mathbf{H}$ becomes the quantization axis so that the direction $\mathbf{d}$ (that specifies the direction along which the magnetic quantum number vanishes, $S_3 = 0$) is orthogonal to $\mathbf{H}$.

2. *Walls*
   Since l denotes the direction of the orbital angular momentum of the Cooper pairs, one expects l to stand orthogonal to the walls of the container since a plane of orbital motion parallel to the walls should energetically be favored over the orthogonal configuration. This expectation is verified by calculations.

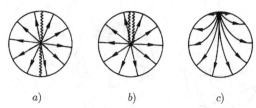

a)   b)   c)

FIGURE 4.7 The $\mathbf{l}\|\mathbf{d}$-field lines in a spherical container. There are necessarily two flux quanta associated with a singular point either in the form of two vortex lines (a) or of one with double circulation (b). Since vortex lines store condensation energy, they act approximately like a rubber band and draw the point to the wall (c), thereby generating a flower-like texture called a *boojum* [24, 25].

Apart from this, currents and probably also electric fields act as directional agents upon $\mathbf{l}$.

Let us now discuss what is called an open system. It is defined by a liquid in a container which is large compared to the dipole length ξ_d (i.e., much bigger than 10^3 cm) and with no magnetic field being present. In order to avoid the pile-up of dipole energy, $\mathbf{d}$ and $\mathbf{l}$-vectors will stay aligned over most of the volume. Only in the neighborhood with a radius ξ_d around the line-like singularities, where the bending energies become comparable with a dipole energy, alignment may be destroyed. Such singularities will be present in any sample prepared carelessly. Moreover, even with the most delicate cooling into the superfluid phase, the geometry of most containers will enforce the existence of some singularities. This will now be discussed separately.

4.8.1 Monopoles

If a sphere is cooled smoothly through the transition region, the field lines of $\mathbf{l}(\mathbf{x})$ will be planted uniformly orthogonal to the walls and develop towards the inside like the spines of a hedgehog. At some place there has to be a point-like singularity. Moreover, since the Euler characteristic of the sphere is $E = 2$, any closed surface orthogonal to the $\mathbf{l}$-field inside the liquid has to be passed by two vortex quanta. Possible field configurations are shown in Fig. 4.7. In the first case, two separate vortex lines of strength one emerge from the singularity, one running to the north, the other to the south pole. In the second case, there is, instead, one single line of vortex strength two at the north pole. In the third case the singularity has settled at the boundary forming a flower-like structure, a texture called a *boojum* [25]. The last case is apparently favored energetically since there is considerable condensation energy stored in the vortex line inside of which the liquid is normal. The vortex line acts like a rubber band (compare the next section on vortex lines) pulling the singularity to the boundary. The first situation corresponds to the field lines of $\phi^{(1)}$ and $\phi^{(2)}$ running along the lines of equal longitude or latitude like on a globe, the $\mathbf{l}$-vector pointing, of course, radially outward. North and south poles are singular.

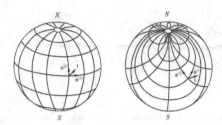

FIGURE 4.8 Two possible parametrizations of a sphere, either with two singularities, as the standard geographic coordinate frame on the globe, or with one singularity, as shown in the right-hand figure. The geographic parametrization corresponds to liquid ³He having two singular vortex lines, one emerging at the north pole and one at the south pole. The vectors $\boldsymbol{\phi}^{(1)} \times \boldsymbol{\phi}^{(2)}$ are tangential to the coordinates. The vector $\mathbf{l} = \boldsymbol{\phi}^{(1)} \times \boldsymbol{\phi}^{(2)}$ points radially outwards. The lower parametrization corresponds to one vortex line with double circulation emerging at the north pole. The south pole is a regular point.

The two other situations correspond to a parametrization of the globe with only one singularity at the north pole (see Fig. 4.8).

In order to estimate the energies let us parametrize the field lines as[10]

$$\mathbf{l} = \mathbf{e}_r, \quad \boldsymbol{\phi} = (\mathbf{e}_\theta + i\mathbf{e}_\varphi)\,e^{i\chi}. \tag{4.165}$$

Then the superfluid velocity is:

$$\mathbf{v}_s = \frac{1}{2m}\left(\boldsymbol{\nabla}\chi - \frac{\cot\theta}{r}\mathbf{e}_\varphi\right) \tag{4.166}$$

with a vorticity

$$2m\,(\boldsymbol{\nabla}\times\mathbf{v}_s) = \frac{1}{r^2}\mathbf{e}_r. \tag{4:167}$$

Integrating this over a spherical closed surface gives $4\pi = 2 \times 2\pi$, corresponding to the passing of *two* vortex units. Choosing $\chi \equiv 0$ we see $\mathbf{v}_s$ to be singular at $\theta = 0$ and π, so that two vortices of one quantum run from the center upwards or downwards [see Fig. 4.7a)], respectively. If χ is chosen to be $\chi = \varphi$, the singularity on the north (south) pole is cancelled with the other one being doubled [see Fig. 4.7(b)]. Inserting these configurations into the energy (4.126) with $2K_1 \sim K_{23} = 2K$ one has [26]

$$E = \frac{\rho_s}{m^2}\frac{\pi}{4}R\log\left(\frac{2R}{\xi} - \frac{5}{2}\right) \tag{4.168}$$

in the first case. Recall that $\rho_s \approx 2\rho\,(1 - T/T_c)$.

The energy of the second case is obtained by replacing $\log\,(2p/\xi - 5/2)$ by the larger value $2\log\,(2R/\xi - 7/4)$. The volume integration has to be cut off at the

---

[10]We neglect, for simplicity, all energy terms involving the **d**-vector.

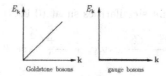

FIGURE 4.9 Spectra of Goldstone bosons versus gauge bosons. Goldstone bosons have energies going to zero with increasing wavelength due to an underlying symmetry. Gauge bosons have no energy for any vector case since their fields correspond to *local* symmetry transformations under which a gauge theory is invariant.

coherence distance $\xi = \xi_0/\sqrt{1 - T/T_c}$ away from the singularity. This is physically the correct procedure. Closer than ξ, the liquid cannot support the large bending energies coming from the directional change of $\phi^{(1)}$ and $\phi^{(2)}$ around the $\mathbf{v}_s$-vortex line and it escapes by Δ leaving the valley of minimal action and returning to the normal liquid point where $\Delta = 0$. At that point, $\mathbf{d}$ and $\boldsymbol{\phi}$ in (4.118) lose their meaning and the singularity is avoided. Since the energy is proportional to Δ^2 and Δ^4, it vanishes in the normal region so that the integration can be cut off there. Remember, though, that the complete energy consists of the sum of the bending energy e of (4.119) and the negative condensation energy $f_{\min}$ of (4.92). When comparing this structure to the monopole-like solutions in gauge theories coupled with Higgs fields [29] there is an essential difference: The energy increases with the radius of the sphere. The energy of a monopole, on the other hand, is constant. The reason for this is simple: In a σ-type of model, a field configuration which is radial asymptotically has a bending energy

$$\left(\nabla_i \frac{x_j}{r}\right)^2 = \left(\frac{\delta_{ij} - x_i x_j/r^2}{r}\right)^2 \sim \frac{1}{r^2}. \tag{4.169}$$

Hence, the integral diverges with R. In a gauge field theory, on the contrary, the vector potential is oriented radially, but the bending energy measures only the gauge invariant derivative $F_{\mu\nu}^2 = (\nabla_\mu A_\nu - \nabla_\nu A_\mu)^2$. This vanishes asymptotically very fast and all the energy is concentrated around the origin.

The situation can also be described in the particle language. In the σ-model, the nontrivial vacuum consists of a coherent superposition of static off-shell Goldstone bosons with many $\mathbf{k}$-vectors. Their energy increases with k^2 and even in the asymptotic region there is a considerable amount of energy. In gauge theory, the asymptotic region contains only longitudinal gauge particles which, by gauge invariance, correspond to Goldstone bosons with energies that vanish identically for all $\mathbf{k}$-vectors (see Fig. 4.9).

Therefore, the asymptotic region is free of energy. Since it is the curvature of the container walls which enforces asymptotic bending energy (or the presence of Goldstone bosons close to the walls) the growth of energy with the radius of a sphere cannot be avoided, even if one patches together the field of a monopole with that of another monopole and forms what may be called a monopolium [26]. In order to study the situation, the point singularities sit at $(0, 0, C)$ and $(0, 0, -C)$. Then an ansatz

$$\mathbf{l} = \frac{r^2 - C^2}{\lambda} \cos\theta \; \mathbf{e}_r + \frac{r^2 + C^2}{\lambda} \sin\theta \; \mathbf{e}_\theta \qquad (4.170)$$

with

$$\lambda \equiv \left[\left(r^2 - C^2 \right)^2 + 4C^2 r^2 \sin^2\theta \right] \qquad (4.171)$$

can be used to construct a vector field ϕ so that the superfluid velocity is

$$\mathbf{v}_s = \frac{1}{2M} \left\{ \boldsymbol{\nabla}\chi - \frac{r^2\cos^2\theta - C^2}{\lambda r \sin\theta} \mathbf{e}_\theta \right\}. \qquad (4.172)$$

The energy becomes

$$E = \frac{56}{24}\pi \frac{\rho_s}{m^2} R + \frac{\pi}{2}\frac{\rho_s}{m^2} C \left[\ln \frac{2C}{\xi} - \left(\frac{9}{4} + \frac{3\pi^2}{32} \right) \right] \qquad (4.173)$$

within a sphere of radius R. Thus, in addition to the energy proportional to R enforced by the spherical container, there is a linear binding energy with a logarithmic correction which stems from the bending energy in the neighborhood of the vortex line. The vortex line pulls the point singularities together according to an almost constant force.

Note, that apart from the first term in the energy caused by geometry, there is an essential difference of this σ-model result with what one expects for string like solutions of pure gauge theories. There, color is supposed to be screened completely in the vacuum so that the color field does not leave the vortex line. This is the reason why the force is purely linear! The monopolium state can be stabilized by placing ions of equal charge at both ends.

4.8.2 Line Singularities

If a cylindrical container is cooled, the l-lines will develop radially inwards. One therefore expects a singular line along the axis. At this line, the liquid would have to be in its normal state since the l-vectors are undefined. This amounts to the accumulation of a large condensation energy, which can be avoided by the l-lines flaring upwards like in a chimney [26] (see Fig. 4.10). Then the entire liquid can remain superfluid and contain only bending energies.

Quantitatively, the energy can be minimized by an l-field

$$\mathbf{l} = \mathbf{e}_z \cos\beta + \mathbf{e}_\rho \sin\beta, \qquad (4.174)$$

with

$$\beta(\rho = 0) = 0, \quad \beta(\rho = R) = \frac{\pi}{2}. \qquad (4.175)$$

There are many complex vectors ϕ which can be constructed with this l, for example:

$$\phi = e^{im\varphi} \left[-\sin\beta \; \mathbf{e}_z + \cos\beta \; \mathbf{e}_\rho + i\mathbf{e}_\varphi \right]. \qquad (4.176)$$

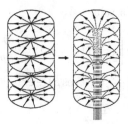

FIGURE 4.10 Cylindrical container with the $\mathbf{l} \parallel \mathbf{d}$ field lines spreading outwards when moving upwards. The line singularity on the left stores condensation energy. The curved configuration on the right contains only bending energy, which is preferable in a large container. Small containers and magnetic fields, on the other hand, may give preference to the singular line.

They lead to a superfluid velocity

$$\mathbf{v}_s = \frac{1}{2M\rho} \left(m - \cos\beta\right) \mathbf{e}_\rho. \tag{4.177}$$

At $m = 1$, there is no vortex line along the axis. This situation is favored energetically. Inserting $\mathbf{v}_s$ and $\mathbf{l}$ into the energy (4.125) and extremizing with respect to $\delta\beta(\rho)$, one finds the solution for $\beta(\rho)$ from the integral:

$$\frac{\rho}{R} = \exp\left(\int_{\beta(\rho)}^{T/2} \left\{ \frac{K_s \cos^2\beta + K_b \sin^2\beta}{K_s \sin^2\beta + \frac{\rho_s}{4m^2}\left(1 - \cos\beta\right)^2} \right\}^{1/2} d\beta \right). \tag{4.178}$$

The total energy of this configuration is

$$E \approx 1.145\pi \frac{\rho_s}{m^2} L, \tag{4.179}$$

where L is the length of the cylinder. Here, the weak coupling equalities (4.120) have been used.

Note that from (4.177) there is an azimuthal current flowing in this field configuration, which therefore may have a nonvanishing orbital angular momentum. In order to calculate this, consider the second, convective, part of the current

$$\nabla \times \mathbf{l} = (\mathbf{l} \cdot \nabla)\,\beta\mathbf{e}_\varphi = -(\cos\beta)'\mathbf{e}_\varphi. \tag{4.180}$$

This part also circulates around the axis but with a different radial dependence. The total angular momentum is then, due to (4.94),

$$\mathbf{L} = \int d^3x\,(\mathbf{x} \times \mathbf{j})\,. \tag{4.181}$$

It is directed along the z-axis with a value

$$
\begin{aligned}
L_z &= 2\pi \int dz d\rho \rho \left[\frac{\rho_s}{2M\rho}(1 - \cos\beta) - c\frac{\partial}{\partial\rho}(\cos\beta) \right] \\
&= 2\pi \int dz d\rho \left[\frac{\rho_s}{2M}(1 - \cos\beta) + c\cos\beta \right] \\
&\approx R^2 \frac{\rho_s}{2m} \left(1 - \frac{1}{\pi} + \frac{2}{\pi^2} \right) L.
\end{aligned} \tag{4.182}
$$

For the last equation, we have again inserted the weak-coupling value $c = \rho_s/4$. During the phase transition, the angular momentum must manifest itself in a recoil imparted upon the container. It would be interesting to detect this effect experimentally.[11]

There is also a way to prepare the singular vortex line. For this, a magnetic field has to be turned on along the z- axis which drives the **d**-vectors into the xy-plane. This enforces a singularity in the **d**-field lines along the axis causing the liquid to be normal there. Once the condensation energy is spent, the weak dipole force is sufficient to pull also the l-field into the radial direction [30].

4.8.3   Solitons

Let us now turn to planar textures in an open geometry [26, 28]. A direction may be defined by magnetic field pointing, say, along the z-axis. Then, the **d**-vectors will be forced to lie in the xy-plane:

$$
\mathbf{d} = \sin\psi\,\hat{\mathbf{x}} + \cos\psi\,\hat{\mathbf{y}}. \tag{4.183}
$$

The bending energy is minimized by a constant ψ in space. The dipole force pulls the l vector in the same or in the opposite directions. Since this force is very weak, there will be some regions where l is parallel and others where l is anti-parallel to d. The wall separating the different domains is stabilized by the competition between bending and dipole energy. If the thickness of the domain wall, a, shrinks, the bending energy density grows like $\frac{\rho_s}{m^2}\frac{1}{a^2} \times a$, while the corresponding dipole term falls off like $\frac{\rho_s}{m^2}\frac{1}{\xi d^2} \times a$. Conversely, a large domain accumulates an overwhelming dipole energy. Equilibrium is reached at $a \approx \xi_d^2$. If one studies, for simplicity, only configurations with a pure z-dependence and with l in the xy-plane

$$
\mathbf{l} = \sin\chi\,\hat{\mathbf{x}} + \cos\chi\,\hat{\mathbf{y}}, \tag{4.184}
$$

the most general $\mathbf{\Phi}$-vector is

$$
\mathbf{\Phi} = e^{i\varphi}\left(-\cos\chi\,\hat{\mathbf{x}} + \sin\chi\,\hat{\mathbf{y}} + i\hat{\mathbf{z}} \right), \tag{4.185}
$$

---

[11]Also, the boojum in a sphere has an angular momentum which would set the sphere into rotation when cooling through the transition point.

and the bending energy becomes, for the weak-coupling values of the parameters,

$$f_{\text{bend}} = \frac{\rho_s^{\|}}{8m^2} \left\{ 2(\boldsymbol{\nabla}\psi)^2 + 2(\boldsymbol{\nabla}\varphi)^2 + \frac{1}{2}(\boldsymbol{\nabla}\chi)^2 \right\}. \tag{4.186}$$

The dipole energy contributes

$$f_{\text{dip}} = \frac{\rho_s^{\|}}{8m^2} \frac{2}{\xi_d^{\perp 2}} \sin^2(\chi - \psi). \tag{4.187}$$

The phase φ occurs only in the bending energy and is uniform, in equilibrium. The remaining dependence on the fields χ, ψ can be diagonalized by setting

$$v \equiv \chi - \psi, \qquad u \equiv \chi + 4\psi. \tag{4.188}$$

Then, the energy takes the sine-Gordon form

$$f = f_{\text{bend}} + f_{\text{dip}} = \frac{\rho_s^{\|}}{4m^2} \left(\frac{1}{20}u_z^2 + \frac{1}{5}v_z^2 + \frac{1}{\xi_d^{\perp 2}}\sin^2 v \right). \tag{4.189}$$

This is minimized by a constant u and a soliton in the variable v:

$$\sin v_{\text{sol}} = \cosh^{-1}(z/\xi_{\text{sol}}), \qquad \tan \frac{v_{\text{sol}}}{2} = e^{\pm z/\xi_{\text{sol}}}, \tag{4.190}$$

where the width of the soliton is of the order of the dipole length

$$\xi_{\text{sol}} \equiv \frac{1}{\sqrt{5}}\xi_s^{\perp}, \tag{4.191}$$

as expected. The energy per unit area of the domain wall is

$$\frac{E}{\sigma} = \frac{\rho_s^{\|}}{4m^2} \frac{2}{\xi_d^2} \int_{-\infty}^{\infty} dz \cosh^{-1}(z/\xi_{\text{sol}}) = \frac{\rho_s^{\|}}{m^2} \frac{\xi_{\text{sol}}}{\xi_d^2} = \frac{\rho_s}{m^2} \frac{1}{\sqrt{5}} \frac{1}{\xi_d}. \tag{4.192}$$

The soliton corresponds to d- and l-vectors twisting in opposite directions inside the domain wall with l moving four times as far as d (see Fig. 4.11).

The presence of such a domain wall can be detected in the laboratory via a nuclear magnetic resonance experiment (NMR). Suppose a vibrating field is turned on along the z-axis (in addition to the static orienting field H^{ext}). This is what is done in a so-called longitudinal resonance experiment. Then the vector l associated with the spin fluctuates *around* the z-direction (see Ref. 16), so that its azimuthal angle is

$$\psi = \psi_{\text{sol}} + \delta, \tag{4.193}$$

and consequently

$$u = u_{\text{sol}} + 4\delta,$$
$$v = v_{\text{sol}} - \delta. \tag{4.194}$$

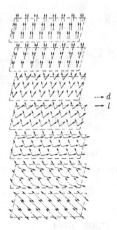

FIGURE 4.11 Field vectors in a composite soliton. At $z = +\infty$, $\mathbf{l}$ and $\mathbf{d}$ are parallel, at $z = -\infty$ they are antiparallel. Inside the domain wall of size ξ_d the vectors change their direction, $\mathbf{l}$ four times as much as $\mathbf{d}$.

This gives an additional vibrational energy

$$\delta^2 f = \frac{\rho_s^\|}{4m^2} \times \left[\delta_z^2 + \frac{1}{\xi_d^{\perp 2}} \left(1 - \frac{2}{\cosh^2(z/\xi_{\text{sol}})} \right) \delta^2 \right]. \tag{4.195}$$

The extrema of this energy correspond to the bound states of the Schrödinger equation[12]

$$\left[-\nabla_z^2 + \frac{1}{\xi_d^{\perp 2}} \left(1 - \frac{2}{\cosh^2(z/\xi_{\text{sol}})} \right) \right] \delta(z) = \lambda \delta(z). \tag{4.196}$$

This is a standard soluble problem (see the textbook on quantum mechanics by Landau-Lifshitz, ch. 23). The ground state is

$$\delta(z) \propto \frac{1}{\cosh^2(z/\xi_{\text{sol}})}, \tag{4.197}$$

with

$$s \equiv \frac{1}{2} \left[-1 + \sqrt{1 + 4\frac{2}{\xi_d^{\perp 2}} \xi_{\text{sol}}^2} \right] = \frac{1}{2} \left[-1 + \sqrt{1 + 4\frac{2}{5}} \right] \approx 0.306. \tag{4.198}$$

Since $s \leq 1$ there is only one bound state. This bound state has an energy

$$\lambda = \frac{1}{2} \left(\sqrt{65} - 7 \right) \frac{1}{\xi_d^{\perp 2}}. \tag{4.199}$$

---

[12]The time driving term can be shown to go as $\frac{1}{2}\delta^2$, so that λ corresponds to a frequency square.

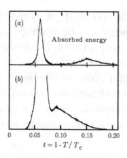
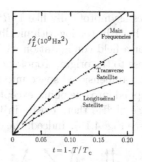

FIGURE 4.12 Nuclear magnetic resonance frequencies in a superfluid ^{3}He-A sample in an external magnetic field, as measured in Ref. [27]. The large peak corresponds to the main absorption line, the small peak to the right is a satellite frequency line attributed to the trapping of spin waves in planar domain walls. The lower part of the figure shows the position of these lines for different external frequencies of the longitudinal magnetic field. The ratio of the satellite frequency to the main frequency agrees with the theoretical calculation.

The continuum has a spectrum

$$\lambda = k^2 + \frac{1}{\xi_{\mathrm{d}}^{\perp 2}}. \tag{4.200}$$

Experimentally, the vibrating field is homogeneous so that in the continuum only the $\mathbf{k} = 0$ value is excited. This leads to the main NMR resonance absorption line. If now a soliton is present, this is the only observed signal. The bound state trapped by the soliton has the effect of creating a line whose frequency lies by a factor $\frac{1}{2}\left(\sqrt{65} - 7\right) \approx (0.728)^2$ below the main line. Such a "satellite" frequency has indeed been observed experimentally (see Fig. 4.12).

Note that the satellite line provides a good test for the weak-coupling values of the coefficients $K_{1,2,3}$ in the bending energies. If κ denotes the ratio

$$\kappa \equiv 2K_1/(K_2 + K_3), \tag{4.201}$$

the frequency should be found at

$$\frac{1}{2\kappa}\left[\sqrt{(3\kappa + 2)(\kappa + 2)} - (5\kappa + 2)\right], \tag{4.202}$$

instead of $(0.728)^2$. The experimental value $(0.74)^2$ implies that κ is close to 1, in agreement with the weak-coupling result.

If the parameter s had been a positive number, there would have been more bound states, one for each $n = 0, 1, 2, \ldots, s$.

4.8.4 Localized Lumps

We have argued before that the energies of point- and line-like singularities are necessarily not localized. In a hedgehog-like field structure, the σ-model bending energy behaves asymptotically like $1/R^2$, so that the spherical (or cylindrical) integral diverges linearly (or logarithmically). The energy can be confined to a small region only for a field configuration which is asymptotically flat but contains some knots, say, close to the origin. Topologically, one has to find a nontrivial mapping of the entire three-dimensional space into the parameter space of the liquid where all points are mapped into one point, except those in a small neighborhood of the origin. In the A-phase of ^{3}He there exists, in fact, such a mapping with l- and d-vectors aligned [35] (see Fig. 4.13). Indeed, the covering space of the parameter space SO(3)

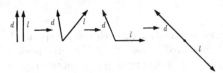

FIGURE 4.13 Vectors of orbital and spin orientation in the A-phase of superfluid ^{3}He.

is SU(2), which is equivalent to S_3, the surface of a sphere in four dimensions. Since the ordinary space corresponds to S_3, which is the space S_3 with the north pole removed, one has a nontrivial mapping $S_3 \to S_3'$ with a large neighborhood of the north pole of S^3 mapped into one point of S_3, accounting for asymptotic uniformity. This corresponds to a diffuse smoke-ring type of configuration which moves through the liquid with a velocity $\mathbf{v} = \hbar/mR$, a momentum $P \approx \hbar \rho_s R^2/m$, and an energy $E \approx \hbar^2 \rho_s R/m^2$, respectively. Note that the velocity is inversely proportional to the radius R of the smoke-ring.

Actually, the topological stability does not prevent this object from having only a small lifetime. While it moves through the liquid, orbital friction eats up the energy and decreases the size. Once the object has shrunk to the size of the order of the dipole length ξ_d, the locking and $\mathbf{d}$ and l-vectors will be overcome, and the parameter space is increased to

$$R = S^2 \times SO(3)/Z_2. \tag{4.203}$$

Then the topological stability is lost. The knot in field space unwinds and disappears.

4.8.5   Use of Topology in the A-Phase

In the A-phase of ^{3}He, as in gauge theories, topological considerations are helpful in classifying the different stable field configurations. In the superfluid, topological stability means that there is no continuous deformation to a lower energy state within the hydrodynamic limit. Since this limit is an approximation, the stability is not perfect. The size of the order parameter Δ which, in the hydrodynamic limit, is assumed to be pinned down at the value of minimal energy, does in fact fluctuate. On rare occasions it will arrive at the point $\Delta = 0$, where the liquid becomes normal. This process is called nucleation. For example, there is topological stability in a superconductor contained in a torus with the phase of the order parameter changing by $e^{in\varphi}$ when going once around the circle. There is no continuous way to relax the ensuing supercurrent in the hydrodynamic limit. But the supercurrent *does* decay, albeit it may take years. The reason is that, at some place on the inner boundary, the size of the order parameter may, by fluctuations, climb up from the valley of lowest energy into the normal phase with $\Delta = 0$. There the phase φ loses its meaning and can unwind by one unit of 2π. This point may lie at the inside of the torus and can develop into a thin flux tube. This tube can carry one unit of electric flux away from the supercurrent. Such a process is facilitated by putting together two superconductors in a Josephson junction where the diffusion of such units can be observed in the clearest fashion. Thus topological stability in the hydrodynamic regime really amounts to metastability with extremely long life times. For most purposes, such life times can be assumed to be infinite. Then the topological classification provides us with good quantum numbers of field configurations.

What is the connection between two field configurations of the same topological class? They can be deformed into each other by continuous changes only of the directions of the fields with the magnitude being fixed. If initial and final states are both dynamically stable, there is an energy barrier to be crossed during such a deformation. Its energy density is only due to the bending of the field lines and, therefore, extremely small as compared with the condensation energy which enforces topological stability.

Consider now the topology in the parameter space of ^{3}He [32]. In the A-phase, the vacuum is determined by the product of the vectors d_a and ϕ_i. The vector d_a covers the surface of the unit sphere in 3 dimensions, S^2, the complex vector $\phi = \phi^{(1)} + i\phi^{(2)}$ is a three-parameter space equivalent to the space $SO(3)$, i.e., a sphere of radius $r = 1$ with diametrically opposite points at the surface identified. Every point is determined by the direction of the vector $\mathbf{l} = \phi^{(1)} \times \phi^{(2)}$ and the length which characterizes the azimuthal angle of $\phi^{(1)}$ in the plane orthogonal to $\mathbf{l}$. Due to the occurance of a product $d_a\phi_i$, a sign change of d_a can be absorbed in ϕ_i so that the total parameter space is

$$R = S^2 \times SO(3)/Z_2. \tag{4.204}$$

Stable singular points exist if the homotopy group $\pi_2(R)$ of mappings of the sphere S^2 in three dimensions into this parameter space is nontrivial. But it is well-known [33] that, for the above space $\pi_2(R) = Z$, the group consists of the integer numbers. Thus each point singularity is characterized by an integer number. There can be infinitely many different stable classical field configurations of the monopole type. This purely topological argument is based on the independence of the vectors $\mathbf{d}$ and $\mathbf{l}$. We know, however, that the dipole force tries to align the $\mathbf{d}$ and $\mathbf{l}$-vectors. For this reason, as soon as the size of the container exceeds the dipole length $\xi_d \approx 10^{-3}$ cm, $\mathbf{d}$ and $\mathbf{l}$ will stay parallel asymptotically thereby reducing the parameter space to

$$R = SO(3) \quad (\mathbf{d} \parallel \mathbf{l}). \tag{4.205}$$

Then the homotopy group is $\pi_2(\mathrm{R}) = 0$ and there are no monopoles.

Thus monopoles could be created only in very small regions ($r << 10^{-3}$, cm) of the liquid. Their $\mathbf{d}$ and $\mathbf{l}$ field lines are non-aligned. As a consequence, their neighborhood contains considerable dipole energy. If the volume of the neighborhood becomes much larger than the dipole length, fluctuations in the liquid cause nucleation to the normal phase with the monopole vanishing in favor of a $\mathbf{d} \parallel \mathbf{l}$ alignment and no dipole energy. Quantitatively, the transition point to the configuration with $\mathbf{d} \parallel \mathbf{l}$ is determined by the competition of the small dipole energy density $f_d \sim (\rho/m^2)(1/\xi_d^2)$, stored in a finite volume, with the large condensation energy density $f_c \sim (\rho/m^2)(1/\xi^2)$, stored in the immediate neighborhood vlume ξ^3 of the singularity of size $\xi(\approx 1000\,\text{Å})$. The relaxation occurs at

$$\frac{R^3}{\xi_d^2} > \frac{\xi^3}{\xi^2}, \qquad (4.206)$$

or

$$R > \sqrt[3]{\xi \xi_d^2} \approx 10^{-4}\,\text{cm}. \qquad (4.207)$$

For line singularities we have to consider $\pi_1(R)$. If $\mathbf{d}$ and $\mathbf{l}$ are independent of each other, this homotopy group is

$$\pi_1(\mathrm{R}) = Z_4. \qquad (4.208)$$

Hence there are four types of line singularities which can be labelled by their vortex strengths $s = \pm\frac{1}{2}, \pm 1$. Examples:

$$\pm\frac{1}{2} \;\; : \;\; \boldsymbol{\Phi} = e^{\pm i\xi/2}\left(\mathbf{e}_x + i\mathbf{e}_y\right), \quad \mathbf{d} = \mathbf{e}_x \cos\frac{\gamma}{2} \mp \mathbf{e}_y \sin\frac{\gamma}{2}, \quad \mathbf{l} = \mathbf{e}_z \qquad (4.209)$$

$$\pm 1 \;\; : \;\; \boldsymbol{\Phi} = \left(\mathbf{e}_z + i\mathbf{e}_\varphi\right), \qquad\qquad \mathbf{d} = \mathbf{e}_\rho, \quad \mathbf{l} = \mathbf{e}_\rho. \qquad (4.210)$$

As the volume increases, $R \gg \xi_d$, the dipole force leads again to alignment of $\mathbf{d}$ and $\mathbf{l}$, reducing the parameter space to

$$\pi_1(\mathrm{R}) = Z_2 \quad (\mathbf{l} \parallel \mathbf{d}). \qquad (4.211)$$

Thus only two types of singular lines survive and one sees from (4.209) that it is the ± 1 vortex lines which survive.

4.8.6   Topology in the B-Phase

The discussion of the hydrodynamic limit can be extended to the B-phase. Consider the parametrization (4.104) of the degenerate ground state

$$A_{ai} = \Delta_B R^{ai}\left(\mathbf{n}, \theta\right) e^{i\varphi} \qquad (4.212)$$

with Δ_B pinned at the point (4.107) of minimal energy density (4.107). The matrix R may be written explicitly as

$$R_{ai}\left(\mathbf{n}, \theta\right) = \cos\theta\,\delta_{ai} + (1 - \cos\theta)\,n_a n_i + \sin\theta\,\epsilon_{aik} n_k. \qquad (4.213)$$

Inserting this into the collective action (4.78), (4.80) the energy becomes the sum of bending energies involving gradients of $\theta, \mathbf{n}$ and φ.

The parameter space of (4.212) consists of the direct product of a phase (which is isomorphic to the circle S^1) and the group space $SO(3)$. As the dipole force is turned on, the angle θ is pined at the value $\theta \approx 104°$ and the space $SO(3)$ is narrowed down to the different *directions* of $\mathbf{n}$ only, covering the surface of a sphere S^2. The point- and line-like singularities are classified by considering the homotopy groups $\pi_2(R)$ and $\pi(R)$ of the parameter spaces $R=S^2\times SO(3)$ for small configurations, $r \ll \xi_d$, and $R = S^2 \times S^2$ for large ones.

In the first case one has

$$\pi_2(R) = 0, \quad \pi_1(R) = Z + Z_2. \tag{4.214}$$

Thus, there are no topologically stable point singularities while there are two types of vortex lines: One set has its origin in the pure phase $e^{i\phi}$ of the parametrization and is characterized by an arbitrary integer N. These vortex lines are of exactly the same type as those of superfluid ^{4}He. In addition, there are singular lines in the $\mathbf{n}$, θ parameter space, two of which can annihilate each other (due to Z_2). For large samples where $\theta = 104°$, the homotopy groups are

$$\pi_2(R) = Z, \quad \pi_1(R) = Z. \tag{4.215}$$

Thus, there are stable point like solutions of arbitrary integer charge, the simplest being a hedgehog with the $\mathbf{n}$-vector pointing radially. The line singularities are all due to the phase $e^{i\varphi}$ and therefore again of the same nature as in superfluid bosonic helium.

The B-phase possesses also interesting planar structures. In order to classify them one has to map the line $z \in (-\infty, \infty)$ into the parameter space $SO(3)$. In an open geometry any such mapping can be deformed into the identity. Stable configurations arise if a magnetic field is turned on along the z-direction which aligns the $\mathbf{n}$ vector parallel or anti-parallel. Note, however, that, contrary to the A-phase, the directional energy of the magnetic field is quite weak: Since the B-phase corresponds to $J = 0$ -Cooper pairs, only the small distortion of the wave function caused by the dipole coupling which leads to a net magnetic energy of the order of

$$f_{\text{mg}}(H) \sim g_{\text{d}} \left(\frac{\gamma}{\Delta}\right)^2 (\mathbf{n} \cdot \mathbf{H})^2 \tag{4.216}$$

Thus, the characteristic length over which bending and magnetic energies are comparable is much larger than ξ_d, namely

$$\xi_{\text{mg}}(H) \sim \frac{\Delta}{\gamma H}\xi_d. \tag{4.217}$$

With $\Delta \sim 1\,\text{m}^0\text{K}$ and $\gamma H \sim 0.156 \times 10^3\,\text{m}^0\text{K/gauss}$, this is, at 100 gauss, of the order of one mm. At large distances, however, this weak-coupling does result in the

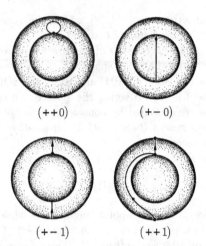

$$(++0) \qquad\qquad (+-0)$$

$$(+-1) \qquad\qquad (++1)$$

FIGURE 4.14 Parameter space of ^{3}He-B containing the parameter space of the rotation group. Thus, points which lie diametrically opposite on the surface are identical. The dipole energy (hyperfine coupling between the spins) favors a spherical shell within this sphere. It corresponds to rotations around any axis by 104°. Planar textures (solitons) have to start and end asymptotically on this shell. The figure shows the four topologically distinct classes of paths starting and ending on this spherical shell.

n-vector lying parallel or anti-parallel to **H**. By the same token, also the dipole force is active and the angle θ between **d** and **l** settles at the value $\theta \approx 104°$.

We can visualize this asymptotic situation by drawing a sphere of radius π and by specifying, within this, the surface of fixed radius forming at a given $\theta \approx 104°$. Then, any planar field configuration corresponds to a line starting and ending at the north or south pole of the $\theta \approx 104°$ -surface. Thus, the asymptotic space is Z_2. There are eight classes of mappings, four of which are the mirror images of the others. They are shown in Fig. 4.14.

The first class $(+ + 0)$ is trivial and can be deformed continuously into the uniform field configuration. The second, $(+ - 0)$, is a θ soliton where θ starts out and ends at $\theta \sim 104°$. The third soliton, $(+-1)$, has the angle θ run from 104° to π and come back from the identical point at the south pole into 104° with **n** pointing in the opposite direction. The last class, $(+ + 1)$, is topologically equivalent to the sum of the previous two and can, in fact decay into them.

In order to imagine the different energies of these field configurations remember that the dipole force makes the radial shells have constant dipole energy with a minimal valley at the shell $\theta \approx 104°$. The magnetic force, on the other hand, pulls **n** into the z-direction, thus creating a potential valley running through the sphere from north to south. Since the magnetic force is much weaker than the dipole force, however, this valley is extremely flat. Let us now follow the movement of the field configuration as z runs from $+\infty$ to $-\infty$. Clearly, the order parameter θ likes to

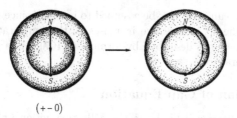

$(+-0)$

FIGURE 4.15 Possible path followed by the order parameter in a planar texture (soliton) when going from $z = -\infty$ to $z = +\infty$. The path tries to stay all the way on the spherical shell preferred by the dipole energy.

stay close to the north and south poles for the largest possible portion of the z-axis. The crossing over to the other takes place on a small piece only. The dipole energy is the strongest effect at hand, the value of θ stays fixed at 104°. Thus, the curve representing the field moves as shown in Fig. 4.15.

While crossing to the other side, θ moves through the valley $\theta \approx 104°$ and has to overcome only the magnetic energy. Correspondingly, the soliton $(+ - 0)$ has the size determined by ξ_{mg}, which is quite large. This is in contrast to the soliton $(+ - 1)$, which always has to cross the dipole barrier and has the much smaller size ξ_d.

Finally, the last configuration $(+ + 1)$ can lower its energy by deforming the line as shown in Fig. 4.16. By inspecting this figure it is obvious that such a soliton can decay into the previous two, one with dipole and one with the much lower magnetic energy.

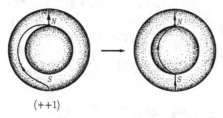

$(++1)$

FIGURE 4.16 Another possible class of solitons has an order parameter which starts at the north pole (say) of the spherical shell, goes to the surface of the sphere, re-emerges at the diametrically opposite (identical) point and ends up at the point it started out. Along the way it tries to stay as much as possible on the spherical shell preferred by the dipole energy.

4.9   Hydrodynamic Properties at All Temperatures $T \leq T_c$

Until now, the discussion has been limited to temperatures in the vicinity of T_c. Only then can the expansion of the collective action (4.66) in A_{ai} converge. For

$T \ll T_c$ the field A_{ai} can no longer be assumed to be small since it fluctuates around a finite average value A_{ai}^0 whose size increases as the temperature decreases. For $T \approx T_c$, such a behavior is described by formula (4.107). Here we shall extend the results to all temperatures.

4.9.1   Derivation of Gap Equation

We separate the collective field A_{ai} in the collective action (4.66) into a constant background field A_{ai}^0, plus fluctuations A'_{ai} around it, by setting

$$A_{ai} \equiv A_{ai}^0 + A'_{ai}. \tag{4.218}$$

Then the collective action becomes a functional of A'_{ai}:

$$\mathcal{A}_{\text{coll}}[A'_{ai}] = -\frac{i}{2} \text{Tr} \log \begin{pmatrix} i\partial_t - \xi(-i\boldsymbol{\nabla}) & i\tilde{\nabla}_i \sigma_a(A_{ai}^0 + A'_{ai}) \\ i\tilde{\nabla}_i \sigma_a(A_{ai}^{0*} + A'^*_{ai}) & i\partial_t + \xi(i\boldsymbol{\nabla}) \end{pmatrix}$$

$$-\frac{1}{3g} \int d^4x A_{ai}^{0*} A_{ai}^0 - \frac{1}{3g} \int d^4x \left(A_{ai}^{0*} A'_{ai} + \text{c.c.} \right) - \frac{1}{3g} \int d^4x A_{ai}'^* A'_{ai}. \tag{4.219}$$

Here we have introduced the dimensionless derivative

$$\tilde{\nabla}_i \equiv \frac{1}{2p_F} \overset{\leftrightarrow}{\boldsymbol{\nabla}}_i. \tag{4.220}$$

The trace log part can be rewritten by analogy with (4.70) as

$$\mathcal{A}_{\text{coll}}[A'_{ai}] = -\frac{i}{2} \text{Tr} \log \begin{pmatrix} i\partial_t - \xi(-i\boldsymbol{\nabla}) & i\tilde{\nabla}_i \sigma_a A_{ai}^0 \\ i\tilde{\nabla}_i \sigma_a A_{ai}^{0*} & i\partial_t + \xi \end{pmatrix} - \frac{1}{3g} \int d^4x A_{ai}^{0*} A_{ai}^0$$

$$- \frac{i}{2} \text{Tr} \log \left\{ 1 - iG_{A^0} \begin{pmatrix} 0 & i\tilde{\nabla}_i \sigma_a A'_{ai} \\ i\tilde{\nabla}_i \sigma_a A'^*_{ai} & 0 \end{pmatrix} \right\}$$

$$- \frac{1}{3g} \int d^4x \left(A_{ai}^{0*} A'_{ai} + \text{c.c.} \right) - \frac{1}{3g} \int d^4x |A'_{ai}|^2, \tag{4.221}$$

where

$$G_{A^0} \equiv i \begin{pmatrix} i\partial_t - \xi(-i\boldsymbol{\nabla}) & i\tilde{\nabla}_i \sigma_a A_{ai}^0 \\ i\tilde{\nabla}_i \sigma_a A_{ai}^{0*} & i\partial_t + \xi(i\boldsymbol{\nabla}) \end{pmatrix}^{-1} \tag{4.222}$$

is the propagator in the presence of the constant A^0 field.

The first two terms can be dropped since they are an irrelevant constant due to their lack of depending on the fluctuating field A'. Expanding in powers of A', we have

$$\mathcal{A}_{\text{coll}}[A'_{ai}] = \sum_{n=1}^{\infty} \mathcal{A}_n[A'_{ai}] \tag{4.223}$$

with a linear term

$$\mathcal{A}_1[A'_{ai}] = \frac{1}{2} \text{Tr} \left(G_{A^0} i\tilde{\nabla}_i \sigma_a A'_{ai} \frac{\tau^+}{2} \right) - \frac{1}{3g} \int d^4x A_{ai}^{0*} A'_{ai} + \text{c.c.} \tag{4.224}$$

where $\tau^+/2$ is the same 2×2-matrix as $\sigma^+/2 = \begin{pmatrix} 0 & 1 \\ 0 & 0 \end{pmatrix}$ but acting on the two field components of (4.65). The quadratic term is

$$\mathcal{A}_2[A'_{ai}] = \frac{i}{4} \mathrm{Tr} \left[G_{A^0} \begin{pmatrix} 0 & i\tilde{\nabla}_i \sigma_a A'_{ai} \\ i\tilde{\nabla}_i \sigma_a A'^{*}_{ai} & 0 \end{pmatrix} \right]^2 - \frac{1}{3g} \int d^4x A'^{*}_{ai} A'_{ai}. \quad (4.225)$$

The linear term is eliminated by the requirement that $\mathcal{A}$ is stationary under fluctuations in A'_{ai}. This condition yields the *gap equation*:

$$A^0_{ai} = \frac{3g}{2} \mathrm{Tr} \left(\sigma_a i\tilde{\nabla}_i G_{A^0}(x, y) \frac{\tau^-}{2} \right) \Bigg|_{x=y-\epsilon}. \quad (4.226)$$

The propagator (4.222) can be calculated most easily for the case of a unitary matrix

$$\Delta_{\alpha\beta}(\tilde{p}) \equiv \tilde{p}_i(\sigma_a)_{\alpha\beta} A_{ai}, \quad (4.227)$$

where $\tilde{p}$ denotes the dimensionless vector $\mathbf{p}/p_F$. Then

$$\Delta_{\alpha\beta} \Delta^{\dagger}_{\beta\gamma} = \frac{1}{2} \mathrm{Tr}(\Delta\Delta^{\dagger}) \delta_{\alpha\gamma}. \quad (4.228)$$

The condition is satisfied if A^0_{ai} has the form (4.103) in the A-phase or (4.104) in the B-phase [not, however, for the A_1-phase (4.105)]. In A- and B-phases the right-hand side becomes

$$\left\{ \begin{matrix} \Delta^2_A \sin^2\theta \\ \Delta^2_B \end{matrix} \right\} \tilde{p}^2 \equiv \Delta^2 \tilde{p}^2, \quad (4.229)$$

where θ is the angle between $\mathbf{l}$ and the momentum vector $\tilde{p}$. In momentum space, the propagator is

$$G_{A^0}(\omega, p) = \frac{1}{\omega^2 + \xi^2(p) + \Delta^2_A \sin^2\theta \ \tilde{p}^2} \begin{pmatrix} (i\omega + \xi(\mathbf{p}))\delta_{\alpha\beta} & -\Delta_{\alpha\beta}(\tilde{p}) \\ -\Delta^{\dagger}_{\alpha\beta}(\tilde{p}) & (i\omega - \xi(\mathbf{p}))\delta_{\alpha\beta} \end{pmatrix} (4.230)$$

for the A-phase, with $\Delta^2_A \sin^2\theta$ replaced by Δ^2_B in the B-phase. This matrix can be diagonalized via a so-called *Bogoliubov-transformation*. This is a 2×2-matrix in which the diagonal values display pure propagators of the energy

$$E(\mathbf{p}) = \pm \sqrt{\xi(\mathbf{p})^2 + \left\{ \begin{matrix} \Delta^2_A \sin^2\theta \\ \Delta^2_B \end{matrix} \right\}}. \quad (4.231)$$

The energies show a gap $\Delta^2_A \sin^2\theta$ or Δ^2_B. In the B-phase, the gap is isotropic just as in a superconductor. In the A-phase, on the other hand, there is an anisotropy along the $\mathbf{l}$-axis with the gap vanishing for momenta along $\mathbf{l}$.

The size of the gap is found by solving the gap equation (4.226). Inserting (4.230), this takes the form

$$
A_{ai}^0 = 3g \sum_j T \sum_{\omega_n,\mathbf{p}} \tilde{p}^i \tilde{p}^j \frac{1}{\omega_n^2 + E^2(\mathbf{p})} A_{aj}^0
$$

$$
= 3g \sum_j \sum_{\mathbf{p}} \tilde{p}^i \tilde{p}^j \frac{1}{2E(\mathbf{p})} \tan \frac{E(\mathbf{p})}{2T} A_{aj}^0, \tag{4.232}
$$

or

$$
\frac{\delta_{ij}}{3g} = \sum_{\mathbf{p}} \tilde{p}^i \tilde{p}^j \frac{1}{2E(\mathbf{p})} \tan \frac{E(\mathbf{p})}{2T}. \tag{4.233}
$$

The momentum integration can be split into size and direction [recall (3.61)]

$$
\int \frac{d^3p}{(2\pi)^3} \approx \mathcal{N}(0) \int \frac{d\hat{\mathbf{p}}}{4\pi} \int d\xi, \tag{4.234}
$$

where $\mathcal{N}(0)$ is the density of states (3.62) at the surface of the Fermi sea. Since the integration over $d\xi$ is cut off at a value $\omega_{\text{cutoff}} \approx \frac{1}{10}T_F$, the momenta stay sufficiently close to the Fermi sphere forcing the dimensionless vectors $\tilde{\mathbf{p}} \equiv \mathbf{p}/p_F$ to be approximately equal to the unit vectors $\hat{\mathbf{p}} \equiv \mathbf{p}/|\mathbf{p}|$. Then (4.233) becomes

$$
\frac{1}{g} \int \frac{d\hat{\mathbf{p}}}{4\pi} \hat{p}_i \hat{p}_j \approx \mathcal{N}(0) \sum_j \int \frac{d\hat{\mathbf{p}}}{4\pi} \hat{p}_i \hat{p}_j \left(\int_{-\omega_{\text{cutoff}}}^{\omega_{\text{cutoff}}} d\xi \frac{1}{2E} \tan \frac{E}{2T} \right). \tag{4.235}
$$

We may eliminate the coupling constant g in favor of the critical temperature T_c by using (4.76). This gives

$$
\int_{-\omega_{\text{cutoff}}}^{\omega_{\text{cutoff}}} d\xi \frac{1}{2\xi} \tan \frac{\xi}{2T_c} = \frac{3}{4} \int_{-1}^{1} dz \left(1 - z^2\right) \int_{\omega_{\text{cutoff}}}^{\omega_{\text{cutoff}}} d\xi \frac{1}{2\sqrt{\xi^2 + \Delta^2}} \tan \frac{\sqrt{\xi^2 + \Delta^2}}{2T}. \tag{4.236}
$$

In order to extract the finite content, one may subtract

$$
\int_{\omega_{\text{cutoff}}}^{\omega_{\text{cutoff}}} d\xi \frac{1}{2\xi} \tan \frac{\xi}{2T}
$$

on both sides. Then ξ-integral converges and we can remove the cutoff which leads to

$$
\log \frac{T}{T_c} = \frac{3}{4} \int_{-1}^{1} dz \left(1 - z^2\right) \int_{-\infty}^{\infty} d\xi \left[\frac{1}{2\sqrt{\xi^2 + \Delta^2}} \tan \left(\sqrt{\xi^2 + \Delta^2}/2T \right) - \frac{1}{2\xi} \tan \frac{\xi}{2T} \right]. \tag{4.237}
$$

From this we may calculate T/T_c as a function of Δ_{AB}/T_c as follows:

For small T, the integral increases as $\log T$ due to the small-ξ behavior of the second term. The finite term is determined by setting $\tan\left(\sqrt{\xi^2 + \Delta^2}/2T\right) \approx 1$ (which is good to fit experimental data, except in the A-phase for $z \approx \pm 1$) and integrating the first while partially integrating the second term:

$$
\begin{aligned}
\log \frac{T}{T_c} &\approx \frac{3}{4}\int_{-1}^{1} dz(1 - z^2)\left\{\left(\log \frac{\sqrt{\xi^2 + \Delta^2} + \xi}{2T} - \log \frac{\xi}{2T}\tan\frac{\xi}{2T}\right) + \int_{0}^{\infty} dx \frac{\log x}{\cosh^2 x}\right\} \\
&= \frac{3}{4}\int_{-1}^{1} dz(1 - z^2)\left(\log \frac{4T}{\Delta} - \log \frac{4e^\gamma}{\pi}\right) \\
&= \log \frac{T}{\Delta_{\max}e^\gamma/\pi} - \frac{3}{8}\int_{-1}^{1} dz(1 - z^2)\log \frac{\Delta^2}{\Delta_{\max}^2}.
\end{aligned}
\tag{4.238}
$$

In the B-phase, $\Delta \equiv \Delta_{\max} = \Delta_B$, and

$$
\Delta_B/T_c = \pi e^{-\gamma} \approx 1.76, \quad T \approx 0. \tag{4.239}
$$

In the A-phase, $\Delta = \Delta_A \sin\Theta$, and the integral becomes

$$
-\frac{3}{8}\int_{-1}^{1} dz(1 - z^2)\log(1 - z^2) = \frac{5}{6} - \log 2 \approx \log 1.15, \tag{4.240}
$$

so that

$$
\Delta_A/T_c = \pi e^{-\gamma}\frac{e^{5/6}}{2} \approx 2.03. \tag{4.241}
$$

For small T, this value is approached exponentially fast $\sim e^{-\Delta_B/T}$ for the B-phase and with a power law T^4 for the A-phase (due to the vanishing of $\Delta_A \sin\theta$ along the anisotropy axis l).

For arbitrary T, the calculation of (4.237) is done (as in the case of superconductivity in Chapter 3) by using the expansion into Matsubara frequencies

$$
\frac{1}{2E}\tan\frac{E}{2T} = \frac{1}{2E}T\sum_{\omega_n}\left(\frac{1}{i\omega_n + E} - \frac{1}{i\omega_n - E}\right) = T\sum_{\omega_n}\frac{1}{\omega^2 + \xi^2 + \Delta^2}. \tag{4.242}
$$

This can be integrated over ξ to find, for the gap equation (4.237):

$$
\log \frac{T}{T_c} = 2\pi \frac{3}{4}\int_{-1}^{1} dz(1 - z^2)\, T\sum_{\omega_n > 0}\left(\frac{1}{\sqrt{\omega_n^2 + \Delta^2}} - \frac{1}{\omega_n}\right). \tag{4.243}
$$

At this place we introduce the auxiliary dimensionless quantity

$$
\delta = \frac{\Delta}{\pi T}, \tag{4.244}
$$

and a reduced version of the Matsubara frequencies:

$$
x_n \equiv (2n + 1)/\delta. \tag{4.245}
$$

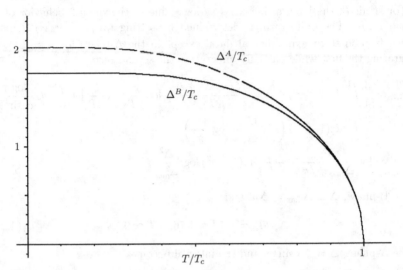

FIGURE 4.17 Fundamental hydrodynamic quantities of superfluid ³He-B and -A, shown as a function of temperature. The superscript FL denotes the Fermi liquid corrected values.

Then the gap equation (4.237) takes the form

$$\log \frac{T}{T_c} = \frac{2}{\delta} \left[\frac{3}{4} \int_{-1}^{1} dz (1 - z^2) \right] \sum_{n=0}^{\infty} \left(1 \Big/ \sqrt{x_n^2 + \left\{ \begin{matrix} 1 \\ 1 - z^2 \end{matrix} \right\}} - 1/x_n \right) \qquad (4.246)$$

in the B- and in the A-phase, respectively. In the B-phase, the angular integral of the parentheses gives a factor 1, so that

$$B: \quad \log \frac{T}{T_c} = \frac{2}{\delta} \sum_{n=0}^{\infty} \left(\frac{1}{\sqrt{x_n^2 + 1}} - \frac{1}{x_n} \right). \qquad (4.247)$$

In the A-phase it leads to

$$A: \quad \log \frac{T}{T_c} = \frac{2}{\delta} \sum_{n=0}^{\infty} \left\{ \frac{3}{4} \left[(1 - x_n^2) \arctan \frac{1}{x_n} + x_n \right] - \frac{1}{x_n} \right\}. \qquad (4.248)$$

The curves $\Delta_{A,B}/T_c$ are plotted in Fig. 4.17.

The $T \approx T_c$ behavior can be extracted from (4.246) by expanding the sum for large x_n. The leading term is

$$\left\{ \begin{matrix} 1 \\ 1 - z^2 \end{matrix} \right\} \sum_{n=0}^{\infty} \frac{1}{2x_n^3} = \left\{ \begin{matrix} 1 \\ 1 - z^2 \end{matrix} \right\} \frac{\delta^3}{2} \sum_{n=0}^{\infty} \frac{1}{(2n+1)^2} = \left\{ \begin{matrix} 1 \\ 1 - z^2 \end{matrix} \right\} \frac{\delta^3}{2} \frac{7}{8} \zeta(3),$$

$$(4.249)$$

so that

$$\Delta_B/T_c = \pi\delta = \pi\sqrt{\frac{8}{7\zeta(3)}}\left(1 - \frac{T}{T_c}\right)^{1/2} \approx 3.063\left(1 - \frac{T}{T_c}\right)^{1/2},$$

$$\Delta_A/T_c = \pi\delta = \pi\sqrt{\frac{10}{7\zeta(3)}}\left(1 - \frac{T}{T_c}\right)^{1/2}, \tag{4.250}$$

in agreement with the determination (4.107).

4.9.2 Ground State Properties

The superfluid densities do not only characterize the hydrodynamic bending energies. They also appear in the description of the thermodynamic quantities of the ground state. Close to the critical temperature T_c, these can be extracted directly from the Ginzburg-Landau free energies (4.108). These limiting results can be used to cross-check the general properties to be calculated now.

Free Energy

Since the ground state field A_{ai}^0 is constant in space and time the first two terms in Eq. (4.221) can be calculated explicitly. In energy momentum space the matrix inside the trace log is diagonal

$$\begin{pmatrix} \epsilon - \xi(\mathbf{p}) & \tilde{p}_i\sigma_a A_{ai}^0 \\ \tilde{p}_i\sigma_a A_{ai}^{0*} & \epsilon + \xi(\mathbf{p}) \end{pmatrix} \tag{4.251}$$

in the functional indices $\epsilon, \mathbf{p}$. In the 4×4 matrix space this can be diagonalized via a Bogoliubov transformation with the result

$$\begin{pmatrix} (\epsilon - E(\mathbf{p}))\begin{pmatrix} 1 & 0 \\ 0 & 1 \end{pmatrix} & 0 \\ 0 & (\epsilon + E(\mathbf{p}))\begin{pmatrix} 1 & 0 \\ 0 & 1 \end{pmatrix} \end{pmatrix} \tag{4.252}$$

where $E(\mathbf{p})$ are the quasi-particle energies (4.231). Thus the first trace log term in the expression (4.221) can be written as

$$-i(t_b - t_a)V \int \frac{d\epsilon}{2\pi} \frac{d^3p}{(2\pi)^3} \log\left(\epsilon - E(\mathbf{p})\right)\left(\epsilon + E(\mathbf{p})\right). \tag{4.253}$$

The second term contributes simply

$$-\frac{1}{3g}\left\{ \begin{matrix} 3\Delta_B^2 \\ 2\Delta_A^2 \end{matrix} \right\}(t_b - t_a)V. \tag{4.254}$$

After a Wick rotation which replaces $\mathcal{A} \to iE/T$, $t_b - t_a \to -i/T$, $\int_{-\infty}^{\infty} d\epsilon \to iT\sum_{\omega_n}$, this corresponds to the free-energy density

$$f = -\sum_{\omega_n}\sum_{\mathbf{p}} \log[(i\omega_n - E(\mathbf{p}))(i\omega_n + E(\mathbf{p}))] + \frac{1}{g}\left\{ \begin{matrix} \Delta_B^2 \\ \frac{2}{3}\Delta_A^2 \end{matrix} \right\} + \text{const}. \tag{4.255}$$

The constant accounts for the unspecified normalization of the functional integration. It is removed by subtracting the free fermion system with $\Delta = 0$, $g = 0$ [note that $\Delta^2 \sim e^{-1/g\mathcal{N}(0)} \to 0$ for $g \to 0$ due to (4.77), (4.239), (4.241)]. Since the energy of the free fermion system is well-known

$$f_0 = -2T \sum_{\mathbf{p}} \log\left(1 - e^{\xi(\mathbf{p})/T}\right), \tag{4.256}$$

it is sufficient to study only

$$\Delta f = f - f_0 = -T \sum_{\omega_n, \mathbf{p}} \log \frac{i\omega_n - E(p)}{i\omega_n - \xi(p)} + (E \to -E, \ \xi \to -\xi) + \frac{1}{g}\left\{ \begin{matrix} \Delta_B^2 \\ \frac{2}{3}\Delta_A^2 \end{matrix} \right\}. \tag{4.257}$$

This energy difference is the condensation energy associated with the transition into the superfluid phase.

The sum over Matsubara frequency can be performed, by analogy with the treatment of the propagator in Eq. (1.103), using Cauchy's formula:

$$T \sum_{\omega_n} \log\left(1 - \frac{E}{i\omega_n}\right) = -\frac{1}{2\pi i} \int \frac{dz}{e^{z/T} + 1} \log\left(1 - \frac{E}{z}\right), \tag{4.258}$$

where the contour C encircles all poles along the imaginary axis at $z = i\omega_n$ in the positive sense but passes the logarithmic cut from $z = 0$ to E on the left if $E > 0$ (see Fig. 1.1). By deforming the contour C into C' the integral becomes

$$-\int_0^E \frac{dz}{e^{z/T} + 1} = \int_0^E dE \, n_E^f. \tag{4.259}$$

Since

$$\frac{\partial n_E^f}{\partial E} = -n_E^f(1 - n_E^f)/T, \tag{4.260}$$

this can be calculated as

$$-\int_0^E dE n_E^f = T \int_{1/2}^{n_E^f} df' \frac{1}{1 - f'} = -T \log[2(1 - n_E^f)]. \tag{4.261}$$

Therefore the expression (4.257) becomes

$$\Delta f = T \sum_{\mathbf{p}} \left[\log(1 - n_E^f)n_E^f - \log(1 - n_\xi^f)n_\xi^f \right] + \frac{1}{g}\left\{ \begin{matrix} \Delta_B^2 \\ \frac{2}{3}\Delta_A^2 \end{matrix} \right\}, \tag{4.262}$$

where n_ξ^f denotes the free-fermion distribution. Alternatively, we may write

$$\Delta f = 2T \sum_{\mathbf{p}} \left\{ \log(1 - n_E^f)) - (E - \xi) \right\} + \frac{1}{g}\left\{ \begin{matrix} \Delta_B^2 \\ \frac{2}{3}\Delta_A^2 \end{matrix} \right\} - 2T \sum_{\mathbf{p}} \log(1 - n_\xi^f). \tag{4.263}$$

The first two terms give the energy of the superfluid ground state, while the last term is the negative of the energy of the free system.

The explicit calculation can be conveniently done by studying Δf of (4.263) at fixed T as a function of g. At $g = 0$, $\Delta_{AB} = 0$ and $\Delta f = 0$. As g is increased to its physical value, the gap increases to Δ_{AB}. Now, since Δf is extremal in changes of Δ at fixed g and T, all g-dependence comes from the variation of the factor $1/g$, i.e.,

$$\left.\frac{\partial \Delta f}{\partial g}\right|_T = \left\{ \begin{array}{c} \Delta_B^2 \\ \frac{2}{3}\Delta_A^2 \end{array} \right\}. \tag{4.264}$$

We can therefore calculate Δf by simply performing the integral

$$\Delta f = -\int_{1/g}^{\infty} d\,(1/g') \left\{ \begin{array}{c} \Delta_B^2\,(1/g') \\ \frac{2}{3}\Delta_A^2\,(1/g') \end{array} \right\}. \tag{4.265}$$

The $1/g$-dependence of the gap is obtained directly from (4.227), (4.246) as

$$\frac{1}{g\mathcal{N}(0)} - \log\left(2\frac{e^{\gamma}}{\pi}\frac{\omega_c}{T}\right) = \frac{3}{4}\int_{-1}^{1} dz(1 - z^2)A(\delta^2, z), \tag{4.266}$$

with the angle-dependent function

$$A(\delta^2, z) \equiv \frac{1}{\delta}\sum_{n=0}^{\infty}\left[1 \Big/ \sqrt{x_n^2 + \left\{ \begin{array}{c} 1 \\ \frac{1}{1-z^2} \end{array} \right\}} - 1/x_n\right]. \tag{4.267}$$

Differentiating this with respect to δ^2 at fixed T yields

$$\begin{aligned}
\left.\frac{\partial}{\partial \delta^2}\left(\frac{1}{g\mathcal{N}(0)}\right)\right|_T &= -\frac{1}{2\delta}\frac{3}{4}\int_{-1}^{1} dz(1 - z^2)A(\delta^2, z) \\
&= -\frac{1}{2\delta^2}\phi^{B,A}(\delta^2) = -\frac{1}{2\delta^2}\left\{ \begin{array}{c} \rho_s^B/\rho \\ \rho_s/\rho \end{array} \right\},
\end{aligned} \tag{4.268}$$

where ρ_s^B and ρ_S^A are the superfluid densities of B- and A-phases, respectively. Using these we can change variables of integration in (4.265) from g' to δ using (4.266), and write

$$\Delta f = \mathcal{N}(0)\pi^2 T^2 \frac{1}{2}\int_0^{\delta^2} d\delta'^2 \left\{ \begin{array}{c} \phi^B(\delta'^2) \\ \frac{2}{3}\phi^A(\delta'^2) \end{array} \right\}. \tag{4.269}$$

Inserting ϕ^B from the upper part of equation (4.323), we can perform the integral with the result:

$$\frac{1}{\delta^2}\int_0^{\delta^2} d\delta'^2 \phi^B(\delta'^2) = \frac{4}{8}\sum_{n=0}^{\infty}\left[-\frac{1}{\sqrt{x_n^2 + 1}} + 2\left(\sqrt{x_n^2 + 1} - x_n\right)\right]. \tag{4.270}$$

We shall denote this angular average by $\tilde{\phi}^B$. By analogy with the relation $\phi_s^B = \rho_s^B/\rho$ [see (4.268)], we shall also write $\tilde{\phi}^B \equiv \tilde{\rho}_s^B/\rho$, and state the result (4.270) in the form

$$\frac{\tilde{\rho}_s^B}{\rho} \equiv \frac{4}{\delta} \sum_{n=0}^{\infty} \left[-\frac{1}{\sqrt{x_n^2 + 1}} + 2\left(\sqrt{x_n^2 + 1} - x_n\right) \right]. \qquad (4.271)$$

When plotted against temperature, this function starts out like $1 - T/T_c$ for $T \sim T_c$, and is equal to unity at $T = 0$.

Similarly, we may integrate the lower part of Eq. (4.269). The integral

$$\frac{1}{\delta^2} \int_0^\delta d\delta'^2 \phi^A(\delta'^2)$$

produces a further gap function, that will be encountered later in the discussion of the superfluid density as the ratio $\rho_s^\parallel/\rho$ in Eq. (4.335).

The condensation energy can therefore be written in the simple form

$$\Delta f = -\mathcal{N}(0)\pi^2 T^2 \left\{ \begin{array}{c} \frac{1}{2}\tilde{\rho}_s^B/\rho \\ \frac{1}{3}\rho_s^\parallel/\rho \end{array} \right\} \delta^2. \qquad (4.272)$$

For $T \to T_c$, both $\tilde{\rho}_s^B$ and $\rho_s^\parallel$ behave like $(1 - T/T_c)^2$, so that

$$\Delta e \approx -\mathcal{N}\pi^2 T^2 \frac{1}{2}\left(1 - \frac{T}{T_c}\right)^2 \frac{8}{7\zeta(3)} \left\{ \begin{array}{c} 1 \\ \frac{5}{6} \end{array} \right\}, \qquad (4.273)$$

in agreement with our previous calculation (4.108) in the Ginzburg-Landau regime for $T \sim T_c$.

For $T \to 0$, both $\tilde{\rho}^B$ and $\rho_s^\parallel$ tend to ρ, with the approach to that limit like

$$\delta^2 \pi^2 T^2 \to \left\{ \begin{array}{c} 3.111 \\ 4.118 \end{array} \right\} T_c^2. \qquad (4.274)$$

Thus the condensation energies become at zero temperature

$$\Delta f|_{T=0} = -\left\{ \begin{array}{c} 0.236 \\ 0.209 \end{array} \right\} c_n(T_c). \qquad (4.275)$$

The right-hand part of the equation has been normalized with respect to the specific heat of the liquid just above the critical temperature

$$c_n(T_c) = -\frac{2}{3}\pi^2 \mathcal{N}(0) T_c. \qquad (4.276)$$

The full temperature dependence of Δe is shown in Fig. 4.18.

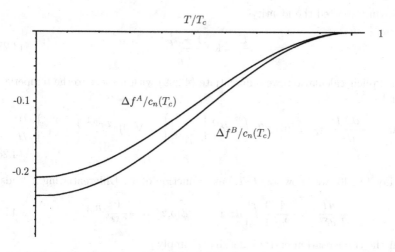

FIGURE 4.18 Condensation energies of A- and B-phases as functions of the temperature.

Entropy

Let us now calculate the entropy. For this it is useful to note that at fixed T and $1/g$ the energy is extremal with respect to small changes in Δ. It is this condition which previously lead to the gap equation (4.226). Thus when forming

$$s = -\frac{\partial f}{\partial T} \qquad (4.277)$$

we do not have to take into account the fact that Δ^2 varies with temperature. Therefore we find

$$\Delta s = -\frac{\partial \Delta f}{\partial T} = -2 \sum_{\mathbf{p}} \left[\log(1 - n_E^f) - \frac{T}{n_E^f(1 - n_E^f)} \frac{\partial n_E^f}{\partial T} \right]. \qquad (4.278)$$

But the derivative is

$$\frac{\partial n_E^f}{\partial T} = n_E^f(1 - n_E^f)\frac{E}{T^2}, \qquad (4.279)$$

so that the entropy becomes

$$\Delta s = -2 \sum_{\mathbf{p}} \left[\log\left(1 - n_{\mathbf{p}}^f\right) - n_E^f \frac{E_{\mathbf{p}}}{T} \right], \qquad (4.280)$$

which can be rewritten in the more familiar form

$$\Delta s = -2 \sum_{\mathbf{p}} \left[(1 - n_E^f) \log(1 - n_E^f) + n_E^f \log n_E^f \right], \qquad (4.281)$$

after having inserted the identity

$$\frac{E}{T} = \log \frac{1 - n_E^f}{n_E^f}. \tag{4.282}$$

For an explicit calculation, we differentiate (4.269) with respect to the temperature and find

$$\Delta s = -\frac{\partial \Delta f}{\partial T} = \mathcal{N}(0)\pi^2 T \int_0^{\delta^2} d\delta'^2 \left\{ \begin{array}{c} \phi^B \\ \frac{2}{3}\phi^A \end{array} \right\} + \mathcal{N}(0)\pi^2 T^2 \frac{1}{2} \left\{ \begin{array}{c} \phi^B \\ \frac{2}{3}\phi^A \end{array} \right\} \frac{\partial \delta^2}{\partial T}. \tag{4.283}$$

From Eq. (4.246) we know $\log(T/T_c)$ as a function of δ^2. Differentiating it leads to

$$\frac{1}{T}\frac{dT}{d\delta^2} = -\frac{1}{2\delta^2}\frac{3}{4}\int_{-1}^{1} dz(1 - z^2)\phi(\delta, z) = -\frac{1}{2\delta^2}\phi^{B,A}, \tag{4.284}$$

so that the condensation entropy density is simply

$$\Delta s^{B,A} = -\mathcal{N}(0)\pi^2 T \left\{ \begin{array}{c} 1 \\ \frac{2}{3} \end{array} \right\} \int_0^{\delta^2} d\delta'^2 \left[1 - \phi^{B,A}(\delta'^2) \right]. \tag{4.285}$$

If we normalize this again with the help of $c_n(T_c)$, it can be written as

$$\Delta s^{B,A}/c_n(T_c) = -\left\{ \begin{array}{c} \frac{2}{3}\left(1 - \tilde{\rho}_B/\rho\right) \\ \left(1 - \rho_s^{\parallel}/\rho\right) \end{array} \right\} \delta^2. \tag{4.286}$$

For $T \to T_c$ this behaves like

$$\Delta s^{A,B}/c_n(T_c) \underset{T \approx T_c}{\approx} -\frac{3}{2}\left\{ \begin{array}{c} 1 \\ \frac{5}{6} \end{array} \right\}\left(1 - \frac{T}{T_c}\right)\frac{8}{7\zeta(3)}. \tag{4.287}$$

In order to calculate the $T \to 0$ limit we consider the expansion (4.269). For $T \to 0$, $\delta \to \infty$ so that the spacings of $x_n = (2n+1)/\delta$ become infinitely narrow and the sum converges towards an integral according to the rule[13]

$$\sum_{n=0}^{\infty} f(x_n) = \frac{\delta}{2} \int dx f(x) - \frac{1}{2! \cdot 3\delta^2} \left(f'(\infty) - f'(0)\right)$$

$$+ \left[\left(\frac{1}{2! \cdot 3}\right)^2 - \frac{1}{4! \cdot 5}\right]\frac{1}{\delta^4}\left(f'''(\infty) - f'''(0)\right) + \dots . \tag{4.288}$$

For $\tilde{\rho}^B$ this implies

$$\left.\frac{\tilde{\rho}_B}{\rho}(\delta^2)\right|_{\delta^2 \to 0} = 2\int_0^{\infty} dx \left[-\frac{1}{\sqrt{x^2+1}} + 2\left(\sqrt{x^2+1} - x\right)\right] - \frac{2}{3\delta^2} + \dots$$

$$= 1 - \frac{2}{3\delta^2} + \dots . \tag{4.289}$$

---

[13]Note that this Euler-MacLaurin expansion misses exponential approaches $e^{-\delta}$.

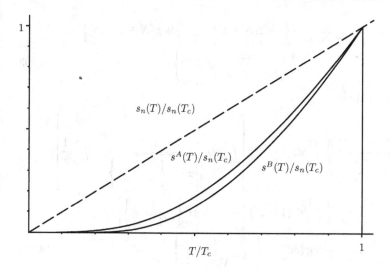

FIGURE 4.19 The temperature behavior of the condensation entropies in B- and A-phases.

Similarly, we can treat the series for $\rho_s^{\parallel}/\rho$ in (4.335):

$$\left.\frac{\rho_s^{\parallel}}{\rho}(\delta^2)\right|_{\delta^2 \to 0} = \frac{3}{2}\int_0^\infty dx\left[-3x + \left(3x^2 + 1\right)\arctan\frac{1}{x}\right] - \frac{1}{\delta^2} + \dots$$

$$= 1 - \frac{1}{\delta^2} + \mathcal{O}\left(\frac{1}{\delta^4}\right). \tag{4.290}$$

Note that for $T \to 0$, the condensation entropy densities are in *both* phases

$$\Delta s^{B,A} = -\frac{2}{3}\mathcal{N}(0)\pi^2 T. \tag{4.291}$$

This is cancelled exactly with the normal entropy

$$s_n = \frac{2}{3}\mathcal{N}(0)\pi^2 T. \tag{4.292}$$

Hence the total entropy vanishes at $T = 0$, as it should. The full T behavior is plotted in Fig. 4.19.

It is worth pointing out that the procedure of going from sums to integrals works only if the integral over the function $f(x)$ has no singularity at $x = 0$. In the $T \to 0$ limit of (4.246), for example, the following more careful limiting procedure becomes necessary

$$\sum_{n=0}^N \frac{1}{x_n} = \delta\sum_{n=1}^N \frac{1}{2n+1} = \delta\left(\sum_{n=1}^{2(N+1)} \frac{1}{n} - \frac{1}{2}\sum_{n=1}^{N+1}\frac{1}{n}\right)$$

$$\underset{N \text{ large}}{\approx} \quad \delta \left\{ \log 2(N+1+\gamma - \frac{1}{2}\left[\log(N+1)+\gamma\right] \right\}$$

$$= \quad \frac{\delta}{2}\left\{ \int_{1/\delta}^{x_N} \frac{dx}{x} + \log(2e^\gamma) \right\}. \tag{4.293}$$

Thus one would obtain

$$\log\frac{T}{T_c} \xrightarrow{T\to 0} \frac{3}{4}\int_{-1}^{1} dz(1-z^2)\left[\int_0^\infty dx \frac{1}{\sqrt{x^2 + \left\{ \dfrac{1}{1-z^2} \right\}}} - \int_{1/\delta}^\infty \frac{1}{x}\right]$$

$$= \quad \frac{3}{4}\int_{-1}^{1} dz(1-z^2)\left[-\gamma - \log\delta\left\{\frac{1}{\sqrt{1-z^2}}\right\}\right]$$

$$= \quad -\log(\delta e^\gamma) + \left\{ \begin{matrix} 0 \\ \log\left(e^{5/6}/2\right) \end{matrix} \right\}, \tag{4.294}$$

in agreement with (4.239), (4.240). The above treatment of the logarithmic divergence is equivalent to applying the mnemonic rule that the $\sum_{n=0}^\infty 1/x_n$ can be replaced by the integral

$$\frac{\delta}{2}\int_0^x \frac{dx'}{x'} \to \frac{\delta}{2}(\log x - \log 0). \tag{4.295}$$

At the lower limit one has to substitute

$$\log 0 \quad \to \quad -\log\left(2\delta e^\gamma\right) = -\log\frac{2\Delta e^\gamma/\pi}{T} = -\log\frac{2\Delta}{\Delta_{\text{BCS}}}\frac{T_c}{T}, \tag{4.296}$$

where Δ_{BCS} denotes the isotropic gap of the B-phase at zero temperature

$$\Delta_{\text{BCS}} = \pi e^{-\gamma}T_c \sim 1.764\, T_c. \tag{4.297}$$

The mnemonic rule can be extracted directly from the relation

$$\int_{-\omega_c}^{\omega_c} \frac{d\xi}{2\xi}\tan\frac{\xi}{2T} = \frac{2}{\delta}\sum_{n=0}^\infty \frac{1}{x_n} = \log\left(\frac{2\omega_c}{T}\frac{e^\gamma}{\pi}\right). \tag{4.298}$$

Specific Heat

By a further differentiation with respect to the temperature we immediately obtain the specific heat

$$\Delta c^{B,A} = T\frac{\partial\Delta s^{B,A}}{\partial T} = \Delta s^{B,A} - \mathcal{N}(0)\pi^2 T\left\{ \begin{matrix} 1 \\ 2 \\ 3 \end{matrix} \right\}\left(1 - \phi^{B,A}(\delta^2)\right)T\frac{\partial\delta^2}{\partial T}$$

$$= \Delta s^{B,A} + 2\mathcal{N}(0)\pi^2 T\frac{1-\phi^{B,A}(\delta^2)}{\phi^{B,A}(\delta^2)}\left\{ \begin{matrix} 1 \\ 2 \\ 3 \end{matrix} \right\}\delta^2. \tag{4.299}$$

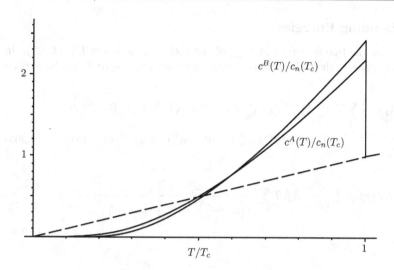

FIGURE 4.20 Specific heat of A- and B-phases as a function of temperature. The dashed line is the contribution of the normal Fermi liquid.

This can be rewritten in terms of the superfluid density function as

$$\Delta c^B / c_n(T_c) = \frac{T}{T_c} \left[-\frac{3}{2} \left(1 - \tilde{\rho}_s^B / \rho \right) + 3 (\rho/\rho_s^B - 1) \right] \delta^2, \qquad (4.300)$$

$$\Delta c^A / c_n(T_c) = \frac{T}{T_c} \left[-(1 - \rho_s^{\parallel}/\rho) + 2 (\rho/\rho_s - 1) \right] \delta^2. \qquad (4.301)$$

At $T = T_c$ there are finite discontinuities

$$\Delta c^B / c_n(T_c) = \frac{3}{2} \frac{8}{7\zeta(3)} = 1.43, \qquad (4.302)$$

$$\Delta c^A / c_n(T_c) = \frac{10}{7\zeta(3)} = 1.19, \qquad (4.303)$$

which can also be derived directly from Ginzburg-Landau expressions in Eqs. (3.20). For the full specific heat one has to add the normal contribution of the normal Fermi liquid to both equations (4.296), which is simply equal to T/T_c.

For $T \to 0$ we use the results (4.289), (4.290) to find

$$\Delta c^{B,A} / c_n(T_c) = -T/T_c. \qquad (4.304)$$

This is exactly the opposite of the specific heat of the normal liquid so that the curves for the total $c^{A,B}/c_n(T_c)$ start out very flat at the origin [exponentially flat for the B-phase due to the nonzero gap (i.e., a finite activation energy) and power-like for the A-phase since the gap vanishes along l]. The full temperature behavior of the specific heat is shown in Fig. 4.20.

Certainly, all these results need strong-coupling corrections which are presently only known in the Ginzburg-Landau regime $T \to T_c$.

4.9.3 Bending Energies

Consider now the free-field part $\mathcal{A}_2[A'_{ai}]$ of the collective action in Eq. (4.225). In momentum space, with the imaginary zeroth component $k_0 = -i\nu$, it can be written in the form

$$\mathcal{A}_2 = \frac{1}{2}\sum_k \left\{ A'_{ai}(k)^* L_{11}^{ij}(k) A'_{aj}(k) + A'_{ai}(-k) L_{22}^{ij}(k) A'_{aj}(-k)^* \right.$$
$$\left. + A'_{ai}(k)^* L_{12}^{ij,ab}(k) A'_{bj}(-k)^* + A'_{ai}(-k) L_{21}^{ij,ab}(k) A'_{bi}(k) \right\}, \quad (4.305)$$

where

$$L_{11}^{ij}(k) = L_{22}^{ij}(k) = \int \frac{d^3p}{(2\pi)^3} \tilde{p}_i \tilde{p}_j T \sum_{\omega_n} \frac{\omega_n^2 - \nu^2/4 + \xi_+\xi_-}{\left[\left(\omega_n - \frac{\nu}{2}\right)^2 + E_+^2 \right]\left[\left(\omega_n + \frac{\nu}{2}\right)^2 + E_-^2 \right]} - \frac{\delta_{ij}}{g},$$

$$(4.306)$$

and

$$L_{12}^{ij,ab}(k) = \left[L_{21}^{ij,ab}(k) \right]^* \qquad\qquad\qquad\qquad\qquad (4.307)$$

$$= -\int \frac{d^3p}{(2\pi)^3} \tilde{p}_i \tilde{p}_j \tilde{p}_{i'} \tilde{p}_{j'} A_{a'i'}^0 A_{b'j}^{0*} \, t_{a'b',ab} \, T \sum_{\omega_n} \frac{1}{\left[\left(\omega_n - \frac{\nu}{2}\right)^2 + E_+^2 \right]\left[\left(\omega_n + \frac{\nu}{2}\right)^2 + E_-^2 \right]},$$

with the tensor

$$t_{a'b'ab} \equiv \frac{1}{2}\mathrm{tr}(\sigma_{a'}\sigma_{b'}\sigma_a\sigma_b) = \delta_{a'a}\delta_{b'b} + \delta_{a'b}\delta_{b'a} - \delta_{a'b'}\delta_{ab}, \qquad (4.308)$$

and the energies (containing the abbreviation $\mathbf{v} \equiv \mathbf{p}/m$):

$$\left\{ \begin{matrix} \xi_+ \\ \xi_- \end{matrix} \right\} \equiv \frac{(\mathbf{p} \pm \mathbf{k}/2)^2}{2m} + \ldots = \frac{\mathbf{p}^2}{2m} \pm \frac{1}{2}\mathbf{v}\cdot\mathbf{k} + \frac{\mathbf{k}^2}{8m} \approx \xi \pm \frac{1}{2}\mathbf{v}\cdot\mathbf{k} + \ldots ,$$

$$\left\{ \begin{matrix} E_+ \\ E_- \end{matrix} \right\} = \sqrt{\left\{ \begin{matrix} \xi_+^2 \\ \xi_-^2 \end{matrix} \right\} + \Delta^2} \approx E \pm \frac{1}{2}\mathbf{v}\cdot\mathbf{k}\frac{\xi}{E} + \frac{1}{8}(\mathbf{v}\cdot\mathbf{k})^2\frac{\Delta^2}{E^3} + \ldots . \quad (4.309)$$

Here ξ and $E = \sqrt{\xi^2 + \Delta^2}$ denote the energy values of ξ_+, ξ_- and E_+, E_-. As usual, the integral over d^3p can be split into size and directional integral, and we can approximate $\mathbf{v} \approx v_F\hat{\mathbf{p}}$, as in (4.76). Compare also with (3.221) and (3.222).

We now rearrange the terms in the sum in such a way that we obtain combinations of single sums of the type

$$T\sum_{\omega_n} \frac{1}{i\omega_n - E}, \qquad\qquad\qquad\qquad (4.310)$$

which lead to the Fermi distribution function [recall (3.199)]

$$T\sum_{\omega_n} \frac{1}{i\omega_n - E} = n_E^f \equiv \frac{1}{e^{E/T} + 1}, \qquad\qquad\qquad (4.311)$$

with the property

$$n_E^f = 1 - n_{-E}^f. \tag{4.312}$$

If we introduce the notation $\omega_\pm \equiv \omega \pm \nu/2$, the decomposition of the different terms $L_{12}^{ijab}(k)$ can be done as in Eq. (3.201). After that we use formula (4.311), and the fact that the frequency shifts ν in $\omega_\pm$ do not appear at the end, since they amount to a mere translation in the infinite sum. Collecting the different terms we find

$$L_{12}^{ij,ab}(k) = \left[L_{21}^{ij,ab}(k)\right]^* = -\int \frac{d^3p}{(2\pi)^3} \tilde{p}_i \tilde{p}_j \tilde{p}_{i'} \tilde{p}_{j'} A_{a'i'}^0 A_{b'j'}^{0*} \frac{t_{a'b',ab}}{2E_-E_+} \tag{4.313}$$

$$\times \left\{ \frac{E_+ + E_-}{(E_+ + E_-)^2 + \nu^2}\left(1 - n_{E_+}^f - n_{E_-}^f\right) + \frac{E_+ - E_-}{(E_+ - E_-)^2 + \nu^2}\left(n_{E_+}^f - n_{E_-}^f\right) \right\}.$$

In the first expression we decompose

$$\frac{\omega_n^2 - \nu^2/4 + \xi_+\xi_-}{[\omega_+^2 + E_+^2][\omega_-^2 + E_-^2]} = \frac{1}{2}\left\{ \frac{1}{\omega_+^2 + E_+^2} + \frac{1}{\omega_-^2 + E_-^2} \right.$$

$$\left. -(E_+^2 + E_-^2 + \nu^2 - 2\xi_+\xi_-)\frac{1}{(\omega_+^2 + E_+^2)(\omega_-^2 + E_-^2)} \right\}. \tag{4.314}$$

For summing up the first two terms we use the formula

$$T\sum_\omega \frac{1}{\omega^2 + E^2} = \frac{1}{2E}(n_{-E}^f - n_E^f) = \frac{1}{2E}\tanh\frac{E}{2T}. \tag{4.315}$$

In the last term of (4.314), the right-hand factor is treated as before in (3.201). Replacing its factor $E_-^2 + E_+^2 + \nu^2$ once by $(E_- + E_+)^2 + \nu^2 - 2E_-E_+$ and once by $(E_- - E_+)^2 + \nu^2 + 2E_-E_+$, and proceeding as in the derivation of Eq. (3.205), we obtain

$$L_{11}^{ij}(k) = L_{22}^{ij}(k) = \int \frac{d^3p}{(2\pi)^3} \tilde{p}_i \tilde{p}_j$$

$$\times \left\{ \frac{E_+E_- + \xi_+\xi_-}{2E_+E_-}\frac{E_+ + E_-}{(E_+ + E_-)^2 + \nu^2}\left(1 - n_{E_+}^f - n_{E_-}^f\right) \right. \tag{4.316}$$

$$\left. - \frac{E_+E_- - \xi_+\xi_-}{2E_+E_-}\frac{E_+ - E_-}{(E_+ - E_-)^2 + \nu^2}\left(n_{E_+}^f - n_{E_-}^f\right) \right\} - \frac{\delta_{ij}}{g}.$$

For the remainder of this chapter we shall specialize on the static case with $k_0 = 0$. We consider only the long-wavelength limit of small $\mathbf{k}$. At $\mathbf{k} = 0$ we find from (4.316) and (4.313)

$$L_{11}^{ij}(0) = \mathcal{N}(0)\int \frac{d\hat{\mathbf{p}}}{4\pi}\hat{p}_i\hat{p}_j \int d\xi \left\{ \frac{E^2 + \xi^2}{4E^3}\left[\tan\frac{E}{2T} + 2n_E^{f'}\right] - \frac{1}{g} \right\}, \tag{4.317}$$

and

$$L_{12}^{ij,ab}(0) = -\frac{1}{2}\mathcal{N}(0)\int \frac{d\hat{\mathbf{p}}}{4\pi}\hat{p}_i\hat{p}_j\hat{p}_i'\hat{p}_j'\frac{\phi(\Delta)}{\Delta^2}A_{a'i'}^0 A_{b'j'}^{0*}t_{a'b'ab}, \tag{4.318}$$

where we have introduced the function

$$\phi(\Delta) = \Delta^2 \left[\int_0^\infty d\xi \frac{1}{E^3} \tan \frac{E}{2T} + 2 \int_0^\infty d\xi \frac{1}{E^2} n_E^{f\prime} \right]. \tag{4.319}$$

The first integral in this equation can be done by parts, after which it turns into the expression (3.214) for the Yoshida function of the superconductor:

$$\phi(\Delta) = 1 - \frac{1}{2T} \int_0^\infty d\xi \frac{1}{\cosh^2(E/2T)}. \tag{4.320}$$

In the A-phase, the gap depends on the direction $\hat{\mathbf{p}}$ of the momentum, so that the gap Δ, and with it the function $\phi(\Delta)$, depend on $z = \hat{\mathbf{p}} \cdot \hat{\mathbf{z}}$.

We now observe that, due to the gap equation (4.235), $L_{11}^{ij}(k)$ can also be expressed in terms of the Yoshida function (3.214) of the superconductor as

$$L_{11}^{ij}(0) = -\frac{1}{2} \mathcal{N}(0) \int \frac{d\hat{\mathbf{p}}}{4\pi} \hat{p}_i \hat{p}_j \phi(\Delta). \tag{4.321}$$

For $T \approx 0$, this function approaches zero exponentially. The full temperature behavior is best calculated by using the Matsubara sum expression for $\phi(\Delta)$ that can be read off from (4.307):

$$\phi(\Delta) = 2T \sum_{\omega_n} \int d\xi \frac{\Delta^2}{(\omega_n^2 + E^2)^2} = -2\Delta^2 T \sum_{\omega_n} \frac{\partial}{\partial \omega_n^2} \int d\xi \frac{1}{\omega_n^2 + \xi^2 + \Delta^2}$$

$$= -2\Delta^2 T \sum_{\omega_n} \frac{\partial}{\partial \omega_n^2} \frac{\pi}{\sqrt{\omega_n^2 + \Delta^2}} = 2T\pi \sum_{\omega_n > 0} \frac{1}{\sqrt{\omega_n^2 + \Delta^2}^3}. \tag{4.322}$$

Using again the variables δ and x_n from (4.244) and (4.245), and the directional parameter z, we may write

$$\phi(\Delta) = \frac{2}{\delta} \left\{ \begin{matrix} 1 \\ 1 - z^2 \end{matrix} \right\} \sum_{n=0}^\infty \frac{1}{\sqrt{x_n^2 + \left\{ \begin{matrix} 1 \\ 1 - z^2 \end{matrix} \right\}}^3} \quad \text{in} \left\{ \begin{matrix} B \\ A \end{matrix} \right\} \text{-phase}. \tag{4.323}$$

For $T \approx T_c$ where $\delta \approx 0$, the Yoshida function has the limiting behaviors

$$\phi(\Delta) \approx 2\delta^2 \left\{ \begin{matrix} 1 \\ 1 - z^2 \end{matrix} \right\} \frac{7\zeta(3)}{8} \quad \text{in} \left\{ \begin{matrix} B \\ A \end{matrix} \right\} \text{-phase}. \tag{4.324}$$

Let us consider the equations (4.317), (4.321) in more detail and rewrite $L_{11}^{ij}(0)$ as follows

$$L_{11}^{ij}(0) = -\frac{1}{4m^2 v_F^2} \rho_{ij}, \tag{4.325}$$

where

$$\rho_{ij} \equiv 3\rho \int \frac{d\hat{\mathbf{p}}}{4\pi} \hat{p}_i \hat{p}_j \phi(\Delta). \tag{4.326}$$

In the B-phase, the angular integral in (4.321) is trivial and yields

$$\rho^B_{ij} = \frac{2}{3} v^2_F m^2 \mathcal{N}(0) \phi(\Delta) \delta_{ij} = \rho \phi(\Delta) \delta_{ij} \equiv \rho^B_s \delta_{ij} \qquad (4.327)$$

where $\phi^B(\Delta)$ is the upper of the functions (4.323) (the isotropic Yoshida function). The invariant ρ^B_s is called the superfluid density of the B-phase. For $T \approx T_c$, we use the expression (4.322) to see that

$$\rho^B_s \approx 2\rho \left(1 - \frac{T}{T_c} \right). \qquad (4.328)$$

For $T = 0$, on the other hand, we better use (4.319) to deduce that $\phi = 1$, which implies that in this limit

$$\rho^B_s = \rho, \qquad T = 0. \qquad (4.329)$$

The full T-dependence of the reduced superfluid density $\tilde{\rho}^B \equiv \rho^B/\rho$ is plotted in Fig. 4.21.

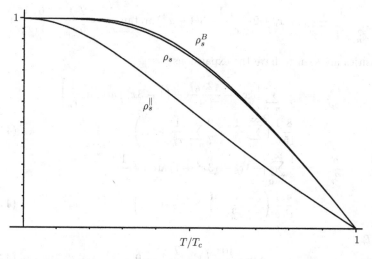

FIGURE 4.21 Temperature behavior of the reduced superfluid densities in the B- and in the A-phase of superfluid ^{4}He.

The figure contains also the corresponding quantities of the A-phase. They are found from the integral

$$\rho^A_{ij} \equiv 3\rho \int \frac{d\hat{\mathbf{p}}}{4\pi} \hat{p}_i \hat{p}_j \phi(\Delta), \qquad (4.330)$$

after expanding it into covariants as

$$\rho^A_{ij} = \rho_s \left(\delta_{ij} - l_i l_j \right) + \rho^{\parallel}_s l_i l_j, \qquad (4.331)$$

whose coefficients are the superfluid densities of the A-phase

$$\frac{\rho_s}{3\rho} \equiv \int \frac{d\hat{p}}{4\pi} p_x^2 \phi(\Delta) = \frac{1}{2} \int_{-1}^{1} \frac{dz}{2}(1-z^2)\phi(\Delta) \equiv \frac{1}{2}\phi^A(\Delta),$$

$$\frac{\rho_s^{\parallel}}{3\rho} \equiv \int \frac{d\hat{p}}{4\pi} p_z^2 \phi(\Delta) = \frac{1}{2} \int_{-1}^{1} \frac{dz}{2}(1-z^2)\phi(\Delta). \tag{4.332}$$

The second function is the same ratio that occurred in the lower expression (4.272) for the free energy.

The spatially averaged ϕ function in the first line will appear repeatedly in the further description of the A-phase and has therefore been given an extra name $\phi^A(\Delta)$. Using the integrals

$$\int_{-1}^{1} dz \frac{1}{\sqrt{x_n^2 + 1 - z^2}^3} = \frac{2}{x_n(x_n^2+1)} = \frac{2}{x_n^3} - \frac{2}{x_n^5} + \dots,$$

$$\int_{-1}^{1} dz \frac{z^2}{\sqrt{x_n^2 + 1 - z^2}^3} = 2\left(\frac{1}{x_n} - \arctan\frac{1}{x_n}\right) = \frac{2}{3}\frac{1}{x_n^3} - \frac{2}{5}\frac{1}{x_n^5} + \dots, \tag{4.333}$$

$$\int_{-1}^{1} dz \frac{1}{\sqrt{x_n^2 + 1 - z^2}} = x_n + 2\frac{x_n^2+1}{x_n} - 3(1+x_n^2)\arctan\frac{1}{x_n} = \frac{2}{5}\frac{1}{x_n^3} - \frac{6}{35}\frac{1}{x_n^5} + \dots,$$

these densities are seen to have the expansions

$$\frac{\rho_s}{3\rho} = \frac{1}{2\delta} \sum_{n=0}^{\infty} \left[3x_n - \frac{2x_n}{x_n^2+1} + \left(1 - 3x_n^2\right)\arctan\frac{1}{x_n}\right]$$

$$= \frac{8}{15}\frac{1}{\delta}\left(\sum_{n=0}^{\infty}\frac{1}{x_n^3} - \frac{9}{7}\sum_{n=0}^{\infty}\frac{1}{x_n^5} + \dots\right), \tag{4.334}$$

$$\frac{\rho_s^{\parallel}}{3\rho} = \frac{1}{\delta} \sum_{n=0}^{\infty} \left[-3x_n - (3x_n^2 + 1)\arctan\frac{1}{x_n}\right]$$

$$= \frac{4}{15}\frac{1}{\delta}\left(\sum_{n=0}^{\infty}\frac{1}{x_n^3} - \frac{6}{7}\sum_{n=0}^{\infty}\frac{1}{x_n^5} + \dots\right). \tag{4.335}$$

For $T \approx T_c$:

$$\delta \approx \frac{10}{7\zeta(3)}\left(1 - \frac{T}{T_c}\right) \to 0. \tag{4.336}$$

The first two sums in each expression can be done with the results:

$$\rho_s \approx \frac{8}{5}c_3\delta^2\left(1 - \frac{9}{7}\frac{c_5}{c_3}\delta^2 + \dots\right) \approx 2\left(1 - \frac{T}{T_c}\right) + \dots, \tag{4.337}$$

$$\rho_s^{\parallel} \approx \frac{4}{5}c_3\delta^2\left(1 - \frac{6}{7}\frac{c_5}{c_3}\delta^2 + \dots\right) \approx \left(1 - \frac{T}{T_c}\right) + \dots, \tag{4.338}$$

where the coefficients

$$c_k = \frac{2^k - 1}{2^k}\zeta(k) \tag{4.339}$$

are the results of the sums $\sum_{n=0}^{\infty}(2n+1)^{-k}$ ($c_3 \approx 1.0518$, $c_5 \approx 1.0045$). The higher terms are omitted on the right-hand sides, although they will be of use later.

For $T = 0$, we have again $\phi(\Delta) = 1$, and from Eqs. (4.332) and (4.333) we find that

$$\rho_s = \rho_s^{\parallel} = \rho, \quad T = 0.$$

The full temperature behavior of the superfluid densities was shown above in Fig. 4.21.

Consider now the function $L_{12}^{ij,ab}(0)$. Here it is useful to introduce a tensor

$$\rho_{ijkl} \equiv \frac{3}{2}\rho \int \frac{d\hat{\mathbf{p}}}{4\pi}\hat{p}_i\hat{p}_j\hat{p}_k\hat{p}_l \; \phi(\Delta)/\Delta^2, \tag{4.340}$$

in terms of which $L_{12}^{ij,ab}(0)$ can be written as

$$L_{12}^{ijab}(0) = -\frac{1}{2mv_F^2}\rho_{ijkl}A_{a'k}^0 A_{b'l}^{0*} \; t_{a'b'ab}, \tag{4.341}$$

with the tensor $t_{a'b'ab}$ of Eq. (4.308). In the B-phase, where the gap is isotropic, the angular integration is trivial and we find from (4.340):

$$\rho_{ijkl} = \frac{1}{10}\left(\delta_{ij}\delta_{kl} + \delta_{il}\delta_{kj} + \delta_{ik}\delta_{jl}\right)\rho^B. \tag{4.342}$$

In the A-phase, this tensor can be expressed in terms of the three covariants

$$\begin{aligned}
\hat{A}_{ijkl} &= \delta_{ij}\delta_{kl} + \delta_{il}\delta_{kj} + \delta_{ik}\delta_{jl}, \\
\hat{B}_{ijkl} &= \delta_{ij}l_kl_l + \delta_{ik}l_jl_l + \delta_{il}l_jl_k + \delta_{jk}l_il_l + \delta_{jl}l_il_k + \delta_{kl}l_il_j, \\
\hat{C}_{ijkl} &= l_il_jl_kl_l,
\end{aligned} \tag{4.343}$$

as follows:

$$\rho_{ijkl} = A\hat{A}_{ijkl} + B\hat{B}_{ijkl} + C\hat{C}_{ijkl}. \tag{4.344}$$

Contracting this with $\delta_{ij}\delta_{kl}$ and $\delta_{ij}l_kl_l$, we find that the coefficients A and B are given by the following combinations of ρ_s and $\rho_s^{\parallel}$:

$$A = \frac{1}{8}\rho_s, \quad A + B = \frac{1}{4}\rho_s^{\parallel}. \tag{4.345}$$

The third covariant leads to another function of the gap parameter $\gamma(\Delta)$ defined by:

$$3A + 6B + C = \frac{3}{8}\gamma(\Delta) \tag{4.346}$$

where $\gamma(\Delta)$ is calculated from the angular integral

$$\gamma(\Delta) \equiv 4\rho \int \frac{d\hat{\mathbf{p}}}{4\pi}\hat{p}_z^4\phi(\Delta)\frac{\Delta_A^2}{\Delta^2}. \tag{4.347}$$

Inserting (4.323) and performing the angular integrals, we find the series representation

$$\frac{\gamma}{\rho} = \frac{4}{\delta} \sum_{n=0}^{\infty} \left[3x_n + \frac{2}{x_n} - 3(x_n^2 + 1) \arctan \frac{1}{x_n} \right]. \tag{4.348}$$

By comparing this series with (4.248) and (4.334), we see that $\gamma(\Delta)$ is not a new gap function. In fact, by adding and subtracting the series for $4\log(T/T_c)$, we find

$$\frac{\gamma}{\rho} \equiv -4 \log \frac{T}{T_c} - 2\frac{\rho_s^{\|}}{\rho}. \tag{4.349}$$

For $T \approx T_c$, γ starts out like

$$\gamma \approx 2 \left(1 - \frac{T}{T_c} \right), \tag{4.350}$$

just as ρ_s/ρ. As T approaches zero, however, there is a logarithmic divergence which is due to the zeros in the gap along the l direction [see Eq. (4.347)].

Let us now turn to the bending energies. For this, we expand $L_{11}(k)$ and $L_{12}(k)$ to lowest order in the momentum $\mathbf{k}$ and find

$$f_{\text{grad}} = \frac{1}{4m^2} \left(\rho_{ijkl}^{11} \partial_k A_{ai}^* \partial_l A_{aj} / \Delta_{A,B}^2 + \text{Re}\, \rho_{ijkl,ab}^{12} \partial_k A_{ai}^* \partial_l A_{bj}^* \right) \text{ in } \left\{ \begin{matrix} \text{A} \\ \text{B} \end{matrix} \right\} \text{–phase.} \tag{4.351}$$

Here we have dropped in the primes on the fields, since in the presence of derivatives, the additional constant A_{ai}^0 does not matter. The tensor coefficients are found by performing similar calculations as in Eqs. (3.194) and (3.195) for the superconductor, except that we must now include directional integrations in momentum space:

$$\rho_{ijkl}^{11} = \frac{3\rho}{2} \int \frac{d\hat{\mathbf{p}}}{4\pi} \hat{p}_i \hat{p}_j \hat{p}_k \hat{p}_l \left[\phi(\Delta) - \frac{1}{2}\bar{\phi}(\Delta) \right] \frac{\Delta_{A,B}^2}{\Delta^2}, \tag{4.352}$$

$$\rho_{ijkl,ab}^{12} = -\frac{3\rho}{2} \int \frac{d\hat{\mathbf{p}}}{4\pi} \hat{p}_i \hat{p}_j \hat{p}_k \hat{p}_l \hat{p}_m \hat{p}_n \frac{1}{2}\bar{\phi}(\Delta) \frac{\Delta_{A,B}^2}{\Delta^4} A_{a'm}^0 A_{b'n}^0 t_{a'b'ab}, \tag{4.353}$$

where $\phi(\Delta)$ is the same as in (4.320) and (4.322), whereas $\bar{\phi}(\Delta)$ denotes another function of the gap:

$$\bar{\phi}(\Delta) \equiv 2\pi T \Delta^4 \sum_{\omega_n > 0} \frac{1}{\sqrt{\omega^2 + \Delta^2}^5} = \frac{2}{\delta}(1 - z^2)^2 \sum_{n=0}^{\infty} \frac{1}{\sqrt{x_n^2 + \left\{ \begin{matrix} 1 - z^2 \\ 1 \end{matrix} \right\}}^5}. \tag{4.354}$$

In the superconductor, this function does not appear in the hydrodynamic limit. We therefore expect a similar cancellation in the B-phase, where the gap is isotropic. In fact, by assuming that

$$A_{ai} = \Delta_B \, e^{i\varphi} R_{ai}(\theta), \tag{4.355}$$

we find that the second term in (4.351) becomes

$$\mathrm{Re}\, \rho^{12}_{ijklab} \partial_b e^{-i\varphi} R_{ai}(\theta) \partial_l e^{i\varphi} R_{bj}(\varphi)$$

$$= -\frac{3\rho}{2} \int \frac{d\hat{p}}{4\pi} \hat{p}_i \hat{p}_j \hat{p}_k \hat{p}_l \frac{1}{2} \bar{\phi}(\Delta) \frac{\Delta^2_B}{\Delta^2} \hat{p}_m \hat{p}_n R_{a'm} R_{b'n} t_{a'b'ab} \left(-\partial_k \varphi \partial_l \varphi R_{ai} R_{bj} + \partial_k R_{ai} \partial_l R_{bj} \right)$$

$$= -\frac{3\rho}{2} \int \frac{d\hat{p}}{4\pi} \hat{p}_k \hat{p}_l \frac{1}{2} \bar{\phi}(\Delta) \frac{\Delta^2_B}{\Delta^2} \left(-\partial_k \varphi \partial_l \varphi - \hat{p}_k \hat{p}_l \partial_k R_{ai} \partial_l R_{aj} \right). \tag{4.356}$$

And this coincides exactly with the $\bar{\phi}$ content in $\rho^{11}_{ijkl} \partial_k e^{-i\varphi} R_{ai} \partial_l e^{i\varphi} R_{aj}$.

Note that the two terms change sign for different reasons: $\partial_k \varphi \partial_l \varphi$ does, because of the equality of the phases $e^{i\varphi}$, and $\partial_k R_{ai} \partial_l R_{bj}$ does, because of symmetry properties of the tensor $t_{a'b'ab}$. Thus, for the B-phase, the result is simply

$$f_{\mathrm{grad}} = \frac{1}{4m^2} \rho_{ijkl} \partial_k A^*_{ai} \partial_l A_{aj} / \Delta^2_B, \tag{4.357}$$

with ρ_{ijkl} being the tensor discussed before in (4.340). This result is exactly the same as for a superconductor except for two additional direction vectors $\hat{p}_i \hat{p}_j$ inserted into the spatial average which are contracted with the vector indices of the fields $A^*_{ai} A_{aj}$. Inserting the decomposition (4.342) we find the energy (see Appendix 4B for details)

$$f_{\mathrm{grad}} = \frac{1}{4m^2} \frac{\rho^B_s}{2} \left[(\boldsymbol{\nabla}\varphi)^2 + \frac{4}{5}(\partial_i \tilde{\theta}_j)^2 - \frac{1}{5}(\boldsymbol{\nabla}\tilde{\theta})^2 - \frac{1}{5}\partial_i \tilde{\theta}_j \partial_j \tilde{\theta}_i \right]. \tag{4.358}$$

For $T \approx T_c$, we insert (4.328) and reobtain the previous Ginzburg-Landau result (4.159), if we use the fact that close to T_c:

$$K_{23} \underset{T \approx T_c}{\approx} 2K_1. \tag{4.359}$$

The superfluid density ρ^B_s was shown before in Fig. 4.21.

In the A-phase, matters are considerably more complicated. This is due to the fact that the gap size varies which prevents the $\bar{\phi}(\Delta)$ function to cancel. Consider the field dependent parts of the ρ^{12} contribution:

$$\mathrm{Re}\, A_{a'm}{}^0 A_{b'n}{}^0 t_{a'b'ab} \partial_k A^*_{ai} \partial_l A^*_{bj} / \Delta^2_A = \tag{4.360}$$

$$\Delta_A{}^2 \mathrm{Re}\, d_{a'} d_{b'} \phi_m \phi_n t_{a'b'ab} \left(\partial_k d_a \partial_l d_b \phi^\dagger_i \phi^\dagger_j + d_a d_b \partial_k \phi_b \phi^\dagger_i \partial_l \phi^\dagger_j \right).$$

Contracting now the indices a' and b', we see that the gradients of **d** appear with opposite signs in the form

$$-\mathrm{Re}\, \Delta_A{}^2 \partial_k d_a \partial_e d_b \phi_m \phi_n \phi^\dagger_i \phi^\dagger_j \tag{4.361}$$

whereas the derivatives $\partial \phi$ keep their sign

$$\Delta^2_A \mathrm{Re}\, \phi_m \partial_k \phi^*_i \partial_l \phi^\dagger_j. \tag{4.362}$$

Using Formula (4A.5) of Appendix 4A, the expression (4.360) can be cast in the form

$$\Delta_A^2 \{ (\partial_k l_m) l_i (\partial_l l_n) l_j \tag{4.363}$$
$$- [(\epsilon_{mpr} l_r \partial_k l_p l_i - 2mv_{sk} (\delta_{mi} - l_m l_i)) \times (\epsilon_{nqs} l_s \partial_q l_q l_j - 2mv_{sl} (\delta_{nj} - l_n l_j))] \}.$$

The calculation is simplified considerably by observing that an expression

$$\Delta_A^2 \mathrm{Re}\, \Phi_m^* \partial_k \Phi_i \Phi_n \partial_e \Phi_j^\dagger, \tag{4.364}$$

instead of (4.362), would give exactly the same result as that in (4.363), except with a plus sign in front of the bracket instead of a minus sign. Thus (4.362) can be written as

$$2\Delta_A^2 (\partial_k l_m) l_i (\partial_e l_n) l_j - \Delta_A^2 \mathrm{Re}\, \Phi_m^* \partial_k \Phi_i \Phi_n \partial_e \Phi_j^\dagger. \tag{4.365}$$

Now, the second term together with (4.353) corresponds to an energy

$$\Delta f^{(2)} = -\frac{1}{4m^2} \frac{3\rho}{2} \int \frac{d\hat{\mathbf{p}}}{4\pi} \hat{p}_i \hat{p}_j \hat{p}_k \hat{p}_l \frac{1}{2} \bar{\phi}(\Delta) \frac{\Delta_A^2}{\Delta^2} \left(\frac{\Delta_A^2}{\Delta^2} \hat{p}_m A_b^* \hat{p}_n A_{bn} \right) \partial_k A_{ai}^* \partial_l A_{aj} \frac{1}{\Delta_A^2}, \tag{4.366}$$

which again cancels the $\bar{\phi}$-part in the ρ^{11}-term. Hence, this part of the energy f has again the form (4.357), thus simply doubling it.

Let us now study the contribution of the first term in (4.365) to the energy:

$$\Delta f = -\frac{1}{4m^2} \frac{3\rho}{2} \int \frac{d\hat{\mathbf{p}}}{4\pi} \hat{p}_i \hat{p}_j \hat{p}_k \hat{p}_l \hat{p}_n \bar{\phi}(\Delta) \frac{\Delta_A^4}{\Delta^4} (\partial_k l_m) l_i (\partial_l l_n) l_j. \tag{4.367}$$

Since $\hat{\mathbf{p}} \cdot \mathbf{l} = z$, we can introduce the tensor

$$\bar{\rho}_{ijkl} = \frac{3\rho}{2} \int \frac{d\hat{\mathbf{p}}}{4\pi} \hat{p}_i \hat{p}_j \hat{p}_k \hat{p}_l \frac{z^2}{1 - z^2} \bar{\phi}(\Delta) \frac{\Delta_A^2}{\Delta^2}, \tag{4.368}$$

so that the additional energy (4.367) can be written as

$$\Delta f = -\frac{1}{4m^2} \bar{\rho}_{ijkl} \partial_k l_i \partial_l l_j. \tag{4.369}$$

Decomposing $\bar{\rho}_{ijkl}$ in the same way as ρ_{ijkl} in (4.346), we find the coefficients

$$\bar{A} = \frac{1}{8} \bar{\rho}_s,$$
$$\bar{A} + \bar{B} = \frac{1}{4} \bar{\rho}_s^\|, \tag{4.370}$$

where $\bar{\rho}_s$, $\bar{\rho}_s^\|$ are auxiliary quantities defined as

$$\bar{\rho}_s \equiv \frac{3}{4} \rho \int_{-1}^{1} dz (1 - z^2) \bar{\phi}(\Delta) \frac{\Delta_A^2}{\Delta^2},$$
$$\bar{\rho}_s^\| \equiv \frac{3}{2} \rho \int_{-1}^{1} dz z^4 \bar{\phi}(\Delta) \frac{\Delta_A^2}{\Delta^2}. \tag{4.371}$$

Inserting the explicit form (4.354) for $\bar{\phi}(\Delta)$, we can partially integrate Eq. (4.371) and find

$$
\begin{aligned}
\bar{\rho}_s &= \frac{3}{4}\rho \int_{-1}^{1} dz (1 - z^2)^2 z^2 \frac{2}{\delta} \sum_{n=0}^{\infty} \frac{1}{\sqrt{x_n^2 + 1 - z^2}^5} \\
&= -\frac{3}{4}\rho \int_{-1}^{1} dz \left(\frac{1}{3} - 2z^2 + \frac{5}{3}z^4 \right) \frac{\phi(\Delta)}{(1 - z^2)}, \quad (4.372) \\
\bar{\rho}_s^{\parallel} &= \frac{3}{4}\rho \int_{-1}^{1} dz (z^4)^2 (1 - z^2) \frac{2}{\delta} \sum_{n=0}^{\infty} \frac{1}{\sqrt{x_n^2 + 1 - z^2}^5} \\
&= -\frac{3}{4}\rho \int_{-1}^{1} dz \left(z^2 + \frac{5}{3}z^4 \right) \frac{\phi(\Delta)}{(1 - z^2)}. \quad (4.373)
\end{aligned}
$$

The auxiliary quantities are therefore expressible in terms of the superfluid densities as follows:

$$
\begin{aligned}
\bar{\rho}_s &= \frac{2}{3}\rho_s^{\parallel} - \frac{1}{3}\rho_s, \\
\bar{\rho}_s^{\parallel} &= -\rho_s^{\parallel} + \frac{1}{2}\gamma. \quad (4.374)
\end{aligned}
$$

If we now perform the contractions of the covariants in (4.357) and (4.369), we find the energy in the form given in (4.125), but now with coefficients (see Appendix 4B for details):

$$
\begin{aligned}
& 2mc = \tfrac{1}{2}\rho_s^{\parallel}, \quad 2c_0 m = \rho_s^{\parallel}, \\
& 4m^2 K_1^d = \rho_s, \quad 4m^2 K_2^d = \rho_0 = \rho_s - \rho_s^{\parallel}, \quad (4.375) \\
& 4m^2 K_s = \rho_s/4, \quad 4m^2 K_t = (\rho_s + 4\rho_s^{\parallel})/12, \quad 4m^2 K_b = (\rho_s^{\parallel} + \gamma)/2.
\end{aligned}
$$

Their temperature dependence is known for all T down to $T = 0$ (see Fig. 4.22). The twist, bend, and splay bending constants are displayed in Fig. 4.22. There is no need to plot K_1^d, K_2^d since K_1^d is equal to $\rho_s/4m^2$, which was plotted in Fig. 4.21. Similarly, the coefficients c, c_0 need no extra plot, since they are proportional to $\rho_s^{\parallel}$ of Fig. 4.21. To see what K_2^d looks like we introduce, by analogy with $\rho_s^{\parallel}$, the longitudinal quantity

$$
K_{\parallel}^d \equiv K_1 - K_2, \quad (4.376)
$$

which is equal to $\rho_s^{\parallel}/4m^2$.

If **d** is locked to **l**, the bending constants K_1^d, K_2^d change K_t, K_b, K_s into

$$
\begin{aligned}
K_s^l &= K_s + \rho_s = 5\rho_s/4, \\
K_t^l &= K_t + \rho_s = (13\rho_s + 4\rho_s^{\parallel})/12, \\
K_b &= K_b + K_{\parallel}^d = (3\rho_s^{\parallel} + \gamma)/2. \quad (4.377)
\end{aligned}
$$

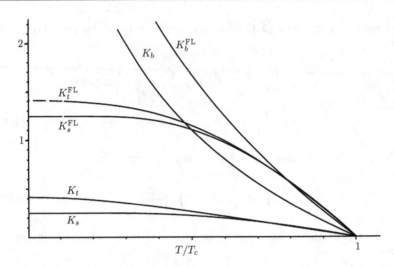

FIGURE 4.22 Superfluid stiffness functions K_t, K_b, K_s of the A-phase, in units of $\rho/4m^2$, as functions of the temperature, once without and once with Fermi liquid corrections, indicated by the superscript FL.

4.9.4   Fermi-Liquid Corrections

In order to compare the above results with experiment at all temperatures below T_c, the pair interaction turns out not to be sufficient. The more T drops below T_c, the more other interactions become important. Here we shall discuss the most relevant of these which is due to a current-current coupling between particle and spin currents.

In Landau's theory of the normal Fermi liquid, these interactions are parametrized with coupling constants F_1, F_1^S as follows:

$$\mathcal{A}_{\text{curr-curr}} = -\frac{1}{2} \int d^4x \left[\frac{F_1}{2\mathcal{N}(0)} \psi^* i\tilde{\nabla}_i\psi\psi^* \tilde{\nabla}_i\psi + \frac{F_1^S}{2\mathcal{N}(0)} \psi^* i\tilde{\nabla}_i\sigma_a\psi\psi^* i\tilde{\nabla}_i\sigma^a\psi \right].$$
(4.378)

Using the particle and spin currents of Eq. (4.100) and the relation $2\mathcal{N}(0)p_F^2 = 3\rho$, this can be written compactly as

$$\mathcal{A}_{\text{curr-curr}} = -\frac{1}{2} \int d^4x \frac{m^2}{\rho} \left(\frac{1}{3}F_1 j_i^2 + \frac{1}{3}F_1^S j_{ai}^2 \right).$$
(4.379)

As in the case of the pair interaction, these fourth order expressions in the fundamental fields ψ^*, ψ can be eliminated in favor of quadratic ones by introducing collective fields φ_i, φ_{ai} and by adding, to the action, the complete squares

$$\frac{1}{2} \int d^4x \frac{m^2}{\rho} \left[\frac{1}{3}F_1 \left(j_i + \frac{\rho}{m^2} \frac{1}{F_1} \varphi_i \right)^2 + \frac{1}{3}F_1^S \left(j_{ai} + \frac{\rho}{m^2} \frac{1}{F_1^S} \varphi_{ai} \right)^2 \right],$$
(4.380)

by analogy with (4.60). Then the current-current interaction becomes

$$\mathcal{A}_{\text{curr}-\text{curr}} = \int d^4x \left[j_i \varphi_i + j_{ai}\varphi_{ai} + \frac{1}{2}\frac{\rho}{m^2}\left(\frac{1}{\frac{1}{3}F_1}\varphi_i^2 + \frac{1}{\frac{1}{3}F_1^S}\varphi_{ai}^2 \right) \right]. \quad (4.381)$$

After integrating out the Fermi fields, the first term in (4.69) is changed to $-i/2$ times Tr log of the matrix

$$\begin{pmatrix} i\partial_t - \xi(-i\boldsymbol{\nabla}) + \frac{i}{2m}\tilde{\boldsymbol{\nabla}}_i\varphi_i + \frac{i}{2m}\tilde{\boldsymbol{\nabla}}_i\sigma_a\varphi_{ai} & i\tilde{\boldsymbol{\nabla}}_i\sigma_a A_{ai} \\ i\tilde{\boldsymbol{\nabla}}_i\sigma_a A_{ai}^* & i\partial_t + \xi(-i\boldsymbol{\nabla}) + \frac{i}{2m}\tilde{\boldsymbol{\nabla}}_i\varphi_i + \frac{i}{2m}\tilde{\boldsymbol{\nabla}}_i\sigma_a\varphi_{ai} \end{pmatrix}, (4.382)$$

depending on φ_i, φ_{ai}.

In the hydrodynamic limit, where only quadratic field dependencies are considered, there is a simple method to find this dependence without going again through loop calculations. For this we observe that a term in the action

$$\int d^4x \left(j_i\varphi_i + j_{ai}\varphi_{ai} \right) \quad (4.383)$$

is equivalent to adding velocity source terms to the energy density, thereby forming quantity that looks like an enthalpy density, except that the roles of pressure and volume are played by momenta and velocities, i.e.,

$$f \to f_{\text{ent}} = f - p_i V_i - p_{ai} V_{ai}. \quad (4.384)$$

Here $p_i \equiv mj_i$, $p_{ai} \equiv mj_{ai}$ are the momentum densities of particle and spin flow. We call $e \to f_{\text{ent}}$ the *flow enthalpy*. The minimum of this quantity determines the equilibrium properties of the system at externally enforced velocities V_i, V_{ai} of particles and spins:

$$V_i \equiv \varphi_i/m, \quad V_{ai} \equiv \varphi_{ai}/m. \quad (4.385)$$

Consider now the energy (4.125) in a planar texture which has all l-vectors parallel. If we want to take into account the effect of the current-current interactions we must extend this expression. Recall that the earlier calculations were all done in a frame in which the normal part of the liquid was at rest. When studying nonzero velocities of the system, as we now do, we must add to the energy density the kinetic contribution of the normal particle and spin flows

$$\frac{\rho_n}{2}\left(v_{n_i}^{\perp \, 2} + v_{n_{ai}}^{\perp \, 2} \right) + \frac{\rho_n^{\parallel}}{2}\left(v_{n_i}^{\parallel \, 2} + v_{n_{ai}}^{\parallel \, 2} \right), \quad (4.386)$$

where $\mathbf{v}^{\perp}$ and $\mathbf{v}^{\parallel}$ are defined by

$$\begin{aligned} \mathbf{v}^{\perp} &= \mathbf{v} - \mathbf{v}^{\parallel}, \\ \mathbf{v}^{\parallel} &= \mathbf{l}\,(\mathbf{l} \cdot \mathbf{v}), \end{aligned} \quad (4.387)$$

with similar definitions for the spin velocities. The corresponding currents are

$$\begin{aligned} \mathbf{p} &= m\mathbf{j} = \rho_s\mathbf{v}_s + \rho_n\mathbf{v}_n, \quad (4.388) \\ \mathbf{p}_a &= m\mathbf{j}_a = \rho_s\mathbf{v}_{sa} + \rho_n\mathbf{v}_{na}. \quad (4.389) \end{aligned}$$

The additional terms (4.386) are necessary to guarantee the correct Galilei transformation properties of the energy density f.

We now study the equilibrium properties of the liquid. First we minimize the flow enthalpy (4.389). If topology does not enforce a nonzero superflow, both velocities $\mathbf{v}_n$ and $\mathbf{v}_s$ are equal to a single velocity $\mathbf{v}$. Thus, in equilibrium, we may rewrite the flow enthalpy density also as

$$f_{\text{ent}} = \frac{\rho}{2} \left(v_i^2 + v_{ai}^2 \right) - p_i V_i - p_{ai} V_{ai}. \tag{4.390}$$

This expression is minimal at

$$v_i = V_i, \quad v_{ai} = V_{ai}, \tag{4.391}$$

where it has the equilibrium value

$$f_{\text{ent}}\Big|_{\text{eq}} = -\frac{\rho}{2} \left(V_i^2 + V_{ai}^2 \right). \tag{4.392}$$

Let us compare this with the calculation of the flow enthalpy from the trace log term of the collective action. The enthalpy density is

$$f_{\text{ent}} = -\log \left(\begin{array}{cc} i\partial_t - \xi(\mathbf{p}) + p_i V_i + p_i \sigma_a V_{ai} & \tilde{p}_i \sigma_a A'_{ai} \\ \tilde{p}_i \sigma_a A_{ai}'^* & i\partial_t + \xi(\mathbf{p}) - p_i V_i - p_i \sigma_a V_{ai} \end{array} \right). \tag{4.393}$$

The quadratic term in the fluctuating field A'_{ai} around the extremum has been calculated before. It has led to the hydrodynamic-limit result

$$f = \frac{\rho_s}{2} \left(v_{si}^{\perp\,2} + v_{sai}^{\perp\,2} \right) + \frac{\rho_s^{\|}}{2} \left(v_{si}^{\|\,2} + v_{sai}^{\|\,2} \right). \tag{4.394}$$

In addition, there are now linear terms

$$\Delta_1 f = -\rho_s \left(v_{si} V_i + v_{s\,ai}^{\perp} V_{ai} \right) - \rho_s^{\|} \left(v_{si}^{\|} V_i + v_{sai}^{\|} V_{ai} \right). \tag{4.395}$$

We would like to find quadratic terms in V_i, V_{ai}. They certainly have the form

$$\Delta_2 f = -\frac{a}{2} \left(V_i^{\perp 2} + V_{ai}^{\perp 2} \right) - \frac{a^{\|}}{2} \left(V_i^{\|2} + V_{ai}^{\|2} \right). \tag{4.396}$$

In order to determine a and $a^{\|}$, we simply minimize the enthalpy in $v_{si}^{\perp,\|}$ and $v_{sai}^{\perp,\|}$, which become equal at $V_i^{\perp,\|}$ and $V_{ai}^{\perp,\|}$, respectively. At these velocities,

$$f_{\text{ent}}\Big|_{\text{eq}} = -\frac{\rho_s + a}{2} \left(V_i^{\perp 2} + V_{ai}^{\perp 2} \right) - \frac{\rho_s^{\|} + a^{\|}}{2} \left(V_i^{\|2} + V_{ai}^{\|2} \right). \tag{4.397}$$

Comparing this with (4.390), we see that

$$\begin{aligned} a &= \rho_n = \rho - \rho_s, \\ a^{\|} &= \rho_n^{\|} = \rho - \rho_s^{\|}, \end{aligned} \tag{4.398}$$

implying that the coefficients in (4.396) are simply the normal-liquid densities. Thus, the hydrodynamic limit of the collective energy density is given by

$$
f = \frac{\rho_s}{2} \left(v_{s\,i}^{\perp\,2} + v_{s\,ai}^{\perp\,2} \right) + \frac{\rho_s^{\parallel}}{2} \left(v_{si}^{\parallel\,2} + v_{sai}^{\parallel\,2} \right) - j_i \varphi_i - j_{ai} \varphi_{ai}
$$
$$
- \frac{1}{2} \frac{m^2}{\rho} \left[\left(\frac{1}{\frac{1}{3} F_1^s} + \frac{\rho_n}{\rho} \right) \varphi_i^{\perp\,2} + \left(\frac{1}{\frac{1}{3} F_1} + \frac{\rho_n^{\parallel}}{\rho} \right) \varphi_i^{\parallel\,2} \right.
$$
$$
\left. + \left(\frac{1}{\frac{1}{3} F_1^S} + \frac{\rho_n}{\rho} \right) \varphi_{ai}^{\perp\,2} + \left(\frac{1}{\frac{1}{3} F_1^S} + \frac{\rho_n^{\parallel}}{\rho} \right) \varphi_{ai}^{\parallel\,2} \right]. \tag{4.399}
$$

We now complete the squares in the fields $\varphi_{ai}^{\perp}$ and $\varphi_{ai}^{\parallel}$, and obtain

$$
f = \frac{1}{2} \rho_s \left(v_{s\,i}^{\perp\,2} + v_{s\,ai}^{\perp\,2} \right) + \frac{\rho_s^{\parallel}}{2} \left(v_{si}^{\parallel\,2} + v_{sai}^{\parallel\,2} \right)
$$
$$
+ \frac{1}{2} \frac{m^2}{s} \left(\frac{\frac{1}{3} F_a^s}{1 + \frac{1}{3} F_1 \frac{\rho_n}{\rho}} j_i^{\perp\,2} + \frac{\frac{1}{3} F_a^a}{1 + \frac{1}{3} F_1^S \frac{\rho_n}{\rho}} j_{ai}^{\perp\,2} + \frac{\frac{1}{3} F_1}{1 + \frac{1}{3} F_a^s \frac{\rho_n}{\rho}} j_i^{\parallel\,2} + \frac{\frac{1}{3} F_1^S}{1 + \frac{1}{3} F_1^S \frac{\rho_n}{\rho}} j_{ai}^{\parallel\,2} \right)
$$
$$
- \frac{1}{2} \frac{\rho}{m^2} \left[\left(\frac{1}{\frac{1}{3} F_1} + \frac{\rho_n}{\rho} \right) \left(\varphi_i^{\perp} - \frac{\frac{1}{3} F_a^s}{1 + \frac{1}{3} F_1 \frac{\rho_n}{\rho}} j_i^{\perp} \right)^2 \right.
$$
$$
\left. + \left(\frac{1}{\frac{1}{3} F_1^S} + \frac{\rho_n}{\rho} \right) \left(\varphi_{ai}^{\perp} - \frac{\frac{1}{3} F_1^S}{1 + \frac{1}{3} F_1^S \frac{\rho_n}{\rho}} j_{ai}^{\perp} \right)^2 + (\perp \to \parallel) \right]. \tag{4.400}
$$

The path integrals over the fields φ_i, φ_{ai} can then be performed, and this makes the harmonic terms in brackets disappear.

Finally, we allow l to vary in space. This produces Fermi liquid corrections to the stiffness constants $K_{s,t,b}$ [recall (4.126)], which are plotted in Fig. 4.22.

In the presence of a nontrivial l texture, the currents acquire additional terms. The particle current density j_i becomes

$$
m\mathbf{j}^{\perp} = \rho_s \mathbf{v}_s + c \left(\boldsymbol{\nabla} \times \mathbf{l} \right)^{\perp}, \qquad 2mc = \frac{\rho_s^{\parallel}}{2},
$$
$$
m\mathbf{j}^{\parallel} = \rho_s^{\parallel} \mathbf{v}_s^{\parallel} - c^{\parallel} \left(\boldsymbol{\nabla} \times \mathbf{l} \right)^{\parallel}, \qquad 2mc^{\parallel} = \frac{\rho_s^{\parallel}}{2}, \tag{4.401}
$$

where we have separated $\boldsymbol{\nabla} \times \mathbf{l}$ into transverse and longitudinal parts:

$$
(\boldsymbol{\nabla} \times \mathbf{l})^{\perp} = (\boldsymbol{\nabla} \times \mathbf{l}) - \mathbf{l} \left[\mathbf{l} \cdot (\boldsymbol{\nabla} \times \mathbf{l}) \right],
$$
$$
(\boldsymbol{\nabla} \times \mathbf{l})^{\parallel} = \mathbf{l} \left[\mathbf{l} \cdot (\boldsymbol{\nabla} \times \mathbf{l}) \right], \tag{4.402}
$$

respectively. The squares of the currents are

$$m^2 j_i^{\perp 2} = \rho_s^2 \mathbf{v}^{\perp 2} + \frac{1}{4m^2}\frac{\rho_s^{\|2}}{4}\left[(\nabla \times \mathbf{l})^2 - (\mathbf{l}\cdot(\nabla\times\mathbf{l}))^2\right] + \frac{1}{2m}\frac{\rho_s\rho_s^{\|}}{2}\mathbf{v}^{\perp}(\nabla\times\mathbf{l})^{\perp}$$

$$= \rho_s^2\left[\mathbf{v}^2 - (\mathbf{l}\cdot\mathbf{v})^2\right] + \frac{1}{4m^2}\frac{\rho_s^2}{4}\left[\mathbf{l}\times(\nabla\times\mathbf{l})\right]^2$$

$$+ \frac{1}{2m}\frac{\rho_s\rho_s^{\|}}{2}\left[\mathbf{v}\cdot(\nabla\times\mathbf{l}) - (\mathbf{v}\cdot\mathbf{l})(\nabla\times\mathbf{l})\right], \qquad (4.403)$$

$$m^2 j_i^{\|2} = \rho_s^{\|2}(\mathbf{l}\cdot\mathbf{v})^2 + \frac{1}{4m^2}\frac{\rho_s^{\|2}}{4}\left[\mathbf{l}\cdot(\nabla\times\mathbf{l})\right]^2 - \frac{1}{2m}\frac{\rho_s^{\|2}}{2}(\mathbf{v}\cdot\mathbf{l})(\nabla\times\mathbf{l}), \qquad (4.404)$$

$$m^2 j_{ai}^{\perp 2} = \rho_s^2 v_{ai}^{\perp 2} = \frac{\rho_s^2}{4m^2}\left\{\epsilon_{abc}d_b\left[\nabla_i - l_i(\mathbf{l}\cdot\nabla)\right]d_c\right\}^2$$

$$= \frac{\rho_s^2}{4m^2}\left[(\nabla_i d_a)^2 - (\mathbf{l}\cdot\nabla d_a)^2\right], \qquad (4.405)$$

$$m^2 j_{ai}^{\|\,2} = \rho_s^{\|2} v_{ai}^{\|\,2} = \frac{\rho_s^{\|2}}{4m^2}\left[l_i\epsilon_{abc}d_b(\mathbf{l}\cdot\nabla)d_c\right]^2$$

$$= \frac{\rho_s^{\|2}}{4m^2}(\mathbf{l}\cdot\nabla d_a)^2. \qquad (4.406)$$

Using these, we find the energy density

$$f = \frac{1}{2}\rho_s\frac{1+\frac{1}{3}F_1}{1+\frac{1}{3}F_1\frac{\rho_n}{\rho}}\left[\mathbf{v}_s^2 - (\mathbf{l}\cdot\mathbf{v}_s)^2\right] + \frac{1}{2}\rho_s^{\|}\frac{1+\frac{1}{3}F_1}{1+\frac{1}{3}F_1\frac{\rho_n^{\|}}{\rho}}(\mathbf{l}\cdot\mathbf{v}_s)^2$$

$$+\frac{1}{2}\rho_s\frac{1+\frac{1}{3}F_1^S}{1+\frac{1}{3}F_1^S\frac{\rho_n}{\rho}}\frac{1}{4m^2}\left[(\nabla_i d_a)^2 - (\mathbf{l}\cdot\nabla d_a)^2)\right] + \frac{1}{2}\rho_s^{\|}\frac{1+\frac{1}{3}F_1^S}{1+\frac{1}{3}F_1^S\frac{\rho_n^{\|}}{\rho}}(\mathbf{l}\cdot\nabla d_a)^2$$

$$+\frac{\rho_s^{\|}}{2}\frac{1+\frac{1}{3}F_a^s}{1+\frac{1}{3}F_1\frac{\rho_n}{\rho}}\frac{1}{2m}\left\{\mathbf{v}_s\cdot(\nabla\times\mathbf{l}) - (\mathbf{l}\cdot\mathbf{v}_s)\left[\mathbf{l}\cdot(\nabla\times\mathbf{l})\right]\right\}$$

$$-\frac{\rho_s^{\|}}{2}\frac{1}{2m}\frac{1+\frac{1}{3}F_1}{1+\frac{1}{3}F_1\frac{\rho_n^{\|}}{\rho}}(\mathbf{l}\cdot\mathbf{v}_s)\left[\mathbf{l}\cdot(\nabla\times\mathbf{l})\right]$$

$$+\frac{1}{2}K_s(\nabla\cdot\mathbf{l})^2 + \frac{1}{2}\left(K_t + \frac{1}{4m^2}\frac{\rho_s^{\|2}}{4\rho^2}\frac{\frac{1}{3}F_1^3}{1+\frac{1}{3}F_1\frac{\rho_n^{\|}}{\rho}}\right)\left[\mathbf{l}\cdot(\nabla\times\mathbf{l})\right]^2$$

$$+\frac{1}{2}\left(K_b + \frac{1}{4m^2}\frac{\rho_s^{\|2}}{4\rho^2}\frac{\frac{1}{3}F_1}{1+\frac{1}{3}F_1\frac{\rho_n}{\rho}}\right)\left[\mathbf{l}\times(\nabla\times\mathbf{l})\right]^2. \qquad (4.407)$$

As discussed in the beginning, the mass parameter m in these expressions is the *effective mass* of the screened quasiparticles in the Fermi liquid. As a consequence, the velocity

$$\mathbf{v}_s = \frac{1}{2m}\Phi^*\overleftrightarrow{\nabla}\Phi \qquad (4.408)$$

is not really the correct parameter of Galilean transformations. To play this role, the phase change in the original fundamental fields would have to be

$$\psi \to e^{im_0\mathbf{v}\cdot\mathbf{x}}\psi, \qquad (4.409)$$

where m_0 is the true physical mass of the ^{3}He atoms. If we introduce the corresponding physical velocity

$$\mathbf{v}_{0s} = \frac{i}{2m_0} \Phi^* \overleftrightarrow{\nabla} \Phi, \tag{4.410}$$

with a similar expression for the spin velocity $\mathbf{v}_{0sai}$, the first term in (4.407) takes the form

$$\frac{1}{2} \left(\rho_s \frac{m_0}{m} \right) \frac{1}{1 + \frac{1}{3}F_1 \frac{\rho_n}{\rho}} \frac{m_0 \left(1 + \frac{1}{3}F_1 \right)}{m} \left[\mathbf{v}_{0s}^2 - (\mathbf{1} \cdot \mathbf{v}_s)^2 \right]. \tag{4.411}$$

The other terms in (4.407) change accordingly.

We now add to the energy density the kinetic energy of the normal component

$$\frac{1}{2} \rho_n \left[\mathbf{v}_{0n}^2 - (\mathbf{1} \cdot \mathbf{v}_{0n})^2 \right] + \frac{1}{2} \rho_n (\mathbf{1} \cdot \mathbf{v}_{0n})^2. \tag{4.412}$$

By Galilei invariance, the sum of the coefficients has to add up to the total density $\rho_0 = nm_0$, where n is the number of particles ($\equiv$ number of quasiparticles) per unit volume:

$$\left(\rho_s \frac{m_0}{m} \right) \frac{1}{1 + \frac{1}{3}F_1 \frac{\rho_n}{\rho}} \frac{m_0 \left(1 + \frac{1}{3}F_1 \right)}{m} + \rho_n = \rho_0 = nm_0. \tag{4.413}$$

At $T = 0$, the normal density ρ_n vanishes, and we obtain

$$\rho_s \Big|_{T=0} = \rho = \rho_0 \frac{m}{m_0}. \tag{4.414}$$

Thus, consistency requires the following relation between the effective mass m and the atomic mass $m_0 \equiv m_{^3\text{He}}$:

$$m = \left(1 + \frac{1}{3}F_1 \right) m_0. \tag{4.415}$$

This brings the term (4.413) to the form

$$\frac{1}{2} \rho_0 \frac{\rho_s}{\rho} \frac{1}{1 + \frac{1}{3}F_1 \frac{\rho_n}{\rho}} \left[\mathbf{v}_{0s}^2 - (\mathbf{1} \cdot \mathbf{v}_{0s})^2 \right]. \tag{4.416}$$

The prefactor can be interpreted as the superfluid density with Fermi liquid corrections:

$$\rho_s^{\text{FL}} \equiv \rho_0 \frac{\rho_s}{\rho} \frac{1}{1 + \frac{1}{3}F_1 \frac{\rho_n}{\rho}}. \tag{4.417}$$

It is now convenient to introduce the dimensionless ratio

$$\tilde{\rho}_s^{\text{FL}} \equiv \frac{\rho_s^{\text{FL}}}{\rho_0}. \tag{4.418}$$

At $T = 0$, this goes to unity just as in the uncorrected case. See Fig. 4.23 for a plot.

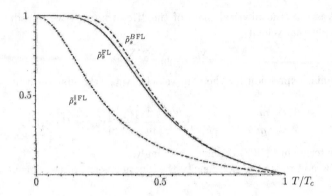

FIGURE 4.23 Superfluid densities of B- and A-phase after applying Fermi liquid corrections, indicated by the superscript FL.

For $T \approx T_c$, however, it receives a strong reduction by a factor

$$\frac{1}{1 + \frac{1}{3}F_1} = \frac{m_0}{m}, \tag{4.419}$$

so that

$$\tilde{\rho}_s^{\mathrm{FL}}\Big|_{T \approx 0} \approx \frac{\rho_s}{\rho} \frac{m_0}{m}. \tag{4.420}$$

Thus near T_c the number of particles in the normal component is equal to the true particle density if it is multiplied by the quasiparticle mass m instead of the atomic mass m_0. Specific-heat experiments [13] determine the effective mass ratios m/m_0 mentioned in the beginning. Inserted into Eq. (4.419), these ratios yield the parameters

$$\frac{1}{3}F_1 = (2.01,\ 3.09,\ 3.93,\ 4.63,\ 5.22) \tag{4.421}$$

at pressures

$$p = (0,\ 9,\ 18,\ 27,\ 34.36)\ \ \mathrm{bar}. \tag{4.422}$$

Similarly, we may go through the Fermi liquid corrections of the spin currents and obtain

$$\tilde{K}_{\mathrm{d}}^{\mathrm{FL}} \equiv \frac{K_{\mathrm{d}}^{\mathrm{FL}}}{\rho_0} = \frac{\rho_s}{\rho} \frac{1 + \frac{1}{3}F_1^S}{1 + \frac{1}{3}F_1^S \frac{\rho_n}{\rho}} \frac{1}{1 + \frac{1}{3}F_1}, \tag{4.423}$$

$$\tilde{K}_{\mathrm{d}}^{\|\mathrm{FL}} \equiv \frac{K_{\mathrm{d}}^{\|\mathrm{FL}}}{\rho_0} = \frac{\rho_s^{\|}}{\rho} \frac{1 + \frac{1}{3}F_1^S}{1 + \frac{1}{3}F_1^S \frac{\rho_n^{\|}}{\rho}} \frac{1}{1 + \frac{1}{3}F_1}, \tag{4.424}$$

while the coefficients c and $c^{\|}$ satisfy

$$\tilde{c}^{\mathrm{FL}} \equiv \frac{2m_0 c^{\mathrm{FL}}}{\rho_0} = \frac{2mc}{\rho} \frac{1}{1 + \frac{1}{3}F_1 \frac{\rho_n}{\rho}}, \tag{4.425}$$

$$\tilde{c}^{\parallel \mathrm{FL}} \equiv \frac{2m_0 c^{\parallel \mathrm{FL}}}{\rho_0} = \frac{2mc^{\parallel}}{\rho} \frac{1}{1 + \frac{1}{3} F_1 \frac{\rho_n^{\parallel}}{\rho}}. \tag{4.426}$$

These are plotted in Fig. 4.24.

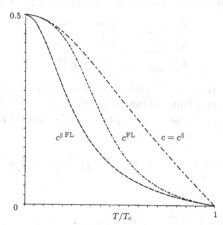

FIGURE 4.24 Coefficients $c = c^{\parallel}$ and their Fermi liquid corrected values in the A-phase, indicated by the superscript FL, in units of ρ/m.

The stiffness coefficients of the pure l-parts of the bending energy receive the Fermi liquid corrections:

$$\tilde{K}_s^{\mathrm{FL}} \equiv \frac{4m_0^2 K_s^{\,\mathrm{FL}}}{\rho_0} = \frac{4m^2 K_s}{\rho} \frac{1}{1 + \frac{1}{3} F_1}, \tag{4.427}$$

$$\tilde{K}_t^{\mathrm{FL}} \equiv \frac{4m_0^2}{\rho_0} K_t^{\mathrm{FL}} = \left(\frac{4m^2 K_t}{\rho} + \frac{1}{4} \frac{\rho_s^{\parallel 2}}{\rho^2} \frac{\frac{1}{3} F_1}{1 + \frac{1}{3} F_1 \frac{\rho_n^{\parallel}}{\rho}} \right) \frac{1}{1 + \frac{1}{3} F_1}, \tag{4.428}$$

$$\tilde{K}_b^{\mathrm{FL}} \equiv \frac{4m_0^2}{\rho_0} K_b^{\mathrm{FL}} = \left(\frac{4m^2 K_b}{\rho} + \frac{1}{4} \frac{\rho_s^{\parallel 2}}{\rho^2} \frac{\frac{1}{3} F_1}{1 + \frac{1}{3} F_1 \frac{\rho_n^{\parallel}}{\rho}} \right) \frac{1}{1 + \frac{1}{3} F_1}. \tag{4.429}$$

They are plotted in Figs. 4.25–4.26. In the sequel, it is convenient to define the momenta

$$\tilde{\mathbf{v}} \equiv 2m_0 \mathbf{v}. \tag{4.430}$$

Then the energy density can be written in the following final form

$$
\begin{aligned}
\frac{4m_0^2}{\rho_0} e &\equiv \frac{1}{2} \tilde{\rho}_s^{\mathrm{FL}} \left[\tilde{\mathbf{v}}_s^2 - (\mathbf{l} \cdot \tilde{\mathbf{v}}_s)^2 \right] + \frac{1}{2} \frac{\tilde{1}}{2} \tilde{\rho}_s^{\parallel \mathrm{FL}} (\mathbf{l} \cdot \tilde{\mathbf{v}}_s)^2 \\
&+ \frac{1}{2} \tilde{K}_d^{\mathrm{FL}} \left[(\nabla_i d_a)^2 - (\mathbf{l} \cdot \nabla_a)^2 \right] + \frac{1}{2} \tilde{K}_d^{\parallel \mathrm{FL}} (\mathbf{l} \cdot \nabla d_a)^2 \\
&+ \tilde{c}^{\mathrm{FL}} \left\{ \tilde{\mathbf{v}} \cdot (\nabla \times \mathbf{l}) - (\tilde{\mathbf{v}}_s \cdot \mathbf{l}) [\mathbf{l} \cdot (\nabla \times \mathbf{l})] \right\}
\end{aligned}
$$

$$- \tilde{c}^{\|FL} (\tilde{\mathbf{v}}_s \cdot \mathbf{l}) [\mathbf{l} \cdot (\boldsymbol{\nabla} \times \mathbf{l})]$$
$$+ \frac{1}{2} \tilde{K}_s^{FL} (\boldsymbol{\nabla} \cdot \mathbf{l})^2 + \frac{1}{2} \tilde{K}_t^{FL} [\mathbf{l} \cdot (\boldsymbol{\nabla} \times \mathbf{l})]^2 + \frac{1}{2} \tilde{K}_b^{FL} [\mathbf{l} \times (\boldsymbol{\nabla} \times \mathbf{l})]^2 . \quad (4.431)$$

In large containers, where $\mathbf{l}$ and $\mathbf{d}$ are locked to each other, the $\tilde{K}_d^{FL}$, $\tilde{K}_d^{\|FL}$ terms can be absorbed into $\tilde{K}_{s,t,b}^{FL}$ which then take the *dipole-locked* values

$$\tilde{K}_s^{FL}\big|_{\text{lock}} = \tilde{K}_s^{FL} + \tilde{K}_d^{FL}, \qquad (4.432)$$
$$\tilde{K}_t^{FL}\big|_{\text{lock}} = \tilde{K}_t^{FL} + \tilde{K}_d^{FL}, \qquad (4.433)$$
$$\tilde{K}_b^{FL}\big|_{\text{lock}} = \tilde{K}_t^{FL} + \tilde{K}_d^{FL}. \qquad (4.434)$$

The temperature dependence of all these quantities is shown in Figs. 4.25 and 4.26 for the experimental Fermi liquid parameters $\frac{1}{3}F_1 = 5.22$ and $\frac{1}{3}F_1^S = -.22$.

The Fermi liquid corrections in the B-phase can be applied in completely analogous manner. There the energy becomes

$$\frac{4m_0^2}{\rho_0} f = \frac{1}{2} \tilde{\rho}_s^{BFL} (\boldsymbol{\nabla}\varphi)^2 + \lambda(4 + \delta)(\nabla_i \theta_j)^2 - (1 + \delta)\nabla_i \theta_j \nabla_j \theta_i - (\nabla_i \theta_i)^2 , \quad (4.435)$$

with the dimensionless parameters

$$\tilde{\rho}_s^{BFL} = \frac{\rho_s^B}{\rho} \frac{1}{1 + \frac{1}{3}F_1 \frac{\rho_n^B}{\rho}}, \qquad (4.436)$$

$$\lambda = \frac{1 + \frac{1}{3}F_1^S}{1 + \frac{1}{3}F_1^S \frac{\rho_n^B}{\rho} + \frac{1}{3}F_1^S \frac{2}{15}\frac{\rho_s^B}{\rho}} \frac{1}{1 + \frac{1}{3}F_1}, \qquad (4.437)$$

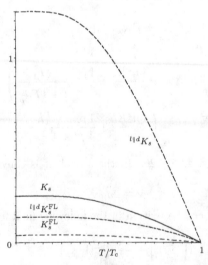

FIGURE 4.25 Coefficient K_s for splay deformations of the fields, and its Fermi liquid corrected values, indicated by the superscript FL, in the A-phase in units of $\rho/4m^2$ as functions of the temperature.

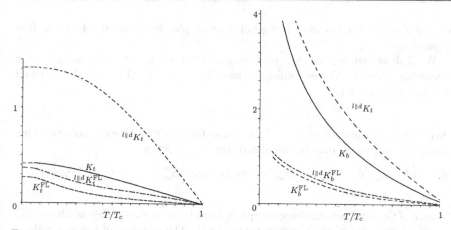

FIGURE 4.26 Remaining hydrodynamic parameters for twist and bend deformations of superfluid ^{3}He-A, together with their Fermi liquid corrected values, indicated by the superscript FL, in units of $\rho/4m^2$, as functions of the temperature.

$$\delta = \frac{\frac{1}{3}F_1^S \frac{\rho_s^B}{\rho}}{1 + \frac{1}{3}F_1^S \frac{\rho_n^B}{\rho}}. \tag{4.438}$$

the first being plotted in Fig. 4.23. The combinations

$$\tilde{K}_1^{BFL} = \rho_s^B \lambda (4 + \delta)/5, \tag{4.439}$$
$$\tilde{K}_2^{BFL} = \rho_s^B \lambda (1 + \delta)/5, \tag{4.440}$$

are plotted in Fig. 4.27.

Certainly, all these results need strong-coupling corrections which are presently only known in the Ginzburg-Landau regime $T \to T_c$.

4.10   Large Currents and Magnetic Fields in the Ginzburg-Landau Regime

The properties of superflow are most easily calculated close to the critical temperature. In this regime, thermodynamic fluctuations are governed by the Ginzburg-Landau form of the energy. There the depairing critical currents have been derived quite some time ago. For the sake of a better understanding of our general results to follow later, we find it useful to review the well-known results.

Suppose a uniform current is set up in a container along the z direction. Since the bending energies tend to straighten out textural field lines it may be expected that, in equilibrium, and in uniform currents, also the textures are uniform. It will be shown later in a detailed study of local stability in Section 4.13, that this assumption is indeed justified in the B-phase. In the A-phase, on the other hand, we shall see that the textural degrees of freedom play an essential role in the flow dynamics.

We shall at first neglect this complication and proceed with a discussion of flow in uniform textures. Correspondingly, the collective field will for now be assumed to have the simple form

$$\Delta_{ai}(z) = \Delta_{ai}^0 e^{i\varphi(z)},\tag{4.441}$$

where Δ_{ai}^0 is a constant matrix. The phase factor $e^{i\varphi(z)}$ allows a non-vanishing matter current, which may be calculated from Eq. (4.88) as

$$j_i = i\left\{K_1|\Delta_{ak}^0|^2\delta_{ij} + K_2\left[\Delta_{aj}^{0*}\Delta_{ai}^0 - (i \leftrightarrow j)\right] + K_3\left[\Delta_{ai}^{0*}\Delta_{aj}^0 - (i \leftrightarrow j)\right]\right\}\partial_j\varphi(z).\tag{4.442}$$

Because of the smallness of strong-coupling corrections on the coefficients $K_i(\leq 3\%)$ we may assume for K the common value (4.82). The presence of a non-vanishing gradient of $\partial_j\varphi$ requires a new minimization of the energy. This will in general modify the normal forms (4.111)–(4.114) of the gap parameters in equilibrium.

4.10.1 B-Phase

For a first crude estimate of the effect of a current we shall assume only the overall size of the gap parameter (4.112) of the B-phase to be changed by the current.

Neglecting Gap Distortion

If the current runs along the z-axis, we find from (4.103)

$$f = \frac{K}{2}5\left[a^2(\partial_z\varphi)^2 + (\partial_z a)^2\right]\Delta_B^2 - 3\alpha\mu a^2\Delta_B^2 + 9\beta_B\beta_0\Delta_B^4.\tag{4.443}$$

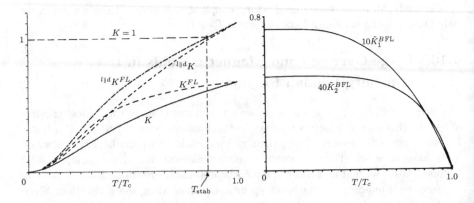

FIGURE 4.27 Hydrodynamic parameters of superfluid ³He-B , together with their Fermi liquid corrected values, indicated by the superscript FL, in units of $\rho/4m^2$, as functions of the temperature.

For the discussions to follow, it is convenient to measure the energy densities in terms of the condensation energy of the B-phase in the weak-coupling limit. In the Ginzburg-Landau regime this is

$$f_c = f_c^B = \frac{\rho}{4m^2\xi_0^2} \left(1 - \frac{T}{T_c}\right)^2. \tag{4.444}$$

By using the definition (4.105) and the *temperature-dependent coherence length*

$$\xi(T) = \frac{\xi_0}{\sqrt{1 - T/T_c}}, \tag{4.445}$$

with ξ_0 from (4.75), we find the simple form

$$\frac{f}{2f_c} = a^2\xi^2 \left[(\partial_z\varphi)^2 + (\partial_z a)^2\right] - \alpha a^2 + \frac{1}{2}\left(\frac{6}{5}\beta_B\right)a^4. \tag{4.446}$$

If one wants to study the system in the presence of a non-vanishing current, one may eliminate the cyclic variable φ in favor of the canonical momentum-like variable

$$j \equiv \frac{1}{2\xi}\frac{\partial}{\partial\partial_z\varphi}\frac{f}{2f_c} = a^2\xi\partial_z\varphi. \tag{4.447}$$

This has the virtue of being z-independent, as follows from the equation of motion for φ.

The associated Legendre transformed energy

$$g = \frac{f}{2f_c} - 2\xi j\partial_z\varphi \tag{4.448}$$

can then be used to study the remaining problem in only one variable $a(z)$

$$g = (\partial_z a)^2 - \alpha a^2 + \frac{1}{2}\left(\frac{6}{5}\beta_B\right)a^4 - \frac{j^2}{a^2}. \tag{4.449}$$

By comparing (4.446) and (4.442) we see that the physical current $j_3 \equiv J$ is determined in terms of the dimensionless quantity j up to a factor two:

$$
\begin{aligned}
J &= 2\frac{\partial f}{\partial\partial_z\varphi} = 10\Delta_B^2 K\partial_z\varphi \\
&= j\frac{\hbar}{2m\xi_0}\rho 2\left(1 - \frac{T}{T_c}\right)^{3/2} \equiv j J_0 \left(1 - \frac{T}{T_c}\right)^{3/2}.
\end{aligned} \tag{4.450}
$$

Thus the quantity j measures the physical mass current in units of

$$J_0 \left(1 - \frac{T}{T_c}\right)^{3/2} = v_0 \left(1 - \frac{T}{T_c}\right)^{1/2} 2\rho \left(1 - \frac{T}{T_c}\right), \tag{4.451}$$

where v_0 is the following reference velocity

$$v_0 \equiv \frac{\hbar}{2m\xi_0}, \tag{4.452}$$

at which the de Broglie wavelength of the quasiparticles equals the coherence length ξ_0. Consequently, we shall refer to v_0 as the *coherence velocity* and to J_0 as *coherence current*.

By analogy with the definition of $\xi_0(T)$ from ξ_0 in (4.445), we introduce also here temperature-dependent quantities that contain the Ginzburg-Landau factor $(1 - T/T_c)$, for instance the temperature-dependent coherence velocity and the current

$$v_0(T) \equiv v_0 \left(1 - \frac{T}{T_c}\right)^{1/2}, \qquad J_0(T) \equiv J_0 \left(1 - \frac{T}{T_c}\right)^{3/2}, \tag{4.453}$$

respectively. Using the superfluid velocity

$$v_s = \frac{\hbar}{2m}\partial_z\varphi, \tag{4.454}$$

we can identify the superfluid density ρ_s via the definition

$$J \equiv \rho_s v_s, \tag{4.455}$$

where

$$\rho_s = a^2\,2\rho \left(1 - \frac{T}{T_c}\right). \tag{4.456}$$

By writing (4.454) in the form

$$v_s = v_0(T)\xi\partial_z\varphi, \tag{4.457}$$

we see that the quantity

$$\kappa \equiv \xi\partial_z\varphi = j/a^2 \tag{4.458}$$

measures the superflow velocity in units of the temperature-dependent coherence velocity $v_0(T)$:

$$\kappa \equiv \frac{v_s}{v_0(T)} = \frac{v_s}{v_0}\left(1 - \frac{T}{T_c}\right)^{-\frac{1}{2}}. \tag{4.459}$$

In order to be able to compare the forthcoming results with experiments we may use the parameters of Wheatley [13], which are listed in Table 4.2, to calculate

$$v_0 = \frac{1}{2m^*\xi_0} = \sqrt{\frac{48}{7\zeta(3)}}\pi\frac{k_B T_c}{p_F} = 7.504\frac{k_B T_c}{p_F}$$

$$\approx \left\{ \begin{array}{ll} 6.25 & \text{cm/sec} \\ 15 & \text{cm/sec} \end{array} \right\} \text{ for } \left\{ \begin{array}{ll} p = 0, & T_c = 1\,\text{mK}, \\ p = 34.36\,\text{bar}, & T_c = 2.7\,\text{mK}. \end{array} \right\} \tag{4.460}$$

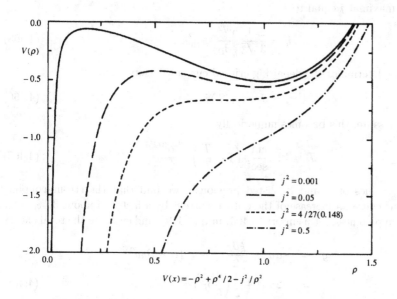

$$V(x) = -\rho^2 + \rho^4/2 - j^2/\rho^2$$

FIGURE 4.28 Shape of potential determining stability of superflow.

It is now quite simple to study the equilibrium gap configuration for a given current density j. According to (4.449) the energy looks like the Lagrangian of a mass point at position a moving as a function of "time" z in a potential which is turned upside down:

$$-V(a) = -\alpha a^2 + \frac{1}{2}\left(\frac{5}{6}\beta_B\right)a^4 - \frac{j^2}{a^2}. \tag{4.461}$$

The shape of this potential is displayed in Fig. 4.28. For a small enough current, there is a constant solution satisfying $\partial V/\partial a = 0$:

$$a(z) \equiv a_0 \tag{4.462}$$

This amounts to a current density

$$j^2 = a_0^4\left[\alpha - \left(\frac{6}{5}\beta_B\right)a_0^2\right]. \tag{4.463}$$

Obviously, this solution can exist only as long as j stays below the maximal value allowed by (4.463). By differentiation we find

$$a_c^2 = \frac{2}{3}\frac{\alpha}{\frac{6}{5}\beta_B}, \tag{4.464}$$

with the maximal j equal to

$$j_c = \frac{2}{3}\frac{1}{\sqrt{3}}\frac{\alpha^{3/2}}{\frac{6}{5}\beta_B} \equiv a_0^2 \kappa_c. \tag{4.465}$$

This value determines the depairing critical current density

$$J_c = J_0(T)j_c. \tag{4.466}$$

At zero pressure, this becomes numerically

$$J_c = 12.5\frac{\mathrm{cm}}{\mathrm{sec}}\rho\left(1 - \frac{T}{T_c}\right)^{3/2}\frac{\alpha^{3/2}}{\frac{6}{5}\beta_B}. \tag{4.467}$$

With the values of α and β_B listed previously we find that the strong-coupling corrections cause an increase of the critical current by a factor of about 30%.

For completeness, let us insert (4.463) into (4.449), and evaluate the total energy

$$\begin{aligned}
g &= -\alpha a^2 + \frac{\frac{5}{6}\beta_B}{2}a^4 - a^2\left(\alpha - \beta_B a^2\right),\\
&= -2\alpha a^2 + \frac{3}{2}\left(\frac{6}{5}\beta_B\right)a^4. \tag{4.468}
\end{aligned}$$

It is cumbersome to express this analytically as a function of j since this would involve solving the cubic equation (4.449). However, if we do not try to express it in terms of the current density j but, instead, in terms of the parameter κ, i.e., the superfluid velocity in natural units, we have [see (4.447), (4.448), (4.449)]

$$\kappa^2 = \alpha - \beta_B a^2, \tag{4.469}$$

and the current dependence of the energy is written explicitly as:

$$g = -\frac{1}{2\beta_B}\left(\alpha - \kappa^2\right)\left(\alpha + 3\kappa^2\right). \tag{4.470}$$

Note that the free energy itself is simply

$$\frac{f}{2f_c} = -\frac{1}{2\beta_B}\left(\alpha - \kappa^2\right)^2. \tag{4.471}$$

Including a Magnetic Field

The critical currents in the B-phase depend sensitively on external magnetic fields. In order to see this consider the additional field energy

$$f_{\mathrm{mg}} = g_z|H_a\Delta_{ai}|^2, \tag{4.472}$$

where g_z was calculated microscopically to be

$$g_z = \frac{3}{2}\frac{\rho}{p_F^2}\frac{\xi_0^2}{v_F^2}\gamma^2\left(1 + \frac{Z_0}{4}\right)^{-2}, \tag{4.473}$$

with γ being the magnetic dipole moment of the ^{3}He atoms

$$\gamma \approx 2.04 \times 10^4 \frac{1}{\text{gauss sec}}, \tag{4.474}$$

and $Z_0 = F_0^S$ is the Fermi liquid parameter of the spin density coupling.

According to Table 4.2 its value is, at zero pressure, $-Z_0 \approx 2.69$. It will be useful to rewrite f_{mg} in a dimensionless form as

$$\frac{f_{\text{mg}}}{2f_c} = h^2 |\hat{H}_a \Delta_{ai}/\Delta_B|^2, \tag{4.475}$$

where $\hat{H}$ is the unit vector in the direction of the field and $h \equiv H/H_0(T)$ measures the magnetic field in terms of the following natural units

$$H_0(T) \equiv \sqrt{\frac{\frac{3}{2}\rho p_F^2}{g_z}} \sqrt{1 - \frac{T}{T_c}} = \left(1 + \frac{Z_0}{4}\right) \frac{v_F}{\xi_0 \gamma} \sqrt{1 - \frac{T}{T_c}}$$

$$\equiv H_0 \sqrt{1 - \frac{T}{T_c}} \approx 16.4K \text{ gauss} \sqrt{1 - \frac{T}{T_c}}. \tag{4.476}$$

For the undistorted gap parameter (4.441), the additional magnetic energy is simply

$$\frac{f_{\text{mg}}}{2f_c} = h^2 a^2. \tag{4.477}$$

This enters into the expression for the equilibrium current (4.463) in the form

$$j^2 = a_0^4 \left[\alpha - \left(\frac{6}{5}\beta_B\right) a_0^2 - h^2\right], \tag{4.478}$$

so that the current is now maximal at

$$a_0^2 = \frac{2}{3} \frac{1}{\frac{5}{6}\beta_B} \left(\alpha - h^2\right), \tag{4.479}$$

with values j_c, κ_c:

$$j_c = \frac{2}{3} \frac{1}{\sqrt{3}} \frac{1}{\frac{5}{6}\beta_B} \left(\alpha - h^2\right)^{3/2}, \qquad \kappa_c = \frac{1}{\sqrt{3}} \frac{1}{\frac{5}{6}\beta_B} \sqrt{\alpha - h^2}. \tag{4.480}$$

Thus, at higher magnetic fields, the liquid supports less superflow. For

$$h_c{}^2 = \alpha, \tag{4.481}$$

the liquid becomes normal. Note that this result is independent of the direction of **H** with respect to the texture of the B-phase.

Let us also calculate the changes in the total energies. With (4.447) and (4.458) we have now

$$\kappa^2 = \alpha - \frac{6}{5}\beta_B a_0^2 - h^2, \tag{4.482}$$

and

$$g^B = -\frac{1}{2\left(\frac{6}{5}\beta_B\right)}\left(\alpha - \kappa^2 - h^2\right)\left(\alpha + 3\kappa^2 + 3h^2\right),\qquad(4.483)$$

or

$$\frac{f^B}{2f_c} = -\frac{1}{2\left(\frac{6}{5}\beta_B\right)}\left(\alpha - \kappa^2 - h^2\right)2.\qquad(4.484)$$

Allowing Gap Distortion

Certainly, the assumption of a purely multiplicative modification of the gap was an over-simplification. For, if we look at the energy for a general gap parameter

$$\Delta^0_{ai} = \Delta_B a_{ai} e^{i\varphi(z)},\qquad(4.485)$$

we find from the energy (4.84) (neglecting the g_d-terms)

$$\frac{f}{2f_c} = \frac{1}{5}(a^*_{ai}a_{ai} + 2a^*_{az}a_{az})(\partial_z\varphi)^2 + \beta_1 a^*_{ai}a_{bj}a^*_{ai}a_{bj} + \beta_2\left(a^*_{ai}a_{ai}\right)^2$$
$$+\beta_3 a^*_{ai}a_{aj}a^*_{bi}a_{bj} + \beta_4 a^*_{ai}a_{bi}a^*_{bj}a_{aj} + \beta_5 a^*_{ai}a_{bi}a^*_{aj}a_{bj}.\quad(4.486)$$

In the absence of magnetic field and current, the energy is invariant under the full group of independent rotations on spin and orbital indices (apart from a phase invariance $a_{ai} \to e^{i\varphi}a_{ai}$).

As a field and a current are turned on, two specific directions in these spaces are singled out. ·Due to the original invariance, however, the energy at the extremum cannot depend on which directions are chosen. Therefore we may pick for both, **H** and the current **J**, the z-direction. Given the solution for the order parameter a_{ai} in this particular case, the general result is obtained by simply performing an appropriate $SO(3)_{\text{spin}} \times SO(3)_{\text{orbit}}$ rotation relevant for the actual directions of **H** and **J**.

We shall now determine the functional direction in which the deformation of the gap parameter has to take place. Consider at first a small current density j. Then the gap parameter can be assumed to be close to the equilibrium value in the B-phase:

$$a_{ai} = a_0\delta_{ai} + a'_{ai} \equiv a_0\delta_{ai} + r_{ai} + i\,i_{ai}\qquad(4.487)$$

with

$$a_0 = \sqrt{\frac{\alpha}{\frac{6}{5}\beta_B}}.\qquad(4.488)$$

Inserting this into (4.444) we can pick up all terms up to quadratic order and find:

$$\frac{\delta^2 f}{2f_c} = \frac{1}{5}\left\{\left(|a'_{ai}|^2 + 2|a'_{az}|^2\right) + 2a_0\left(r_{11} + r_{22} + 3r_{33}\right)\right\}(\partial_3\varphi)^2$$

$$+\frac{1}{15}\frac{\alpha}{\frac{6}{5}\beta_B}\left\{2\left(2\beta_{124} + \beta_{35}\right)R^1 + 4\beta_{12}R^3 + 2\beta_{345}R^5\right.$$

$$\left.-8\beta_1 I' + 8\beta_1 I^3 - 6\beta_1 I^5 - 2\left(3\beta_1 + \beta_{35} - \beta_4\right)I^6\right\}. \qquad (4.489)$$

Here $R^{1,3,5,6}$ represent the following quadratic forms

$$R^1 \equiv r_{11}^2 + r_{22}^2 + r_{33}^2,$$

$$R^3 \equiv 2\left(r_{11}r_{22} + r_{22}r_{33} + r_{33}r_{11}\right),$$

$$R^{5,6} \equiv \left(r_{12} \pm r_{21}\right)^2 + \left(r_{23} \pm r_{32}\right)^2 + \left(r_{31} \pm r_{13}\right)^2, \qquad (4.490)$$

with $I^{1,3,5,6}$ being the same expressions in terms of the imaginary parts i_{ai}.

Now, the linear term involves only the real diagonal elements r_{11}, r_{22} and r_{33}. Thus only these will develop new equilibrium values. Moreover, since r_{11} and r_{22} enter symmetrically, their new values are equal. For small currents we are just led to a new gap parameter

$$\Delta_{ai} = \Delta_B \begin{pmatrix} a & & \\ & a & \\ & & c \end{pmatrix} e^{i\varphi(z)}. \qquad (4.491)$$

We shall now assume that this form of the distortion is also present in stronger currents, up to its critical value J_c.

In order to ensure this, we have to examine the stability of this form under small oscillations for any current. This will be done in Sections 4.13.5 and 4.15, where local stability of the form (4.491) will indeed be found (up to J_c). As a side result, the analyses will provide us with the energies of all collection excitations in the presence of superflow.

In order to study the problem with a distorted gap (4.491) let us, at first, neglect the strong-coupling corrections. Then the energy (4.446) takes the simple form

$$\frac{f}{2f_c} = \frac{1}{5}\xi^2\left[\left(2a^2 + 3c^2\right)(\partial_z\varphi)^2 + 2a_z^2 + 3c_z^2\right]$$

$$-\frac{1}{3}\left(2a^2 + c^2\right) + \frac{1}{15}\left(4a^4 + 2a^2c^2 + \frac{3}{2}c^4\right) + h^2c^2, \qquad (4.492)$$

where we have included the magnetic field. The current density is now

$$j = \frac{1}{5}\left(2a^2 + 3c^2\right)\xi\partial_z\varphi = \frac{2a^2 + 3c^2}{5}\kappa, \qquad (4.493)$$

so that the superfluid density becomes

$$\rho_s^\| = \frac{1}{5}\left(2a^2 + 3c^2\right)2\rho\left(1 - \frac{T}{T_c}\right). \qquad (4.494)$$

Note that this is valid only parallel to the flow, which is why we have added a superscript $\parallel$ to ρ_s. Since a and c are different, an additional small gradient of φ orthogonal to the flow would be associated with a different current density

$$j = \frac{1}{5} \left(4a^2 + c^2 \right) \xi \partial_z \varphi, \tag{4.495}$$

i.e., the transverse superfluid density would rather be

$$\rho_s^\perp = \frac{1}{5} \left(4a^2 + c^2 \right) 2\rho \left(1 - \frac{T}{T_c} \right). \tag{4.496}$$

The Legendre transformed energy reads

$$\begin{aligned}
g &= \frac{1}{5}\xi^2 \left(2a_z^2 + 3c_z^2 \right) - \frac{1}{3} \left(2a^2 + c^2 \right) \\
&+ \frac{1}{15} \left(4a^4 + 2a^2 c^2 + \frac{3}{2}c^4 \right) - \frac{5j^2}{2a^2 + 3c^2} + h^2 c^2.
\end{aligned} \tag{4.497}$$

Minimizing this with respect to a and c we find two equations

$$-\frac{2}{3} + \frac{1}{15} \left[2 \left(2a^2 + c^2 \right) + 4a^2 \right] + \frac{10j^2}{\left(2a^2 + 3c^2 \right)^2} = 0, \tag{4.498}$$

$$-\frac{1}{3} + \frac{1}{15} \left[\left(2a^2 + c^2 \right) + 2c^2 \right] + 15\frac{j^2}{\left(2a^2 + 3c^2 \right)^2} + h^2 = 0. \tag{4.499}$$

They are solved by

$$a_0^2 = 1 + \frac{3}{2}h^2, \tag{4.500}$$

$$j^2 = \frac{1}{25} \left(2 + 3c_0^2 + 3h^2 \right)^2 \frac{1}{3} \left(1 - c_0^2 + 6j^2 \right). \tag{4.501}$$

Thus, in the absence of a magnetic field, the gap parameter orthogonal to the flow is not distorted after all,

$$\Delta^\perp \equiv \Delta_B\, a_0 = \Delta_B, \tag{4.502}$$

whereas the gap parallel to the flow is reduced to

$$\Delta^\parallel \equiv \Delta_B\, c_0, \tag{4.503}$$

with c_0 satisfying (4.502).

The current has a maximal size for

$$c_c^2 = \frac{4}{9} - \frac{h^2}{13}, \tag{4.504}$$

where j_c, κ_c take the values

$$j_c = \frac{2}{9}\sqrt{\frac{5}{3}} \left(1 - 3h^2 \right)^{3/2}, \qquad \kappa_c = \frac{1}{3}\sqrt{\frac{5}{3}} \left(1 - 3h^2 \right)^{1/2}. \tag{4.505}$$

The critical current is smaller than the previously calculated value by a factor of about $3/4$.

The energy can be expressed most simply as a function of κ. From (4.493) and (4.502) we identify

$$\kappa^2 = \frac{1}{3}\left(1 - c^2 - 6h^2\right). \tag{4.506}$$

Inserting this into (4.492), which in terms of κ reads

$$\begin{aligned}\frac{f}{2f_c} &= \frac{1}{5}\left(2a^2 + 3c^2\right)\kappa^2 \\ &\quad - \frac{1}{3}\left(2a^2 + c^2\right) + \frac{1}{15}\left(4a^4 + 2a^2c^2 + \frac{3}{2}c^4\right) + h^2c^2,\end{aligned} \tag{4.507}$$

we may evaluate only half of the quadratic terms according to the general rule that in equilibrium, the fourth-order part is half the opposite of the quadratic one. In this way we easily find

$$\frac{f}{2f_c} = -\frac{1}{2}\left(1 - \kappa^2 - h^2\right)^2 - \frac{5}{2}\left(h^2 + \frac{2}{5}\kappa^2\right)^2, \tag{4.508}$$

so that

$$\begin{aligned}g &= \frac{f}{2f_c} - 2j\kappa \\ &= -\frac{1}{2}\left(1 - \kappa^2 - h^2\right)^2 - 2\kappa^2\left(1 - 3h^2 - \frac{9}{5}\kappa^2\right).\end{aligned} \tag{4.509}$$

Let us now see how strong-coupling corrections modify this result. It is straightforward to calculate that then the free energy reads

$$\begin{aligned}g &= \frac{1}{5}\left(2a_z^2 + 3c_z^2\right) - \frac{1}{3}\left(2a^2 + c^2\right) + \frac{1}{15}\left[\beta_{12}\left(2a^2 + c^2\right)^2 + \beta_{345}\left(2a^4 + c^4\right)\right] \\ &\quad - \frac{5j^2}{2a^2 + 3c^2} + h^2c^2.\end{aligned} \tag{4.510}$$

The local minimum is given by the extrema in a and c:

$$-\frac{2}{3} + \frac{4}{15}\left[\beta_{12}\left(2a^2 + c^2\right) + \beta_{345}a^2\right] - 10\frac{j^2}{\left(2a^2 + 3c^2\right)^2} = 0, \tag{4.511}$$

$$-\frac{1}{3} + \frac{2}{15}\left[\beta_{12}\left(2a^2 + c^2\right) + \beta_{345}c^2\right] - 15\frac{j^2}{\left(2a^2 + 3c^2\right)^2} + h^2 = 0, \tag{4.512}$$

so that c and a are now related by

$$\left(4\beta_{12} + 3\beta_{345}\right)a^2 + \left(2\beta_{12} - \beta_{345}\right)c^2 = 5\left(1 + \frac{3}{2}h^2\right). \tag{4.513}$$

From this we find

$$\kappa^2 = \frac{25j^2}{(2a^2 + 3c^2)^2} = \frac{5\beta_{345}}{3(\beta_{12} + 3\beta_{345})}\left(\alpha - \frac{6}{5}\beta_B c^2 - \frac{2\beta_{12} + \beta_{345}}{\beta_{345}}3h^2\right), \quad (4.514)$$

and obtain the longitudinal gap parameter

$$\frac{\Delta^{\|2}}{\Delta_B^2} = c^2 = \frac{1}{\frac{6}{5}\beta_B}\left[\alpha - \frac{9}{5}\left(\frac{4\beta_{12}}{3\beta_{345}}\right)\kappa^2 - \left(1 + \frac{2\beta_{12}}{\beta_{345}}\right)3h^2\right]. \quad (4.515)$$

Similarly, we find for the transversal direction

$$\frac{\Delta^{\perp 2}}{\Delta_B^2} = a^2 = \frac{1}{\frac{6}{5}\beta_B}\left[\alpha + \frac{3\beta_{12}}{\beta_{345}}h^2 - \left(1 - \frac{2\beta_{12}}{\beta_{345}}\right)\frac{3}{5}\kappa^2\right]. \quad (4.516)$$

The current is now maximal at

$$\kappa_c^2 = \frac{5}{9}\frac{5\beta_{345}}{8\beta_{12} + 11\beta_{345}}\left(\alpha - \frac{4\beta_{12} + 3\beta_{345}}{5\beta_{345}}3h^2\right), \quad (4.517)$$

where j_c, κ_c become

$$j_c = \frac{1}{\frac{6}{5}\beta_B}\frac{2\sqrt{5}}{9}\sqrt{\frac{5\beta_{345}}{8\beta_{12} + 11\beta_{345}}}\left(\alpha - \frac{4\beta_{12} + 3\beta_{345}}{5\beta_{345}}3h^2\right)^{3/2},$$

$$\kappa_c = \sqrt{\frac{5}{9}}\sqrt{\frac{5\beta_{345}}{8\beta_{12} + 11\beta_{345}}}\left(\alpha - \frac{4\beta_{12} + 3\beta_{345}}{5\beta_{345}}3h^2\right)^{1/2}. \quad (4.518)$$

The free energy density is found by the same method as before

$$\frac{f}{2f_c} = -\frac{1}{2}\left\{(\alpha - \kappa^2 - h^2)^2\frac{1}{\frac{6}{5}\beta_B} + \frac{5}{\beta_{345}}\left(h^2 + \frac{2}{5}\kappa^2\right)^2\right\}. \quad (4.519)$$

This result is rather simple since some of the strong-coupling corrections cancel in the first quadratic term of (4.508)

$$2a^2 + c^2 = \frac{3}{\frac{6}{5}\beta_B}(\alpha - \kappa^2 - h^2). \quad (4.520)$$

The free energy density g, on the other hand, looks more complicated because of the awkward form of the longitudinal superfluid density

$$\frac{\rho_s^{\|}}{2\rho\left(1 - \frac{T}{T_c}\right)} = \frac{1}{5}(2a^2 + 3c^2) = \frac{1}{\frac{6}{5}\beta_B}\left\{\alpha - \frac{4\beta_{12} + 3\beta_{345}}{5\beta_{345}}3h^2 - \frac{8\beta_{12} + 11\beta_{345}}{5\beta_{345}}\frac{3}{5}\kappa^2\right\}$$

$$(4.521)$$

entering the additional term

$$-2j\kappa = -2\kappa^2\frac{2a^2 + 3c^2}{5}. \quad (4.522)$$

4.10.2   A-Phase

Before discussing the result in the B-phase further, it is useful to compare them with the A-phase. As before, we shall first assume a uniform texture

$$\Delta_{ai} = \Delta_{ai}^0 e^{i\varphi(z)}. \tag{4.523}$$

Later we shall see that this ansatz is stable only for very small currents. Still, it is instructive to go through the same calculation as in the B-phase.

The kinetic energy has the form

$$f = \frac{\kappa}{2}\xi^2 \left(|\Delta_{ai}^0|^2 + 2|\Delta_{a3}^0|^2 \right) \kappa^2. \tag{4.524}$$

If we suppose again that the gap parameter suffers only from a change of size, we may write

$$\Delta_{ai}^0 = \Delta_A \hat{\Delta}_{ai} = \Delta_B a\, \hat{\Delta}_{ai}, \tag{4.525}$$

and we assume for $\hat{\Delta}_{ai}$ the standard form up to spin and orbital rotation. From (4.524) we see that the bending energy is minimal if $\hat{\Delta}_{a3}$ is chosen to vanish implying that the field $l(\mathbf{x})$ points in the direction of flow. Then the total energy has the form

$$f = \frac{\kappa}{2}\Delta_B^2 2a^2 (\partial\varphi)^2 - 3\alpha\mu a^2 \Delta_B^2 + 4\beta_0\beta_A a^4 \Delta_B^4.$$

The current has now the form

$$J = 4m\kappa\Delta_B^2 a^2 \xi \partial_z\varphi. \tag{4.526}$$

In order to compare with the previously derived results for the B-phase it is useful to measure again all energies in units of $2f_c$ of the B-phase. Then we obtain[14]

$$\frac{f}{2f_c} = \frac{2}{5}a^2 \left(\partial_z\varphi\right)^2 - \frac{2}{3}\alpha a^2 + \frac{2}{9}\frac{6}{5}\beta_A a^4, \tag{4.527}$$

for which the dimensionless current density is now

$$j = \frac{2}{5}a^2 \xi \partial_z\varphi = \frac{2}{5}a^2 \kappa. \tag{4.528}$$

Therefore the Legendre transformed energy

$$g = \frac{f}{2f_c} - 2j\kappa = -\frac{5j^2}{2a^2} - \frac{2}{3}\left(\alpha a^2 - \frac{2}{5}\beta_A a^4\right) \tag{4.529}$$

---

[14]This result agrees, of course, with energy (4.492) of the distorted B-phase if one inserts $c = 0$ and takes $\alpha = \beta_A$, i.e., the weak-coupling limit, since the planar phase and the A-phase are energetically the same.

is extremal at

$$j^2 = \frac{4}{15}a^4 \left(\alpha - a^2 \frac{4}{5}\beta_A\right). \tag{4.530}$$

By comparison with (4.525) we find the gap parameter as a function of the velocity κ as

$$a^2 = \frac{5}{4\beta_A} \left(\alpha - \frac{3}{5}\kappa^2\right), \tag{4.531}$$

from which we may calculate

$$\Delta_A{}^2 = \Delta_B{}^2 a^2. \tag{4.532}$$

The current has a maximum at $a^2 = 5\alpha/6\beta$ with the critical values

$$j_c = \frac{\sqrt{5}}{9} \frac{\alpha^{3/2}}{\beta_A},$$

$$\kappa_c = \frac{\sqrt{5}}{3}\alpha^{1/2}. \tag{4.533}$$

In terms of κ the energies take the simple explicit forms:

$$\frac{f}{2f_c} = -\frac{1}{\frac{6}{5}\beta_A} \left(\alpha - \frac{3}{5}\kappa^2\right)^2,$$

$$g = \frac{f}{2f_c} - \frac{1}{\beta_A} \left(\alpha - \frac{3}{5}\kappa^2\right)\kappa. \tag{4.534}$$

It is important to realize that all these results are true, irrespective of the presence of a magnetic field: The **d**-texture can always lower its energy by orienting itself orthogonal to **H**, because of the absence of magnetic energy.

4.10.3   Critical Current in Other Phases for $T \sim T_c$

For completeness let us analyze the energies of the Ginzburg-Landau expansion in the presence of superflow in all the above possible phases. It could happen that the presence of superflow induces a transition into a phase with zero current which is unphysical because of its high energy. In order to eliminate this possibility we shall carry out an analysis for all known phases found in the analysis of the Ginzburg-Landau energy by Barton and Moore [21]. For each of these the order parameter may be written as

$$A_{ai} = \Delta \hat{A}_{ai}, \tag{4.535}$$

where $\hat{A}_{ai}$ is sometimes normalized to unity

$$\text{tr}\left(\hat{A}_{ai}\hat{A}_{ai}\right) = 1. \tag{4.536}$$

The energy is

$$f = -\mu\Delta^2 + \beta\hat{\beta}\Delta^4, \tag{4.537}$$

where β is a combination of various β_i's for the phase under consideration. This is minimal at

$$\Delta_0^2 = \frac{\mu}{2\beta_0\hat{\beta}}, \tag{4.538}$$

with $f = -f_c$ and the condensation energy density

$$f_c = \frac{\mu^2}{4\beta_0}\frac{1}{\hat{\beta}}. \tag{4.539}$$

Now let there be an equilibrium current flowing through a uniform texture. The order parameter may be normalized as

$$A^{ai} = \Delta\hat{A}_{ai}e^{i\varphi}, \tag{4.540}$$

so that the bending energies are

$$f = \frac{K}{2}\Delta^2\left[(\partial_i\varphi)^2 + 2(\partial_x\varphi)^2\left(A_{ax}\right)^2 + 2(\partial_y\varphi)^2(A_{ay})^2 + 2(\partial_z\varphi)^2(A_{az})^2\right]. \tag{4.541}$$

In the presence of the velocity $(\partial_i\varphi)/2m$, the energy does not minimize any longer a gap value (4.538), but it is minimal at a new modified order

$$\Delta = \Delta_0 a, \tag{4.542}$$

so that the energy can be written as

$$f = -\frac{K}{2}\Delta_0^2 a^2\left[(\partial_i\varphi)^2 + 2(\partial_x\varphi)^2 + 2(\partial_x\varphi)^2(A_{ax})^2 + 2(\partial_y\varphi)^2(A_{ay})^2 + 2(\partial_z\varphi)^2(A_{az})^2\right]$$
$$- \mu\Delta_0^2 a^2 + \beta_0\beta a^4\Delta_0^4. \tag{4.543}$$

It is again convenient to divide out the condensation energy of the phase in the absence of a current by substituting

$$\begin{aligned}\mu\Delta_0^2 &= 2f_c, \\ \hat{\beta}\beta_0\Delta_0^4 &= f_c.\end{aligned} \tag{4.544}$$

In addition, we have from (4.82)

$$\frac{K\Delta_0^2}{2} = \frac{3}{5}\mathcal{N}(0)\mu\frac{1}{2\beta_0\beta}\xi_0^2 = \frac{6}{5}f_c. \tag{4.545}$$

Therefore the energy has the generic reduced form

$$\frac{f}{f_c} = \frac{6}{5}a^2\xi_0^2(\partial\varphi)^2 - 2a^2 + a^4, \tag{4.546}$$

with

$$\alpha = 1 + 2|\hat{A}_{ai}\hat{j}|^2 \tag{4.547}$$

and $\hat{j}$ being the direction of the current. The physical current is

$$\begin{aligned}
J &= f_c 4m \frac{6}{5} a^2 \xi_0{}^2 (\partial\varphi)\alpha \\
&= \rho \left(1 - \frac{T}{T_c}\right)^{7/2} \frac{1}{\xi_0 m^2} \frac{1}{2\beta} a^2 \zeta(\partial\varphi) \\
&= jJ_0. \tag{4.548}
\end{aligned}$$

Here J_0 is the same quantity as introduced in (4.451), and j is the dimensionless reduced current density

$$j = \frac{1}{2\beta} a^2 \xi(\partial\varphi). \tag{4.549}$$

At a fixed j, we have to minimize

$$\begin{aligned}
\frac{g}{f_c} &= \frac{f - \frac{24}{5} j(\partial\varphi)}{f_c} \\
&= -\frac{24}{5} \frac{j^2}{a^2} \frac{\beta^2}{\alpha} - 2a^2 + a^4. \tag{4.550}
\end{aligned}$$

The equilibrium value of a lies at

$$j^2 = Ra^4 \left(1 - a^2\right), \tag{4.551}$$

where R is the quantity

$$R = \frac{5}{12} \frac{\alpha}{\beta^2}. \tag{4.552}$$

Since β and α are independent of a, the current is maximal for $a^2 = 2/3$, where it is given by

$$j^2 = \frac{1}{3} \frac{4}{9} R. \tag{4.553}$$

Let us now calculate the parameters α and R for each of the different superfluid phases. The results are displayed in Table 4.3. Since α depends on the direction of the current with respect to the texture, the energy has to be minimized for each of the standard forms A_{ai} listed in [21]. In the second column we have therefore marked the possible directions of the equilibrium current. Clearly, in the presence of strong-coupling corrections, R is modified by a suitable factor. The last column contains the condensation energy as compared to that of the B-phase. At the critical current, the energy $-g$ is lower than $-f_c^B$ by a factor $\frac{4}{3}$. Thus it might, in principle, happen

TABLE 4.3 Parameters of the critical currents in all theoretically known phases

Phases	α	direction of current	β	β_{GL}	R_{GL}	$f_c/f_c{}^B$	$g_c/f_c{}^B$
B	$\frac{5}{3}$	x,y,z	$\beta_{12} + \frac{1}{3}\beta_{345}$	$\frac{5}{6}$	1	1	$-\frac{4}{3}$
planar	1	z	$\beta_{12} + \frac{1}{2}\beta_{345}$	1	$\frac{5}{9}$	$\frac{5}{6}$	$-\frac{10}{9}$
polar	1	y,z	$\beta_{12} + \beta_{345}$	$\frac{3}{2}$	$\frac{5}{12}$	$\frac{5}{9}$	$-\frac{20}{27}$
α	$\frac{5}{3}$	x,y,z	$\beta_2 + \frac{1}{3}\beta_{345}$	$\frac{4}{3}$	$\frac{5}{27}$	$\frac{5}{18}$	$-\frac{10}{27}$
bipolar	1	z	$\beta_2 + \frac{1}{2}\beta_{345}$	$\frac{3}{2}$	$\left(\frac{5}{8}\right)^2$	$\frac{5}{9}$	$-\frac{20}{27}$
axial	1	z	β_{245}	1	$\frac{5}{12}$	$\frac{5}{6}$	$-\frac{10}{9}$
β	1	y,z	β_{234}	3	$\frac{5}{108}$	$\frac{6}{15}$	$-\frac{8}{15}$
γ	1	z	β_{124}	2	$\frac{5}{48}$	$\frac{5}{12}$	$-\frac{5}{9}$

that one of the higher-lying phases drops underneath a lower one when increasing the current. It can be checked, however, that such a crossover does not take place. For this we compare g at the critical currents

$$g = -4a^2 + 3a^4 = -\frac{4}{3}\left(\frac{f_c}{f_c^B}\right)f_c^B. \tag{4.554}$$

for the different phases. Starting out with the B-phase, the energy drops from -1 to $-\frac{4}{3}$. In the A-phase it starts out at $-\frac{5}{6}$ and drops down to $-\frac{10}{9}$. This value is underneath -1 so that there is, in principle, the possibility of a crossover, but we can check that the energy of the B-phase drops fast enough to avoid a collision. Similar arguments can be applied to any other pair of phases. In order to study this behavior in detail one has to plot the energy g as a function of the current density j. As a function of a, the energy g is most easily determined by solving the cubic equation (4.551) in a geometric way, writing

$$a^2 = \frac{1}{3} + \frac{1}{3}\cos\frac{2}{3}\varphi - \frac{1}{\sqrt{3}}\sin\frac{2}{3}\varphi, \tag{4.555}$$

where

$$\cos\varphi = \frac{3}{2}\sqrt{3}\frac{j}{\sqrt{R}}. \tag{4.556}$$

At $j = 0$ and $j = j_c$, the angle ϕ and the size a are given by

$$j = 0, \quad \varphi = \frac{\pi}{2}, \quad a^2 = 1,$$

$$j = j_c, \quad \varphi = 0, \quad a^2 = \frac{2}{3}. \tag{4.557}$$

Consider now the non-inert phases. Then the coefficients α and β contain one more parameter, for instance an angle θ. In addition to

$$\frac{\partial g}{\partial a} = 0 \tag{4.558}$$

which leads, as before, to

$$\frac{j^2}{R} = a^4 \left(1 - a^2\right). \tag{4.559}$$

Now we have to minimize g also with respect to θ :

$$\frac{\partial g}{\partial \theta} \equiv g' = 0. \tag{4.560}$$

Therefore we also have

$$\left(-2\frac{j^2}{a^2}\frac{1}{R} - 2a^2 + a^4\right) f_c' + \frac{R}{R^2}\frac{j^2}{a^2} f_c = 0, \tag{4.561}$$

where, according to (4.944), the θ-dependence is ruled by the differential equation

$$\frac{R'}{R} = -\left(2\frac{\beta'}{\beta} - \frac{\gamma'}{\gamma}\right). \tag{4.562}$$

From (4.932) we see that f_c depends on θ only via $1/\beta(\theta)$. Hence

$$f_c' = -\frac{\beta'}{\beta} f_c, \tag{4.563}$$

so that (4.561) becomes

$$\frac{j^2}{R} = a^4 \left(1 - \frac{a^2}{2}\right)\frac{1}{1 - \tau}, \tag{4.564}$$

where we have abbreviated

$$\tau \equiv \frac{\beta \alpha'}{\beta' \alpha}. \tag{4.565}$$

By equating (4.564) and (4.559) we find

$$a^2 = \frac{\tau}{\tau - \frac{1}{2}}, \tag{4.566}$$

and the relation between current density and angle θ becomes

$$j^2 = \frac{1}{2}\frac{\tau^2(\theta)}{[\tau(0) - 1/2]^3} R(\theta). \tag{4.567}$$

This current density is maximal if θ solves the equation

$$\left(\frac{\alpha'}{\alpha} + \frac{\beta'}{\beta}\right)\left(\frac{\alpha''}{\alpha'} - \frac{\beta''}{\beta'}\right) = \frac{\alpha'}{\alpha}\left(2\frac{\alpha'}{\alpha} - \frac{5}{2}\frac{\beta'}{\beta}\right). \tag{4.568}$$

If $\alpha'' = 0$, this can also be written in the more convenient form

$$-2\beta \left(\alpha'^2 \beta' + \frac{1}{2} \alpha' \beta'' \right) = \beta' \alpha \left(\beta'' \alpha - \frac{5}{2} \alpha' \beta' \right). \tag{4.569}$$

As an example consider the ζ-phase with

$$A_{ai} = \frac{\Delta}{\sqrt{s}} \begin{pmatrix} \sin\theta\cos\phi & -i\sin\theta\sin\phi & 0 \\ i\sin\theta\sin\phi & \sin\theta\cos\phi & 0 \\ 0 & 0 & \sqrt{2}\cos\theta \end{pmatrix}. \tag{4.570}$$

Actually, this parametrization interpolates between several phases:

$$\text{polar} \quad : \quad \text{all } \phi, \ \theta = 0, \quad A_{ai} = \Delta \begin{pmatrix} 0 & & \\ & 0 & \\ & & \sqrt{2} \end{pmatrix},$$

$$\text{planar} \quad : \quad \text{all } \phi, \ \theta = \tfrac{\pi}{2}, \quad A_{ai} = \Delta \begin{pmatrix} 1 & & \\ & 1 & \\ & & 0 \end{pmatrix}, \tag{4.571}$$

$$B \quad : \quad \phi = 0, \ \sin\theta = \sqrt{23}, \quad A_{ai} = \frac{\Delta}{\sqrt{3}} \begin{pmatrix} 1 & & 0 \\ & 1 & \\ 0 & & 1 \end{pmatrix},$$

and, certainly, the non-inert phase ζ itself. The potential energy is

$$f_p = -\mu\Delta^2 + \Delta^4 \beta_0 \beta_\zeta, \tag{4.572}$$

where

$$\begin{aligned} \beta_\zeta &= \beta_1 \, 4 \left(1 - 2\sin^2\phi\sin^2\theta \right)^2 + \beta_2 \, 4 \\ &\quad + \beta_{35} \left[2\sin^4\theta \left(1 - \sin^2 2\phi \right) + \varphi\cos^4\theta \right] + \beta_4 \left[2\sin^4\theta \left(1 + \sin^2 2\phi \right) + \varphi\cos\varphi\theta \right] \\ &= (4\beta_1 + 2\beta_{345})\sin^4\theta + (4\beta_1 + 4\beta_{345})\cos^4\theta + 4\beta_2 \\ &\quad + (\beta_4 - \beta_{35} - 2\beta_1)\, 2\sin^4\theta\sin^2 2\phi + 8\beta_1 \sin^2\theta\cos^2\theta\cos^2\phi. \end{aligned} \tag{4.573}$$

Minimizing this with respect to ϕ gives either

$$\tan^2\theta\cos^2\phi = T \equiv \frac{2\beta_1}{\beta_4 - \beta_{135} - \beta_1} \tag{4.574}$$

or the trivial solution

$$\phi = 0, \pi. \tag{4.575}$$

In the latter case, A_{ai} interpolates only between the three phases (4.571). In particular, the previously discussed distorted B-phase is contained in it.

In either case, the function β becomes:

$$\beta_\zeta = \beta_4 \sin^4\theta + (\beta_{1345} + \beta_1 T)\cos^4\theta + \beta_2, \tag{4.576}$$

$$\beta_\zeta^{\phi=0} = \beta_{12} + \frac{1}{2}\beta_{345}\left(\sin^4\theta + 2\cos^4\theta\right). \tag{4.577}$$

Consider now the bending energy. Inserting (4.571) into (4.80) gives

$$f_{\text{bend}} = \frac{K}{2}\Delta^2\left[(\partial_i\varphi)^2(1+\sin^2\theta) + (\partial_z\varphi)^2(2-3\sin^2\theta)\right]. \tag{4.578}$$

The orientation of the current with respect to the texture depends on the equilibrium value of θ. If

$$\sin^2\theta \lessgtr \frac{2}{3}, \tag{4.579}$$

the current points in the x, y-plane, or in the z-direction, respectively. In these two cases the bending energies are

$$f_{\text{bend}} = \frac{K}{2}\Delta^2(\partial_i\varphi)^2\left\{ \begin{matrix} 1+\sin^2\theta \\ 3-2\sin^2\theta \end{matrix} \right\}. \tag{4.580}$$

Therefore we identify

$$\alpha = \left(\begin{matrix} 1+\sin^2\theta \\ 3-2\sin\theta \end{matrix} \right), \quad \sin^2\theta \begin{matrix} < & 2/3, \\ > & 2/3. \end{matrix} \tag{4.581}$$

In the absence of a current the extremal value for θ is given by

$$\tan^2\theta = \frac{T\beta_1 + \beta_{1345}}{\beta_4}. \tag{4.582}$$

In the Ginzburg-Landau domain we find

$$T = -\frac{1}{2}, \quad \tan^2\theta = \frac{3}{4}, \quad \sin^2\theta = \frac{3}{7}. \tag{4.583}$$

At the value where $\sin\theta = 2/3$, both equations are solved at equal θ. Setting more generally $\sin^2\theta = x$, we can easily calculate the critical current. For simplicity we use only the weak-coupling values of β_i and have

$$\alpha = 1+x, \quad \alpha' = 1, \quad \alpha'' = 0,$$
$$\beta = x^2 + \frac{3}{4}(1-x)^2 + 1 = \frac{7}{4}\left(x^2 - \frac{6}{7}x + 1\right),$$
$$\beta' = \frac{7}{4}\left(2x - \frac{6}{7}\right), \quad \beta'' = \frac{7}{4}\cdot 2. \tag{4.584}$$

With these values, our equation (4.569) becomes linear and is solved by

$$x = \frac{1}{4}. \tag{4.585}$$

At that point, the parameter $\tau = \beta \alpha' / \beta' \alpha$ is equal to

$$\tau = \frac{19}{10}, \tag{4.586}$$

implying via Eq. (4.566) the equilibrium value of a

$$a^2 = \frac{19}{24}. \tag{4.587}$$

The corresponding critical current density is then from (4.567):

$$j_c = \sqrt{\frac{5}{6 \cdot 27}}. \tag{4.588}$$

Note that this current density is smaller than that of the B-phase by a factor $\sqrt{5/24} \sim 1/2$.

For consistency, we convince ourselves that at critical current the value of x is smaller than at $j = 0$ so that the direction of the current with respect to the texture, and therefore the choice of the bending energy with $\alpha = 1 + x$, remains valid for all equilibrium currents.

As a cross check of this method let us confirm the critical current of the B-phase with gap distortion by using the parametrization (4.577) in the weak-coupling limit:

$$\beta^{\phi=0} = \frac{1}{2} \left[1 + x^2 + 2(1-x)^2 \right] = \frac{3}{2} x^2 - 2x + \frac{3}{2}. \tag{4.589}$$

Here we start out with the B-phase where

$$x = \sin^2 \theta = \frac{2}{3}. \tag{4.590}$$

From our previous calculation we know that $c \leq a$, which says that in all currents the value of θ stays above the value implied by (4.590). Then we have to use the bending energy with

$$x = 3 - 2x. \tag{4.591}$$

Inserting $\beta, \gamma, \beta', \gamma', \beta'' \gamma''$ into (4.569), we find the linear equation

$$x = \sin^2 \theta = \frac{9}{11} \tag{4.592}$$

which is indeed larger than (4.590). The associated values of τ, a^2, and R are $-\frac{42}{15}$, $\frac{28}{33}$, and $\frac{11^3}{4 \cdot (21)^2}$, so that the critical current density becomes

$$j^2 = R a^4 \left(1 - a^2 \right) = \frac{20}{3} \frac{1}{81}, \tag{4.593}$$

as obtained before.

4.11 Is ^{3}He-A a Superfluid?

Equipped with the calculations of the last section and the topological arguments of Sections 4.8.5 and 4.8.6, we are now ready to address an important question: Does superfluid ^{3}He really deserve the prefix "super" in its name (apart from the similarity in the formalism with that of the superconductor)? In order to answer this question one usually performs a gedanken experiment of putting the liquid in a long and wide torus, stirring it up to a uniform rotation along the axis, cooling it down into the A- or B-phase, and waiting whether the liquid will slow down after a finite amount of time. Superconductors and the bosonic superfluid ^{4}He will preserve the rotation for a long time. Mathematically. the reason is that the order parameter describing the condensate is $\Delta_0 e^{i\varphi}$ with φ varying from zero to $2\pi N$ (where N is a very large number) when going around the torus. The liquid can slow down only if N decreases stepwise unit by unit. In order to do so the order parameter has to vanish in a finite volume, for example by the formation of a narrow vortex ring. This may form from a roton on the axis. With time, the radius increases until it reaches the surface where it annihilates, thereby reducing N by one unit (see Fig. 4.29).

FIGURE 4.29 Superflow in a torus which can relax by vortex rings. Figure shows ther formation and their growth until they finally meet their death at the surface. In a super-conductor or superfluid ^{4}He, these rings have to contain a core of normal liquid and are therefore very costly in energy. This assures an extremely long lifetime of superflow. In ^{3}He-A, on the other hand, there can be coreless vortices which could accelerate the decay.

Since such a vortex ring contains a rather large amount of energy (the condensation energy), the probability of this relaxation process is extremely small. Only at a very narrow place (e.g., at a Josephson junction of two superconducting wires) can this process be accelerated so that the relaxation takes place within minutes or seconds.

The maximal size of a current which is stable against this type of decay is reached when the kinetic energy density of the superfluid reaches the order of magnitude of the condensation energy density. Then the liquid can use up the kinetic energy, via fluctuations, to become normal, and the phase $e^{i\varphi}$ can unwind. Obviously, the existence of a macroscopic superflow hinges on the possibility of having large flux numbers conserved topologically along the torus.

In the B-phase, this is indeed the case. According to (4.214) and (4.215), the homotopy group describing the mapping of the axis of a torus into the parameter space of the B-phase contains the group of integer numbers Z which can pile up a macroscopic superflow. In the A-phase, on the other hand, one has in a large torus $\pi_1 = Z_2$ [see (4.211)]. Hence, there is only one nontrivial mapping. The associated flux is of unit strength and therefore necessarily microscopic. Thus it appears as if the liquid ³He is not really "super" at all in comparison with superconductors and superfluid ⁴He.

We shall now show that this is, fortunately, not completely true. Although in a much weaker sense, i.e., with much smaller critical currents and shorter lifetimes, ³He-A does *support* a stable superflow. Moreover, as the temperature drops below a certain value, say T_{stab}, there are even *two separate* supercurrents, which both are topologically conserved [39]. Thus in the weaker sense, ³He-A is really a *double superfluid*.

In order to understand this, one has to observe that in the bulk it is not really necessary to have an overwhelming potential barrier of condensation energy guaranteeing the stability at a macroscopic time scale. A barrier with a moderately large energy density, say $\rho_s/m^2\xi_b^2$, can also prevent a state from decaying, and the length scale characterizing the size of the barrier ξ_b can be quite a bit larger than the coherence length ξ_0. This can make a phase metastable, if the volume is sufficiently large. As argued above, such a decay can only proceed via the nucleation of a vortex tube of length L and diameter d with the energy $(\rho_s/m^2\xi_b^2) \cdot d^2 L$ (for a potential barrier of the order of the dipole force $\xi_d \sim 1000\xi_0$, this energy corresponds to $\approx 10^{-6}$ mK per Cooper pair). The diameter d will adapt itself to the characteristic length scale of the potential barrier, i.e., $d \approx \xi_b$. Thus the energy of the vortex tube is $(\rho_s/M_2)\,L$. It is this number that enters the exponent in the Boltzmann factor dominating the decay rate

$$\frac{1}{\tau} \sim \frac{1}{\tau_0} \exp\left\{ -\left(\frac{\rho_s}{m^2}\frac{1}{\xi_b^2} - f_{\text{curr}} \right) \xi_b^2 \frac{L}{T} \right\}, \tag{4.594}$$

where f_{curr} is the energy density of the current flow, and τ_0 is the characteristic time of vortex motion. This parameter varies for the decay mechanism associated with different barriers, but not by many orders of magnitude. The main effect of the smaller barrier energy lies in the significant reduction of the critical energy density which can be accumulated in the current (note that the barrier strength parameter $1/\xi_b$ can be small enough to be cancelled by the energy density f_{curr} in the exponential). As a consequence, if we are satisfied with a rather small critical current, the potential barrier does not need to be completely unsurmountable to allow the use of topological arguments to classify stable flow configurations.

The important property of ³He-A is that a current, once established, attracts the l-vector into its direction via the second term in the energy (4.125):

$$-\rho_0 \left(\mathbf{l} \cdot \mathbf{v}_s \right)^2. \tag{4.595}$$

It is this term which creates a potential barrier permitting a supercurrent to accumulate. In order to simplify the discussion we shall assume the torus to be sufficiently long and wide to neglect its curvature and boundaries. Thus, the fields in the energy

(4.125) can be assumed to depend only on the variable z (if we assume the z-axis to coincide with the axis of the torus). In order to avoid the use of constraints for respecting the curl condition (4.147) it is convenient to work directly with the parametrization of $\mathbf{l}$ and $\boldsymbol{\phi}$ in terms of Euler angles (4.120), (4.121), so that $\mathbf{v}_s$ of Eq. (4.124) becomes

$$\mathbf{v}_s = -\frac{1}{2m}\left(\boldsymbol{\nabla}\alpha + \cos\beta\,\boldsymbol{\nabla}\gamma\right). \tag{4.596}$$

We shall also express $\mathbf{d}$ in terms of directional angles as

$$\mathbf{d} = (\sin\theta\cos\varphi, \sin\theta\sin\varphi, \cos\theta). \tag{4.597}$$

For pure z-variations of the fields we calculate the derivatives

$$
\begin{aligned}
\boldsymbol{\nabla}\cdot\mathbf{l} &= -\sin\beta\,\beta_z; & \boldsymbol{\nabla}\times\mathbf{l} &= -\gamma_z\mathbf{l}^\perp - \cos\beta\,\beta_z\frac{\mathbf{e}_z\times\mathbf{l}}{|\mathbf{e}_z\times\mathbf{l}|}, \\
\mathbf{l}\cdot(\boldsymbol{\nabla}\times\mathbf{l}) &= -\sin^2\beta\,\gamma_z; & [\mathbf{l}\times(\boldsymbol{\nabla}\times\mathbf{l})]^2 &= \cos^2\beta\left(\beta_z^2 + \sin^2\beta\,\gamma_z^2\right), \quad (4.598)\\
(\boldsymbol{\nabla}_i d_a)^2 &= \theta_z^2 + \sin^2\theta\phi_z^2; & (\mathbf{l}\cdot\boldsymbol{\nabla}d_a)^2 &= \cos^2\beta\left(\theta_z^2 + \sin^2\theta\,\phi_z^2\right),
\end{aligned}
$$

and find the energy density

$$2f = A(s)\alpha_z^2 + G(s)\gamma_z^2 + 2M(s)\alpha_z\gamma_z + B(s)\beta_z^2 + T(s)\left(\theta_z^2 + s\phi_z^2\right) + 2\frac{\rho_s^\parallel}{\xi_d^{\perp 2}}\left[1 - (\mathbf{l}\cdot\mathbf{d})^2\right], \tag{4.599}$$

where the coefficients are the following functions of $s \equiv \sin^2\beta$:

$$
\begin{aligned}
A(s) &\equiv \rho_s^\parallel + \rho_0 s; & \rho_s^\parallel \equiv \rho_s - \rho_0, \\
B(s) &\equiv K_b + (K_s - K_b)s, \\
G(s) &\equiv \rho_s^\parallel + (K_b - 2c_0 + \rho_0 - \rho_s^\parallel)s + (K_t - K_b + 2c_0 - \rho_0)s^2, \\
M(s) &\equiv [\rho_s^\parallel + (\rho_0 - c_0)s]\sqrt{1-s}, \\
T(s) &\equiv K_1^d - K_2^d + K_2^D s.
\end{aligned} \tag{4.600}
$$

Here we have dropped several factors $2m$ by going to time units t_0 in which $2m \equiv 1$, i.e., where

$$t_0 = \frac{v_F}{2p_F} = \frac{1}{2m}$$

is a unit of time. The energy possesses two mass currents

$$J_1 \equiv -\frac{1}{2m}\frac{\partial f}{\partial\alpha_z} = -\frac{1}{2m}\left[A(s)\alpha_z + M(s)\gamma_z\right], \tag{4.601}$$

$$J_2 \equiv -\frac{1}{2m}\frac{\partial f}{\partial[(\alpha+\gamma)_z/2]} = -\frac{1}{2m}\left[G(s)\gamma_z + M(s)\alpha_z + T(\beta)s\phi_z\right], \tag{4.602}$$

which are separately conserved:

$$\partial_z J_1 = \partial_2 J_2 = 0.$$

Note that such a conservation law is certainly not enough to stabilize a superflow since small dissipative effects neglected in (4.599) will ruin the time independence and swallow up momentum and energy. To make the following discussion as transparent as possible, let us go to units which are most natural for the problem at hand: We shall measure all lengths in units of $l_d \equiv \xi_d$, the energy in units of $f_d \equiv \rho_s^\parallel/(4m^2\xi_d^2)$, and the current density as multipoles of $J_d \equiv \rho_s^\parallel/(2m\xi_d)$, respectively. Physically, the f_d is the energy density which the system would have if all **d** and l-vectors were orthogonal, contrary to the dipole alignment force. The second current component in (4.602) is the current which flows if the Bose condensate moves with "dipole velocity" $v_d \equiv 1/2m\xi_d$ parallel to l. Now, the energy $2f$ has again the form (4.599) except that all coefficients are divided by $\rho_s^\parallel$ and there is no $\rho_s^\parallel/\xi_d^2$ in front of the dipole coupling. In the Ginzburg-Landau regime, in which the parameters of the liquid satisfy the identities (4.127), the coefficients simplify to

$$A(s) = 1 + s, \quad B(s) = \frac{1}{2}(3 - 2s), \quad G(s) = 1 - \frac{1}{2}s,$$
$$M(s) = \sqrt{1 - s}, \quad T(s) = 1 + s. \tag{4.603}$$

Since we are interested in the system at a fixed current we study the energy

$$
\begin{aligned}
2g &\equiv 2(f - j\gamma_z) \\
&= A_g j^2 + G_g \gamma_z^2 + 2M_g \gamma_z j + B\beta_z^2 + T\left(\theta_z^2 + s\phi_z^2\right) + 2\left(1 - [\mathbf{l} \cdot \mathbf{d}]^2\right), (4.604)
\end{aligned}
$$

where

$$A_g \equiv -A^{-1}, \quad M_g \equiv M/A, \quad G_g \equiv G - M^2/A \equiv \Delta(s)/A, \tag{4.605}$$

and

$$
\begin{aligned}
\Delta(s) &\equiv GA - M^2 \\
&= \frac{s}{\rho_s^{\parallel 2}}\left\{\rho_s^\parallel K_b + \left[\rho_s^\parallel(K_t - K_b) + (\rho_0 K_b - c_0^2)\right]s + \left[\rho_0(K_t K_b) + c_0^2\right]s^2\right\} \\
&\equiv s\left(\Delta_0 + \Delta_1 s + \Delta_2 s^2\right).
\end{aligned}
\tag{4.606}
$$

In the Ginzburg-Landau regime, this becomes simply (see Appendix 4C)

$$\Delta(s) \underset{\text{GL}}{=} \frac{s}{2}(3 - s). \tag{4.607}$$

In order to gain as much experimental flexibility as possible let us also add a magnetic field

$$g \rightarrow g + g_z\left(\mathbf{d} \cdot \mathbf{H}\right)^2. \tag{4.608}$$

It is convenient to bring this to a form in which it can be compared most easily with the dipole energy. Let H_{dd} be the magnetic field ($H_{dd} \approx 300\ \emptyset$) at which

$$g_z H_{dd}^2 = \rho_s^\parallel/4m^2\xi_d^2. \tag{4.609}$$

If we measure H in terms of these units, say via the dimensionless quantity

$$\mathbf{h} \equiv \mathbf{H}/H_{\mathrm{dd}}, \tag{4.610}$$

we have

$$f = \frac{\rho_s^{\|}}{4m^2\xi_{\mathrm{d}}^2}\,(\mathbf{d}\cdot\mathbf{h})^2, \tag{4.611}$$

which for the energy (4.604) amounts to simply adding

$$2g \to 2g - 2h^2 s. \tag{4.612}$$

In order to obtain a first estimate of the stability properties let us assume j and h to be much smaller than one (i.e., current and field energies are much smaller than the characteristic dipole values). Then the $\mathbf{d} \parallel \mathbf{l}$ alignment force causes a complete locking of these two vectors and we may set $\tau \equiv \beta$, $\phi \equiv \gamma$. Now the energy $2g$ reads

$$2g^l = A_g j^2 + G_g^l \gamma_z^2 + 2M_g \gamma_z j + B^l \beta_z^2 - 2h^2 s \tag{4.613}$$

where G_g^l, B^l have the same form as those in (4.605), but with K_s, K_t, K_b replaced by

$$\begin{aligned}
K_s^l &\equiv K_s + K_1^d, \\
K_t^l &\equiv K_t + K_1^d, \\
K_b^l &\equiv K_b + K_1^d - K_2^d,
\end{aligned} \tag{4.614}$$

as shown in Appendix 4A. In the Ginzburg-Landau regime, their values are, if we divide out the factor $\rho_s^{\|}$:

$$\begin{aligned}
K_s^l &= \frac{1}{2} + 2 = \frac{5}{2}, \\
K_t^l &= \frac{1}{2} + 2 = \frac{5}{2}, \\
K_b^l &= \frac{3}{2} + 2 - 1 = \frac{5}{2}.
\end{aligned} \tag{4.615}$$

Consider now the problem of stability of the $\mathbf{d}\parallel\mathbf{l}\parallel\mathbf{j}\parallel\mathbf{h}$ configuration with $s = 0$. Expanding the energy up to the first power in s gives[15]

$$2g = j^2 + 2j\gamma_z + \frac{K_b}{g_s^{\|}}\beta_z^2 + \left(\frac{\rho_0}{\rho_s^{\|}}j^2 - 2h^2\right)s + \frac{K_b}{\rho_s^{\|}}s\gamma_z^2 - 2\frac{c_0 + \frac{1}{2}\rho_s^{\|}}{\rho_s^{\|}}s\gamma_z j \tag{4.616}$$

$$= j^2 + 2j\gamma_z + \frac{K_b}{\rho_s^{\|}}\beta_z^2 + \left[\frac{\rho_0}{\rho_s^{\|}}\left(1 - \frac{\left(c_0 + \frac{\rho_s^{\|}}{2}\right)^2}{\rho_0 K_b}\right)j^2 - 2h^2\right]s + \frac{K_b}{\rho_s^{\|}}s\left(\gamma_z - \frac{c_0 + \frac{\rho_s^{\|}}{2}}{K_b}j\right)^2.$$

---

[15]Here we omit the superscript l, and understand all K's as locked values (4.614).

Note that the term linear in γ_z is a pure surface term and does not influence the stability. Let us introduce the quantity

$$K \equiv \frac{\rho_0 K_b}{\left(c_0 + \frac{\rho_s^{\parallel}}{2}\right)^2}.$$ (4.617)

For small β, the term proportional to β^2 is

$$2\left[\frac{1}{2}\frac{\rho_0}{\rho_s^{\parallel}}\left(1 - K^{-1}\right) - \frac{h^2}{j^2}\right]\beta^2 j^2.$$ (4.618)

Hence the position at $\beta = 0$ is stable if and only if

$$\frac{h^2}{j^2} \leq \frac{h_c^2}{j^2} \equiv \frac{1}{2}\frac{\rho_0}{\rho_s^{\parallel}}(1 - K^{-1}).$$ (4.619)

In the absence of a magnetic field, stability implies [36]

$$K > 1.$$

In the Ginzburg-Landau regime, this is barely satisfied:

$$K \underset{\text{GL}}{=} \frac{10}{9}.$$ (4.620)

For decreasing temperature, however, ρ_0 is known to vanish. Hence we expect K to cross eventually the line $K = 1$. If one uses the energy parameters (4.126), but with the Fermi liquid corrections of Subsection 4.9.4 [37], one can argue that this will happen well within the A-phase at a temperature [38]

$$T_{\text{stab}} \equiv T(K = 1) \approx 0.86\,T_c.$$ (4.621)

Thus we can conclude: For $T \in (T_{\text{stab}}, T_c)$, the presence of a superflow acts self-stabilizing. It creates its own potential well, which prevents the free motion of $\mathbf{d} \parallel \mathbf{l}$ away from the direction of the current. In the parameter space SO(3) of the $\mathbf{d} \parallel \mathbf{l}$ - phase, this corresponds to a potential mountain around the equatorial region (see Fig. 4.30). This mountain is sufficient to prevent the deformation of contours to the two basic ones (corresponding to integer and half-integer spin representation). For these deformations, the passage of the equator would have to be unhindered (see Fig. 4.31).

It is easy to convince oneself that the SO$_3$-sphere with forbidden equatorial regions allows an infinity of inequivalent paths: The allowed type within the SO(3) sphere has its upper face coinciding with the lower one (except for a reflection on the axis). The parameter space becomes equivalent to a torus and $\pi_1 = Z$. Therefore *there are* again large quantum numbers which are conserved topologically in the weaker sense discussed above. Thus there exists indeed superflow in ^{3}He-A.

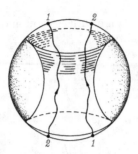

FIGURE 4.30 In the presence of a superflow in ^{3}He-A, the l-vector is attracted to the direction of flow. In the parameter space of ^{3}He-A this force corresponds to forbidding the equator of the sphere, thereby favoring a conical section. Since diametrally opposite points are identical, the topology is infinitely connected. The figure shows an example of a closed curve with two breaks.

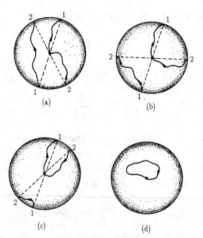

FIGURE 4.31 Doubly connected parameter space of the rotation group corresponding to integer and half-integer spin representations. Note that the continuous deformation of arbitrary contours to the two fundamental ones (either a point or a line running from a point at the surface to the diametrally opposite point) always has to pass via the equator of the sphere. An alignment force between l and the current which forbids the equator of the sphere therefore changes drastically the topology to being infinitely connected.

Note that in the dipole-locked regime with $\beta_0 = 0$ both currents (4.602) and (4.601) coincide and are equal to

$$j_1 \equiv J_1/J_{\mathrm{d}} = j_2 \equiv J_2/J_{\mathrm{d}} = -(\alpha_z + \gamma_z). \tag{4.622}$$

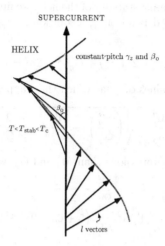

FIGURE 4.32 Helical texture in the presence of a supercurrent. The vectors show the directions of $\mathbf{l}$ which rotate around the axis of superflow when preceding along the z-axis. The angle of inclination has a constant value β_0. The pitch of the helix is constant with a ratio $\gamma_z/j \approx (c_0 + \rho_s^{\parallel})/K_b \approx \frac{3}{5}$.

Topological conservation in a torus implies that $\langle \alpha_z + \gamma_z \rangle$ is pinned down at $2\pi/L$ times an integer number, say N, when going once around the axis. Hence both currents are topologically stable at a value

$$j_1 = j_2 = 2\pi N/L, \qquad (4.623)$$

where L is the length of the torus.

What happens when the temperature drops below T_{stab}? Then the quadratic term becomes negative and β starts moving away from the forward direction. We can then show that the higher orders in β stop this motion at a value $\beta_0 \neq 0$. In this case the coefficient of the last term in (4.616) becomes finite so that γ_z will be driven to an average value

$$\langle \gamma_z \rangle \approx \gamma_z^0 \equiv \frac{c_0 + \frac{\rho_s^{\parallel}}{2}}{K_b} j \underset{\text{GL}}{=} \frac{3}{5} j. \qquad (4.624)$$

A texture with fixed angle of inclination β_0 and $\gamma_z = \gamma_z^0$ looks like a helix with constant pitch γ_z^0 (see Fig. 4.32).

It is in this helical texture that the two currents (4.601) and (4.602) no longer coincide and, moreover, become both conserved topologically [39].

In order to prove the dynamic stability of the helix we first consider all stationary solutions. Since $2g$ does not depend on γ, a solution at $s \equiv s_0$ is stationary if and only if[16]

$$2g' = A'_g j^2 + 2M'_g \gamma_z j + G'_g \gamma_z^2 - 2h^2 = 0. \tag{4.625}$$

For every s_0 there are two values of γ_z at which this happens:

$$\frac{\gamma_z^{\pm}}{j} = -\frac{M'_g}{G'_g} \pm \sqrt{\left(\frac{M'_g}{G'_g}\right)^2 - \frac{A'_g}{G'_g} + 2\frac{h^2}{j^2}\frac{1}{G'_g}}. \tag{4.626}$$

Since M and G are simpler expressions than M_g and G_g, we use

$$M'_g = \frac{M'A - AM'}{A^2} \underset{\text{GL}}{=} -\frac{1}{2\sqrt{1-s}}, \tag{4.627}$$

$$G'_g = G' - 2MM'/A + M^2 A'/A^2 \underset{\text{GL}}{=} \left(5 + 6s + 9s^2 + 4s^3\right)/4(1+s)^2, \tag{4.628}$$

which satisfy

$$M_g'^2 - G'_g A'_g = \left(M'^2 - G'A'\right)/A^2, \tag{4.629}$$

to write [see (4.605) and the forthcoming Appendix 4C]

$$\frac{\gamma_z^{\pm}}{j} = \left(A^2 G' - 2MM'A + M^2 A'\right)^{-1} \tag{4.630}$$

$$\times \left[-(M'A - MA') \pm A\sqrt{M'^2 - G'A' + A^2 G_g'^2 h^2/j^2}\right]$$

$$\underset{\text{GL}}{=} \left[\sqrt{1-s}\left(5 + 6s + 9s^2 + 4s^3\right)\right]^{-1} \tag{4.631}$$

$$\times \left[3 - s \pm \sqrt{(3-s)^2 - 2[1 - (1+s)^2 2h^2/j^2](5 + 6s + 9s^2 + 4s^3)(1-s)}\right].$$

This equation has two solutions if

$$M'^2 - G'A' - A^2 G'_g 2h^2/j^2 > 0. \tag{4.632}$$

Consider at first the case $h = 0$. After a somewhat tedious calculation one can write

$$M'^2 - G'A' = \alpha(s - s^+)(s - s^-)/4(1-s), \tag{4.633}$$

with

$$s^{\pm} \equiv \frac{\beta}{\alpha}\left(1 \pm \sqrt{1 + 4\alpha K_b \rho_0 (1 - K^{-1})/\beta \rho_s^{\|2}}\right), \tag{4.634}$$

where

$$\alpha \equiv \left[\rho_0^2 2\rho_0 c_0 + gc_0^2 + 8(K_t - K_b)\rho_0\right]/\rho_s^{\|2} \underset{\text{GL}}{=} 8, \tag{4.635}$$

---

[16] As we noted after Eq. (4.616), a linear term in γ_z does not influence the field equation of the system since it is a pure surface term. It is a mere constant after integration over z.

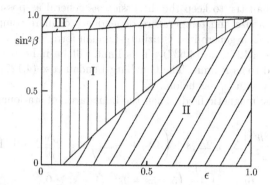

FIGURE 4.33 Three different regions in which there are equilibrium configurations of the texture at $H = 0$ (schematically).

$$\beta = \frac{2\rho_0}{(c_0 + \frac{1}{2}\rho_s^{\parallel})\rho_s^{\parallel 2}}\left[3c_0 K_b(K^{-1} - 1) + (c_0 + \frac{1}{2}\rho_s^{\parallel})(2K_t - \frac{1}{2}\rho_s^{\parallel}) - \frac{3}{2}\rho_s^{\parallel}K_b\right],$$

$$(4.636)$$

and K is the ratio (4.617). In the absence of a magnetic field, $s^{\pm}$ give the boundaries of stationary solutions. Due to our incomplete knowledge of the parameters of the liquid we shall estimate the regions in the following fashion: Since the passage of K through unity is eventually enforced by the vanishing of ρ_0, we shall assume, for simplicity, that all coefficients have their Ginzburg-Landau value in the list (4.127), except for ρ_0 which we assume to be equal to

$$\rho_0 = \rho_s^{\parallel}(1 - \epsilon) = \rho_s^{\parallel}\frac{9}{10}K. \qquad (4.637)$$

Then

$$K = \frac{10}{9}(1 - \epsilon), \quad \epsilon = 1 - \frac{9}{10}K, \qquad (4.638)$$

and

$$\alpha = 8 + \epsilon^2, \quad \beta = 3 + 6\epsilon, \qquad (4.639)$$

so that

$$s^{\pm} \underset{\mathrm{GL}}{=} \left[3 + 6\epsilon \pm (1 - \epsilon)\sqrt{17 - 10\epsilon}\right]/(8 + \epsilon^2). \qquad (4.640)$$

The curves $s^{\pm}(\epsilon)$ are shown in Fig. 4.33.

The regions above the upper and below the lower curve correspond to stationary solutions. As the lower curve drops underneath the axis ($\epsilon < 1/10$), the solution becomes meaningless. But this is precisely the region discussed before in which the $\beta = 0$ solution is stable.

In the following we shall try to keep the discussion as general as possible but find it useful to indicate a size and temperature dependence of more complicated expressions by exhibiting their generalized Ginzburg-Landau form in which only ρ_0 deviates from the values (4.127) via (4.637). This limit will be indicated with a symbol $\underset{L}{=}$ and be referred to as L-limit. The Ginzburg-Landau case (4.127) will be exhibited with an equality sign $\underset{GL}{=}$, as before.

Let us now include the magnetic field. Then the boundaries of stationary solutions are

$$\alpha(s - s^{\dagger})(s - s^{-}) + \frac{8h^2}{j^2}\left[\Delta_0 + 2\Delta_1 s + \left(3\Delta_2 + \frac{\rho_0}{\rho_s^{\parallel}}\Delta_1\right)s^2 + 2\frac{\rho_0}{\rho_s^{\parallel}}\Delta_2 s^3\right](1 - s)$$

$$\underset{GL}{=} 8s^2 - 6s - 1 + 4\frac{h^2}{j^2}\left(5 + 6s + 9s^2\right)(1 - s) \geq 0. \qquad (4.641)$$

This equation is no longer quadratic in s and its solution is complicated. It is gratifying to note that the physically interesting regions can easily be studied with a good approximation. First observe that at $s \geq 0$ there are stationary solutions if the magnetic field is larger than the value given by

$$\alpha s^{\dagger}s^{-} + 8\frac{h_c^2}{j^2}\Delta_0 = 0. \qquad (4.642)$$

This implies [see (4.233)]:

$$\frac{h_c^2}{j^2} = -\frac{\alpha s^{\dagger}s^{-}}{8K_b}\rho_s^{\parallel}. \qquad (4.643)$$

But from (4.634) one has

$$\alpha s^{\dagger}s^{=} - 4\frac{K_b\rho_0}{\rho_s^{\parallel 2}}\left(1 - K^{-1}\right), \qquad (4.644)$$

so that the value of h_c from (4.643) coincides with the critical value determined previously from the stability of the texture with $\beta = 0$ (see (4.619). Thus as h exceeds h_c, the aligned solution destabilizes in favor of a new extremal solution. The new equilibrium position can be calculated to lowest order in $\Delta h^2 \equiv h^2 h_c^2$ by expanding formula (4.641):

$$\left[8\frac{h_c^2}{j^2}(2\Delta_1 - \Delta_0) - (s^{\dagger} + s^{-})\alpha\right]s + 8\frac{\Delta h^2}{j^2} > 0, \qquad (4.645)$$

which amounts to

$$s < s_h^{-} \equiv \frac{4}{\beta - 4\frac{h_c^2}{j^2}(2\Delta_1 - \Delta_c)}\frac{\Delta j^2}{j^2}. \qquad (4.646)$$

Using the limiting value

$$\frac{h_c^2}{j^2} \underset{L}{=} (1 - 10\epsilon)/20, \qquad (4.647)$$

we estimate the prefactor to be

$$\frac{4}{\beta - 4\frac{h_c^2}{j^2}(2\Delta_1 - \Delta_0)} \underset{L}{=} \frac{10}{3 + 6\epsilon - (1 - 10\epsilon)^2/10} \tag{4.648}$$

$$= \begin{cases} 100/29 \\ 100/36 \end{cases} \text{for} \quad \begin{matrix} \epsilon = 0, & T = T_c, \\ \epsilon = 1/10, & T = T_{\text{stab}}, \end{matrix} \tag{4.649}$$

which is therefore $\approx 1/3$ for all temperatures between T_c and T_{stab}.

Within this small-s region we can now solve (4.626) for $\gamma_z^{\pm}$. Since γ_z goes with the square root of $s - s_h^-, s_h$, it is sufficient to keep, for small Δh^2, only the constants in the other terms and we find

$$\frac{\gamma_z^{\pm}}{j} \approx \frac{c_0 + \frac{1}{2}\rho_s^{\parallel}}{K_b} \pm \frac{\rho_s^{\parallel}}{2K_b}\sqrt{\alpha(s - s_h^-)(s_h^- - s_h^{\dagger})} \tag{4.650}$$

$$= \frac{c_0 + \frac{1}{2}\rho_s^{\parallel}}{K_b} \pm \frac{\rho_s^{\parallel}}{K_b}\sqrt{\frac{\beta}{2}}\left[1 + \frac{\alpha}{\beta^2}\frac{4K_b\rho_0}{\rho_s^{\parallel 2}}\left(1 - K^{-1}\right)\right]^{1/4}\sqrt{s_h^- - s}.$$

If we choose, in addition, also $K \approx 1$, we have

$$\frac{\gamma_z}{j} \approx \frac{\gamma_z^0}{j} \pm \frac{\rho_s^{\parallel}}{K_b}\sqrt{\frac{\beta}{2}}\sqrt{s_h^- - s}, \tag{4.651}$$

which in the L-limit reads explicitly

$$\frac{\gamma_z}{j} \underset{L}{\approx} \frac{3}{5}\left(1 \pm \sqrt{\frac{5}{4}\left[\frac{20}{9}\frac{h^2}{j^2} - (K - 1)\right] - s}\right). \tag{4.652}$$

As the magnetic field increases one can solve this equation for the external positions only numerically. The results are shown in Figs. 4.34 a)–c) for three different values of ϵ : $\epsilon = 0$, $\epsilon = 0.1$, $\epsilon = 0.2$. Note that the small-s regions coincide if the magnetic field lines are labelled by $\Delta h^2/j^2$ rather than h^2/j^2. Let us now find out which of these positions correspond to stable extrema. The energy density can be written in the form

$$2g = \bar{B}s_z^2 + V(s, \gamma_z). \tag{4.653}$$

The stationary points were determined from

$$\frac{\partial V}{\partial s}\left(s_0, \gamma_z^{\pm}\right) = 0. \tag{4.654}$$

If we now assume linear oscillations around this value we have

$$2\delta^2 g = \bar{B}(\delta s_z)^2 + \frac{\partial V}{\partial \gamma_z}(s_0, \gamma_z^{\pm})(\delta\gamma_z)$$

$$+ \frac{\partial^2 V}{\partial s^2}(s_0, \gamma_z^{\pm}) + 2\frac{\partial^2 V}{\partial s\partial\gamma_z}(s_0, \gamma_z^{\pm}) + \frac{\partial^2 V}{\partial\gamma_z}(s_0, \gamma_z^{\pm}). \tag{4.655}$$

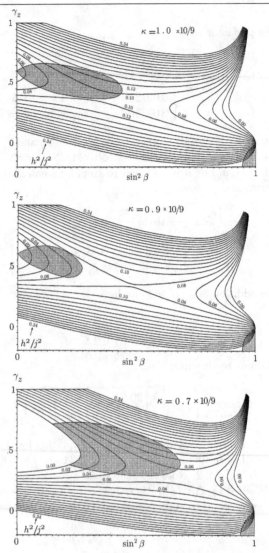

FIGURE 4.34 Pitch values for stationary helical solutions as a function of the angle of inclination β_0. The curves are lines of constant ratio between magnetic field and current. The shaded areas are regions of stability for the helical texture, the left one has l close to the direction of flow, the right one has l transverse to the flow. a) The temperature lies close to the transition point; b) at the lower temperature $T = T_{\text{stab}}$, at which the helix begins forming in a zero magnetic field; c) at a temperature below $T = T_{\text{stab}}$. The temperature dependence of the hydrodynamic coefficients is simplified assuming that only ρ_0 differs from the Ginzburg-Landau values.

The second term is a pure surface term and can be ignored. The equations of motion of (4.655) are linear. Therefore the superposition principle holds and we can test stability separately by using plane waves of an arbitrary wave vector k. With such an ansatz $2\delta^2 g$ becomes

$$2\delta^2 g = \left(V' + \bar{B}k^2\right)(\delta s)^2 + 2\dot{V}'k(\delta s)/(\delta\gamma) + \ddot{V}k^2(\delta\gamma)^2. \tag{4.656}$$

This is positive-definite for all k if

$$\ddot{V} \equiv \frac{\partial^2 V}{\partial \gamma_{lz}^2} > 0, \tag{4.657}$$

and

$$V''\ddot{V} - \dot{V}'^2 > 0. \tag{4.658}$$

In terms of the functions defined in (4.605), these conditions read

$$G_g > 0, \tag{4.659}$$

$$2\left(G_g''\gamma_z^{\pm 2} + 2M_g''j\gamma_z^{\pm} + A_g''j^2\right)G_g - 4\left(G_g' + M_g'j\gamma_z^{\pm}\right)^2 > 0. \tag{4.660}$$

Using (4.626), the second condition takes the alternative form

$$D^{\pm} = \left(G_g''\gamma_z^{\pm 2} + 2M_g''j\gamma_z^{\pm} + A_g''j^2\right)\frac{G_g}{2G_g'^2} - \frac{(\gamma_z^+ - \gamma_z^-)^2}{4j^2} > 0. \tag{4.661}$$

Now it is easy to see that $G_g > 0$ for all s, and only (4.661) remains to be tested. Analytically, the small-s region is simple to study. Since

$$\frac{G_g}{2G_g'^2} \approx \frac{\rho_s^{\|}}{2K_b}s, \tag{4.662}$$

we have to satisfy

$$\frac{\rho_s^{\|}}{2K_b}\left\{-2\frac{\rho_0^2}{\rho_s^{\|2}} + \left(c_0 - \frac{\rho_s^{\|}}{4} + 2\frac{\rho_0 c_0}{\rho_s^{\|}}\right)\frac{1}{\rho_s^{\|}}\frac{c_0 + \frac{1}{2}\rho_s^{\|}}{K_b}\right.$$
$$\left. + \left[2\left(K_t - K_b\right) + 2c_0 + \frac{\rho_s^{\|}}{2} - 2\frac{(c_0 + \frac{1}{2}\rho_s^{\|})^2}{\rho_s^{\|}}\right]\left(\frac{c_0 + \frac{1}{2}\rho_s^{\|}}{K_b}\right)^2\right\}s$$
$$> \frac{\rho_s^{\|2}}{K_b^2}\frac{\beta}{2}\sqrt{1 + \frac{\alpha}{\beta^2}\frac{4K_b\rho_0}{\rho_s^{\|2}}}(1 - K^{-1})(s_h^- - s). \tag{4.663}$$

For $K \approx 1$, we can keep only terms linear in $K - 1$, s_h^-, s. Using the generalized Ginzburg-Landau values for the parameters gives

$$\frac{1}{5}\left[\frac{1}{2}\frac{36}{25} + \frac{1}{4}\left(\frac{9}{10}\right)^2(K - 1)\right]s > \frac{36}{125}(s_h^- - s). \tag{4.664}$$

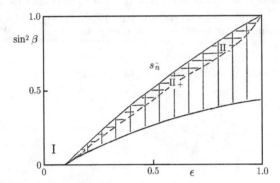

FIGURE 4.35 Regions of stable helical texture, II- and II+. In the region of overlap there are two possible pitch values γ_z^+, γ_z^- for which a helix can be stable.

But on the left-hand side, $K - 1$ can be neglected since it contributes higher orders in s. Thus we find that the extremal solutions, which exist for

$$s < s_h^-, \tag{4.665}$$

are stable if

$$s > \frac{2}{3}s_h^-. \tag{4.666}$$

Using (4.652), this result can also be phrased in the form

$$\left(\frac{\gamma_z - \gamma_z^0}{\gamma_z^0}\right)^2 < \frac{1}{3}(1 - K). \tag{4.667}$$

In Fig. 4.35, this statement amounts to the upper third portion underneath the curve s_h^- to be stable at $s \approx 0$. In general, the stability can be decided only numerically. In Figs. 4.34 a–c we have encircled the stable regions with a dashed line. Note that for fixed h^2, the instability sets in as γ_z^+ and γ_z^- become separated. Looking at the expression (4.661) the reason is clear: The second derivative V' is positive but not very large. If the branches separate too much, the positivity cannot be maintained. We see that as h^2 increases, the helix is stable only up to $s \approx 0.3 - 0.45$. Beyond this it collapses. For completeness, we have also indicated the stable regions in Fig. 4.36.

Note that the existence of the dipole force is essential for stability. First of all, the position $s = 0$ is never stable if the vectors $\mathbf{d}$ and $\mathbf{l}$ are not coupled at all. To see this remember that the constant K of (4.617) would be, in the Ginzburg-Landau region [compare (4.614)], considerably smaller than unity:

$$K = \frac{K_b\rho_0}{(c_0 + \frac{1}{2}\rho_s^\parallel)^2} = \frac{2}{3} < 1. \tag{4.668}$$

There is no hope that this situation reverses for smaller temperatures (since $\rho_0 \to 0$ for small T). The magnetic field does not help since it couples only to d. Also the

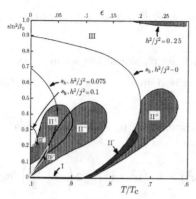

FIGURE 4.36 Regions of a stable helical texture (shaded areas). Contrary to Figs. 4.33 and 4.35, the full temperature dependence of the hydrodynamic coefficients is taken into account, including Fermi liquid corrections. The regions of a stable helical texture are denoted by II- and II+. In the overlap region, there are two possible pitch values γ_z^+, γ_z^- for which a helix can be stable.

hope that a position $s \neq 0$ may be stable is futile, even though there are stationary solutions: If we calculate in the Ginzburg-Landau limit

$$M'^2 - A'G' = \frac{1}{4(1-s)} - (1)\left(-\frac{1}{2}\right) > 0, \qquad (4.669)$$

this is fulfilled for all physical values $s = \sin^2 \beta \in (0, 1)$ with

$$\frac{\gamma_z^\pm}{j} = \frac{1}{\sqrt{1-s}} \frac{1}{3 - 2s - s^2} \left[3 - s \pm (1+s)\sqrt{3 - s}\right]. \qquad (4.670)$$

At $s = 0$, the ratios are 1.577 and 0.4226. For $s \to 1$, the upper branch tends monotonously to infinity with $1/\sqrt{1-s}$, the lower goes to zero with $\sqrt{1-s}$. Thus $(\gamma_z^+ - \gamma_z^-)^2/j^2$ increases rapidly. It is exactly for this reason why there is no hope of making $D > 0$ in (4.661). The second term is too large (remembering that the separation between γ_z^+ and γ_z^- is also the origin of the instability for small s in the dipole-locked regime).

Recognizing this fact we are compelled to study the effect of the dipole force with more sensitivity than implied by the assumption of dipole locking in the above discussion. Certainly, the results gained there will be valid for h, $j \ll 1$, i.e., as long as the dipole force is strong with respect to the other alignment forces. What happens if h, j grow to a comparable size? Consider again first the stability of the forward position where $\mathbf{d} \| \mathbf{l} \| \mathbf{j} \| \mathbf{h}$. For small, θ, β the quadratic part in the energy can be written as

$$2g = \text{const} + \frac{1}{\rho_s^\|}\left[\rho_0 j^2 \beta^2 + K_b\left(\beta_z^2 + \beta^2 \chi_z^2\right) - 2\left(c_0 + \frac{\rho_s^\|}{2}\right)\beta^2 \gamma_z j\right.$$

$$+ \left(K_1^d - K_2^d\right)\left(\theta_z^2 + \beta^2\phi_z^2\right) + 2\left(\theta^2 + \beta^2 - 2\theta\beta\omega(\gamma - \phi)\right) - 2h^2\beta^2\Big].$$
(4.671)

Introducing coordinates which are regular at the origin

$$u = \beta\cos\gamma, \quad v = \beta\sin\gamma, \quad \bar{u} = \theta\cos\phi, \quad \bar{v} = \theta\sin\phi,$$
(4.672)

so that

$$\gamma_z = (uv_z - vu_z)/\sqrt{u^2 + v^2}; \quad \phi_z = (\bar{u}\bar{v}_z - \bar{v}\bar{u}_z)/\sqrt{\bar{u}^2 + \bar{v}^2},$$
(4.673)

the energy becomes

$$2g = \frac{1}{\rho_s^{\parallel}}\left\{\rho_0 j^2\left(u^2 + v^2\right) + K_b\left(u_z^2 + v_z^2\right) - 2\left(c_0 + \frac{\rho_s^{\parallel}}{2}\right)(uv_z - vu_z)\,j\right.$$
$$\left. + \left(K_1^d - K_2^d\right)\left(\bar{u}_z^2 + \bar{v}_z^2\right) + 2\left((u - \bar{u})^2 + (v - \bar{v})^2\right) - 2h^2\left(\bar{u}^2 + \bar{v}^2\right)\right\}.$$

Since the corresponding equations of motion are linear, we can again test the stability of all plane waves

$$u, \bar{u} \sim \sin kx; \quad v, \bar{v} \sim \cos kx,$$
(4.674)

each of which gives

$$2g = \frac{1}{\rho_s^{\parallel}}\left(\rho_0 j^2 + K_b k^2 - 2kj\left(c_0 + \frac{\rho_s^{\parallel}}{2}\right) + 2\rho_s^{\parallel}\right)\left(u^2 + v^2\right)$$
$$+ \left[\left(K_1^d - K_2^d\right)k^2 - 2\left(h^2 - 1\right)\rho_s^{\parallel}\right]\left(\bar{u}^2 + \bar{v}^2\right)$$
$$\times \left\{\frac{1}{\rho_s^{\parallel}}\left(\rho_0 j^2 + K_b k^2 - 2kj\left(c_0 + \frac{\rho_s^{\parallel}}{2}\right) - \frac{4\rho_s^{\parallel 2}}{\left(K_1^d - K_2^d\right)k^2 - 2(h^2 - 1)\rho_s^{\parallel}} + 2\rho_s^{\parallel}\right)\right.$$
$$\times \left(u^2 + v^2\right)\left[\left(K_1^d - K_2^d\right)k^2 - 2(h^2 - 1)\rho_s^{\parallel}\right]$$
$$\left.\times \left[\left(\bar{u} - \frac{2\rho_s^{\parallel}}{\left(K_1^d - K_2^d\right)k^2 - 2(h^2 - 1)\rho_s^{\parallel}}u\right)^2 + (u \to v)\right]\right\}.$$
(4.675)

Since $K_1^d - K_2^d > 0$, the second term is positive definite for all k if:

$$h^2 < 1.$$
(4.676)

Thus we remain with deciding the region in the h, p-plane for which

$$A(k) = \rho_0 j^2 + K_b k^2 + 2\rho^{\parallel} - 2kj\left(c_0 + \frac{\rho_s^{\parallel}}{2}\right) - \frac{4\rho_s^{\parallel 2}}{\left(K_1^d - K_2^d\right)k^2 - 2(h^2 - 1)\rho_s^{\parallel}}$$
$$= \rho_0 j^2 + K_b\left(k - \frac{c_0 + \frac{\rho_s^{\parallel}}{2}}{K_b}j\right)^2 - K_b\left(\frac{c_0 + \frac{\rho_s^{\parallel}}{2}}{K_b}\right)^2$$
$$+ \frac{2\rho^{\parallel}}{s}\left[1 - \frac{2\rho_s^{\parallel}}{\left(K_1^d - K_2^d\right)k^2 - 2\left(h^2 - 1\right)\rho_s^{\parallel}}\right] > 0.$$
(4.677)

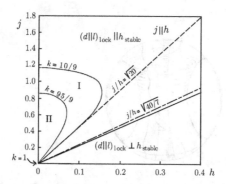

FIGURE 4.37 If the assumption of dipole locking is relaxed, the regions of stability shrink as shown in this figure. The entire region to the left of the line $j/h = \sqrt{20}$ is stable in the dipole-locked limit. The finite strength of dipole locking reduces this to region I. It is reduced to the origin, depending on whether the temperature is $T = T_c$ or $T = T_{\text{stab}}$. Here a a stable helix can begin forming. For completeness, we have given also the region II for a temperature half-way between T_c and T_{stab}. Similarly, if dipole locking would be perfect, the whole region below $j/h = \sqrt{40/7}$ would be stable, with $\mathbf{d} \parallel \mathbf{l}$ pointing orthogonal to the magnetic field. The finiteness of the dipole locking force reduces this region to the solid curve which becomes horizontal for large h.

Only the regions with $k \approx h^2/(K_1^d - K_2^d)$ and $k \approx (c_0 + \frac{\rho_s^\parallel}{2})j/K_b$ are dangerous. If we assume h, $j \ll 1$, then $k \ll 1$ is also a dangerous value, and we expand

$$
\begin{aligned}
A(k) &\approx \rho_0 j^2 + K_b\left(k - \frac{c_0 + \frac{\rho_s^\parallel}{2}}{K_b}j\right)^2 - K_b\left(\frac{c_0 + \frac{\rho_s^\parallel}{2}}{K_b}\right)^2 \\
&\quad + 2\rho_s^\parallel\left(-h^2 + \left(K_1^d - K_2^d\right)k^2/2\rho_s^\parallel\right) \\
&= \rho_0 j^2 + L_b^l\left(k - \frac{c_0 + \frac{\rho_s^\parallel}{2}}{K b^l}j\right)^2 - \rho_0 K^{-1} - 2\rho_s^\parallel h^2 > 0. \quad (4.678)
\end{aligned}
$$

From this we find

$$
\frac{h^2}{j} < -\frac{1}{2}\frac{\rho_0}{\rho_s^\parallel}\left(K^{-1} - 1\right) \underset{\text{L}}{=} \frac{1}{2}\left(\frac{1}{10} - \epsilon\right), \quad (4.679)
$$

in agreement with the dipole-locked result (4.619), as it should. Thus the straight line (4.679) will now be tangential to the stability curve at the origin. For larger values of h, that curve bends upwards and cuts the z axis at some finite value of j. In Fig. 4.37 we have plotted the new stability curves for the generalized Ginzburg-Landau constants with $\rho_0 = \rho_s^\parallel(1 - \epsilon)$ at $\epsilon = 0$, $\epsilon = 0.05$ and $\epsilon = 0.1$. Even at $h = 0$, the forward texture is stable only for $j \leq j_{\max} = 1.17$, 0.83, 0, respectively. The reason for the onset of stability at $h = 0$ is easy to understand: The current tries to

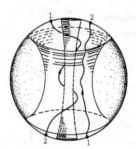

FIGURE 4.38 As a stable helix forms in the presence of a superflow in ^{3}He-A, the parameter space reduces even more. In addition to the equator being forbidden by the alignment force, a narrow cylinder along the axis is outruled as well. The topology of the remainder is doubly-infinitely connected. Continuous paths can either break at the surface and continue from the diametrally opposite point or they can wind an arbitrary number of times around the central one.

curl up the texture in form of a helix (see the second term of (4.678)). The dipole force drags $\mathbf{d}$ behind. But the bending energies of $\mathbf{d}$ favor a uniform $\mathbf{d}$ texture. Thus, if the current is too strong, the $\mathbf{d} \parallel \mathbf{l}$ alignment breaks. As soon as $\mathbf{d}$ and $\mathbf{l}$ are decoupled, the texture destabilizes, as was observed before in the general case.

The full analysis of equilibrium positions in the unlocked case is tedious. However, as long as j, h are small enough, say $j < \frac{3}{4}j_{\max}$, $h < h_c$, the results of the dipole locked situation are perfectly applicable.

Let us now turn to a disussion of the physical content of the helix which was alluded to in the beginning of this chapter. As the helix forms at $h > h_c$, the $\beta = 0$ position turns into a potential mountain which forbids the alignment of $\mathbf{d} \parallel \mathbf{h}$ with $\mathbf{j} \parallel \mathbf{h}$. In the SO(3) parameter space of the dipole-locked A-phase, this amounts to removing a narrow cylindrical region running along the axis (shown in Fig. 4.38). Together with the potential mountain around the equator discussed before, the parameter space becomes now doubly-infinitely connected:

$$\pi_1 = \mathbf{Z} + \mathbf{Z}. \tag{4.680}$$

In addition to paths running from south to north, continuing again at the diametrically opposite point at the south, etc., also those which wind an arbitrary number of times around the narrow cylinder become topologically inequivalent. Physically, this corresponds to the fact that in a torus not only

$$\langle \alpha_z + \gamma_z \rangle = 2\pi N/L, \tag{4.681}$$

but also the average pitch

$$\langle \gamma_z \rangle = 2\pi M/L, \tag{4.682}$$

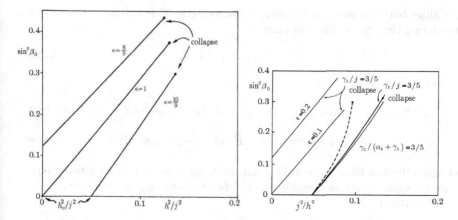

FIGURE 4.39 Angle of inclination as a function of the magnetic field at different temperatures. The values $K = \frac{10}{9}, 1, \frac{8}{9}$ correspond to $T = T_c$, $T = T_{stab}$, $T < T_{stab}$. As the magnetic field is increased, the helix collapses. Then the magnetic field is so strong that it tears apart the stabilizing dipole locking between 1 and d. The solid curves show the behavior if only the temperature dependence of ρ_0 is taken into account. The dashed curves contain the full T-dependence.

are both topological invariants. A consequence of this is that, when increasing the magnetic field beyond h_c, or when decreasing the temperature so that $h_c^2 < 0$, the pitch value $\langle \gamma_z \rangle \approx \left(c_0 + \frac{1}{2}\rho_s^\parallel \right) / K_b j = \gamma_z^0$ at which the helix begins forming [due to the last term in (4.616)] will be frozen. Therefore the angle of inclination β_0 will be pinned down topologically, precisely at the value corresponding to s_h^- [see (4.646)].

In Fig. 4.39 we have displayed the curves of constant $\gamma_z/(\alpha_z + \gamma_z) = \gamma_z^0/j$ for increasing h^2/j^2 at fixed values of ϵ, until the point of collapse. These curves can be deduced from plots like those in Figs. 4.34 a)–c), by following almost a straight line to the right starting from $\gamma_z/j = \frac{3}{5}$. The line is not exactly straight since this would imply $\gamma_z/j = \frac{3}{5}$ rather than $\gamma_z/(\alpha_z + \gamma_2) = \frac{3}{5}$. The relation is, in the Ginzburg-Landau limit,

$$\frac{\gamma_z}{j} \underset{GL}{=} \frac{\gamma_z/(\alpha_z + \gamma_z)}{1 + s - (1 + s - \sqrt{1 - s})\gamma_z/(\alpha_z + \gamma_z)}$$

$$\approx \frac{3}{5}\left(1 - \frac{1}{10}\sin^2 \beta + \dots\right),$$

(4.683)

so that there is very little deviation for small s.

The separate topological conservation of the two currents is intimately related with the fact that ³He-A contains p-wave Cooper pairs. Remembering our discussion of Eq. (4.131) there are two current terms of different physical origin. The helix stabilizes both currents *topologically* and provides, in addition, the perfect tool for measuring their ratio. The pair current

$$J^{\text{pair}} = \rho_s \mathbf{v}_s - \rho_0 \mathbf{l}(\mathbf{l} \cdot \mathbf{v}_s) \qquad (4.684)$$

is a sum of two terms

$$
\begin{aligned}
J^{\text{pair}} &= -\left(\rho_s - \rho_0 \cos^2 \beta\right) \frac{1}{2m} \left(\alpha_z + \cos \beta \gamma_z\right) \mathbf{e}_z + \rho_0 \mathbf{l}^{\perp} \cos \beta \frac{1}{2m} \left(\alpha_z + \cos \beta \gamma_z\right) \\
&= -\frac{1}{2m} A \left(\alpha_z + \cos \beta \gamma_z\right) \mathbf{e}_z + \rho_0 \mathbf{l}^{\perp} \cos \beta \frac{1}{2m} \left(\alpha_z + \cos \beta \gamma_z\right), \qquad (4.685)
\end{aligned}
$$

of which the first flows in the z-direction, while the second forms stratified layers of currents whose direction changes with $\mathbf{l}^{\perp}$ when proceeding along the helix.

The orbital current

$$
\begin{aligned}
J^{\text{orb}} &= c \left(\nabla \times \mathbf{l}\right) - c_0 \mathbf{l} \left[\mathbf{l} \cdot \left(\nabla \times \mathbf{l}\right)\right] \\
&= c_0 \cos \beta \sin^2 \beta \gamma_z \mathbf{e}_z - \left(c - c_0 \sin^2\right) \gamma_z \mathbf{l}^{\perp} + c \cos \beta \beta_z \mathbf{e}_\varphi, \qquad (4.686)
\end{aligned}
$$

on the other hand, is the sum of three terms, the last of which points into the azimuthal direction

$$\mathbf{e}_\varphi \equiv (\mathbf{e}_z \times \mathbf{l}) / |\mathbf{e}_z \times \mathbf{l}|,$$

and vanishes in equilibrium where $\beta_z = 0$. With $\beta = \beta_0$ and α_z, γ_z frozen topologically, both currents are determined. In particular, the ratio of their z-components is

$$\frac{J_z^{\text{orb}}}{J_z^{\text{pair}}} = 2mc_0 \sin^2 \beta_0 \cos \beta_0 \frac{\gamma_z}{(\rho_s - \rho_0 \cos^2 \beta_0)(d_z + \cos \beta_0 \gamma_z)}. \qquad (4.687)$$

There is a simple way to measure $\sin^2 \beta_0$. As is well-known, sound attenuation is sensitive to the angle between the l-vector and the direction of propagation of the sound [43]. In fact, if δ denotes this angle, the attenuation constant is given by

$$
\begin{aligned}
\alpha &\equiv \alpha_\perp \cos^4 \delta + 2\alpha_c \sin^2 5 \cos^2 \delta + \alpha_\perp \sin^4 \delta, \\
&= (\alpha_{11} - 2\alpha_c + \alpha_\perp) \cos^4 \delta + 2 (\alpha_c \alpha_\perp) \cos^2 \delta + \alpha_\perp. \qquad (4.688)
\end{aligned}
$$

If the helix is probed with a transverse signal, the angle δ becomes

$$\cos \delta = \sin \beta \cos \gamma. \qquad (4.689)$$

Therefore one has the averages

$$\langle \cos^2 \delta \rangle = \sin^2 \beta_0 \langle \cos^2 \gamma \rangle = \frac{1}{2} \sin^2 \beta, \quad \langle \cos^4 \delta \rangle = \sin^4 \beta_0 \langle \cos^4 \gamma \rangle = \frac{3}{8} \sin^4 \beta, \quad (4.690)$$

so that

$$\alpha = \alpha_\perp + (\alpha_c - \alpha_\perp) \sin^2 \beta_0 + \frac{3}{8} \sin^4 \beta. \qquad (4.691)$$

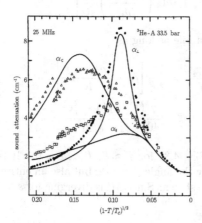

FIGURE 4.40 Sound attenuation can be parametrized in terms of three constants. Experimental measurements are shown here and compared with theoretical calculations of Ref. [43]. The most sensitive test for a helical texture can be performed in the region of largest difference between $\alpha_\perp$ and α_{11}.

The experimental values for the coefficients are displayed in Fig. 4.40 (taken from Ref. [43]). Thus, if one goes into a region of large $|\alpha_c - \alpha_\perp|$, and turns on a magnetic field, α will stay constant for $h < h_c$ [from (4.619) and (4.647)]. For $h > h_c$ it will begin to drop linearly in $\Delta h^2/j^2$ (if $\alpha_c \alpha_\perp < 0$) with a slope $\approx (\alpha_c - \alpha_\perp) 3\Delta h^2/j^2$. It appears that this effect has been seen at the University of California in La Jolla [42].[17]

Until now we have focussed our discussion on helical textures which may develop from a previously aligned $\mathbf{d} \parallel \mathbf{l} \parallel \mathbf{j} \parallel \mathbf{h}$ configuration. A look at Figs. 4.34 a–c shows that there is another domain of stability for $s \approx 1$ (open helices), as h^2/j^2 exceeds some critical value $h_{c_2}^2/j^2$. The reason for this is obvious: If h is large enough, a potential valley is created for the $\mathbf{d}$-vector at $\theta \approx \pi/2$. Dipole locking stabilizes also $\mathbf{l}$ in this position. In order to calculate the boundary in the dipole-locked regime, consider the energy for $s \approx 1$:

$$
\begin{aligned}
2g &= \text{const} + 2\left(\frac{h^2}{j^2} - \frac{1}{2}\frac{\rho_0 \rho_s^{\parallel}}{\rho_s^2}\right)(1-s)j^2 + 2\sqrt{1-s}\frac{\rho_s - c_0}{\rho_s}\gamma_z j \\
&\quad + \frac{K_t}{\rho_s^{\parallel}}\gamma_z^2 \\
&= \text{const} + 2\left[\frac{h^2}{j^2} - \frac{\rho_0 \rho_s^{\parallel}}{2\rho_s^2}\left(\frac{\rho_s c_0}{\rho_s}\right)^2 \frac{\rho_s^{\parallel}}{2K_t}\right]j^2
\end{aligned}
$$

---

[17]I thank Prof. Kazumi Maki for a discussion of this point, and of the experiment performed by R.L. Kleinberg at La Jolla [42].

$$+\frac{K_t}{\rho_s^{\parallel}}\left[\gamma_z + \sqrt{1s}\left(\frac{\rho_s - c_0}{\rho_s}\right)\frac{\rho_s^{\parallel}}{K_t}\right]^2. \tag{4.692}$$

Thus the $\beta \approx \pi/2$ -position is stable as long as

$$\frac{h^2}{j^2} > \frac{h_{c_2}^2}{j^2} = \frac{\rho_s^{\parallel}}{2\rho_s^2}\left[\rho_0 + \frac{(\rho_s - c_0)^2}{K_t}\right]$$

$$\underset{L}{=} \frac{7}{40}\frac{(1-\epsilon)\left(1-\frac{2}{7}\epsilon\right)}{(1-\epsilon/2)^2}. \tag{4.693}$$

This boundary was shown in Fig. 4.37 for $T \approx T_c$ (i.e., $\epsilon \approx 0$). It is important to realize that for $h > h_{c_2}$, not only the angle $\beta = \pi/2$, but also an entire neighborhood of it is stable. This can easily be shown: Since M'_g is diverging for $s \approx 1$ as $1/\sqrt{1-s}$, the solution of (4.630) becomes simply

$$\frac{\gamma_z^-}{j} \approx \frac{1}{M'_g}\left[\frac{h^2}{j^2} - \frac{A'_g}{2}\right]\bigg|_{s\approx 1} \underset{L}{=} -2\frac{2-\epsilon}{1-\epsilon}\sqrt{1-s}\left[\frac{h^2}{j^2} - \frac{1-\epsilon}{8(1-\epsilon/2)^2}\right],$$

$$\frac{\gamma_z^+}{j} \approx -2\frac{M'_g}{G'_g}\bigg|_{s\approx 1} \underset{L}{=} \frac{1}{\sqrt{1-s}}\frac{1-\epsilon}{6-\frac{5}{2}\epsilon}, \tag{4.694}$$

which can be compared with Fig. 4.34 a)–c).

Now the first stability criterion (4.659) is fulfilled trivially:

$$G_g|_{s=1} = \frac{K_t}{\rho_s^{\parallel}} > 0. \tag{4.695}$$

The second criterion on the determinant in Eq. (4.661) is, on the other hand, dominated by the singularity in M''_g where

$$2M''_g j\gamma_z^{\pm}\frac{G_g}{2G'^2_g} \gtrsim \frac{\gamma_z^{+2}}{4j^2}. \tag{4.696}$$

Inserting (4.694) we see that only the negative values of γ_z^- on the lower branch can satisfy this under the condition:

$$\frac{h^2}{j^2} - \frac{A'_g}{2} > \frac{M'^2_g}{M''_g}\frac{1}{G_g}. \tag{4.697}$$

Inserting the parameters of the liquid, this becomes exactly the same condition as (4.693), but now it guarantees stability of *all* positions in the neighborhood of $s \approx 1$. Note that, contrary to $s = 0$, where $s = 0$ and $s \neq 0$ correspond to two different parameter manifolds, the point $s = 1$ is in no way special as compared to its neighborhood.

If dipole locking is relaxed, the straight boundary (4.693) in the j, h plane will curve for larger values of j (see Fig. 4.37), and will eventually approach an asymptotic line $j = j_{\max}$. In order to find $j_{\max}$, consider the terms of the energy quadratic in $\tilde{\beta} \equiv \beta - \pi/2$, $\tilde{\theta} \equiv \theta - \pi/2$, γ, ϕ:

$$2g = \text{const} + 2\left(\frac{h^2}{j^2}\tilde{\theta} - \frac{\rho_0\rho_s^{\parallel}}{2\rho_s^2}\tilde{\beta}^2\right)j^2 + K_s\tilde{\beta}_z^2 + 2\tilde{\beta}\frac{\rho_s c_0}{\rho_s}\gamma_z j$$

$$+\frac{K_t}{\rho_s^{\parallel}}\gamma_z^2 + \frac{K_1^d}{\rho_s^{\parallel}}\left(\theta_z^2 + \phi_z^2\right) + 2\left(\tilde{\theta} - \tilde{\beta}\right)^2 + 2\left(\gamma - \phi\right)^2, \qquad (4.698)$$

where now K_s, K_t are the unlocked values ($K_s \underset{\text{GL}}{=} K_t \underset{\text{GL}}{=} \frac{1}{2}$ also $K_1^d \underset{\text{GL}}{=} 2$). For a plane wave ansatz this becomes

$$2g = \text{const} \quad +\tilde{B}\tilde{\beta}^2 + \tilde{T}\tilde{\theta}^2 - 4\tilde{\theta}\tilde{\beta} - \tilde{G}\gamma^2 + 2Mjk\tilde{\beta}\gamma + \tilde{F}\phi^2 - 4\gamma\phi, \qquad (4.699)$$

with

$$\tilde{B} = K_s k^2 - \frac{\rho_0\rho_s^{\parallel}}{\rho_s^2}j^2 + 2,$$

$$\tilde{G} = \frac{K_t}{\rho_s^{\parallel}}k^2 + 2, \qquad\qquad (4.700)$$

$$\tilde{T} = \frac{K_1^d}{\rho_s^{\parallel}}k^2 + 2h^2 + 2,$$

$$\tilde{F} = \frac{K_1^d}{\rho_s}k^2 + 2.$$

After a few quadratic completions one finds

$$2g = \text{const} + \bar{B}\tilde{\beta}^2 + \tilde{T}\tilde{\theta}^2 + \bar{G}\bar{\gamma}^2 + \tilde{F}\bar{\phi}^2, \qquad (4.701)$$

with

$$\bar{G} \equiv \tilde{G} - 4/\tilde{F} = \frac{K_t^l}{\rho_s^{\parallel}}k^2\left(1 + \frac{K_t K_1^d}{2K_t^l\rho_s^{\parallel}}k^2\right)\bigg/\left(1 + \frac{1}{2}\frac{K_1^d}{\rho_s^{\parallel}}k^2\right),$$

$$\bar{B} = \tilde{B} - \tilde{M}^2/\bar{G}j^2 - 4/\tilde{T} \qquad\qquad (4.702)$$

$$= \frac{K_s}{\rho_s^{\parallel}}k^2 - \frac{\rho_0\rho_s^{\parallel}}{\rho_s^2}j^2 + 2 - \left(\frac{\rho_s - c_0}{\rho_s}\right)^2\frac{k^2}{\bar{G}}j^2 - \frac{2}{\frac{K_1^d}{2\rho_s^{\parallel}}k^2 + h^2 + 1},$$

and new angles

$$\bar{\theta} = \tilde{\theta} - \tilde{\beta}/\tilde{T},$$

$$\bar{\gamma} = \gamma - \left(Mjk/\bar{G}\right)\beta, \qquad\qquad (4.703)$$

$$\bar{\phi} = \phi - \gamma/\tilde{F}.$$

We now observe that $\tilde{F} \geq 0$, $\tilde{T} \geq 0$, $\bar{G} \geq 0$, and that the quadratic form (4.701) is positive-definite for all k if

$$\bar{B}(k) > 0. \tag{4.704}$$

For small h^2, j^2, we can expand in h^2, j^2, and k^2 and recover the dipole-locked result (4.693). As h increases, the stability curve approaches the line $j = j_{\max}$ determined by the $h = \infty$ version of (4.704) which renders

$$j_{\max}^2 - \frac{2\rho_0^2/\rho_s^{\parallel}}{\rho_0 + \frac{(\rho_s - c_0)^2}{K_t^l}} \underset{\mathrm{GL}}{=} \frac{40}{7}. \tag{4.705}$$

In fact, if the coefficients are close enough to their Ginzburg-Landau values, the value $\bar{B}(k = 0)$ is the most dangerous one, yielding the boundary curve,

$$j^2 \leq \frac{2\rho_s^2}{\rho_s^{\parallel}} \frac{1}{\rho_0 + \frac{(\rho_s - c_0)^2}{K_t^l}} \frac{h^2}{h^2 + 1}, \tag{4.706}$$

which starts out as (4.693) and becomes horizontal for $h \gg 1$ (see Fig. 4.37).

A final remark concerns the possibility that the stability discussion presented here becomes invalid due to transverse oscillations which we have neglected. Certainly, these oscillations must be included to believe the above stability criteria. This makes the discussion of the energy much more tedious. Until now, only oscillations with very small transverse momentum have been tested. Fortunately it turns out that at least for this limit the transverse oscillations have higher energies than the longitudinal ones, so that the instabilities are always triggered along the z-direction. Since the discussion of this point is very technical, the reader is referred to Ref. [40].

A similar helical texture exists in an external magnetic field, as was found soon after the above discovery of the helix in a flowing superfluid [41].

4.11.1   Magnetic Field and Transition between A- and B-Phases

At zero flow we can observe a transition between A- and B-phases at a magnetic field which satisfies

$$\frac{\alpha}{\frac{6}{5}\beta_B} \left(\alpha - h^2\right)^2 + \frac{5}{\beta_{345}} h^4 = \frac{\alpha}{\frac{6}{5}\beta_A}. \tag{4.707}$$

The solution is

$$h_{AB}^2 = \frac{\alpha}{3} \frac{\beta_{345}}{2\beta_{12} + \beta_{345}} \left[1 - f(\beta)\right], \tag{4.708}$$

where

$$f(\beta) \equiv \sqrt{\frac{3\beta_B}{\beta_A \beta_{345}}} \sqrt{2\beta_{13} - \beta_{345}}. \tag{4.709}$$

In the weak-coupling limit, the result is

$$h_{AB}^2 = \frac{1}{6}. \tag{4.710}$$

Recalling the definition of h in Eq. (4.476), this translates into the relation between the physical magnetic field H, and the temperature between A- and B-phase:

$$\frac{H^2}{H_0^2} = \frac{1}{6}\left(1 - \frac{T_{AB}}{T_c}\right), \tag{4.711}$$

from which we see that the transition temperature is shifted downwards quadratically with an increasing magnetic field. At the polycritical pressure p_{pc} one has

$$\beta_A = \beta_B, \tag{4.712}$$

and

$$3\beta_{13} = 2\beta_{345}, \quad f(\beta) = 1, \tag{4.713}$$

so that the transition occurs at zero magnetic field, as it should.

Let T_{AB}^0 be the temperature of the transition in the absence of a magnetic field. We may expand the right-hand side of Eq. (4.708) around this temperature by setting

$$1 - f(\beta) = f'\Big|_{T_{AB}^0}\left(1 - \frac{T_{AB}^0}{T}\right). \tag{4.714}$$

Inserting this into (4.708) we find that a magnetic field shifts the transition from T_{AB}^0 to T_{AB}^h according to the approximate formula

$$h_{AB}^2 = \frac{\alpha}{3}\frac{\beta_{345}}{2\beta_{12} + \beta_{345}}f'\Big|_{T_{AB}^0}\left(T_{AB}^0 - T_{AB}^h\right), \tag{4.715}$$

or in terms of physical magnetic fields:

$$\frac{H_{AB}^2}{H_0^2} = \frac{\alpha}{3}\frac{\beta_{345}}{2\beta_{12} + \beta_{345}}f'\Big|_{T_{AB}^0}\left(T_{AB}^0 - T_{AB}^h\right)\left(1 - \frac{T_{AB}^0}{T_c} + \frac{T_{AB}^0 - T_{AB}^h}{T_c}\right). \tag{4.716}$$

Thus we see that far away from the polycritical pressure p_{pc} there is a quadratic response of T_{AB}^h to H. Close to p_{pc}, on the other hand, the response is linear. This explains why the experimental curve of phase transition shows the most significant dependence on H in the neighborhood of the polycritical pressure p_{pc}.

It should be noted that, in the absence of strong-coupling corrections, the order parameter of the B-phase is distorted continuously into that of the A-phase as H reaches H_{AB}. There the order parameters $a^2 = 1 + \frac{3}{2}h^2$, $c^2 = 1 - 6h^2$ become $a^2 = \frac{5}{4}$, $c^2 = 0$.

Thus the transition is of second order. Since it is sometimes believed that strong-coupling corrections become small for $p \to 0$, this amounts to a decreasing latent heat.

4.12   Large Currents at Any Temperature $T \leq T_c$

4.12.1   Energy at Nonzero Velocities

For general temperatures $T \leq T_c$ we shall confine our discussion to the weak-coupling regime. Fermi liquid correction will be included at a later stage.

Adding the external source

$$\mathbf{vJ} = -\mathbf{v}\psi^*(x)\frac{i}{2}\overset{\leftrightarrow}{\nabla}\psi(x) \tag{4.717}$$

to the action (4.75) gives rise, in the 2×2 matrix M of (4.80), to the additional entries

$$\begin{pmatrix} \mathbf{vp} & 0 \\ 0 & \mathbf{vp} \end{pmatrix}. \tag{4.718}$$

Therefore, the final collective action (4.83) becomes simply

$$\mathcal{A}^v = -\frac{i}{2}\mathrm{Tr}\log\begin{pmatrix} i\partial_t - \xi(\mathbf{p}) + \mathbf{vp} & A_{ai}\sigma_a\overset{\leftrightarrow}{\nabla}_i/2 \\ A^{ai*}\sigma_a\overset{\leftrightarrow}{\nabla}_i/2 & i\partial_t + \xi(p) + \mathbf{vp} \end{pmatrix} - \frac{1}{3g}\int d^3x|A_{ai}|^2. \tag{4.719}$$

For constant field configurations, $A_{ai} \equiv A^0_{ai}$, this results in the free energy density at velocity v:

$$g^v \equiv -\frac{T}{V}\mathcal{A}^v \tag{4.720}$$

$$= -T\sum_{\omega_n,\mathbf{p}}[\log(i\omega_n + \mathbf{vp} - E(\mathbf{p})) + (E - E)] + \frac{1}{3g}|A^0_{ai}|^2 + \mathrm{const},$$

with

$$E(\mathbf{p}) = \sqrt{\xi^2(\mathbf{p}) + \Delta^2_\perp + (1 - r^2z^2)}. \tag{4.721}$$

As usual in such expressions, it is convenient to subtract from this the free energy of the free Fermi liquid, now with the external source $\mathbf{vp}$. Then

$$g^v_0 = -T\sum_{\omega_n,\mathbf{p}}[\log(i\omega_n + \mathbf{vp} - \xi(\mathbf{p})) + (\xi \to -\xi)] + \mathrm{const}. \tag{4.722}$$

For $\mathbf{v} = 0$ this quantity was calculated earlier [recall Eq. (4.264)]. For $\mathbf{v} \neq 0$ we observe that we may perform a quadratic completion

$$\mp\mathbf{vp} - \xi(\mathbf{p}) = -\frac{(\mathbf{p} \pm m\mathbf{v})^2}{2m} + \mu + \frac{m}{2}\mathbf{v}^2. \tag{4.723}$$

The first term gives the same g_0 as the $\mathbf{v} = 0$ formula since the integration over $\mathbf{p}$ is merely shifted by $m\mathbf{v}$. As far as the additional kinetic energy $m\mathbf{v}^2/2$ is concerned we may assume it to be very much smaller than $p_F^2/2m$ so that we can expand

$$
\begin{aligned}
g_0^v &= g_0^0 + T \sum_{\omega_n, \mathbf{p}} \left(\frac{e^{i\omega_n \mu}}{i\omega_n - \xi(\mathbf{p})} - \frac{e^{-i\omega_n \mu}}{i\omega_n + \xi(\mathbf{p})} \right) \frac{m}{2} \mathbf{v}^2 \\
&= g_0^0 - \sum_{\mathbf{p}} n(\xi) \frac{m}{2} \mathbf{v}^2 \\
&= g_0^0 - \frac{g}{2} \mathbf{v}^2,
\end{aligned}
\tag{4.724}
$$

thus arriving at the usual form of a Galilean transformed energy.

4.12.2   Gap Equations

We shall now allow for anisotropic gaps (4.104) of the same distorted form (4.491) as discussed previously in the Ginzburg-Landau limit, i.e.,

$$
A_{ai}^0 = \Delta^0 \begin{pmatrix} a & & \\ & a & \\ & & c \end{pmatrix} = \begin{pmatrix} \Delta_\perp & & \\ & \Delta_\perp & \\ & & \Delta_\parallel \end{pmatrix},
\tag{4.725}
$$

where $\Delta_\perp$ and $\Delta_\parallel$ are the gaps orthogonal and parallel to the flow. Introducing the gap distortion parameter

$$
r \equiv 1 - \frac{\Delta_\parallel^2}{\Delta_\perp^2},
\tag{4.726}
$$

and the directional cosine z of the quasiparticle momentum with respect to the preferred axis, which lies parallel to the current for symmetry reasons, we may write the anisotropic gap as

$$
|A_{ai}^0 \hat{p}_i|^2 = \Delta^2(z) = \Delta_\perp^2 \left(1 - z^2 \right) + \Delta_\parallel^2 z^2 \equiv \Delta_\perp^2 \left(1 - r^2 z^2 \right).
\tag{4.727}
$$

This paramerization of the gap permits a simultaneous discussion of B-, A-, planar, and polar phases. With the form (4.725) the last term in the free energy g^v becomes

$$
\frac{1}{3g} |A_{ai}|^2 = \frac{1}{3g} \left(2\Delta_\perp^2 + \Delta_\parallel^2 \right) = \frac{1}{g} \Delta_\perp^2 \left(1 - \frac{r^2}{3} \right).
\tag{4.728}
$$

Minimizing g^v with respect to $\Delta_\parallel^2$ and $\Delta_\perp^2$ we find the two conditions:

$$
\left[\frac{1}{g} - T \sum_{\omega_n, \mathbf{p}} 3z^2 \frac{1}{(\omega_n - i p_F z)^2 + E^2(\mathbf{p})} \right] \Delta_\parallel = 0,
\tag{4.729}
$$

$$
\left[\frac{1}{g} - T \sum_{\omega_n, \mathbf{p}} \frac{3}{2} \left(1 - z^2 \right) \frac{1}{(\omega_n - i v p_F z)^2 + E^2(\mathbf{p})} \right] \Delta_\perp = 0.
\tag{4.730}
$$

If we assume both gaps $\Delta_\parallel$ and $\Delta_\perp$ to be nonzero, there are two nontrivial gap equations. Sometimes it is useful to compare the general result with the hypothetical case of a gap that is free of distortion, $\Delta_\parallel = \Delta_\perp$ or $r = 0$. Then only the average gap equation [$\frac{1}{3}$ (longitudinal $+2$ transverse)] survives with no directional factor z in the integration and with $r = 0$ inserted. Moreover, since the polar phase in which $\Delta_\perp$ vanishes (corresponding to $r \to -\infty$) is physically rather uninteresting, due to its small condensation energy, we shall henceforth work with the average gap equation together with the transverse one (4.730). From the latter we shall often draw comparison with the A-phase by inserting $r = 1$.

In the two gap equations, the sums over Matsubara frequencies may be performed in the standard fashion using Formula (4.242) to find

$$T \sum_{\omega_n} \frac{1}{2E} \left(\frac{1}{i\omega_n + vp_F z - E(p)} - \frac{1}{i\omega_n + vp_F z + E(p)} \right)$$
$$= \frac{1}{4E} \left[\tanh \frac{E - vp_F z}{2T} + (v \to -v) \right]. \tag{4.731}$$

Decomposing the integral over momenta according to direction and size

$$\int \frac{d^3 p}{(2\pi)^3} \approx \mathcal{N}(0) \int_{-1}^1 \frac{dz}{2} \int_{-\infty}^\infty d\xi, \tag{4.732}$$

the average and the transverse gap equations become

$$\frac{1}{g\mathcal{N}(0)} = \int_{-1}^1 \frac{dz}{2} \gamma(z),$$
$$\frac{1}{g\mathcal{N}(0)} = \int_{-1}^1 \frac{dz}{2} \frac{3}{2} \left(1 - z^2\right) \gamma(z), \tag{4.733}$$

where $\gamma(z)$ denotes the function

$$\gamma(z) \equiv T \sum_{\omega_n} \int_{-\infty}^\infty d\xi \frac{1}{(\omega_n - vp_F z)^2 + E^2(\mathbf{p})}$$
$$= \int_{-\infty}^\infty d\xi \frac{1}{4E} \left[\tanh \frac{E - vp_F z}{2T} + (v \to -v) \right], \tag{4.734}$$

which is logarithmically divergent. It may be renormalized via the critical temperature which satisfies

$$\frac{1}{g\mathcal{N}(0)} = \int_{-\infty}^\infty d\xi \frac{1}{2\xi} \tanh \frac{\xi}{2T_c} = \log \left(\frac{\omega_c}{T_c} 2e^{-\gamma}/\pi \right)$$
$$= \log \frac{T}{T_c} + \int_{-\infty}^\infty d\xi \frac{1}{2\xi} \tanh \frac{\xi}{2T}. \tag{4.735}$$

Subtracting this expression on both sides the gap equations take the form

$$\log \frac{T}{T_c} = \int_{-1}^1 \frac{dz}{2} \gamma(z), \tag{4.736}$$

$$\log \frac{T}{T_c} = \int_{-1}^1 \frac{dz}{2} \frac{3}{2} (1 - z^2) \gamma(z), \tag{4.737}$$

with the subtracted finite function

$$\gamma(z) = \int_{-\infty}^{\infty} d\xi \left\{ \frac{1}{4E} \left[\tanh \frac{E - v p_F z}{2T} + (v \to -v) \right] - \frac{1}{2\xi} \tanh \frac{\xi}{2T} \right\}. \quad (4.738)$$

For calculations it is more convenient to return to the Matsubara sum forms (4.729) and (4.730). Then the integrals over d can be performed and with the above renormalization procedure, and we find the simple expression

$$\begin{aligned}
\gamma(z) &= \pi \sum_{\omega_n} \left(\frac{1}{\sqrt{(\omega_n - iv p_F z)^2 + \Delta_\perp^2 (1 - r^2 z^2)}} - \frac{1}{\omega_n} \right) \\
&= \frac{1}{\delta} \sum_{n=-\infty}^{\infty} \left(\frac{1}{\sqrt{(x_n - i\nu z)^2 + (1 - r^2 z^2)}} - \frac{1}{x_n} \right). \quad (4.739)
\end{aligned}$$

In this and many formulas to come we have introduced the following dimensionless variables:

$$\begin{aligned}
\delta &= \frac{\Delta_\perp}{\pi T}, \\
\nu &= \frac{v p_F}{\Delta_\perp}, \quad (4.740) \\
x_n &= \frac{\omega_c}{\Delta_\perp}.
\end{aligned}$$

In order to check the gap equations we first re-derive the previous Ginzburg-Landau results by going to the limit $T \to T_c$. Then the variables x_n become very large and we may approximate the two equations (4.736) and (4.737) for $\log T/T_c \approx 1 - T/T_c$ as

$$\begin{aligned}
1 - \frac{T}{T_c} &\approx \frac{2}{\delta} \int_{-1}^{1} \frac{dz}{2} \left\{ \begin{array}{c} 1 \\ \frac{3}{2}(1 - z^2) \end{array} \right\} \sum_{n=0}^{\infty} \left[1 + \left(2\nu^2 - r^2 \right) z^2 - 2i\nu_z x_n \right] / 2x_n^3 \\
&= \delta^2 \left[1 + \left\{ \begin{array}{c} \frac{1}{3} \\ \frac{1}{5} \end{array} \right\} \left(2\nu^2 - r^2 \right) \right] \frac{7\zeta(3)}{8}. \quad (4.741)
\end{aligned}$$

Combining the two equations we see that near T_c, the gap distortion grows with the square of the velocity:

$$r^2 = 1 - \frac{\Delta_\parallel^2}{\Delta_\perp^2} = 1 - \frac{c^2}{a^2} = 2\nu^2. \quad (4.742)$$

Inserting this into (4.741), the transverse gap behaves like

$$\Delta_\perp^2 = \pi^2 T^2 \delta^2 \approx \frac{8}{7\zeta(3)} \pi^2 T_c^2 \left(1 - \frac{T}{T_c} \right) \approx 3.063^2 \left(1 - \frac{T}{T_c} \right), \quad (4.743)$$

i.e., it is independent of the current velocity. Note that it follows the same near-T_c-equation as the gap in a superconductor [compare (3.121)]. Inserting this into

Eq. (4.740), we find an equation for the reduced current velocity $\kappa \equiv v/v_0$ [see (4.459)]:

$$
\begin{aligned}
\nu^2 &= \frac{v^2 p_F^2}{\Delta_\perp^2} = \frac{v^2 p_F^2}{\dfrac{8}{7\zeta(3)}\pi^2 T_c^2 \left(1 - \dfrac{T}{T_c}\right)} \\
&= \frac{3}{2}\frac{v^2}{\left(\dfrac{1}{2m\xi_0}\right)^2}\frac{1}{1 - \dfrac{T}{T_c}} = \frac{3}{2}\frac{v^2}{v_0^2(T)} = \frac{3}{2}\kappa^2.
\end{aligned}
\tag{4.744}
$$

Via the gap distortion (4.742), this determines the ratio of logitudinal versus trasverse gap as a function of κ:

$$
\frac{\Delta_\parallel^2}{\Delta_\perp^2} = \frac{c^2}{a^2} = 1 - 3\kappa^2.
\tag{4.745}
$$

These results agree with the Ginzburg-Landau formulas (4.515) and (4.516), if we insert the appropriate expansion coefficients β_i of the corresponding state.

In the opposite limit of zero temperature the distance between neighboring values x_n goes to zero so that we may replace the sum over x_n in (4.739) by an integral according to the rule

$$
\sum_{x_n} \xrightarrow[T\to 0]{} \frac{2}{\delta}\int dx_n.
\tag{4.746}
$$

As in Chapter 3 and in Eq. (4.293), care is necessary to treat the last sum $\sum_n 1/x_n$ in (4.739), since each term diverges at $T = 0$. As in the case of the superconductor gap equation (3.159), leading to (3.161) and (3.162), the proper replacement is

$$
\frac{2}{\delta}\sum_{n=0}^{\infty}\frac{1}{x_n} \to \int_{1/\delta}^{x_N}\frac{dx}{x} + \log(2e^\gamma).
\tag{4.747}
$$

Therefore we obtain

$$
\begin{aligned}
\log\frac{T}{T_c} &\xrightarrow[T\to 0]{} \mathrm{Re}\int_{-1}^{1}\frac{dz}{2}\left\{\frac{1}{\frac{3}{2}(1 - z^2)}\right\} \\
&\times \left[\int_0^\infty dx\,\frac{1}{\sqrt{(x - i\nu z)^2 + 1 - r^2 z^2}} - \int_{1/\delta}^\infty\frac{1}{x} - \log 2 - \gamma\right]
\end{aligned}
\tag{4.748}
$$

$$
= -\mathrm{Re}\int_{-1}^{1}\frac{dz}{2}\left\{\frac{1}{\frac{3}{2}(1 - z^2)}\right\}\log\left(\sqrt{1 - (\nu^2 + r^2)z^2 - i\epsilon z} - i\nu z\right) - \log\delta - \gamma.
$$

Taking γ and $\log\delta$ to the other side, the $\log T$-divergence cancels and we find

$$
\log\frac{\Delta_\perp(T = 0)}{\Delta_{\mathrm{BCS}}} = -\mathrm{Re}\int_{-1}^{1}\frac{dz}{2}\left\{\frac{1}{\frac{3}{2}(1 - z^2)}\right\}\log\left(\sqrt{1 - (\nu^2 + r^2)z^2 - i\epsilon z} - i\nu z\right),
\tag{4.749}
$$

where we have introduced the $T = 0$ gap of the BCS theory [compare (3.153)]

$$\Delta_{\mathrm{BCS}} = \pi T_c e^{-\gamma} \sim 1.7638\, T_c. \tag{4.750}$$

When calculating the logarithm, we have to be careful to use the correct square root. Taking the branch cut, as usual, to the left, this is specified by the $i\epsilon$ prescription.

As a cross check we see that for $\nu = 0$, $r = 0$ (B-phase) the orthogonal gap becomes $\Delta_\perp = \Delta_{\mathrm{BCS}}$, while for $r = 1$ (A-phase) the lower equation gives

$$\log \frac{\Delta_\perp}{\Delta_{\mathrm{BCS}}} = -\int_{-1}^{1} \frac{dz}{2} \frac{3}{2} \left(1 - z^2\right) \log \sqrt{1 - z^2} = \frac{5}{6} - \log 2, \tag{4.751}$$

such that

$$\Delta_\perp = \Delta_{\mathrm{BCS}} \frac{e^{5/6}}{2} \sim 2.03 T_c. \tag{4.752}$$

While the full solution of the gap equation (4.749) can only be found numerically, we can see directly that at $T = 0$, the gap distortion parameter r vanishes for all $\nu = v p_F / \Delta_\perp \leq 1$, so that $\Delta_\perp = \Delta_{\mathrm{BCS}}$ is a solution of both equations (4.748): in fact, the real part of the logarithm vanishes identically for $r = 0$.

The full T-behavior of the gaps can be found from the average [i.e., $\frac{1}{3}$ (longitudinal $+2$ transverse)] between the two gap equations (4.736) and (4.737). Then there is no dependence on z, and the result is

$$\log \frac{\Delta_\perp(T = 0)}{\Delta_{\mathrm{BCS}}} = -\mathrm{Re} \int_{-1}^{1} \frac{dz}{2} \log \left(\sqrt{1 - \nu^2 z^2} - i\epsilon z - i\nu z \right). \tag{4.753}$$

There exists a real part only for $\nu > 1$, which is

$$-\int_{1/\nu}^{1} dz \log \left(\sqrt{\nu^2 z^2 - 1} + \nu z \right) = -\int_{1/\nu}^{1} dz\, \mathrm{acosh}\, \nu z = \left[z\, \mathrm{acosh}\, \nu z - 1/\nu \sqrt{\nu^2 z^2 - 1} \right]_{1/\nu}^{1}, \tag{4.754}$$

which is why the $T = 0$ -gap without distortion (using a superscript u for "undistorted") is given by[18]

$$\log \frac{\Delta_B^u(T = 0)}{\Delta_{\mathrm{BCS}}} = -\theta(\nu - 1) \left(\mathrm{acosh}\nu - \frac{1}{\nu} \sqrt{\nu^2 - 1} \right). \tag{4.755}$$

The gap remains equal to Δ_{BCS} up to $\nu = 1$. From there on it drops rapidly to zero. The place where it vanishes is found from (4.755) by inserting the limit $\nu = v\, p_F / \Delta_B \to \infty$, which yields in this limit

$$-\log \Delta_{\mathrm{BCS}} = -\log 2 v p_F + 1, \tag{4.756}$$

---

[18] Carrying the Euler-Maclaurin expansion one step further, the right-hand side in (4.755) would have an additional $-\frac{1}{6}\delta^2 \nu \sqrt{\nu^2 - 1}$, which in (4.757) would give a factor $\exp\{-\frac{1}{6}\frac{\pi^2 T^2}{(p_F v)^2}\} \sim 1 - \frac{2}{3}e^{2(\gamma-1)} (T/T_c)^2 \sim 1 - 0.29 (T/T_c)^2$.

or

$$\frac{v p_F}{\Delta_{\text{BCS}}} = \frac{e}{2} \approx 1.359. \tag{4.757}$$

In physical units this amounts to

$$\frac{v}{v_0} = v\, 2m\xi_0 = v\frac{2p_F}{\pi T_c}\sqrt{\frac{7\zeta(3)}{48}} = \frac{v p_F}{\pi e^{-\gamma} T_c} 2e^{-\gamma}\sqrt{\frac{7\zeta(3)}{48}}$$

$$= \frac{p_F v}{\Delta_{\text{BCS}}} 0.47 \approx 0.64. \tag{4.758}$$

For comparison we see that in the A-phase

$$\log \frac{\Delta_\perp(T=0)}{\Delta_{\text{BCS}}} = -\text{Re} \int_{-1}^{1} \frac{dz}{2}\frac{3}{2}\left(1-z^2\right)\log\left(\sqrt{1-z^2-\nu^2 z^2} - i\epsilon z - i\nu z\right)$$

$$= -\int_{-1}^{1} \frac{dz}{2}\frac{3}{2}\left(1-z^2\right)\log\sqrt{1-z^2}$$

$$-2\int_{1/\sqrt{1+\nu^2}}^{1} \frac{dz}{2}\frac{3}{2}\left(1-z^2\right)\log\frac{\sqrt{(1+\nu^2)z^2-1}+\nu z}{\sqrt{1-z^2}}, \tag{4.759}$$

implying that[19]

$$\log \frac{\Delta_\perp(T=0)}{\Delta_{\text{BCS}} e^{5/6}/2} = -\int_{1/\sqrt{1+\nu^2}}^{1} \frac{3}{2}\left(1-z^2\right)\text{acosh}\frac{\nu z}{\sqrt{1-z^2}}$$

$$= \frac{1}{2}\frac{\nu^2}{1+\nu^2} - \frac{1}{2}\log\left(1+\nu^2\right). \tag{4.760}$$

Here the gap decreases smoothly and hits zero at

$$\frac{v p_F}{\Delta_{\text{BCS}}} = \frac{e^{5/6}}{2}\sqrt{e} \approx 1.897, \tag{4.761}$$

or

$$\frac{v}{v_0} \approx 0.892. \tag{4.762}$$

The full solution of the gap equations are shown in Fig. 4.41. For comparison we also have displayed the solutions in the B-phase and the A-phase neglecting the gap distortion [i.e., using $r \equiv 0$ or $\Delta_\perp \equiv \Delta_\parallel$, and either of the two equations (4.729), (4.730)].

All curves as functions of v are double-valued. It will be seen later that this behavior is an artifact of neglecting Fermi liquid corrections. Once included, these will turn the lower branches anticlockwise into the region of higher velocities while

---

[19]To next order in $1/\delta$, the Euler-Maclaurin expansion gives $-\nu^2/\left(1+\nu^2\right)^2 2\delta^2$ which enters in (4.761) as a factor $e^{-\pi^2 T^2/2(p_F \nu)^2} \approx 1 - 2e^{8/3+2\gamma}\left(T/T_c\right)^2 \approx 1 - 0.44\left(T/T_c\right)^2$.

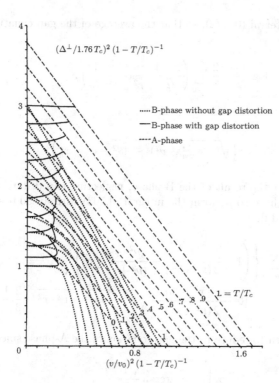

FIGURE 4.41 Velocity dependence of the gap in the A- and B-phases.

distorting only slightly the upper branches at lower velocities. In this way the curves become single-valued.

For numerical calculation we have used formulas (4.736)–(4.738), after having performed the integrals over dz analytically: The average gap equation requires the integral

$$\mathrm{Re}\int_{-1}^{1} \frac{dz}{2} \frac{1}{\sqrt{(x - i\nu z)^2 + 1 - r^2 z^2}} = \mathrm{Re}\int_{-1}^{1} \frac{dz}{2} \frac{1}{\sqrt{1 + x^2 - 2i\nu x z - (r^2 + \nu^2) z^2}}. \quad (4.763)$$

The square root has to be taken with a positive real part, i.e., with the standard choice of the branch cut running to the left from zero to $-i\infty$. The result is

$$\frac{\alpha_n}{\sqrt{\nu^2 + r^2}} \equiv \frac{1}{\sqrt{\nu^2 + r^2}} \arcsin \frac{\nu^2 + r^2 + i\nu x_n}{\sqrt{\nu^2 + r^2 + r^2 + x_n^2}}, \quad (4.764)$$

or

$$\alpha_n = \frac{1}{2} \arccos \left(\left\{ \left(\nu^2 + r^2\right) \sqrt{(1 + x_n^2 - \nu^2 - r^2)^2 + 4\nu^2 x_n^2} - \left(\nu^2 + r^2\right)^2 - \nu^2 x_n^2 \right\} \right), \quad (4.765)$$

which lies in the interval $(0, \pi/2)$, so that the average of the gap equations (4.748) becomes

$$\log \frac{T}{T_c} = \frac{2}{8} \sum_{n=0}^{\infty} \left(\frac{1}{\sqrt{\nu^2 + r^2}} \alpha_n - \frac{1}{x_n} \right). \tag{4.766}$$

For $r = 0$, we have

$$\alpha_n = \frac{1}{2} \cos \left(\sqrt{(1 + x_n^2 + \nu^2)^2 - 4\nu^2} - \nu^2 - x_n^2 \right), \tag{4.767}$$

and thus we recover the result of the B-phase, neglecting gap distortion. For the transverse gap, we have to perform the integral (4.763) with an additional weight factor $\frac{3}{2}(1 - z^2)$ and find

$$
\log \frac{T}{T_c} = \frac{3}{2} \frac{2}{\delta} \sum_{n=0}^{\infty} \left\{ \left(1 - \frac{1}{2(\nu^2 + r^2)} + \frac{\nu^2 - \frac{r^2}{2}}{(\nu^2 + r^2)^2} x_n^2 \right) \frac{\alpha_n}{\sqrt{\nu^2 + r^2}} \right.
$$
$$
\left. + \mathrm{Re} \frac{\nu^2 + r^2 - 3i\nu x_n}{2(\nu^2 + r^2)^2} \sqrt{(x_n - i\nu)^2 + 1 - r^2} - \frac{2}{3} \frac{1}{x_n} \right\}. \tag{4.768}
$$

At $r = 1$, this is seen to reduce to the gap equation of the A-phase since

$$\alpha_n \Big|_{r=1} = \arctan \frac{\sqrt{1 + \nu^2}}{x_n}, \tag{4.769}$$

and

$$\mathrm{Re} \frac{\nu^2 + 1}{2(\nu^2 + 1)^2} \frac{3i\nu x_n}{}(x_n - i\nu) = \frac{1 - 2\nu^2}{2(\nu^2 + 1)}. \tag{4.770}$$

The second term in the sum may be evaluated explicitly by defining

$$\gamma_n \equiv \arctan \frac{3\nu x_n}{\nu^2 + r^2} \in (0, \pi/2),$$
$$\beta_n \equiv \arctan \frac{2\nu x_n}{1 + x_n^2 - \nu^2 - r^2} \in (0, \pi), \tag{4.771}$$

which brings it to the form

$$
\frac{\sqrt{(\nu^2 + r^2)^2 + 9\nu^2 x_n^2}}{2(\nu^2 + r^2)^2} \left[\left(\nu^2 + r^2 - 1 - x_n^2 \right)^2 + 4\nu^2 x_n^2 \right]^{\frac{1}{4}} \cos \left(\gamma_n + \frac{\beta_n}{2} \right)
$$
$$
= \frac{\sqrt{2\nu x_n}}{2(\nu^2 + r^2)} \frac{1}{\sqrt{\sin \beta_n}} \frac{1}{\cos \gamma_n} \cos \left(\gamma_n + \frac{\beta_n}{2} \right). \tag{4.772}
$$

4.12.3 Superfluid Densities and Currents

By construction, the current density in the presence of the external source (4.717) is

$$-\frac{\partial g^v}{\partial v}. \tag{4.773}$$

This, however, is not the full current density of the system. In calculating g^v we have assumed the field A_{ai}^0 to be a constant in space (and time). In this way the Cooper pairs have been forced artificially to remain immobile. In full thermal equilibrium also these follow the drag of the external source, and in the ground state the gap A_{ai}^0 acquires a phase modulation

$$e^{2im\mathbf{v}\mathbf{x}} A_{ai}^0. \tag{4.774}$$

The gradient of this order parameter accounts for the flow of the condensate, i.e., the ^{3}He quasiparticles bound in Cooper pairs.

If calulated at an unmodulated gap A_{ai}^0, the gradient (4.773) accounts only for the movement of the quasiparticles which are *not* bound in Cooper pairs. These make up the normal component of the superfluid. If the current associated with this flow is written with a subscript n it reads:

$$J_n \equiv -\left.\frac{\partial g^v}{\partial v}\right|_{A_{ai}^0} = \text{const.} \tag{4.775}$$

From this we deduce the mass density of the normal component via the relation

$$J_n \equiv \rho_n v. \tag{4.776}$$

Since the full current would be

$$J = \rho v, \tag{4.777}$$

we may attribute the difference entirely to the flow of Cooper pairs and write

$$J_s \equiv J - J_n = (\rho - \rho_n)\, v = \rho_s v. \tag{4.778}$$

This quantity defines the superfluid density ρ_s from the density of flowing pairs in the liquid. Since $J = \rho v$ is obtained from $-\partial g_0^v / \partial v$, J_s is obtained directly from the derivative of the condensation energy

$$J_s = \frac{\partial g_c^v}{\partial v} \equiv \frac{\partial g^v}{\partial v} - \frac{\partial g_0^v}{\partial v} = \frac{\partial g^v}{\partial v} - (\Delta = 0)\,. \tag{4.779}$$

Using (4.722) we perform the differentiation and find

$$
\begin{aligned}
J_s &= -T \sum_{\omega_n, \mathbf{p}} p_F z \left(\frac{1}{i\omega_n + v p_F - E} + (E \to -E) \right) - (\Delta = 0) \\
&= \frac{3}{2} \int_{-1}^{1} \frac{dz}{2} 2\mathcal{N}(0) \int_{-\infty}^{\infty} d\xi \left\{ \left[\tanh \frac{E - v p_F z}{2T} - (v \to -v) \right] - [\Delta = 0] \right\},
\end{aligned} \tag{4.780}
$$

where we have replaced the momentum sum as in (4.732), and added a factor 2 for the two spin orientations. For numerical evaluations it is more convenient to keep the Matsubara sum (4.780), but perform analytically the integration over ξ. Then we find

$$
\begin{aligned}
J_s &= \frac{3\rho}{p_F} \int_{-1}^{1} \frac{dz}{2} \int_{-\infty}^{\infty} d\xi\, T \sum_{\omega_n} \frac{i\omega_n + v p_F z}{(\omega_n - i v p_F z)^2 + \Delta_\perp^2 (1 - r^2 z^2) + \xi^2} \\
&= \frac{3\rho}{p_F} \pi T \int_{-1}^{1} \frac{dz}{2} z \operatorname{Re} \sum_{\omega_n} \frac{i\omega_n + v p_F z}{\sqrt{(\omega_n - v p_F z)^2 + \Delta_\perp^2 (1 - r^2 z^2)}} \\
&= \left[\frac{6}{\delta \nu} \int_{-1}^{1} \frac{dz}{2} z \operatorname{Re} \sum_{n=0}^{\infty} \frac{i x_n + \nu z}{\sqrt{(x_n - i\nu z)^2 + 1 - r^2 z^2}} \right] \rho v = \rho_s^\parallel v,
\end{aligned}
\tag{4.781}
$$

where we have inserted $2\mathcal{N}(0) = \frac{3}{2}\rho/p_F^2$ from (4.16). Thus we can identify the superfluid density parallel to the flow as:

$$
\begin{aligned}
\frac{\rho_s^\parallel}{\rho} &= \frac{3}{2} \frac{1}{p_F v} \int_{-1}^{1} \frac{dz}{2} \int_{-\infty}^{\infty} d\xi \left\{ \left[\tanh \frac{E - v p_F z}{2T} - (v \to -v) \right] - [\Delta = 0] \right\} \\
&= \frac{6}{\delta \nu} \int_{-1}^{1} \frac{dz}{2} z \operatorname{Re} \sum_{n=0}^{\infty} \frac{i x_n + \nu z}{\sqrt{(x_n - i\nu)^2 + 1 - r^2 z^2}}.
\end{aligned}
\tag{4.782}
$$

The integral over z yields

$$
\begin{aligned}
\frac{\rho_s^\parallel}{\rho} &= \frac{3\rho}{\delta \nu} \sum_{n=0}^{\infty} \left\{ \frac{\nu}{(\nu^2 + r^2)^2} \left(\nu^2 + r^2 + 3 r^2 x_n^2 \right) \frac{\alpha_n}{\sqrt{\nu^2 + r^2}} \right. \\
&\quad \left. - \operatorname{Re} \frac{2}{(\nu^2 + r^2)} \left[i x_n + \frac{1}{2}\nu \left(1 - 3i \frac{\nu x_n}{\nu^2 + r^2} \right) \right] \sqrt{(x_n - i\nu)^2 + 1 - r^2} \right\}.
\end{aligned}
\tag{4.783}
$$

If we neglect gap distortion by setting $r = 1$, we recover the result of previous calculations

$$
\frac{\rho_s^\parallel}{\rho} \overset{r=0}{\equiv} \frac{3}{\delta \nu^3} \sum_{n=0}^{\infty} \left\{ \alpha_n - \frac{1}{\sqrt{2}} \left[\left(\nu^2 - 9 x_n^2 \right) \left(1 - \nu^2 + x_n^2 \right) - 12 \nu^2 x_n^2 \right. \right.
$$
$$
\left. \left. - 9 x_n^2 \nu^2 \sqrt{(1 + \nu^2 + x_n^2)^2 - 4\nu^2} \right] \right\}.
\tag{4.784}
$$

The last term becomes simply $-3 x_n \nu / (\nu^2 + 1)^2$, so that $\rho_s^\parallel$ at $r = 1$ reduces to the expression for the A-phase:

$$
\frac{\rho_s^\parallel}{\rho} \overset{\text{A-phase}}{\equiv} \frac{3}{\delta} \frac{1}{(\nu^2 + 1)^{\frac{3}{2}}} \sum_{n=0}^{\infty} \left[\left(\nu^2 + 1 + 3 x_n^2 \right) \arctan \frac{\sqrt{\nu^2 + 1}}{x_n} - 3\sqrt{\nu^2 + 1}\, x_n \right].
\tag{4.785}
$$

For general r, the real part in (4.783) consists of two terms, of which the second coincides with -2ν times the corresponding term in the transverse gap equation. The first may be rewritten in terms of the angle β of (4.771) as

$$
\begin{aligned}
-\operatorname{Re} \frac{2 i x_n}{\nu^2 + r^2} \sqrt{(x_n - i\nu)^2 + 1 - r^2} &= -\operatorname{Re} \frac{2 i x_n}{\nu^2 + r^2} \left[\left(1 + x_n^2 - \nu^2 - r^2 \right)^2 + 4\nu^2 x_n^2 \right]^{\frac{1}{4}} e^{-i\beta_n/2} \\
&= -\frac{2 x_n}{\nu^2 + r^2} \sqrt{\nu x_n \tan \beta_n / 2}.
\end{aligned}
\tag{4.786}
$$

Let us compare the result with our Ginzburg-Landau calculation in Section 4.10.2. For $T \sim T_c$, we may take the limit $x_n \to \infty$ and remain with

$$\frac{\rho_s^{\parallel}}{\rho} \approx \frac{6}{\delta} \int_{-1}^{1} \frac{dz}{2} z^2 \left(1 - r^2 z^2\right) \sum_{n=0}^{\infty} \frac{1}{x_n^3} = 6\delta^2 \left(\frac{1}{3} - \frac{r^2}{5}\right) \frac{7\zeta(3)}{8}. \qquad (4.787)$$

This coincides with our previous result if we insert (4.744):

$$\frac{\rho_s^{\parallel}}{\rho} = \frac{3}{\nu} \int_{-1}^{1} \frac{dz}{2} \int_{0}^{\infty} dx \frac{ix + \nu z}{\sqrt{(x - i\nu z)^2 + 1 - r^2 z^2}}$$

$$= \frac{3}{\nu} \int_{-1}^{1} \frac{dz}{2} z \left[\nu z - \mathrm{Re}\, i\sqrt{1 - (\nu^2 + r^2) z^2} - i\epsilon\nu z\right]. \qquad (4.788)$$

The square root gives a contribution only for $z^2 > 1/\nu^2 + r^2$, implying that $\rho_s^{\parallel}$ remains equal to ρ until $\nu^2 = 1 - r^2$.

Since the upper branch of the gap is isotropic up to $\nu = 1$, there is also an upper branch with $\rho_s^{\parallel} \equiv \rho$ up to $\nu = 1$. On the lower branch one has $\nu^2 > 1 - r^2$ and

$$\left.\frac{\rho_s^{\parallel}}{\rho}\right|_{T=0} = 1 - \frac{3}{\nu} \mathrm{Re} \int_{1/(\nu^2 + r^2)}^{1} dz z \sqrt{(\nu^2 + r^2) z^2 - 1}$$

$$= 1 - \Theta\left(\nu^2 + r^2 - 1\right) \frac{1}{\nu^2 (\nu^2 + r^2)} \sqrt{\nu^2 + r^2 - 1}^3, \qquad (4.789)$$

where $\Theta(z)$ is the Heaviside function. This result agrees with those of B- and A-phases for $r = 0$ and 1, respectively.

4.12.4 Critical Currents

In the Ginzburg-Landau regime, the critical currents are known from Section 4.10. These results agree with the present calculation since $\rho_s^{\parallel}$ of (4.788) is the same as before. In the opposite limit $T \to 0$, an exact calculation is difficult but the current can be fixed to a high accuracy by the following consideration:

Due to the distortion of the gap, the current J_s as a function of v must be below the current calculated by neglecting distortion. Now, up to $\nu = 1$, both currents are identical since the gap distortion was derived to be zero for $\nu \le 1$. Thus $J_s(v)$ is known to reach the value

$$J_s(v)\big|_{\nu=1} = \rho v_{\nu=1} = \rho \frac{\Delta_{\nu=1}}{p_F}. \qquad (4.790)$$

Since $\Delta_{\nu=1} = \Delta_{\mathrm{BCS}}$ also at $T = 0$ [see the text after (4.752)], we have the lower bound on the critical current

$$J_s(v) \ge \rho \frac{\Delta_{\mathrm{BCS}}}{p_F} \approx 0.47 J_0. \qquad (4.791)$$

As an upper bound we may use the maximum of $J_s^{Bu}(v)$ which can easily be calculated exactly. We shall see in a moment that the critical velocity is determined by

$$\nu_c = \frac{1}{\sqrt{1 - \left(2^{1/3} - 1\right)^2}} \approx 1.036. \tag{4.792}$$

Inserting this into the superfluid density (4.789) at $r = 0$, we find

$$\left.\frac{\rho_s^{Bu}}{\rho}\right|_{\nu_c} = 1 - \left(2^{1/3} - 1\right)^3 \approx 0.982. \tag{4.793}$$

This leads to a critical current

$$J_c^{Bu} = \rho \left[1 - \left(2^{1/3} - 1\right)^3\right] \frac{1}{\sqrt{1 - \left(2^{1/3} - 1\right)^2}} \frac{\Delta^{Bu}|_{\nu_c}}{p_F}. \tag{4.794}$$

But the gap at ν_c can be evaluated from (4.755), with the result

$$\begin{aligned}
\log\left.\frac{\Delta_B^{\parallel}}{\Delta_{\mathrm{BCS}}}\right|_{\nu_c} &= -\log\left(\nu_c + \sqrt{\nu_c^2 - 1}\right) + \sqrt{1 - \frac{1}{\nu_c^2}} \\
&= \log \nu_c + \log\left[1 - \left(2^{1/3} - 1\right)\right] + \left(2^{1/3} - 1\right),
\end{aligned} \tag{4.795}$$

so that

$$\left.\Delta_B\right|_{\nu_c\, T=0} = \Delta_{\mathrm{BCS}}\, e^{2^{1/3}-1} \left[1 - \left(2^{1/3} - 1\right)\right] \nu_c. \tag{4.796}$$

Thus we find, altogether, a critical current

$$\begin{aligned}
J_c^{Bu} &= \left[1 - \left(2^{1/3} - 1\right)^3\right] 2^{-1/3} e^{\left(2^{1/3}-1\right)} \frac{\Delta_{\mathrm{BCS}}}{p_F} \rho \\
&\approx 1.0112 \frac{\Delta_{\mathrm{BCS}}}{p_F} \rho \approx 0.486 J_0.
\end{aligned} \tag{4.797}$$

This lies only slightly above the lower bound. Therefore the true critical current, including the effect of gap distortion, is determined extremely well by the upper and lower bounds

$$0.470 J_0 \le J_c^B \le 0.486 J_0. \tag{4.798}$$

Note that the critical velocity in the B-phase is

$$\begin{aligned}
v_c &= \nu_c \frac{\Delta}{\Delta_{\mathrm{BCS}}} \frac{\Delta_{\mathrm{BCS}}}{p_F} = 2^{1/3} e^{\left(2^{1/3}-1\right)} \frac{\Delta_{\mathrm{BCS}}}{p_F} \\
&\approx 1.029 \frac{\Delta_{\mathrm{BCS}}}{p_F} \approx 0.48\, v_0,
\end{aligned} \tag{4.799}$$

i.e., it is reached immediately above $\nu = 1$. Thus the critical velocity v_c lies between $0.47 v_0$ and $0.48 v_0$.

Let us now derive (4.792). Certainly, the maximum of the current is determined by

$$\frac{d}{dv} J_s = \frac{d}{dv} \rho_s(v) + \rho_s(v) = 0. \tag{4.800}$$

In general, ρ_s is a function of ν, δ, T, where ρ is itself a function of ν and T via gap equation:

$$\log \frac{T}{T_c} = \gamma(\delta, \nu). \tag{4.801}$$

We can therefore express the derivative at fixed T as

$$\frac{\partial}{\partial v} = \frac{\partial \nu}{\partial v} \left(\frac{\partial \delta}{\partial \nu} \frac{\partial}{\partial \delta} + \frac{\partial}{\partial \nu} \right). \tag{4.802}$$

But since $\nu = v p_F / \Delta_\perp$ we have

$$\frac{\partial \nu}{\partial v} = \frac{p_F}{\pi T} \left(\frac{1}{\delta} - \frac{1}{\delta^2} \frac{\partial \nu}{\partial v} \frac{\partial \delta}{\partial \nu} \right), \tag{4.803}$$

or

$$\frac{\partial \nu}{\partial v} = \frac{\nu}{v} \frac{1}{1 + \frac{v}{\delta} \frac{\partial \delta}{\partial v}}. \tag{4.804}$$

In this way, the extremal condition (4.802) may be written in terms of the natural variables as

$$\left(\frac{\partial \delta}{\partial \nu} \frac{\partial}{\partial \delta} + \frac{\partial}{\partial \nu} \right) \rho_s + \left(\frac{1}{\delta} \frac{\partial \delta}{\partial \nu} + \frac{1}{\nu} \right) \rho_s = 0. \tag{4.805}$$

The derivative $\partial \delta / \partial \nu$ may be taken from (4.804) as $-\left(\partial \gamma / \partial \nu \right) / \left(\partial \gamma / \partial \delta \right)$. Let us evaluate condition (4.805) at zero temperature. Then ρ_s becomes independent of δ, and the first term in (4.805) is absent. The gap equation (4.804), on the other hand, has the form

$$\log \frac{\Delta}{\Delta_{BCS}} = \gamma_0(\nu), \tag{4.806}$$

so that

$$\frac{1}{\delta} \frac{\partial \delta}{\partial \nu} = \frac{\partial \gamma_0}{\partial \nu}. \tag{4.807}$$

The critical velocity at $T = 0$ is therefore obtained from the simple relation

$$\left[\nu \frac{\partial \rho_s}{\partial \nu} + \left(\nu \frac{\partial \gamma_\nu}{\partial \nu} + 1 \right) \rho \right]_{\nu_c} = 0. \tag{4.808}$$

For the B-phase, neglecting gap distortion, we see from (4.789)

$$\frac{\partial \rho_s^{Bu}}{\partial \nu} = -\theta(\nu - 1)\frac{3}{\nu^4}\sqrt{\nu^2 - 1}. \tag{4.809}$$

The gap function γ_0 is taken from (4.755), so that

$$\frac{\partial \gamma_0}{\partial \nu} = -\frac{\sqrt{\nu^2 - 1}}{\nu^2}\theta(\nu - 1). \tag{4.810}$$

The condition (4.808) becomes

$$-3\nu\frac{\sqrt{\nu^2 - 1}}{\nu^4} + \left(1 - \frac{\sqrt{\nu^2 - 1}}{\nu}\right)\left(1 - \frac{\sqrt{\nu^2 - 1}^3}{\nu^3}\right) = 0. \tag{4.811}$$

This awkward equation is solved by setting $y \equiv \sqrt{1 - \frac{1}{\nu^2}}$ and rewriting

$$y^3 + 3y^2 + 3y - 1 = 0, \tag{4.812}$$

which has the only real solution

$$y = 2^{1/3} - 1, \tag{4.813}$$

which verifies the critical current of the previous discussion (4.792)–(4.799).

For comparison we may use (4.808) to derive also the depairing critical current for the A-phase. From (4.789) with $r = 1$ and (4.760) we see that

$$\frac{\rho_s^{\parallel}}{\rho} = \frac{1}{1 + \nu^2}, \tag{4.814}$$

$$\log\frac{\Delta_\perp}{\Delta_{BCS}} = \gamma_0(v) = \log\frac{e^{5/6}}{2} - \frac{1}{2}\log\left(1 + \nu^2\right) + \frac{\nu^2}{2\left(1 + \nu^2\right)}, \tag{4.815}$$

which is inserted into (4.808) to give

$$-\frac{2\nu^2}{(1 + \nu^2)^2} + \frac{1}{(1 + \nu^2)}\left(1 - \frac{\nu^4}{(1 + \nu^2)^2}\right) = 0. \tag{4.816}$$

This is solved by $\nu_c^2 = 1/\sqrt{2}$. From this we obtain the critical current

$$\begin{aligned}
J_c^A &= \frac{\rho_s^{\parallel}}{\rho}\nu_c\frac{\Delta_\perp}{\Delta_{BCS}}\frac{\Delta_{BCS}}{p_F}\rho \\
&= \left(\frac{\sqrt{2}}{\sqrt{2} + 1}\right)\frac{1}{2^{1/4}}\left(\frac{e^{5/6}}{2}\sqrt{\left(\sqrt{2} - 1\right)\sqrt{2}e^{\frac{\sqrt{2}-1}{2}}}\right)\frac{\Delta_{BCS}}{p_F}\rho \\
&= \sqrt{2}\left(\sqrt{2} - 1\right)^{3/2}e^{\frac{\sqrt{2}-1}{2}}\frac{e^{5/6}}{2}\frac{\Delta_{BCS}}{p_F}\rho \\
&\approx 0.534\frac{\Delta_{BCS}}{p_F}\rho \approx 0.25J_0,
\end{aligned} \tag{4.817}$$

implying a critical velocity

$$v_c = \frac{J_c}{\rho_s^\parallel} = \sqrt{\sqrt{2} - 1} e^{\frac{\sqrt{2}-1}{2}} \frac{e^{5/6}}{2} \frac{\Delta_{\mathrm{BCS}}}{p_F}$$

$$\approx 0.911 \frac{\Delta_{\mathrm{BCS}}}{p_F} \approx 0.428 v_0. \tag{4.818}$$

4.12.5 Ground State Energy at Large Velocities

Let us now consider the superfluid in motion. As before, we imagine bringing the liquid adiabatically from $v = 0$ to its actual velocity. This will result in an additional energy

$$g_c^{(v)} = g_c|_{v=0} + \int_0^v dv' \rho_s^\parallel(v') v', \tag{4.819}$$

where $g|_{v=0}$ is the previously calculated condensation energy f_c, and $\rho_s^\parallel$ is the superfluid density parallel to the flow. Alternatively, we may take the energy of the freely moving fermions

$$g_0 = f_0 - \frac{\rho}{2} v^2, \tag{4.820}$$

and form the total energy as a combination

$$g = f_0 + f_c - \int_0^v dv' \rho_n^\parallel(v') v', \tag{4.821}$$

where

$$\rho_n^\parallel = \rho - \rho_s^\parallel \tag{4.822}$$

is the density of the normal component of the liquid.

4.12.6 Fermi Liquid Corrections

With (4.821), the expression for the energy reaches a convenient form which permits the inclusion of the quantitatively very important Fermi liquid corrections due to the current-current coupling (4.379). In Eq. (4.393) we have seen that the associated molecular fields φ_i enter the collective action on the same footing as the velocity v of the liquid.

In equilibrium we expect a constant nonzero mean molecular field. For symmetry reasons, only the field parallel to the flow can contribute. Therefore we may substitute simply $\mathbf{v} \to \mathbf{v} + \varphi$ in (4.821) and add, after this, the quadratic term as in (4.392). Then the energy, corrected by the constant mean molecular field, may be written as

$$g^* = \min_\varphi \left[f_0 + f_c - \int_0^{\mathbf{v}+\varphi} dv' \rho_n^\parallel(v') v' - \frac{1}{2} \rho \frac{1}{F_1/3} \varphi^2 \right]. \tag{4.823}$$

Differentiating with respect to φ we see that the minimum lies at the mean field

$$\varphi = -\frac{F_1}{3}\frac{\rho_n(v+\varphi)}{\rho}(\mathbf{v}+\varphi). \tag{4.824}$$

Inserting this back into (4.823), the energy becomes an explicit function of the quantity

$$v^* \equiv v + \varphi. \tag{4.825}$$

This may be interpreted as the local fluid velocity felt by the quasi-particles including the effects of the molecular field. In terms of v^* it reads:

$$\begin{aligned}
g^* &= f_0 + f_c - \int_0^{v^*} dv'\rho_n^\parallel(v')v' - \frac{1}{2}\frac{F_1/3}{\rho}\rho_n^{\parallel\,2}(v^*)v^{*2} \\
&= f_0 + f_c - \int_0^v dv' J_n(v') - \frac{1}{2}\frac{F_1/3}{\rho}J_n^2(v^*). \tag{4.826}
\end{aligned}$$

Given an arbitrary physical velocity v, the quantity v^* may be found from Eq. (4.824), which can be rewritten in the form

$$\left(1 + \frac{F_1}{3}\frac{\rho_n(v^*)}{\rho}\right)v^* = v. \tag{4.827}$$

Expression (4.826) allows a calculation of the Fermi-liquid-corrected supercurrent and superfluid density. By differentiation with respect to v we find

$$\begin{aligned}
J_n^*(v) &= -\frac{\partial g^*}{\partial v} = J_n(v^*)\frac{\partial v^*}{\partial v} + \frac{F_1/3}{\rho}J_n(v^*)\frac{\partial J_n(v^*)}{\partial v} \\
&= J_n(v^*)\left\{\left[1 + \frac{F_1/3}{\rho}\frac{\partial J_n}{\partial v^*}\right]\frac{\partial v^*}{\partial v}\right\}. \tag{4.828}
\end{aligned}$$

By writing (4.827) in the form

$$v^* + \frac{F_1/3}{\rho}J_n(v^*) = v, \tag{4.829}$$

we see that the factor in curly brackets of (4.828) is unity. Hence the Fermi-liquid-corrected current equals the uncorrected one, except for its evaluation at the local velocity v^* rather than the physical v:

$$J_n^*(v) \equiv \rho_n^*(v)v = J_n(v^*) = \rho_n(v^*)v^*. \tag{4.830}$$

As in Section 4.9.4 we have found it convenient to introduce ρ_n^* as the corrected density of the normal component

$$\rho_n^*(v) \equiv \rho_n(v^*)\frac{v^*}{v} = \frac{\rho_n(v^*)}{1 + \frac{F_1}{3}\frac{\rho_n}{\rho}(v^*)}, \tag{4.831}$$

which is reduced with respect to ρ_n by the ratio of v^* and v. The same reduction appears in the superfluid density. Here we have to subtract the normal current from the total one, ρv. In order to do so we have to remember that ρ contains the true mass of the ^{3}He atoms $m = m_{3\text{He}}$, while all quantities derived from the original action involve the effective mass $m^* = (1 + F_1/3)\, m$.

Therefore the supercurrent is given by

$$
\begin{aligned}
J_s^*(v) &= \rho v - J_n^*(v) \\
&= \rho v - \rho_n\left(v^*\right) v^* \\
&= \rho v - \rho \frac{m^*}{m} \frac{\rho_n(v^*)}{\rho} \frac{v}{1 + \frac{F_1}{3}\frac{\rho_n(v^*)}{\rho}},
\end{aligned}
\tag{4.832}
$$

which can further be brought to the form:

$$
\begin{aligned}
J_s^*(v) &= \rho v - \left(1 + \frac{F_1}{3}\right)\frac{\rho_n(v^*)}{\rho}\frac{1}{1 + \frac{F_1}{3}\frac{\rho_n(v^*)}{\rho}}\frac{1}{1 + \frac{F_1}{3}\frac{\rho_n(v^*)}{\rho}} \\
&= \rho v \frac{\rho_s\left(v^*\right)}{\rho}\frac{1}{1 + \frac{F_1}{3}\frac{\rho_n(v^*)}{\rho}}.
\end{aligned}
\tag{4.833}
$$

The effect of Fermi-liquid corrections is to reduce the superfluid fraction ρ_s/ρ by a factor $1/[1 + \frac{F_1}{3}\frac{\rho_n(v^*)}{\rho}]$, and to change the velocity coordinate from v to v^*:

$$
\frac{\rho_s^{\text{FL}}(v)}{\rho} = \frac{\rho_s\left(v^*\right)}{\rho}\frac{1}{1 + \frac{F_1}{3}\frac{\rho_n(v^*)}{\rho}}.
\tag{4.834}
$$

Note that for small velocities the integral in (4.826) may be performed, which brings g^* to the simple form

$$
\begin{aligned}
g^* &= f_0 + f_c - \frac{\rho}{2}v^2 + \frac{1}{2}\rho\frac{\rho_s\left(v^*\right)}{\rho}\frac{1}{1 + \frac{F_1}{3}\frac{\rho_n(v^*)}{\rho}}v^2 \\
&= f_0 + f_c - \frac{\rho_n^*\left(v\right)}{2}v^2.
\end{aligned}
\tag{4.835}
$$

As far as our Figure 4.41 is concerned we learn that in Δ, $\Delta^{\parallel}$, $\Delta^{\perp}$, the curves remain the same except that the v axis has to be read as v^*. The same statement holds for $\rho^{\parallel}$ and $\rho^{\perp}$ which are, in addition, reduced by the factors $1/[1 + \frac{F_1}{3}\frac{\rho_n(v^*)}{\rho}]$.

Sometimes in an experiment, the velocity v is given, rather than the current. The corresponding quantity v^* is easily extracted graphically from a plot of J_s versus v^* (see Fig. 4.42). By rewriting Eq. (4.833) as

$$
\frac{J_s\left(v^*\right)}{\rho} = \frac{1 + \frac{F_1}{3}v^*}{F_1/3} - \frac{v}{F_1/3}
\tag{4.836}
$$

we see that for any value of $F_1/3$, we may draw a straight line of unit slope and intercept $(1 - v)/(F_1/3)$. It intercepts the curves of J_s/ρ at v^*. The same statement

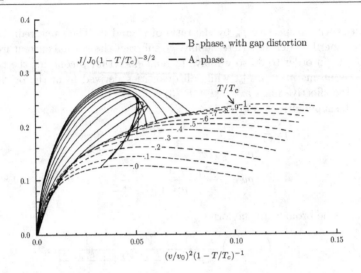

FIGURE 4.42 Current as a function of velocity.

holds for the reduced quantities $(J/J_0)(1 - T/T_c)^{-3/2}$, $(v/v_0)(1 - T/T_c)^{-\frac{1}{2}}$, except that the result carries a factor $1/[2(1 - T/T_c)]$.

The Fermi liquid corrections have the pleasant property of removing the double valuedness of the variables when plotted as a function of v rather than v^*. The reason is that the lower branch of $\rho_s^\parallel(v^*)$ corresponds, via (4.829), to a higher physical velocity v at the same v^*. This has the effect of rotating all lower branches with positive slope anticlockwise, until their slopes are negative. In this all curves become single-valued even at zero pressure where F_1 takes its smallest value where $F_1/3 \approx 2$.

4.13   Collective Modes in the Presence of Current at all Temperatures $T \leq T_c$

In Subsection 4.10 we have seen that, in the neighborhood of the critical temperature, the distorted gap parameter (4.725) is stable under small space- and time-independent fluctuations. Here we want to extend this consideration to all temperatures below T_c. For simplicity we shall only consider the weak-coupling limit in which the hydrodynamic properties were discussed (recall Section 4.9).

4.13.1   Quadratic Fluctuations

As in Section 4.9. we parametrize the fluctuations around the extremal field configuration A_{ai}^0 by

$$A_{ai}' = A_{ai} - A_{ai}^0. \tag{4.837}$$

Inserting this into the collective action (4.221) and expanding in powers of A'_{ai} up to quadratic order, we find

$$\delta^2 \mathcal{A}^v = -\frac{i}{4}\text{Tr}\left[G^v \begin{pmatrix} 0 & A'_{ai}\sigma_a i\tilde{\nabla}_i \\ A'^*_{ai}\sigma_a i\tilde{\nabla}_i & 0 \end{pmatrix} G^v \begin{pmatrix} 0 & A'_{ai}\sigma_a i\tilde{\nabla}_i \\ A'^*_{ai}\sigma_a i\tilde{\nabla}_i & 0 \end{pmatrix}\right]$$
$$-\frac{1}{3}\int d^4x |A'_{ai}|^2, \tag{4.838}$$

where G^v is the generalization of the 4×4 matrix (4.222):

$$G^v = i \begin{pmatrix} i\partial_t + \mathbf{v}\boldsymbol{\nabla} - \xi & A^0_{ai}\sigma_a\tilde{\nabla}_i \\ A^0_{ai}{}^*\sigma_a\tilde{\nabla}_i & i\partial_t + \mathbf{v}\cdot\boldsymbol{\nabla} + \xi \end{pmatrix}^{-1}. \tag{4.839}$$

This contains the source term (4.717) to guarantee the velocity $\mathbf{v}$ of the fluid. as in the collective action (4.719). Assuming that A^0_{ai} is extremal, i.e., that it satisfies gap equations like (4.729), (4.730), there are no linear terms in A'_{ai}.

The explicit form of the matrix G^v in energy-momentum space reads

$$G^v(\epsilon, \mathbf{p}) = \frac{i}{-(\epsilon + \mathbf{p}v)^2 + E^2} \begin{pmatrix} \epsilon + \mathbf{v}\mathbf{p} - \xi(\mathbf{p}) & -A^0_{ai}\sigma_a\tilde{p}_i \\ -A^{0*}_{ai}\sigma_a\tilde{p}_i & \epsilon + \mathbf{v}\mathbf{p} + \xi(\mathbf{p}) \end{pmatrix}. \tag{4.840}$$

It is the propagator of the pair of Fermi field $f = (\psi, c\psi^*)$ in the presence of a velocity v and a constant pair field A^0_{ai}.

We now pass from quantum mechanics to quantum statistics at constant temperature T by replacing everywhere ϵ by $i\omega_n = i(2n+1)\pi T$, and integrals over energies $\int d\epsilon/2\pi$ by sums over Matsubara frequencies

$$\int \frac{d\epsilon}{2\pi} \to iT \sum_{\omega_n}. \tag{4.841}$$

Correspondingly, we decompose the fluctuations of the pair field as in (3.53):

$$A'_{ai}(\tau, \mathbf{x}) = T\sum_{\nu_n}\int \frac{d^3k}{(2\pi)^3} e^{-i(\tau\nu_n - \mathbf{kx})} A'_{ai}(\nu_n, \mathbf{k}), \tag{4.842}$$

with bosonic Matsubara frequencies

$$\nu_n = 2n\pi T. \tag{4.843}$$

With the short notation (3.55),

$$T\sum_{\nu_n}\int \frac{d^3k}{(2\pi)^3} f(\nu_n, \mathbf{k}) = T\sum_k f(k), \tag{4.844}$$

the fluctuation action (4.838) may be written as

$$i\delta^2 \mathcal{A}^v \approx T\sum_k \left\{ T\sum_p \frac{1}{2}\left[G\left(p - \frac{k}{2}\right) \begin{pmatrix} 0 & A'(-k)\sigma_a\hat{p}_i^- \\ A'^*(k)\sigma_a\hat{p}_i^- & 0 \end{pmatrix} \right.\right.$$
$$\left.\left. \times G\left(p + \frac{k}{2}\right) \begin{pmatrix} 0 & A'(k)\sigma_a\hat{p}_i^+ \\ A'^*(-k)\sigma_a\hat{p}_i^+ & 0 \end{pmatrix}\right] - \frac{1}{3g}|A'_{ai}(k)|^2 \right\}. \tag{4.845}$$

Here we have collected again frequency ν_n and momentum $\mathbf{k}$ in a single four-vector symbol k. Also, by restricting our consideration to long wavelengths with $k \ll p_F$ only, we have set $(p \pm k) \approx \hat{p}_i^\mp$. For small k compared to the Fermi energy ε_F and momentum $\mathbf{p}_F$, this can be approximated by $(p \pm k) \approx \hat{p}_i$. After a little matrix algebra, the fluctuation action can be written as

$$i\delta^2 \mathcal{A}^v \approx -\frac{1}{2}T \sum_k \left(A'^*_{ai}(k), A'_{ai}(-k)\right) L^{v\,ij,ab}(k)(k) \begin{pmatrix} A'(k) \\ A'^*(-k) \end{pmatrix}_{bj} \qquad (4.846)$$

with the matrix

$$L^{v\,ij,ab}(k) \equiv \begin{pmatrix} L^{v\,ij}_{11}(k)\delta^{ab} & L^{v\,ij,ab}_{12}(k) \\ L^{v\,ij,ab}_{21}(k) & L^{v\,ij}_{22}(k)\delta^{ab} \end{pmatrix} \qquad (4.847)$$

whose coefficients $F_{\alpha\beta}(k)$ involve the four 2×2 submatrices of G^v

$$G^v(k) = \begin{pmatrix} G^v_{11}(k) & G^v_{12}(k) \\ G_{21}(k) & G^v_{22}(k) \end{pmatrix} \qquad (4.848)$$

as follows:

$$L^{v\,ij}_{11}(k)\delta^{ab} \approx \frac{1}{2}\operatorname*{tr}_{2\times2} T \sum_p G^v_{22}\left(p - \frac{k}{2}\right)\sigma_a \hat{p}_i G^v_{11}\left(p + \frac{k}{2}\right)\sigma_b \hat{p}_j,$$

$$L^{v\,ij}_{22}(k)\delta^{ab} \approx \frac{1}{2}\operatorname*{tr}_{2\times2} T \sum_p G^v_{11}\left(p - \frac{k}{2}\right)\sigma_a \hat{p}_i G^v_{22}\left(p + \frac{k}{2}\right)\sigma_b \hat{p}_j, \qquad (4.849)$$

$$L^{v\,ij}_{12}(k)\delta^{ab} \approx \frac{1}{2}\operatorname*{tr}_{2\times2} T \sum_p G^v_{12}\left(p - \frac{k}{2}\right)\sigma_a \hat{p}_i G^v_{12}\left(p + \frac{k}{2}\right)\sigma_b \hat{p}_j,$$

$$L^{v\,ij}_{21}(k)\delta^{ab} \approx \frac{1}{2}\operatorname*{tr}_{2\times2} T \sum_p G^v_{21}\left(p - \frac{k}{2}\right)\sigma_a \hat{p}_i G^v_{21}\left(p + \frac{k}{2}\right)\sigma_b \hat{p}_j.$$

Using (4.840) we find

$$L^{v\,ij}_{11}(k) = T \sum_p \frac{(i\tilde{\omega}_+ - \xi_+)(i\tilde{\omega}_- + \xi_-)}{(\tilde{\omega}_+^2 + E_+^2)(\tilde{\omega}_-^2 + E_-^2)}\hat{p}_i\hat{p}_j + \frac{1}{3g}\delta_{ij}, \qquad (4.850)$$

$$L^{v\,ij}_{22}(k) = L^{v\,ij}_{11}(-k), \qquad (4.851)$$

$$L^{v\,ij,ab}_{12}(k) = T \sum_p \frac{1}{(\tilde{\omega}_+^2 + E_+^2)(\tilde{\omega}_-^2 + E_-^2)}\hat{p}_i\hat{p}_j\hat{p}_k\hat{p}_l \, t_{aba'b'} A^0_{a'k} A^0_{b'l}, \qquad (4.852)$$

$$L^{v\,ij,ab}_{21}(k) = L^{v\,ij,ak}_{12}(-k)^*. \qquad (4.853)$$

with the tensor $t_{aba'b'}$ being the trace

$$t_{aba'b'} \equiv \frac{1}{2}\operatorname{tr}\left(\sigma^{a'}\sigma^a\sigma^{b'}\sigma^{b'}\right) = \delta_{aa'}\delta_{bb'} + \delta_{ab'}\delta_{ba'} - \delta_{ab}\delta_{a'b'}, \qquad (4.854)$$

and $\xi_\pm$, $\tilde{\omega}_\pm$,$E_\pm$ abbreviating

$$\xi_\pm = \xi \pm v_F \hat{\mathbf{p}} \mathbf{k},$$

$$\tilde{\omega}_\pm \equiv \omega_n \pm \nu/2 - i v \hat{\mathbf{p}} p_F, \tag{4.855}$$

$$E_\pm = E(p \pm k/2) \equiv \sqrt{(\xi \pm v_F \hat{\mathbf{p}} \mathbf{k})^2 + |A_{ai}\hat{p}_i|^2}.$$

If we split the energy-momentum summation into size and angular parts using the density of states $\mathcal{N}(0) = \frac{3}{4}\rho/p_F^2$, then $L_{12}^{v\,ij,ab}$ can be written as an angular average

$$L_{12}^{v\,ij,ab}(k) = \frac{\rho}{2p_F^2} 3 \int_{-1}^{1} \frac{d\hat{\mathbf{p}}}{4\pi} L^v(k_0, \hat{\mathbf{p}}\mathbf{k}) \hat{p}_i \hat{p}_j \hat{p}_k \hat{p}_l \, t_{aba'b'} A_{a'k}^0 A_{b'l}^{0*}, \tag{4.856}$$

where $L^v(k_0, \hat{\mathbf{p}}\mathbf{k})$ is the following function:

$$L^v(k_0, \hat{\mathbf{p}}\mathbf{k}) = T \sum_\omega \int_{-\infty}^{\infty} d\xi \frac{1}{(\tilde{\omega}_+^2 + E_+^2)(\tilde{\omega}_-^2 + E_-^2)}. \tag{4.857}$$

It is a generalization of the Yoshida function $\phi(\Delta)$ in Eq. (4.320) to the nonzero flow situation. For $\mathbf{v} = 0$, $k = 0$:

$$L^v(k_0, \hat{\mathbf{p}}\mathbf{k})\Big|_{\mathbf{v}=0,k=0} = T \sum_\omega \int_{-\infty}^{\infty} d\xi \frac{1}{(\omega^2 + E^2)^2} = \frac{1}{2\Delta^2} \phi(\Delta^2). \tag{4.858}$$

4.13.2 Time-Dependent Fluctuations at Infinite Wavelength

Let us now specialize on those fluctuations which depend only on time and not on space. Then the only preferred spatial direction is the anisotropy axis l and we may decompose the tensor $L_{11}^{v\,ij}$ into components parallel and orthogonal to l, i.e.,

$$L_{11}^{v\,ij}(\nu) = (\delta_{ij} - l_i l_j) L_{11}^{v\perp}(\nu) - l_i l_j L^{v\parallel}(\nu). \tag{4.859}$$

Alternatively we shall decompose

$$L_{11}^{v\,ij}(\nu) = \delta_{ij} L_{11}^{v\perp}(\nu) - l_i l_j F_{11}^{v0}(\nu). \tag{4.860}$$

Using the general decomposition formula for an integral

$$3 \int \frac{d\hat{\mathbf{p}}}{4\pi} f(\mathbf{pl}) \hat{p}_i \hat{p}_j = \left[\int_{-1}^{1} \frac{dz}{2} \frac{3}{2} (1-z^2) f(z) \right] (\delta_{ij} - l_i l_j) + \left[\int_{-1}^{1} \frac{dz}{2} 3z^2 f(z) \right] l_i l_j$$

$$= \left[\int_{-1}^{1} \frac{dz}{2} \frac{3}{2} (1-z^2) f(z) \right] \delta_{ij} - \left[\int_{1}^{1} \frac{dz}{2} \frac{3}{2} (1-3z^2) f(z) \right] l_i l_j, \tag{4.861}$$

which may be verified by contraction with δ_{ij} and $l_i l_j$, we identify

$$L_{11}^{v\perp}(\nu) = \frac{\rho}{2p_F^2} \int_{-1}^{1} \frac{dz}{2} \frac{3}{2} (1-z^2) \left[T \sum_\omega \int_{-\infty}^{\infty} d'\xi \frac{(i\tilde{\omega}_+ - \xi_+)(i\tilde{\omega}_- + \xi_-)}{(\tilde{\omega}_+^2 + E_+^2)(\tilde{\omega}_-^2 + E_-^2)} + \gamma \right],$$

$$L_{11}^{v0}(\nu) = \frac{\rho}{2p_F^2} \int_{-1}^{1} \frac{dz}{2} \frac{3}{2} (1-3z^2) \left[T \sum_\omega \int_{-\infty}^{\infty} d'\xi \frac{(i\tilde{\omega}_+ - \xi_+)(i\tilde{\omega}_- + \xi_-)}{(\tilde{\omega}_+^2 + E_+^2)(\tilde{\omega}_-^2 + E_-^2)} + \gamma \right]$$

$$- \frac{\rho}{2p_F^2} \int_{-1}^{1} \frac{dz}{2} \frac{3}{2} (1-3z^2) \gamma, \tag{4.862}$$

where γ is the gap function introduced earlier in (4.734):

$$\gamma = T \sum_\omega \int_{-\infty}^\infty d\xi \frac{1}{(\tilde\omega^2 + E^2)}. \tag{4.863}$$

Note that in $L_{11}^0(\nu)$, the function γ cancels out. To keep the expressions for both coefficients as similar as possible, however, we have left γ in the equation. The advantage of this is that the square bracket can be simplified, due to the fact that γ may also be summed in terms of variables $\tilde\omega_\pm$ and $E_\pm$ instead of $\tilde\omega, E$, a replacement that merely amounts to a translation of the infinite sum. Thus, taking the average of both forms, we may write

$$\gamma = T \sum_\omega \int_{-\infty}^\infty d\xi \frac{1}{2} \frac{\tilde\omega_+^2 + \tilde\omega_-^2 + E_+^2 + E_-^2}{(\tilde\omega_+^2 + E_+^2)(\tilde\omega_-^2 + E_-^2)}. \tag{4.864}$$

Now the numerators in (4.862) can be combined to

$$L_{11}^{v\perp}(\nu) = \frac{\rho}{2p_F^2} \int_{-1}^1 \frac{dz}{2} \left(\Delta^2 + \frac{\nu^2}{2} \right) L^v(\nu) \equiv \frac{\rho}{4p_F^2} \varphi^{v\perp}(\nu), \tag{4.865}$$

$$L_{11}^{v0}(\nu) = \frac{\rho}{2p_F^2} \int_{-1}^1 \frac{dz}{2} \frac{3}{2}(1 - 3z^2) \left[\left(\Delta^2 + \frac{\nu^2}{2} \right) L^v(\nu) - \gamma \right] \equiv \frac{\rho}{4p_F^2} \varphi^{v0}(\nu), \tag{4.866}$$

with the function $L^v(k_0, \hat{\mathbf{p}}\mathbf{k})$ of Eq. (4.857). On the right-hand side of (4.865) we have introduced convenient dimensionless quantities $\varphi^{v\perp}$, φ^{v0} associated with $L_{11}^{v\perp}$, L_{11}^{v0}.

It is a pleasant feature of the B-phase that the γ term in (4.866) does not contribute due to the simultaneous validity of the longitudinal and the transversal gap equations (4.603) and (4.604). Thus the B-phase acts as if there is no gap distortion at all. This is not so in the A-phase where only the transversal gap equation is available and γ does contribute!

Let us now perform a tensorial decomposition of L_{12}^v. Generalizing (4.861), we may decompose

$$3 \int \frac{d\hat{\mathbf{p}}}{4\pi} F(\mathbf{pl}) \, \hat{p}_i \hat{p}_j \hat{p}_k \hat{p}_l = A \left(\delta_{ij}\delta_{kl} + \delta_{ik}\delta_{jl} + \delta_{il}\delta_{jk} \right)$$
$$+ B \left(\delta_{ij}l_k l_l + \delta_{ik}l_j l_j + \delta_{il}l_j l_k + \delta_{ik}l_i l_l + \delta_{jl}l_i l_k + \delta_{kl}l_i l_j \right) + C l_i l_j l_k l_l, \tag{4.867}$$

where A, B, C are the following angular projections of F:

$$A = \frac{3}{8} \int_{-1}^1 \frac{dz}{2} (1 - z^2)^2 F(z),$$

$$B = -\frac{3}{8} \int_{-1}^1 \frac{dz}{2} (1 - z^2)(1 - 5z^2) F(z),$$

$$C = \frac{3}{8} \int_{-1}^1 \frac{dz}{2} (3 - 30z^2 + 35z^4) F(z),$$

$$= -7B - \frac{3}{8} \int_{-1}^1 \frac{dz}{2} (1 - 3z^2) F(z). \tag{4.868}$$

For the purpose of obtaining the final results in the simplest possible form it is convenient to use the alternative dimensionless projections

$$\sigma_1(\nu) \equiv 2(A+B)\Delta_\perp^2 = \int_{-1}^1 \frac{dz}{2} 3z^2(1-z^2)F^v(\nu)\Delta_\perp^2,$$

$$\sigma_2(\nu) \equiv 4A\Delta_\perp^2 = \int_{-1}^1 \frac{dz}{2}\frac{3}{2}(1-z^2)^2 F^v(\nu)\Delta_\perp^2, \qquad (4.869)$$

$$\sigma_3(\nu) \equiv 2(3A+6B+C)\Delta_\perp^2 = 6\int_{-1}^1 \frac{dz}{2}z^4 F^v(\nu)\Delta_\perp^2.$$

Note that due to (4.238), the functions A and B at $\nu = 0$ contain the information on the orthogonal superfluid density, since

$$\rho_s^\perp = \left[8A + 2c^2(A+B)\right]_{\nu=0}\Delta_\perp^2 = 2\sigma_2(0) + c^2\sigma_1(0). \qquad (4.870)$$

We now evaluate the full tensor $L_{12}^{vijab}(\nu)$ in terms of $\sigma_{1,2,3}$. Contracting (4.867) with $t_{aba'b'}$ of (4.854), we find

$$A\{2\left(\delta_{ai}\delta_{bj} + \delta_{aj}\delta_{bi} - 3\delta_{ab}\delta_{ij}\right)$$
$$+(1-c)\left[2(2l_a l_b - \delta_{ab}) + 2(\delta_{ai}l_b l_j + \delta_{bi}l_a l_j + (i \leftrightarrow j)) - 4\delta_{ab}l_i l_j\right]$$
$$+ (1-c)^2\left[-(\delta_{ab} - 2l_a l_b)(\delta_{ij} + 2l_i l_j)\right]\}$$
$$+B\{\left[(2l_a l_b - \delta_{ab})\delta_{ij} + 2(\delta_{ai}l_b l_j + \delta_{bi}l_a l_j + (i \leftrightarrow j)) - 5\delta_{ab}l_i l_j\right]$$
$$+(1-c)\left[2(2l_a l_b - \delta_{ab})(3l_i l_j + \delta_{ij}) + 2(\delta_{ai}l_b l_j + \delta_{bi}l_a l_j + (i \leftrightarrow j)) - 4\delta_{ab}l_i l_j\right]$$
$$+(1-c)^2\left[(2l_a l_b - \delta_{ab})\delta_{ij} + 5(2l_a l_b - \delta_{ab})l_i l_j\right]\} + C\{c^2(2l_a l_b - \delta_{ab})l_i l_j\}. \quad (4.871)$$

Collecting terms of equal tensorial properties this becomes

$$2A[\delta_{ai}\delta_{bj} + (i \leftrightarrow j)] + \left[-2A - c^2(A+B)\right]\delta_{ab}\delta_{ij} + \left[-2A + 2c^2(A+B)\right]l_a l_b \delta_{ij}$$
$$+ \left[2A + c^2(A+B) - c^2(3A+6B+C)\right]\delta_{ab}l_i l_j$$
$$+ [-2A + 2(A+B)][\delta_{ai}l_b l_j + \delta_{ai}l_a l_j + (i \leftrightarrow j)]$$
$$+ \left[4(1-c)^2(A+B) - 6(1-c^2)B + 2c^2C\right]l_a l_b l_i l_j. \qquad (4.872)$$

Multiplying L_{11}^v, L_{12}^v with the pair of fluctuating fields $A'_{ai} = \Delta_\perp d_{ai}$, and taking account of the fact that the contributions from L_{22}^v, L_{21}^v are complex conjugate to each other, leads to an action

$$i\delta^2 \mathcal{A}^v = -\frac{\Delta_\perp^2 \rho}{4p_F^2}VT\sum_{\nu_n}\left\{\varphi^\perp(\nu_n)\operatorname{tr}\boldsymbol{d}^\dagger\boldsymbol{d} - \varphi^0(\nu_n)|\boldsymbol{dl}|^2\right.$$

$$+ \frac{\sigma_2}{2}(d_{aa}d_{bb} + d_{ai}d_{ia} + \text{h.c.}) - \frac{1}{2}(c^2\sigma_i + \sigma_2)(d_{ia}d_{ia} + \text{h.c.})$$

$$+ \frac{1}{2}(2c^2\sigma_1 - \sigma_2)\left[(1^T\boldsymbol{d})_a(1^T\boldsymbol{d})_a + \text{h.c.}\right] + \frac{1}{2}(\sigma_1 + \sigma_2 - c^2\sigma_3)\left[(1^T\boldsymbol{d})_a(1^T\boldsymbol{d})_a + \text{h.c.}\right]$$

$$+ 2(2\sigma_1 - \sigma_2)\left[(\boldsymbol{d}^T1)_a(1^T\boldsymbol{d})_a + \text{h.c.}\right]$$

$$+ \left.\left[-(1+4c+c^2)\sigma_1 + \frac{3}{2}\sigma_2 + c^2\sigma_3\right]\left[(1^T\boldsymbol{d}1)(1^T\boldsymbol{d}1) + \text{h.c.}\right]\right\}, \qquad (4.873)$$

where $\boldsymbol{d}$ denotes the matrix whose matrix elements are d_{ai}. As a cross-check of the result we verify that this expansion reduces in the static case $\nu_n = 0$, and the Ginzburg-Landau limit $T \sim T_c$, to the previous expression (4.125). Indeed, if we insert $d(0) = d/T$ and $i\delta^2 \mathcal{A} = -\delta^2 fV/T$, and express

$$\frac{\Delta_\perp^2 \rho}{4p_F^2} \approx \frac{1}{6}\frac{p_F^2}{m^2\xi^2}\frac{\rho}{4p_F^2} \approx 2f_c\frac{1}{6}\frac{1}{1 - T/T_c}, \tag{4.874}$$

we observe that for $T \sim T_c$:

$$L^v(0)\Delta_\perp^2 \approx \frac{1}{2}\phi(\Delta^2) \approx \left(1 - \frac{T}{T_c}\right), \tag{4.875}$$

so that $\sigma_{1,2,3}$ have the extremely simple Ginzburg-Landau limits

$$\sigma_i \approx i \cdot \frac{2}{5}\left(1 - \frac{T}{T_c}\right). \tag{4.876}$$

4.13.3   Normal Modes

It is pleasant to realize that also the new formula (4.873), which is valid for all $T \leq T_c$ and $\nu_n \neq 0$, can be diagonalized on the same subspaces of real and imaginary parts of $d_{ai}(\nu_n) \equiv r_{ai}(\nu_n) + i\,i_{ai}(\nu_n)$

$$r_{11}, r_{22}, r_{33}; \quad r_{12}, r_{21}; \quad r_{13}, r_{31}; \quad r_{23}, r_{32};$$
$$i_{11}, i_{22}, i_{33}; \quad i_{12}, i_{21}; \quad i_{13}, i_{31}; \quad i_{23}, i_{32}. \tag{4.877}$$

On these two 3+2+2+2 -dimensional subspaces, the curly brackets in the energy (4.873) can be written as a quadratic form

$$i\delta^2 \mathcal{A}^v = -\frac{\Delta_\perp^2\rho}{4p_F^2}VT\sum_{\nu_n}\Big\{\big[(r_{11}, r_{22}, r_{33})R(r_{11}, r_{22}, r_{33})^T + (r_{12}, r_{21})R^{12}(r_{12}, r_{21})^T$$
$$+ (r_{13}, r_{31})R^{13}(r_{13}, r_{31})^T + (r_{23}, r_{32})R^{23}(r_{23}, r_{32})^T\big]$$
$$+ (R \to I, r_{ai} \to i_{ai})\Big\}, \tag{4.878}$$

in which the real parts are a sum of three matrices

$$R = \begin{pmatrix} \lambda^\perp + c^2\sigma_1 - \sigma_2 & -\sigma_2 & -2c\sigma_1 \\ -\sigma_2 & \lambda^\perp + c^2\sigma_1 - \sigma_2 & -2c\sigma_1 \\ -2c\sigma_1 & -2c\sigma_1 & \lambda^\perp - \lambda^0 + 2\sigma_1 - c^2\sigma_2 \end{pmatrix},$$

$$R^{12} = \begin{pmatrix} \lambda^\perp + c^2\sigma_1 - \sigma_2 & \sigma_2 \\ \sigma_2 & \lambda^\perp + c^2\sigma_1 - \sigma_2 \end{pmatrix}, \tag{4.879}$$

$$R^{13}_{23} = \begin{pmatrix} \lambda^\perp - \lambda^0 - c^2\sigma_3 & 2c\sigma_1 \\ 2c\sigma_1 & \lambda^\perp + c^2\sigma_1 - 2\sigma_2 \end{pmatrix}.$$

Similar matrices are found for the imaginary parts, except that the σ-terms appear with reversed sign.

These matrices serve two purposes. On the one hand, we can now verify the stability under static fluctuations for all temperatures $T \neq T_c$ by finding the eigenvalues at $\nu_n = 0$. On the other hand, the matrices contain information on the energy of collective excitations at infinite wavelengths: By continuing analytically from the discrete values ν_n to physical frequencies

$$\nu_n \to -i(\omega + i\epsilon) \tag{4.880}$$

these energies are given by the frequency ω at which the matrices become singular. The corresponding eigenvectors are the normal modes of the order parameter fluctuations $d_{ai}(\omega)$.

To embark on this calculation it is useful to express the functions $\varphi^\perp$, φ^0 of (4.865), (4.866) in terms of the functions σ_i as follows

$$\varphi^{v\perp}(\nu_n) = \int_{-1}^{1} \frac{dz}{2} \frac{3}{2}(1-z^2)(1-r^2z^2)2F\Delta_\perp^2 + \frac{\nu_n^2}{2\Delta_\perp^2} \int \frac{dz}{2}\frac{3}{2}\left(1-z^2\right)2F\Delta_\perp^2. \tag{4.881}$$

Using (4.869), this becomes

$$\varphi^{v\perp}(\nu_n) = c^2\sigma_1 + 2\sigma_2 + 2\frac{\nu_n^2}{4\Delta_\perp^2}(\sigma_1 + 2\sigma_2). \tag{4.882}$$

Similarly, we find

$$\begin{aligned}\varphi^{v\parallel}(\nu_n) &= 2\sigma_1 + \sigma_3 + 2\frac{\nu_n^2}{4\Delta_\perp^2}(2\sigma_1 + \sigma_3) \\ &\equiv \varphi^{v\parallel} - \varphi^{v0}(\nu_n).\end{aligned} \tag{4.883}$$

Inserting these relations into (4.879) we obtain

$$R = \begin{pmatrix} 3\sigma_2 - 2w^2(\sigma_1 + 2\sigma_2) & \sigma_2 & 2c\sigma_1 \\ \sigma_2 & 3\sigma_2 - 2w^2(\sigma_1 + 2\sigma_2) & 2c\sigma_1 \\ 2c\sigma_1 & 2c\sigma_1 & 2c^2\sigma_3 - 2w^2(2\sigma_1 + \sigma_3) \end{pmatrix},$$

$$R^{12} = \begin{pmatrix} \sigma_2 - 2w^2(\sigma_1 + 2\sigma_2) & \sigma_2 \\ \sigma_2 & \sigma_2 - 2w^2(\sigma_1 + 2\sigma_2) \end{pmatrix}, \tag{4.884}$$

$$R^{13}_{23} = \begin{pmatrix} 2\sigma_1 - 2w^2(2\sigma_1 + \sigma_3) & 2c\sigma_1 \\ 2c\sigma_1 & 2c^2\sigma_1 - 2w^2(\sigma_1 + 2\sigma_2) \end{pmatrix}.$$

For the imaginary parts of d_{ai} the fluctuation matrices are

$$I = \begin{pmatrix} \sigma_2 + 2c^2\sigma_1 - 2w^2(\sigma_1 + 2\sigma_2) & -\sigma_2 & -2c\sigma_1 \\ -\sigma_2 & \sigma_2 + 2c^2\sigma_1 - 2w^2(\sigma_1 + 2\sigma_2) & -2c\sigma_1 \\ -2c\sigma & -2c\sigma_1 & 4\sigma_1 - 2w^2(2\sigma_1 + \sigma_3) \end{pmatrix},$$

$$I^{12} = \begin{pmatrix} 3\sigma_2 + 2c^2\sigma_1 - 2w^2(\sigma_1 + 2\sigma_2) & -\sigma_2 \\ -\sigma_2 & 3\sigma_2 + 2c^2\sigma_1 - 2w^2(\sigma_1 + 2\sigma_2) \end{pmatrix}, \tag{4.885}$$

$$I^{13}_{23} = \begin{pmatrix} 2(\sigma_1 + c^2\sigma_3) - 2w^2(2\sigma_1 + \sigma_3) & -2c\sigma_1 \\ -2c\sigma_1 & 4\sigma_2 - 2w^2(\sigma_1 + 2\sigma_3) \end{pmatrix}.$$

For brevity, we have introduced the dimensionless frequency variable

$$w^2 \equiv -\frac{\nu_n^2}{4\Delta_\perp^2} = \frac{(\omega + i\epsilon)^2}{4\Delta_\perp^2}. \tag{4.886}$$

It is now straightforward to determine the places of vanishing determinants: For R^{12}, I^{12} we immediately find

$$w_{\frac{1}{2}}^2 = \left\{ \begin{array}{c} 0 \\ \frac{\sigma_2}{\sigma_1 + 2\sigma_2} \end{array} \right\} \quad \text{with} \quad \left\{ \begin{array}{c} (1, -1)^T \\ (1, 1)^T \end{array} \right\},$$

$$w_{\frac{1}{2}}^2 = \left\{ \begin{array}{c} \frac{c^2\sigma_1 + 2\sigma_2}{\sigma_1 + 2\sigma_2} \\ \frac{c^2\sigma_1 + \sigma_2}{\sigma_1 + 2\sigma_2} \end{array} \right\} \quad \text{with} \quad \left\{ \begin{array}{c} (1, -1)^T \\ (1, 1)^T \end{array} \right\}, \tag{4.887}$$

respectively. Behind each eigenvalue we have written down the corresponding eigenvector. For R^{13}_{23}, the eigenvalues are given

$$w_{\frac{1}{2}}^2 = \left\{ \begin{array}{c} 0 \\ \frac{\sigma_1}{(\sigma_1 + 2\sigma_2)(2\sigma_1 + \sigma_3)}[(2c^2 + 1)\sigma_1 + 2\sigma_2 + c^2\sigma_3] \end{array} \right\} \text{with} \left\{ \begin{array}{c} (1, -1/c)^T \\ (\sigma_1 + 2\sigma_2, c(2\sigma_1 + \sigma_3))^T \end{array} \right\}. \tag{4.888}$$

For I^{13}_{23} the roots can no longer be written down explicitly. Here we find

$$w_{\frac{1}{2}}^2 = \frac{\sigma_2}{\sigma_1 + 2\sigma_2} + \frac{\sigma_1 + c^2\sigma_3}{2(2\sigma_1 + \sigma_3)} \pm \frac{1}{2(\sigma_1 + 2\sigma_2)(2\sigma_1 + \sigma_3)} \tag{4.889}$$
$$\times \sqrt{[(2\sigma_1 + \sigma_3)2\sigma_2 - (\sigma_1 + c^2\sigma_3)(\sigma_1 + 2\sigma_2)] + 4c^2\sigma_1^2(2\sigma_1 + \sigma_3)(\sigma_1 + 2\sigma_2)}.$$

In the 3×3 subspaces the eigenvalues look simple only for the imaginary components of d_{ai}. First we observe that $(1, 1, c)^T$ is an eigenvector of I with eigenvalue

$$w_1^2 = 0 \qquad (1, 1, c)^T. \tag{4.890}$$

For $c = 1$, this is the pure phase oscillation of zero sound. By adding the second and the last column times c to the first, the determinant of I can be written as

$$|I| = -2w^2(\sigma_1 + 2\sigma_2)$$
$$\times \begin{vmatrix} 1 & 0 & -2c\sigma_1 \\ 1 & \sigma_2 + 2c^2\sigma_1 - 2w^2(\sigma_1 + 2\sigma_2) & -2c\sigma_1 \\ c & -2c\sigma_1 & 4\sigma_1 - 2w^2(2\sigma_1 + \sigma_3) \end{vmatrix}. \tag{4.891}$$

The remaining determinant has the form

$$4(\sigma_1 + 2\sigma_2)(2\sigma_1 + \sigma_3)w^4 - 4w^2\left[(c^2\sigma_1 + \sigma_2)(2\sigma_1 + \sigma_3) + (2 + c^2)\sigma_1(\sigma_1 + 12\sigma_2)\right]$$
$$+ 4(c^2\sigma_1 + \sigma_2)(2 + c^2)\sigma_1 = 0, \tag{4.892}$$

so that the remaining two eigenvalues are

$$w_2^2 = \frac{(2 + c^2)\sigma_1}{2\sigma_1 + \sigma_3},$$
$$w_3^2 = \frac{c^2\sigma_1 + \sigma_2}{\sigma_1 + 2\sigma_2}. \tag{4.893}$$

For the real 3×3 matrix R, finally, we can find a trivial eigenvector $(1, -1, 0)^T$ with eigenvalue

$$w_1^2 = \frac{\sigma_2}{\sigma_1 + 2\sigma_2}. \tag{4.894}$$

It is degenerate with the second of the eigenvalues of R^{12}. By subtracting in the determinant of R the second from the first row we obtain

$$|R| = \left[2\sigma_2 - 2W^2(\sigma_1 + 2\sigma_2)\right]$$
$$\times \begin{vmatrix} 1 & -1 & 0 \\ \sigma_2 & 2\sigma_2 - 2w^2(\sigma_1 + 2\sigma_2) & 2c\sigma_1 \\ 2c\sigma_1 & 2c\sigma_1 & 2c^2\sigma_3 - 2w^2(\sigma_1 + \sigma_3) \end{vmatrix},$$

so that the remaining two roots are found from the secular equation

$$w^4(\sigma_1 + 2\sigma_2)(2\sigma_1 + \sigma_3) - w^2\left[(\sigma_1 + 2\sigma_2)c^2\sigma_3 + (2\sigma_1 + \sigma_3)2\sigma_2\right]$$
$$+ 2c^2(\sigma_2\sigma_3 - \sigma_1^2) = 0, \tag{4.895}$$

which is solved by

$$w_{\frac{2}{3}}^2 = \frac{1}{2}\frac{c^2\sigma_3}{2\sigma_1 + \sigma_3} + \frac{\sigma_2}{\sigma_1 + 2\sigma_2} \pm \frac{1}{2(\sigma_1 + 2\sigma_2)(\sigma_1 + \sigma_3)} \tag{4.896}$$
$$\times \sqrt{\left[(\sigma_1 + 2\sigma_2)c^2\sigma_3 - (2\sigma_1 + \sigma_3)2\sigma_2\right]^2 + 8c^2\sigma_1^2(\sigma_1 + 2\sigma_2)(2\sigma_1 + \sigma_3)}.$$

All these equations are transcendental since the right-hand sides depend again on w^2. They can, however, be solved quite simply in an iterative fashion.

4.13.4 Simple Limiting Results at Zero Gap Deformation

Before attempting a numerical solution of these equations we may extract several results right away: For small current the asymmetry parameter r vanishes at all temperatures. As a consequence, $\sigma_{1,2,3}$ becomes independent of z and we find immediately, from integrating (4.869), that the functions $\sigma_1 : \sigma_2 : \sigma_3$ have a fixed ratio

$1\!:\!2\!:\!3$. As a result we obtain the well-known collective frequencies of the B phase, at all temperatures:

$$R^{12}: \quad \left\{ \begin{matrix} 0 \\ \frac{2}{5} \end{matrix} \right\}, \quad \omega^2 = \left\{ \begin{matrix} 0 \\ \frac{8}{5}\Delta_\perp^2 \end{matrix} \right\} \quad \text{with} \quad \left\{ \begin{matrix} (1,-1)^T \\ (1,1)^T \end{matrix} \right\},$$

$$R_{23}^{13}: \quad \left\{ \begin{matrix} 0 \\ \frac{2}{5} \end{matrix} \right\}, \quad \omega^2 = \left\{ \begin{matrix} 0 \\ \frac{8}{5}\Delta_\perp^2 \end{matrix} \right\} \quad \text{with} \quad \left\{ \begin{matrix} (1,-1)^T \\ (1,1)^T \end{matrix} \right\}, \qquad (4.897)$$

$$R: \quad \left\{ \begin{matrix} \frac{2}{5} \\ 1 \\ \frac{2}{5} \end{matrix} \right\}, \quad \omega^2 = \left\{ \begin{matrix} \frac{8}{5}\Delta_\perp^2 \\ 4\Delta_\perp^2 \\ \frac{8}{5}\Delta_\perp^2 \end{matrix} \right\} \quad \text{with} \quad \left\{ \begin{matrix} (1,-1,0)^T \\ (1,1,1)^T \\ (1,1,-2)^T \end{matrix} \right\},$$

and

$$I^{12}: \quad \left\{ \begin{matrix} 1 \\ \frac{3}{5} \end{matrix} \right\}, \quad \omega^2 = \left\{ \begin{matrix} 4\Delta_\perp^2 \\ \frac{12}{5}\Delta_\perp^2 \end{matrix} \right\} \quad \text{with} \quad \left\{ \begin{matrix} (1,-1)^T \\ (1,1)^T \end{matrix} \right\},$$

$$I_{23}^{13}: \quad \left\{ \begin{matrix} 1 \\ \frac{3}{5} \end{matrix} \right\}, \quad \omega^2 = \left\{ \begin{matrix} 4\Delta_\perp^2 \\ \frac{12}{5}\Delta_\perp^2 \end{matrix} \right\} \quad \text{with} \quad \left\{ \begin{matrix} (1,-1)^T \\ (1,1)^T \end{matrix} \right\}, \qquad (4.898)$$

$$I: \quad \left\{ \begin{matrix} \frac{3}{5} \\ \frac{3}{5} \\ 0 \end{matrix} \right\}, \quad \omega^2 = \left\{ \begin{matrix} \frac{12}{5}\Delta_\perp^2 \\ \frac{12}{5}\Delta_\perp^2 \\ 0 \end{matrix} \right\} \quad \text{with} \quad \left\{ \begin{matrix} (1,-1,0)^T \\ (1,1,-2)^T \\ (1,1,1)^T \end{matrix} \right\},$$

where the eigenvectors have again been recorded in each case. Moreover, since at $T = 0$ there is no gap deformation for $\nu \leq 1$ these results remain true for all velocities up to $\leq v_n$.

It is useful to classify this symmetric situation in terms of angular momentum. The real and imaginary 3×3 matrices contain a $J = 0$, $J = 1$, and $J = 2$ tensor with the correspondence

$$
\begin{aligned}
\tfrac{1}{\sqrt{3}}(1,1,1) &= |00\rangle, \\
\tfrac{1}{\sqrt{2}}(1,-1,0) &= \tfrac{1}{\sqrt{2}}\left(|2,2\rangle + |2,-2\rangle\right), & R, I \\
\tfrac{1}{\sqrt{6}}(1,1,-2) &= |2,0\rangle, \\
\tfrac{1}{\sqrt{2}}(1,-1) &= |1,0\rangle, \\
\tfrac{1}{\sqrt{2}}(1,1) &= \tfrac{1}{\sqrt{2}}\left(|2,2\rangle - |2,-2\rangle\right), & R^{12}, I^{12} \qquad (4.899) \\
\tfrac{1}{\sqrt{2}}(1,-1) &= \tfrac{1}{\sqrt{2}}\left(|1,1\rangle + |1,-1\rangle\right), & R^{13}, I^{13} \\
\tfrac{1}{\sqrt{2}}(1,1) &= \tfrac{1}{\sqrt{2}}\left(|2,1\rangle + |2,-1\rangle\right), \\
\tfrac{1}{\sqrt{2}}(1,-1) &= \tfrac{1}{\sqrt{2}}\left(|1,1\rangle - |1,-1\rangle\right), & R^{23}, I^{23} \\
\tfrac{1}{\sqrt{2}}(1,1) &= \tfrac{1}{\sqrt{2}}\left(|2,1\rangle - |2,-1\rangle\right),
\end{aligned}
$$

explaining the degeneracies among the 5 real $J = 2$ modes, the three real $\omega^2 = 0$ Goldstone modes with $J = 1$, the 5 imaginary $J = 2$ modes with $\omega^2 = \frac{12}{5}\Delta_\perp^2$, and the 3 imaginary $J = 1$ modes with $\omega^2 = 4\Delta_\perp^2$.

Now, if a current is turned on, the levels of different $|J_3|$ within each multiplet split up. Using the explicit forms of analytically continued σ_i functions, to be discussed in the next section, we find for T close to T_c the level structure displayed in Fig. 4.43.

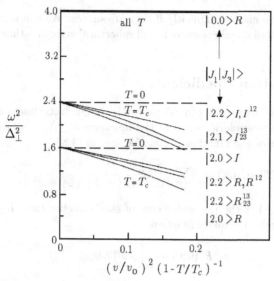

$$(v/v_0)^2 (1 - T/T_c)^{-1}$$

FIGURE 4.43 Collective frequencies of B-phase in the presence of superflow of velocity v at zero and slightly below the critical temperature T_c (Ginzburg-Landau regime). Near T_c, there is a considerable splitting between the levels of different $|J_3|$. The quantum numbers of angular momentum are displayed at the right end of each curve. The gap distortion $r^2 \equiv 1 - \Delta_\parallel^2 / \Delta_\perp^2$ is related to the superfluid velocity v by $r^2 = 3(v/v_0)^2 (1 - T/T_c)^{-1}$ $r^2 = 3(v/v_0)^2 (1 - T/T_c)^{-1}$.

4.13.5 Static Stability

In order to verify static stability we have to take the matrices R, I before analytic continuation at zero Matsubara frequency $\nu_n = 0$ and calculate their eigenvalues. These are found as

$$R \ : \ 2\sigma_2, 2\sigma_2 + c^2\sigma_3 \pm \sqrt{(2\sigma_2 - c^2 r_3)^2 + 8c^2\sigma_1^2},$$

$$R^{12} \ : \ \left\{ \begin{array}{c} 0 \\ 2\sigma_2 \end{array} \right\} \ \text{with} \ \left\{ \begin{array}{c} (1, -1)^T \\ (1, 1)^T \end{array} \right\},$$

$$R^{\,13}_{\,23} \ : \ \left\{ \begin{array}{c} 0 \\ (1 + c^2)\sigma_1 \end{array} \right\} \ \text{with} \ \left\{ \begin{array}{c} (1, -c^{-1})^T \\ (1, c)^T \end{array} \right\}, \tag{4.900}$$

and

$$I \ : \ 2(c^2\sigma_1 + \sigma_2), 2(c^2 + 2)\sigma_1$$

$$I^{12} \ : \ \left\{ \begin{array}{c} 2c^2\sigma_1 + 4\sigma_2 \\ 2c^2\sigma_1 + 2\sigma_2, \end{array} \right\} \ \text{with} \ \left\{ \begin{array}{c} (1, -1)^T \\ (1, 1)^T \end{array} \right\},$$

$$I^{\,13}_{\,23} \ : \ \sigma_1 + 2\sigma_2 + c^2\sigma_3 \pm \sqrt{(\sigma_1 - 2\sigma_2 + c^2)^2 + 4c^2\sigma_1^2}. \tag{4.901}$$

The eigenvectors are marked explicitly if they are simple. We can now easily verify that all nonzero values remain positive for all subcritical velocities thus guaranteeing static stability.

4.14   Fluctuation Coefficients

We have seen in the last section that all properties of quadratic fluctuations at finite wavelengths are expressible in terms of the functions $\sigma_{1,2,3}(\omega^2)$ which in turn are angular projections of the function $F^v(i\nu_n, \hat{\mathbf{p}}\mathbf{k})$ [recall (4.869)]

$$F^v(i\nu_n, \hat{\mathbf{p}}\mathbf{k}) = T\sum_\omega \int_{-\infty}^\infty d\xi \frac{1}{(\tilde{\omega}_+^2 + E_+^2)(\tilde{\omega}_-^2 + E_-^2)} \qquad (4.902)$$

formed at $|\mathbf{k}| = 0$. For the particular case of static fluctuations, $F^v(0,0)$ reduces directly to the standard Yoshida function

$$F^v(0,0) = \frac{1}{2\Delta_\perp^2}\phi^v(0,0). \qquad (4.903)$$

It can then easily be checked that in this case the projection $\sigma_i(0)$ are positive thus guaranteeing the stability of static fluctuation frequencies (4.901): First close to T_c, all nonzero eigenvalues are positive since σ_i have the simple form (4.876). Moreover, as the temperature reaches zero, the gap becomes uniform and

$$F^v\Delta_\perp^2 \to \frac{1}{2} \qquad (4.904)$$

for subcritical velocities so that $\sigma_i(0)$ are positive members with the same ratios $1:2:3$. Inserting this together with $c = 1$ into (4.901) all eigenvalues become again positive. By monotony of the gap distortion at fixed velocity $(v^2/v_0^2)(1 - T/T_c)^{-1}$ (see Fig. 4.41) as a function of temperature, we conclude that there is stability at all temperatures $T \le T_c$ and all subcritical velocities.

For dynamic fluctuations, let us continue F^v analytically in the frequency $-\nu_n$. For this we decompose

$$F^v(i\nu_n, \hat{\mathbf{p}}\mathbf{k}) = T\sum_\omega \int_{-\infty}^\infty d\xi \frac{1}{\tilde{\omega}_-^2 + E_-^2} \frac{1}{\tilde{\omega}_+^2 + E_+^2} \qquad (4.905)$$

as in (3.201), and use the summation formula (3.199) to find

$$\frac{1}{2E}T\sum_\omega \frac{1}{i\tilde{\omega}_\pm \pm E} = \frac{1}{2E}\left[1 \pm \frac{1}{2}\left(\tanh\frac{E + v_Fz}{2T} + \tanh\frac{E - v_Fz}{2T}\right)\right]. \qquad (4.906)$$

Again we have made use of the fact that the frequency shift ν_n in $\omega_\pm$ does not appear in (4.906) since it amounts to a mere translation in the infinite sum. Collecting the different terms we find

$$F^v(i\nu_n, \hat{\mathbf{p}}\mathbf{k}) = T\int_{-\infty}^\infty d\xi$$

$$\times\frac{1}{4E_+E_-}\left\{\frac{E_++E_-}{(E_++E_-)^2+\nu_n^2}\left[\frac{1}{2}\left(\tanh\frac{E_++vp_Fz}{2T}+(v\to-v)\right)+(E_+\leftrightarrow E_-)\right]\right.$$ (4.907)

$$\left.-\frac{E_+-E_-}{(E_+-E_-)^2+\nu_n^2}\left[\frac{1}{2}\left(\tanh\frac{E_++vp_Fz}{2T}+(v\to-v)\right)-(E_+\leftrightarrow E_-)\right]\right\}.$$

with

$$E_\pm^2 = (\xi\pm v_F\hat{\mathbf{p}}\mathbf{k})^2+\Delta^2.$$ (4.908)

At $\mathbf{k}=0$, and $v=0$, we recover the Yoshida function in the presence of superflow (4.903) that governs the superfluid densities:

$$F^v(0,0) = \int_{-\infty}^{\infty}d\xi\frac{1}{E^2}T\sum_\omega\frac{E^2}{(\tilde{\omega}^2+E^2)^2}$$

$$= \phi^v(\Delta^2)/2\Delta_\perp^2.$$ (4.909)

The expression (4.907) can readily be continued analytically to physical frequencies ω by merely replacing

$$i\nu_n\to\omega+i\epsilon,$$ (4.910)

which is independent of the direction $\mathbf{p}$. Let us now turn to the calculation of the functions. For this we consider the continued expression at infinite wavelength

$$F^v(\omega,0)=\int_{-\infty}^{\infty}d\xi\frac{1}{E(4E^2-\omega^2)}\frac{1}{2}\left[\tanh\frac{E+vp_Fz}{2T}+(v\to-v)\right].$$ (4.911)

The temperature region close to T_c is explored most easily by inserting the expansion (4.906). Then the integral over ξ can be done and we find by the same steps as from Eq. (3.201) to (3.202):

$$F^v(\omega,0) = \frac{1}{4}\int_{-\infty}^{\infty}d\xi\,T\sum_\omega\frac{1}{(\omega_n-ivp_Fz)^2+\omega^2/4}$$

$$\times\left[\frac{1}{\xi^2+\Delta^2-\omega^2/4}-\frac{1}{(\omega_n-ivp_Fz)^2+\xi^2+\Delta^2}\right]$$

$$= \frac{\pi}{4}\frac{1}{\sqrt{\Delta^2-\omega^2/4}}\frac{1}{\omega}\left[\tanh\frac{\omega/2+v}{2T}+(v\to-v)\right]$$ (4.912)

$$-\frac{\pi T}{4}\sum_{\omega_m}\frac{1}{(\omega_m-ivp_Fz)^2}+\frac{\omega^2}{4}\frac{1}{\sqrt{(\omega_n-ivp_Fz)^2+\Delta^2}}.$$

Using the previously introduced dimensionless variables (4.886), this may be rewritten as

$$F^v(\omega,0)\Delta_\perp^2 = \frac{\pi}{8}\frac{1}{\sqrt{1-r^2z^2-w^2}}\left\{\tanh\left[\frac{\pi}{2}(w-\nu z)\delta\right]+(\nu\to-\nu)\right\}$$

$$-\frac{1}{2\delta}\mathrm{Re}\sum_{n=0}^{\infty}\frac{1}{(x_n-i\nu z)^2+w^2}\frac{1}{\sqrt{(x_n-i\nu z)^2+1-r^2z^2}},$$ (4.913)

where the square root has to be taken with a positive real part.

In the limit $T \to T_c$, $\delta \to 0$ and the sum is suppressed by one power of δ as compared with the first term so that we may use the simple expression

$$F^v(\omega, 0)\Delta_\perp^2 \underset{T \to T_c}{=} \frac{\pi^2 \delta}{8} \frac{1}{\sqrt{1 - r^2 z^2 - w^2}}. \tag{4.914}$$

For $T \to 0$, the integral is found easily from (4.911) if the velocity v is satisfies $v < \Delta_{\mathrm{BCS}}/p_F \approx v_c$. Then $\tanh[(E \pm p_F z)/2T] = 1$, and we have

$$F^v(\omega, 0) = \frac{1}{4} \int_{-\infty}^{\infty} d\xi \frac{1}{\sqrt{\xi^2 + \Delta^2}} \frac{1}{\xi^2 + \Delta^2(1 - w^2)}. \tag{4.915}$$

It is useful to remove the square root by an auxiliary integration, writing

$$F^v(\omega, 0) = \frac{1}{4\pi} \int_{-\infty}^{\infty} d\xi \int_{-\infty}^{\infty} \frac{1}{\xi^2 + \mu^2 + \Delta^2} \frac{1}{\xi^2 + \Delta^2(1 - w^2)}. \tag{4.916}$$

Using Feynman's formula

$$\frac{1}{AB} = 2 \int_0^1 ds \frac{s}{[sA + (1 - s^2)B]^2} \tag{4.917}$$

this becomes

$$F^v(\omega, 0) = \frac{1}{\pi} \int_0^1 ds \int_{-\infty}^{\infty} \int_{-\infty}^{\infty} d\xi d(\mu s) \frac{1}{[\xi^2 + s^2\mu^2 + \Delta^2 - (1 - s^2)w^2]^2}. \tag{4.918}$$

Due to rotational invariance in the $(\xi, s\mu)$-plane this can be evaluated in polar coordinates to give

$$2 \int_0^1 ds \int_0^{\infty} dr \frac{r}{(r^2 + \Delta^2 - (1 - s^2)w^2)^2} = \int_0^{\infty} ds \frac{1}{sA^2 - (1 - s^2)w^2}. \tag{4.919}$$

Thus we arrive at the simple integral representation

$$F^v(\omega, 0)\Delta_\perp^2 = \frac{1}{2} \int_0^{\infty} ds \frac{1}{s^2 w^2 + 1 - r^2 z^2 - w^2}, \tag{4.920}$$

which can be integrated to

$$F^v(\omega_1, 0)\Delta_\perp^2 = \frac{1}{2} \frac{1}{\sqrt{1 - r^2 z^2 - w^2}} \frac{1}{w} \arcsin \frac{\omega}{\sqrt{1 - r^2 z^2}}. \tag{4.921}$$

We can now proceed to calculate the $\sigma_{1,2,3}$ functions. Consider first the limit $T \to T_c$. Straightforward integration yields, with the overall factor

$$\alpha \equiv \pi^2 \delta/4 = \pi \Delta_\perp/4T, \tag{4.922}$$

the expressions

$$
\sigma_1(w^2) \underset{T \to T_c}{=} \alpha \frac{3}{2} \int_{-1}^{1} \frac{dz}{2} z^2 (1 - z^2) \frac{1}{\sqrt{1 - w^2 - r^2 z^2}} \tag{4.923}
$$

$$
= \frac{3}{4r^5} \left\{ \left[-\frac{3}{4}(1 - w^2)^2 + r^2(1 - w^2) \right] l + \left[\frac{3}{4}(1 - w^2) - \frac{r^2}{2} \right] r \sqrt{1 - w^2 r^2} \right\},
$$

$$
\sigma_2(w^2) \underset{T \to T_c}{=} \alpha \frac{3}{4} \int_{-1}^{1} \frac{dz}{2} (1 - z^2)^2 \frac{1}{\sqrt{1 - w^2 - r^2 z^2}} \tag{4.924}
$$

$$
= \frac{3}{4r^5} \frac{3}{8} \left\{ \left[(1 - w^2)^2 - \frac{8}{3} r^2(1 - w^2 - r^2) \right] l + \left[-(1 - w^2) + 2r^2 \right] r \sqrt{1 - w^2 - r^2} \right\},
$$

$$
\sigma_3(w^2) \underset{T \to T_c}{=} 3 \alpha \int_{-1}^{1} \frac{dz}{2} z^4 \frac{1}{\sqrt{1 - w^2 - r^2 z^2}} \tag{4.925}
$$

$$
= \frac{3}{4r^5} \left\{ \left[-\frac{3}{4}(1 - w^2)^2 + r^2(1 - w^2) \right] l + \left[\frac{3}{4}(1 - w^2) + \frac{r^2}{2} r^2 \right] r \sqrt{1 - w^2 - r^2} \right\}.
$$

Here l is the fundamental integral

$$
l(w^2) \equiv r \int_{-1}^{1} \frac{dz}{2} \frac{1}{\sqrt{1 - w^2 - r^2 z^2}} = \arcsin \frac{r}{\sqrt{1 - w^2}}. \tag{4.926}
$$

This formula may be used as long as $w^2 < 1 - r^2$. For w^2 between $1 - r^2$ and l there is an imaginary part whose sign is controlled by the $i\epsilon$ prescription in w:

$$
l(w^2) = \frac{\pi}{2} + \frac{i}{r} \log \frac{r + \sqrt{w^2 - (1 - r^2)}}{\sqrt{1 - w^2}}. \tag{4.927}
$$

It may in principle give rise to a width of the collective excitation due to pair breaking along directions where the gap is not maximal.

4.15 Stability of Superflow in the B-Phase under Small Fluctuations for $T \sim T_c$

Let us finally investigate the important question whether the ansatz (4.763) for the distorted order parameter is a local minimum of the free energy for all currents up to J_c. Previously, we have shown this form to develop for infinitesimal currents. We shall now study, for all currents up to the critical value J_c, the small fluctuations in the 18 parameter field space A_{ai}.

With the time driving term of the collective action being of the simple pure damping form, it will be sufficient to consider only static fluctuations. It is a disadvantage of the Ginzburg-Landau regime that there are no properly oscillating modes which could easily be detected experimentally. On the other hand, there is the advantage of a simple parametrization of strong-coupling corrections.

Let us parametrize the static fluctuations in the form

$$
A_{ai} = \Delta_B \left[a \left(\delta_{ai} + r l_a l_i \right) + d_{ai} \right], \tag{4.928}
$$

where

$$r \equiv \frac{c}{a} - 1 \tag{4.929}$$

and c and a are the equilibrium values of the gap parameters in the presence of a current (we shall leave out the magnetic field, for simplicity). Inserting (4.928) into the energy we obtain the potential terms for $r = 0$

$$
\left. \delta^2 f / 2 f_c \right|_{\text{pot}} = -\frac{\alpha}{3} |d_{ai}^2| + \frac{a^2}{15} \left[(3\beta_1 + \beta_{35}) \left(d_{ai}^2 + d_{ai}^{*\,2} \right) \right.
$$
$$
+ (6\beta_B - 6\beta_1 + 2\beta_4) |d_{ai}^2| + (4\beta_1 + 2\beta_2) |t|^2 + \beta_2 \left(t^2 + t^{*2} \right)
$$
$$
\left. + 2\beta_{35} |d_{ai}^2| + \beta_4 \left(d_{aj} d_{ja} + \text{h.c.} \right) \right]. \tag{4.930}
$$

Here t denotes the trace of d_{ai}. The linear terms have been left out since they are all of the form $t + t^*$ and cancel at the extremum. Moreover, with the equilibrium value of a^2 being $\alpha / \left(\frac{6}{5} \beta_B \right)$, the first term simply cancels the $6\beta_B$-term inside the bracket.

Neglecting strong-coupling corrections, the expression simplifies to

$$
\left. \delta^2 f / 2 f_c \right|_{\text{pot}} = \frac{a^2}{15} \left\{ 5 |d_{ai}^2| + (d_{aj} d_{ja} + \text{h.c.}) + \left(t^2 + t^{*2} \right) - \frac{3}{2} \left(d_{ai}^2 + \text{h.c.} \right) \right\}. \tag{4.931}
$$

The term containing the gap distortion gives an additional

$$
\frac{a^2}{15} \Big\{ \ \beta_1 \left[4r^2 |d_{33}|^2 + 4r \left(t d_{33}^* + \text{h.c.} \right) \right]
$$
$$
+ \beta_2 \left[2r^2 |d_{33}|^2 + 2r \left(t d_{33}^* + \text{h.c.} \right) + 2r \left(2 + r \right) |d|^2 \right]
$$
$$
+ \beta_3 \left[2r^2 |d_{33}|^2 + 2r \left(d_{i3} d_{3i}^* + \text{h.c.} \right) + 2r(2 + r)|d_{a3}|^2 \right]
$$
$$
+ \beta_4 \left[2r^2 |d_{3i}|^2 + 4r|d_{3i}|^2 + 2r(2 + r)|d_{a3}|^2 \right]
$$
$$
+ \beta_5 \left[2r^2 |d_{33}|^2 + 2r \left(d_{3i} d_{i3}^* + \text{h.c.} \right) + 2r(2 + r)|d_{3i}|^2 \right]
$$
$$
+ \beta_1 r(2 + r) \left(d^2 + \text{h.c.} \right)
$$
$$
+ \beta_2 \left[r^2 \left(d_{33}^2 + \text{h.c.} \right) + 2r \left(t a_{33} + \text{h.c.} \right) \right]
$$
$$
+ \beta_3 \left[r^2 \left(d_{3i}^2 + \text{h.c.} \right) + 2r \left(d_{3i}^2 + \text{h.c.} \right) \right]
$$
$$
+ \beta_4 \left[r^2 \left(d_{33}^2 + \text{h.c.} \right) + 2r \left(d_{i3} d_{3i} + \text{h.c.} \right) \right]
$$
$$
\beta_5 r(2 + r) \left(d_{a3}^2 + \text{h.c.} \right) \Big\} \tag{4.932}
$$

Without strong-coupling effects this simplifies considerably leaving only

$$
\frac{1}{15} \Big\{ 2 \left(c^2 - 1 \right) \left(|d_{ai}^2| + 2|d_{a3}|^2 \right)
$$
$$
- \frac{1}{2} \left(c^2 - 1 \right) \left(d_{ai}^2 + \text{h.c.} \right) + 2 \left(c - 1 \right)^2 \left(d_{33}^2 + \text{h.c.} \right)
$$
$$
+ 2 \left(c - 1 \right) \left(t d_{33} + \text{h.c.} \right) + \left(c^2 - 1 \right) \left(d_{3i}^2 + \text{h.c.} \right)
$$
$$
- \left(c^2 - 1 \right) \left(d_{a3}^2 + \text{h.c.} \right) \Big\} \tag{4.933}
$$

where we have made use of $a = 1$ so that

$$r = c - 1. \tag{4.934}$$

The result can be written in matrix form

$$15\frac{\delta^2 f}{2f_c} = r_{ai} R_{ai,a'i'} \, r_{a'i'} + i_{ai} I_{ai,a'i'} \, i_{a'i'} \tag{4.935}$$

where we have separated d into real and imaginary parts

$$d_{ai} = r_{ai13.33} + i \, i_{ai}. \tag{4.936}$$

The matrix R may be decomposed as $R \times R^{12} \times R^{13} \times R_{23}$ where R is a 3×3 submatrix acting only in the space $\begin{pmatrix} r_{11} \\ r_{23} \\ r_{33} \end{pmatrix}$ while R^{12}, R^{13}, R^{23} are 2×2 blocks in the subspaces

$$\begin{pmatrix} r_{12} \\ r_{21} \end{pmatrix} \begin{pmatrix} r_{13} \\ r_{31} \end{pmatrix} \begin{pmatrix} r_{23} \\ r_{32} \end{pmatrix}. \tag{4.937}$$

An analogous decomposition holds for I. Collecting the different contribution we find

$$\begin{aligned}
R &= \begin{pmatrix} 5 + c^2 & 2 & 2c \\ 2 & 5 + c^2 & 2c \\ 2c & 2c & 9c^2 - 3 \end{pmatrix}, \\
R^{12} &= \begin{pmatrix} c^2 + 1 & 2 \\ 2 & c^2 + 1 \end{pmatrix}, \\
R^{13}_{23} &= \begin{pmatrix} 3c^2 - 1 & 2c \\ 2c & 3c^2 - 1 \end{pmatrix}, \\
I &= \begin{pmatrix} 1 + 3c^2 & -2 & -2c \\ -2 & 1 + 3c^2 & -2c \\ -2c & -2c & 1 + 3c^2 \end{pmatrix}, \\
I^{12} &= \begin{pmatrix} 3c^2 + 5 & -2 \\ -2 & 3c^2 + 5 \end{pmatrix}, \\
I^{13}_{23} &= \begin{pmatrix} -1 + 9c^2 & -2c \\ -2c & 7 + c^2 \end{pmatrix}.
\end{aligned} \tag{4.938}$$

In the absence of a current, we have $c = 1$ and can recover immediately the eigenvalues:

$$\begin{aligned}
R &: \ (10, \, 4, \, 4), \\
R^{12,13,23} &: \ (0, 4), \\
I &: \ (0, \, 6, \, 6), \\
I^{12, \, 13, \, 23} &: \ (6, 10).
\end{aligned} \tag{4.939}$$

We observe the occurrence of 4 Nambu-Goldstone modes corresponding to overall phase oscillations (sound) and three vibrations of the order parameter θ, one for the length and two for the direction.

These correspond to the residual part of the original $SO(3)_{\text{spin}} \times SO(3)_{\text{orbit}} \times U(1)_{\text{phase}}$ symmetry left unbroken by the isotropic parameter A^0_{ai}, of the B-phase.

The strong-coupling corrections change the eigenvalues only slightly. Since the Nambu-Goldstone bosons are a consequence of the symmetry of the action and A^0_{ai}, their eigenvalues remain exactly zero. Collecting the different terms in (4.932) we find the corrected matrices

$$R = 4 \begin{pmatrix} \beta_{12345} & \beta_{12} & \beta_{12} \\ \beta_{12} & \beta_{12345} & \beta_{12} \\ \beta_{12} & \beta_{12} & \beta_{12345} \end{pmatrix} \frac{\alpha}{\frac{6}{5}\beta_B},$$

$$R^{12,13,23} = 2\beta_{345} \begin{pmatrix} 1 & 1 \\ 1 & 1 \end{pmatrix} \frac{\alpha}{\frac{6}{5}\beta_B},$$

$$I = -4\beta_1 \begin{pmatrix} 2 & -1 & -1 \\ -1 & 2 & -1 \\ -1 & -1 & 2 \end{pmatrix} \frac{\alpha}{\frac{6}{5}\beta_B}, \tag{4.940}$$

$$I^{12,13,23} = 2 \begin{pmatrix} -6\beta_1 - \beta_{35} + \beta_4 & \beta_{35} - \beta_4 \\ \beta_{35} - \beta_4 & -6\beta_1 - \beta_{35} + \beta_4 \end{pmatrix} \frac{\alpha}{\frac{6}{5}\beta_B},$$

with eigenvalues

$$R \ : \ (12\beta_B, \ 4\beta_{345}, \ 4\beta_{345}) \frac{\alpha}{\frac{6}{5}\beta_B},$$

$$R^{12,13,23} \ : \ \beta_{345}(0,4) \frac{\alpha}{\frac{6}{5}\beta_B} \tag{4.941}$$

$$I^{12,13,23} \ : \ (-12\beta_1, \ -12\beta_1 + 4(\beta_4 - \beta_{35})) \frac{\alpha}{\frac{6}{5}\beta_B}.$$

Remember that $\frac{\alpha}{\frac{6}{5}\beta_B}\Delta_B$ represents the corrected gap value in the B-phase.

Note that if $\beta_{345} = 0$, there would be two more zero-frequency modes in R. This fact is associated with the accidental degeneracy of polar and planar phase at $\beta_{345} = 0$: the two modes correspond to linear interpolations between these two phases.

Let us now turn on the current. Then we have to add the fluctuations from the term

$$-\frac{5j^2}{|d^2_{ai}| + |d^2_{a3}|}, \tag{4.942}$$

which in equilibrium contributes inside the curly brackets of (4.932):

$$\frac{3\kappa^2}{2a^2 + 3c^2} \left[\left(|d^2_{a1}| + |d^2_{a2}| + 3|d^2_{a3}| \right) \left(2a^2 + 3c^2 \right) \right.$$
$$-4a^2 \left(r^2_{11} + r^2_{22} \right) - 36c^2 r^2_{33}$$
$$\left. -8r_{11}r_{22}a^2 - 24ac \left(r_{11} + r_{22} \right) r_{33} \right]. \tag{4.943}$$

Without strong-coupling corrections $a = 1$ this adds directly

$$
\left(1 - c^2\right) \left\{ \frac{1}{2 + 3c^2} \begin{pmatrix} 3c^2 - 2 & -4 & -12c \\ -4 & 3c^2 - 2 & -12c \\ -12c & -12c & 6 - 27c^2 \end{pmatrix}, \begin{pmatrix} 1 & & \\ & 1 & \\ & & 3 \end{pmatrix}, \right.
$$

$$
\left. \begin{pmatrix} 1 & \\ & 1 \end{pmatrix}, \begin{pmatrix} 1 & \\ & 1 \end{pmatrix}, \begin{pmatrix} 3 & 0 \\ 0 & 1 \end{pmatrix}, \begin{pmatrix} 3 & 0 \\ 0 & 1 \end{pmatrix} \right\} \qquad (4.944)
$$

to R, I, R^{12}, I^{12}, R^{13}_{23}, I^{13}_{23} so that we obtain the new matrices:

$$
R = \begin{pmatrix} 22c^2 + 8 & 2\left(c^2 + 4\right) & 2c\left(9c^2 - 4\right) \\ 2\left(c^2 + 4\right) & 22c^2 + 8 & 2c\left(9c^2 - -4\right) \\ 2c\left(9c^2 - 4\right) & 2c\left(9c^2 - 4\right) & 6c^2\left(9c^2 - 4\right) \end{pmatrix},
$$

$$
R^{12} = \begin{pmatrix} 2 & 2 \\ 2 & 2 \end{pmatrix},
$$

$$
R^{12}_{23} = \begin{pmatrix} 2 & 2c \\ 2c & 2c^2 \end{pmatrix},
$$

$$
I = \begin{pmatrix} 2\left(1 + c^2\right) & -2 & -2c \\ -2 & 2\left(1 + c^2\right) & -2c \\ -2c & -2c & 4 \end{pmatrix},
$$

$$
I^{12} = \begin{pmatrix} 2c^2 + 6 & -2 \\ -2 & 2c^2 + 6 \end{pmatrix},
$$

$$
I^{13}_{23} \begin{pmatrix} 6c^2 + 2 & -2c \\ -2c & 8 \end{pmatrix}. \qquad (4.945)
$$

The eigenvalues are now

$$
R \; : \; \left(\frac{1}{5} \left[\left(27c^4 + 8\right) \pm \frac{1}{3} \sqrt{\left(9c^2\right)^4 - 8\left(9c^2\right)^3 - 48\left(9c^2\right)^2 + 512\left(9c^2\right) + 576} \right], 4c^2 \right),
$$

$$
R^{12} \; : \; (0, 4),
$$

$$
R^{13}_{23} \; : \; \left(0, 2\left(1 + c^2\right)\right),
$$

$$
I \; : \; \left(0, \frac{2}{5}\left(2 + c^2\right), \frac{2}{5}\left(2 + c^2\right)\right),
$$

$$
I \; : \; \left(2c^2 + 4, 2c^2 + 8\right)
$$

$$
I^{13}_{23} \; : \; \left(3c^2 + 5\pm, \sqrt{c^4 - \frac{14}{9}c^2 + 1}\right). \qquad (4.946)
$$

For increasing current, $c^2 = 1 - 3\kappa^2$ decreases and with it also the eigenfrequencies. At the critical current $\kappa_c^2 = 5/27$ the value of c^2 drops to $4/9$ and the eigenvalues become

$$
R \; : \; \left(0, \frac{16}{3}, \frac{16}{9}\right),
$$

$$R^{12} \ : \ (0,4)\,,$$

$$R^{13}_{23} \ : \ \left(0,\frac{26}{9}\right),$$

$$I \ : \ \left(0,\frac{44}{45},\frac{44}{45}\right),$$

$$I^{12} \ : \ \left(\frac{44}{9},\frac{62}{9}\right),$$

$$I^{13}_{23} \ : \ (4.2,7.04)\,. \tag{4.947}$$

The zero eigenvalue in R signalizes the instability for decay into the planar (or A) phase.

Summary

We have presented only a short introduction into the wide field of ^{3}He physics which has been developed over the last forty years. The methods used in describing the physical properties of the superfluid run hand in hand with those which are popular nowadays in particle physics and field theory. For a particle physicist it can be rewarding to study some of the phenomena and their explanations since it may provide him with a more transparent understanding of the σ-type of models. Also, the visualization of functional field spaces in the laboratory may lend a more realistic appeal to topological considerations which have become a current tool in the analysis of solutions of gauge field equations.

Finally, there may even be direct applications of superfluid ^{3}He in particle physics. Since the condensate is characterized by two vectors $\mathbf{L}$ and $\mathbf{S}$, there is a vector $\mathbf{L} \times \mathbf{S}$ which is time-reversal invariant, but parity violating. If there are neutral currents of this symmetry type in weak interactions they may build up a small electric dipole moment in the Cooper pairs. This has to be aligned necessarily with $\mathbf{L} \times \mathbf{S}$. In the condensed phase of the superfluid, this very small dipole moment can pile up coherently and might result in an observable *macroscopic* dipole moment. This could lead to a more sensitive test than those available right now. Unfortunately, the uncertainty in the Cooper pair wave function is, at present, an obstacle to a reliable estimate of the effect. Also, the detection of the resulting macroscopic dipole moment may be hampered by competing orientational effects.

Appendix 4A   Hydrodynamic Coefficients for $T \approx T_c$

Here we give a brief derivation of the hydrodynamic energy (4.125) as it follows from the original form (4.85) which we rewrite as

$$f_{\text{grad}} = \frac{1}{2}\Delta^2_A \left\{ K_1|\partial_i\phi_j|^2 + K_2\left(\partial_i\phi_j^*\partial_j\phi_i\right) + K_3|\boldsymbol{\nabla}\phi|^2 + K_{23}|\phi\boldsymbol{\nabla}d_a|^2 + 2K_1\left(\partial_i d_a\right)^2 \right\} \tag{4A.1}$$

with the notation $K_{12} \equiv K_2 + K_3$. First we process the pure ϕ parts. The first term is decomposed as follows:

$$|\partial_i \phi_j|^2 = (\partial_i \phi^{(1)})^2 + (\partial_i \phi^{(2)})^2.$$ (4A.2)

Observing that the vector $\partial_i \phi^{(1)}$ has only an $\mathbf{l}$ and a $\phi^{(2)}$ component, due to the trivial orthogonality relation $\phi^{(1)} \partial_i \phi^{(1)} = 0$, we write

$$\partial_i \phi^{(1)} = \left(\mathbf{l} \partial_i \phi^{(1)}\right) \mathbf{l} + \left(\phi^{(2)} \partial_i \phi^{(1)}\right) \phi^{(2)}.$$ (4A.3)

In terms of the superfluid velocity

$$v_{si} = \frac{1}{2m} \phi^{(1)} \partial_i \phi^{(2)}$$ (4A.4)

and using the further orthogonality relation $\mathbf{l} \partial_i \phi^{(1,2)} = -(\partial_i \mathbf{l}) \phi^{(1,2)}$ which follows from the orthogonality between $\mathbf{l}$ and $\phi^{(1,2)}$, we have

$$\partial_i \phi^{(1,2)} = -(\phi^{(1,2)} \partial_i \mathbf{l}) \mathbf{l} \mp 2m v_{si} \phi^{(2,1)}.$$ (4A.5)

By squaring this, we obtain

$$(\partial_i \phi^{(1)})^2 = (\phi^{(1)} \partial_i \mathbf{l})^2 + 4m^2 \mathbf{v}_s^2.$$ (4A.6)

Adding once more the same term with $\phi^{(1)}$ and $\phi^{(2)}$ interchanged we obtain

$$
\begin{aligned}
|\partial_i \phi|^2 &= (\phi^{(1)} \partial_i \mathbf{l})^2 + (\phi^{(2)} \partial_i \mathbf{l})^2 + 8m^2 \mathbf{v}s^2 \\
&= (\partial_i \mathbf{l})^2 + 8m^2 \mathbf{v}_s^2,
\end{aligned}
$$ (4A.7)

having dropped a trivially vanishing term $-(\mathbf{l} \partial_i \mathbf{l})^2$. The first term can be decomposed into splay, twist, and bend terms as

$$(\partial_i \mathbf{l})^2 = (\boldsymbol{\nabla} \cdot \mathbf{l})^2 + [\mathbf{l} \cdot (\mathbf{v} \times \mathbf{l})]^2 + [\mathbf{l} \times (\mathbf{v} \times \mathbf{l})]^2$$ (4A.8)

so that we find the final form

$$|\partial_i \phi|^2 = (\boldsymbol{\nabla} \cdot \mathbf{l})^2 + [\mathbf{l} \cdot (\boldsymbol{\nabla} \times \mathbf{l})]^2 + [\mathbf{l} \times (\boldsymbol{\nabla} \times \mathbf{l})]^2 + 8m^2 \mathbf{v}_s^2.$$ (4A.9)

The third derivative term ϕ is treated as follows:

$$
\begin{aligned}
|\boldsymbol{\nabla} \phi|^2 &= (\boldsymbol{\nabla} \phi^{(1)})^2 + (\boldsymbol{\nabla} \phi^{(2)})^2 \\
&= \left[(\mathbf{l} \partial_i \phi^{(1)}) l_i + (\phi^{(2)} \partial_i \phi^{(1)}) \phi_i^{(2)}\right]^2 + (1 \leftrightarrow 2) \\
&= [-(\phi^{(1)} \partial_i \mathbf{l}) l_i - 2m v_{si} \phi_i^{(2)}]^2 + (1 \leftrightarrow 2, v_s \to -v_s) \\
&= (\mathbf{l} \boldsymbol{\nabla} l_j)^2 + 4m v_{sk} [\phi_k^{(2)} \phi_j^{(1)} - (1 \leftrightarrow 2)] (\partial_i l_j) l_i + 4m^2 \left[\mathbf{v}_s^2 - (\mathbf{l} \cdot \mathbf{v}_s)^2\right].
\end{aligned}
$$ (4A.10)

Here the first term is of the pure bend form

$$[\mathbf{l} \boldsymbol{\nabla} l_j]^2 = [\mathbf{l} \times (\boldsymbol{\nabla} \times \mathbf{l})]^2.$$ (4A.11)

The second term can be rewritten using

$$\phi_k^{(2)}\phi_j^{(1)} - \phi_k^{(1)}\phi_j^{(2)} = -\epsilon_{kjm}l_m \tag{4A.12}$$

as

$$-mv_{sk}\epsilon_{kjm}l_m(\partial_i l_j)l_i. \tag{4A.13}$$

With the formula

$$l_i\epsilon_{kjm} = l_k\epsilon_{ijm} + l_j\epsilon_{kim} + l_m\epsilon_{kji} \tag{4A.14}$$

it becomes

$$-4m\left(\mathbf{v}_s\cdot\mathbf{l}\right)\left[\mathbf{l}\cdot\left(\boldsymbol{\nabla}\times\mathbf{l}\right)\right] + 4m\mathbf{v}_s\cdot\left(\boldsymbol{\nabla}\times\mathbf{l}\right). \tag{4A.15}$$

The second gradient term in (4A.1) becomes, finally, by a similar treatment:

$$\partial_i\phi_j^*\partial_j\phi_i = [\mathbf{l}\times(\boldsymbol{\nabla}\times\mathbf{l})]^2 + 4m^2[\mathbf{v}_s^2 - (\mathbf{v}_s\cdot\mathbf{l})^2] - 4m(\mathbf{v}_s\cdot\mathbf{l})[\mathbf{l}\cdot(\boldsymbol{\nabla}\times\mathbf{l})]. \tag{4A.16}$$

Hence, the pure ϕ part of the gradient energy is

$$\begin{aligned}
e^\phi = \ & \frac{1}{2}\Delta_A^2\left\{4m^2(2K_1 + K_{23})\mathbf{v}_s^2 - 4m^2K_{23}\left(\mathbf{l}\cdot\mathbf{v}_s\right)^2\right.\\
& +4mK_3\mathbf{v}_s\cdot(\boldsymbol{\nabla}\times\mathbf{l}) - 4mK_{23}\left(\mathbf{v}_s\cdot\mathbf{l}\right)\left[\mathbf{l}\cdot(\boldsymbol{\nabla}\times\mathbf{l})\right]\\
& \left.+K_1\left(\boldsymbol{\nabla}\cdot\mathbf{l}\right)^2 + K_2\left[\mathbf{l}\cdot(\boldsymbol{\nabla}\times\mathbf{l})\right]^2 + (K_1 + K_{23})\left[\mathbf{l}\times(\boldsymbol{\nabla}\times\mathbf{l})\right]^2\right\}.
\end{aligned} \tag{4A.17}$$

If the $\mathbf{d}$ bending energies are neglected, we find the hydrodynamic energy (4.125) with the coefficients

$$\begin{aligned}
\rho_s &= \Delta_A(2K_1 + K_{23})\,4m^2, \tag{4A.18}\\
\rho_0 &= 2mc_0 = \Delta_A^2 K_{23}\,4m^2, \tag{4A.19}\\
c &= \Delta_A^2 K_3\,2m, \tag{4A.20}\\
c_0 &= \Delta_A^2 K_{23}\,2m, \tag{4A.21}\\
K_s &= K_t = \Delta_A^2 K_1, \tag{4A.22}\\
K_b &= \Delta_A^2(K_1 + K_{23}). \tag{4A.23}
\end{aligned}$$

Inserting the weak-coupling results (4.82) for $K_1, 2, 3$, one has

$$\rho_s = 2\rho\left(1 - \frac{T}{T_c}\right) \tag{4A.24}$$

and the relations

$$\rho_0 = \frac{1}{2}\rho_s = c_0\,2m = 2c\,2m, \tag{4A.25}$$

$$K_s = K_t = \frac{1}{4m^2}\frac{1}{4}\rho_s; \quad K_b = \frac{1}{4m^2}\frac{3}{4}\rho_s. \tag{4A.26}$$

The terms containing the **d**-vectors can be processed similarly. With

$$
\begin{aligned}
|\phi \nabla d_a|^2 &= (\phi^{(1)} \nabla d_a)^2 + (\phi^{(2)} \nabla d_a)^2 \\
&= (\partial_i d_a)^2 - (\mathbf{l} \nabla d_a)^2
\end{aligned}
\tag{4A.27}
$$

we obtain

$$
e^d = \frac{1}{2} \Delta_A^2 \left\{ (2K_1 + K_{23})(\partial_i d_a)^2 - K_{23}(\mathbf{l} \nabla d_a)^2 \right\}
\tag{4A.28}
$$

amounting to the bending constants

$$
K_1^d = \Delta_A^2 (2K_1 + K_{23}), \qquad K_2^d = \Delta_A^2 K_{23}.
\tag{4A.29}
$$

In the case that **d** and **l** are locked to each other by the dipole energy, the general bending energy of the d_a field

$$
e^d = \frac{1}{2} \left\{ K_1^d (\partial_i d_a)^2 - K_2^d (\mathbf{l} \nabla d_a)^2 \right\}
\tag{4A.30}
$$

contributes to the l field an energy

$$
\begin{aligned}
f_{\text{locked}}^d &= \frac{1}{2} (K_1^d \left\{ (\nabla \cdot \mathbf{l})^2 + [\mathbf{l} \cdot (\nabla \times \mathbf{l})]^2 + [\mathbf{l} \times (\nabla \times \mathbf{l})]^2 \right\} \\
&\quad - K_2^d [\mathbf{l} \times (\nabla \times \mathbf{l})]^2).
\end{aligned}
\tag{4A.31}
$$

Adding this to (4.125) we obtain again the general form (4.125), now with the coefficients

$$
K_t^l = K_t + K_1^d, \qquad K_s^l = K_s + K_1^d, \qquad K_b = K_b + K_1^d - K_2^d.
\tag{4A.32}
$$

For the present case with the coefficients (4A.17) and (4A.28) this gives

$$
K_s = K_t = K_b = \Delta_A^2 (3K_1 + K_{23}).
\tag{4A.33}
$$

In the weak-coupling limit these are related to the superfluid density by

$$
K_{s,t,b} = \frac{1}{4m^2} \frac{5}{4} \rho_s.
\tag{4A.34}
$$

Appendix 4B   Hydrodynamic Coefficients for All $T \leq T_c$

For arbitrary temperatures $T \leq T_c$, the hydrodynamic limit is

$$
f = \frac{1}{4m^2} \rho_{ijkl} \partial_k A_{ai}^* \partial_l A_{aj} \frac{1}{\Delta_{AB}^2} - \left\{ \begin{array}{c} \frac{1}{4m^2} \bar{\rho}_{ijkl} \partial_k l_i \partial_l l_j \\ 0 \end{array} \right\}, \qquad \left\{ \begin{array}{c} A \\ B \end{array} \right\} \text{phase}
\tag{4B.1}
$$

with A_{ai} having the forms (4.103), (4.104) but being permitted to contain smooth spatial variations of the direction vectors. We now evaluate this further for the two phases:

A-phase

Here, we have to contract the three covariants of (4.343),

$$\hat{A}_{ijkl} \equiv \delta_{ij}\delta_{kl} + \delta_{il}\delta_{jk} + \delta_{ik}\delta_{jl},$$
$$\hat{B}_{ijkl} \equiv \delta_{ij}l_kl_l + \delta_{ik}l_jl_l + \delta_{il}l_jl_k + \delta_{kj}l_il_l + \delta_{lj}l_il_k + \delta_{kl}l_il_j,$$
$$\hat{C}_{ijkl} \equiv l_il_jl_kl_l, \tag{4B.2}$$

with

$$\partial_k\left(d_a\Phi_i^*\right)\partial_l\left(d_a\Phi_j\right) = \left(\partial_k d_a\partial_l d_a\right)\Phi_i^*\Phi_j + \partial_k\Phi_i^*\partial_l\Phi_j. \tag{4B.3}$$

From $\hat{A}$ we find

$$\hat{A}: \quad |\partial_i\Phi_j|^2 + \partial_i\Phi_j^*\partial_j\Phi_i + |\boldsymbol{\nabla}\boldsymbol{\Phi}|^2 + 2(\boldsymbol{\nabla}d_a)^2 + 2|\boldsymbol{\Phi}\boldsymbol{\nabla}d_a|^2. \tag{4B.4}$$

These gradient terms have been expanded in Appendix 4A in terms of the generic hydrodynamic gradient terms in the energy (4.125). If we use the following short notation for the various invariants in that energy

$$\hat{\rho} \equiv 4m^2\mathbf{v}_s^2, \quad \hat{\rho}_0 \equiv -4m^2\left(\mathbf{1}\cdot\mathbf{v}_s\right)^2,$$
$$\hat{c} \equiv 2m\mathbf{v}_s\cdot(\boldsymbol{\nabla}\times\mathbf{1}), \quad \hat{c}_0 \equiv -2m\left(\mathbf{1}\cdot\mathbf{v}_s\right)[\mathbf{1}\cdot(\boldsymbol{\nabla}\times\mathbf{1})],$$
$$\hat{s} \equiv (\boldsymbol{\nabla}\cdot\mathbf{1})^2, \quad \hat{t} \equiv [\mathbf{1}\cdot(\boldsymbol{\nabla}\times\mathbf{1})]^2, \quad \hat{b} \equiv [\mathbf{1}\times(\boldsymbol{\nabla}\times\mathbf{1})]^2,$$
$$\hat{k}_1^d \equiv (\partial_i d_a)^2, \quad \hat{k}_2^d \equiv -\left(\mathbf{1}\cdot\boldsymbol{\nabla}d_a\right)^2, \tag{4B.5}$$

the hydrodynamic expansion reads

$$f = \frac{1}{2}\left(\frac{\rho_s}{4m^2}\hat{\rho} + \frac{\varrho_0}{4m^2}\hat{\rho}_0 + \frac{c}{2m}\hat{c} + \frac{c_0}{2m}\hat{c}_0 + K_s\hat{s} + K_t\hat{t} + K_b\hat{b} + K_1^d\hat{k}_1^d + K_2^d\hat{k}_2^d\right). \tag{4B.6}$$

With the same invariants we can write (4B.4) as

$$\left(\hat{s} + \hat{b} + \hat{t} + 2\hat{\rho}\right) + \left(\hat{b} + \hat{\rho} + \hat{\rho}_0 + \hat{c}_0\right) + \left(\hat{b} + \hat{\rho} + \hat{\rho}_0 + \hat{c} + \hat{c}_0\right) + 4\hat{K}_1^d + 2\hat{K}_2^d, \tag{4B.7}$$

where parentheses indicate the different terms in (4B.4).

The covariant $\hat{B}_{ijkl}$ has a very simple contribution to the **d** bending energy

$$\hat{B}: \quad 2\left(\mathbf{1}\cdot\boldsymbol{\nabla}d_a\right)^2 = -2\hat{K}_2^d, \tag{4B.8}$$

as follows immediately from $\boldsymbol{\phi}\mathbf{1} = 0$. As far as the gradient terms of the $\boldsymbol{\phi}$ field are concerned we use (4A.5) to rewrite

$$\partial_k\Phi_i^*\partial_l\Phi_i = \left(\boldsymbol{\phi}^{(1)}\partial_k\mathbf{1}\right)l_i\left(\boldsymbol{\phi}^{(1)}\partial_l\mathbf{1}\right)l_j + (1\leftrightarrow 2) - 4m^2 v_{sk}v_{sl}\left[\phi_i^{(1)}\phi_j^{(1)} + (1\leftrightarrow 2)\right]$$
$$+ \left\{\left[2mv_{sl}\phi_j^{(2)}(\boldsymbol{\phi}^{(1)}\partial_k l)l_i + (k\leftrightarrow l, i\leftrightarrow j)\right] - [1\leftrightarrow 2]\right\} \tag{4B.9}$$

and employ (4A.12) to bring the terms in curly brackets to the form

$$-2mv_{sl}\epsilon_{jmn}l_n\partial_k l_m l_i + (k \leftrightarrow l, \quad i \leftrightarrow j). \tag{4B.10}$$

Contracting the pure l terms of (4B.9) with $\hat{B}_{ijkl}$ we find

$$\hat{B}: \quad 5[\phi^{(1)}(\mathbf{l}\cdot\boldsymbol{\nabla})\mathbf{l}]^2 + (\phi^{(1)}\partial_k\mathbf{l})^2 + (1 \leftrightarrow 2) = 5\,(\mathbf{l}\cdot\boldsymbol{\nabla}l_i)^2 + (\partial_k l_i)^2$$
$$= 5\hat{b} + (\hat{s} + \hat{t} + \hat{b}). \tag{4B.11}$$

The first v-terms in (4B.9), on the other hand, contribute

$$\hat{B}: \qquad 4m^2\,(\mathbf{l}\mathbf{v}_s)^2 = -2\hat{\rho}_0 \tag{4B.12}$$

while the others extracted in (4B.10) add to this

$$\hat{B}: \quad -2mv_i\epsilon_{imn}l_n(\mathbf{l}\boldsymbol{\nabla})l_m - 2m(\mathbf{l}v_s)\epsilon_{imn}l_n\partial_i l_m$$
$$\quad - 2m(\mathbf{l}v_s)\epsilon_{imn}l_n\partial_i l_m - 2mv_i\epsilon_{imn}l_n(\mathbf{l}\boldsymbol{\nabla})l_m$$
$$= -4m(\mathbf{l}\cdot\mathbf{v}_s)[\mathbf{l}\cdot(\boldsymbol{\nabla}\times\mathbf{l}) + 4m[\mathbf{v}_s\cdot(\boldsymbol{\nabla}\times\mathbf{l})] - 4m(\mathbf{l}\cdot\mathbf{v}_s)[\mathbf{l}\cdot(\boldsymbol{\nabla}\times\mathbf{l})]$$
$$= \hat{c} + 2\hat{c}_0. \tag{4B.13}$$

The contributions of the third covariant $\hat{C}_{ijkl}$, finally, are obtained by contracting four l-vectors with (4B.8) giving

$$\hat{C}: \quad [\phi'(\mathbf{l}\cdot\boldsymbol{\nabla})\mathbf{l}]^2 + [1 \to 2] = [(\mathbf{l}\cdot\boldsymbol{\nabla})l]^2 = [\mathbf{l}\cdot(\boldsymbol{\nabla}\times\mathbf{l})]^2 = \hat{b}. \tag{4B.14}$$

Collecting all terms we obtain

$$(A\hat{A} + B\hat{B} + C\hat{C})_{ijkl}\partial_k(d_a\Phi_i^*)\partial_l(d_a\Phi_j) = 4A\hat{\rho} + 2(A-B)\hat{\rho}_0 + 4A\hat{K}_1^d + 2(A-B)\hat{K}_2^d$$
$$+(A+B)\hat{C} + 2(A+B)\hat{C}_0 + (A+B)\hat{s} + (A+B)\hat{t} + (3A+6B+C)\hat{b}. \tag{4B.15}$$

Inserting (4.345)-(4.346) we obtain the energy (4.125) with the coefficients

$$\begin{aligned}
2mC &= \tfrac{1}{2}\rho_s^{\|}, & 2mc^{\|} &= 2m(c_0 - c) = \tfrac{1}{2}\rho_s^{\|}, \\
4m^2K_1^d &= \rho_s, & 4m^2K_2^d &= \rho_0, \\
4m^2K_s &= 4m^2K_t = \tfrac{1}{2}\rho_s^{\|}, & 4m^2K_b &= \tfrac{3}{4}\gamma.
\end{aligned}$$

We now turn to the $\bar{\rho}_{ijkl}$-term in the gradient energy (4B.1). This tensor has once more the same expansion into covariants

$$\bar{A}\hat{A}_{ijkl} + \bar{B}\hat{B}_{ijkl} + \bar{C}\hat{C}_{ijk}, \tag{4B.16}$$

with the coefficients $\bar{A}$ and $\bar{B}$ given by (4.370) while $\bar{C}_{ijkl}$ does not contribute when contracting it with $\partial_k l_i\partial_l l_j$ as required by (4.369). In fact, doing this contraction on (4B.16) gives

$$\bar{A}(3\hat{s} + \hat{t} + \hat{b}) + \bar{B}\hat{b}, \quad \bar{A} = \bar{\rho}_s/8, \quad \bar{A} + \bar{B} = \bar{\rho}_s^{\|}/4. \tag{4B.17}$$

This adds $-3\bar{A}, -\bar{A}, -(\bar{A}+\bar{B})$ to the bending constants $\frac{1}{2}4m^2 K_{s,t,b}$, respectively, which therefore become

$$4m^2 K_s = \rho_s/4, \quad 4m^2 K_t = (\rho_s + 4\rho_s^{\parallel})/12, \quad 4m^2 K_b = (\rho_s^{\parallel} + \gamma)/2 \quad (4\text{B}.18)$$

as stated in (4.375).

B-phase

Let the vacuum be given by

$$A_{ai}^0 = \Delta_B R_{ai}(\theta_0)e^{-i\varphi_0}. \quad (4\text{B}.19)$$

We may parametrize the oscillators around this nonzero value by letting

$$R_{ai}(\theta) = R_{aj}(\theta_0)R_{ji}(\tilde{\theta}). \quad (4\text{B}.20)$$

Since the subscripts a of A'_{ai} are always contracted, we may also use

$$\tilde{A}_{ai} \equiv R^{-1}(\theta_0)_{aa'}A_{a'i} \quad (4\text{B}.21)$$

as an order parameter without changing the energy. With this the derivative terms of the field become simply

$$\partial_k \tilde{A}_{ai} = -iL_{ai}^c \partial_k \tilde{\theta}_c = -\epsilon_{cai}\partial_k \tilde{\theta}_c, \quad (4\text{B}.22)$$

where L_{ai}^c are the 3×3 generating matrices of the rotation group $L_{ai}^c = -i\epsilon_{cai}$.

Consider now the expression (4B.1) with coefficient in the B-phase being:

$$\rho_{ijkl} = \frac{3}{2}\rho_s^b \frac{1}{\Delta_B^2}\frac{1}{15}\left(\delta_{ij}\delta_{kl} + \delta_{il}\delta_{jk} + \delta_{ik}\delta_{jl}\right). \quad (4\text{B}.23)$$

The derivatives are

$$\partial_k A_{ai}^* \partial_l A_{aj} = \partial_k \tilde{A}_{ai}^* \partial_l \tilde{A}_{aj}$$
$$= \Delta_B^2 \left(\partial_k \varphi \partial_l \varphi \delta_{ij} + \partial_k \tilde{R}_{ai}\partial_l \tilde{R}_{aj}\right) + \text{mixed terms}. \quad (4\text{B}.24)$$

The mixed terms can be neglected since

$$\Delta_B^2 i \left(\partial_k \tilde{R}_{ai}\tilde{R}_{aj}\partial_l \varphi - \tilde{R}_{ai}\partial_l \tilde{R}_{aj}\partial_k \varphi\right) \quad (4\text{B}.25)$$

is antisymmetric under $(i \leftrightarrow j, \ k \leftrightarrow l)$, while (4B.23) is symmetric. Contracting this with the covariant in (4B.21) gives

$$\Delta_B^2 \left[(3+1+1)(\partial_i\varphi)^2 + \partial_k\tilde{\theta}_c\partial_l\tilde{\theta}_d \epsilon_{cai}\,\epsilon_{caj}\,(\delta_{ij}\delta_{kl} + \delta_{il} + \delta_{ik}\delta_{jl})\right]$$
$$= \Delta_B^2 \left\{5(\partial_i\varphi)^2 + 2(\partial_i\tilde{\theta}_j)^2 + \left[(\partial_i\tilde{\theta}_j)^2 - (\partial_i\tilde{\theta})^2\right] + \left[(\partial_i\tilde{\theta}_j)^2 - \partial_i\tilde{\theta}_j\partial_j\tilde{\theta}_i\right]\right\}, \quad (4\text{B}.26)$$

so that

$$4m^2 f = \frac{\rho_s^B}{2}\left[(\boldsymbol{\nabla}\varphi)^2 + \frac{4}{5}(\partial_i\tilde{\theta}_j)^2 - \frac{1}{5}(\boldsymbol{\nabla}\tilde{\theta})^2 - \frac{1}{5}(\partial_i\tilde{\theta}_j\partial_j\tilde{\theta}_j)\right], \quad (4\text{B}.27)$$

as given in (4.358).

Appendix 4C   Generalized Ginzburg-Landau Energy

If one assumes all temperature dependence to come from $\rho_0 = \rho_s^\parallel (1 - \epsilon) \equiv \rho_s^\parallel \alpha$, the coefficients of the energy are in the dipole-locked regime:

$$A = 1 + \alpha s \underset{\text{GL}}{=} 1 + s, \qquad\qquad\qquad\qquad\qquad (4C.1)$$

$$A_g = -A^{-1},$$

$$A_g' = \alpha A^{-2} \underset{\text{GL}}{=} \frac{1}{(1+s)^2},$$

$$A_g'' = -2\alpha^2 A^{-3} \underset{\text{GL}}{=} -2\frac{1}{(1+s)^3},$$

$$M_g = \left(1 - sA^{-1}\right)\sqrt{1-s} \underset{\text{GL}}{=} \frac{\sqrt{1-s}}{1+s},$$

$$M_g' = -\left[1 + (2 - 3s)A^{-1} - 2(1 - s)\alpha s A^{-3}\right]/2\sqrt{1-s} \underset{\text{GL}}{=} -\frac{1}{2\sqrt{1-s}}\frac{3-s}{(1+s)^2},$$

$$M_g'' = -\left[1 - (4 - 3s)A^{-1} - 4(3s^2 - 5s + 2)A^{-2} + 8(1-s)^2 s\alpha A^{-3}\right]/4(1-s)^{3/2}$$
$$\underset{\text{GL}}{=} \frac{1}{4(1-s)^{3/2}}\frac{1}{(1+s)^4}\left(11 - 7s - 15s^2 + 3s^3\right),$$

$$G_g = \frac{s}{2}\left(5 - 2s(1-s)A^{-1}\right) \underset{\text{GL}}{=} \frac{c}{2}\frac{5 + 3s + 2s^2}{1+s},$$

$$G_g' = \frac{5}{2} - s\left[2 - 3s - s(1-s)A^{-1}\right]A^{-1} \underset{\text{GL}}{=} \frac{1}{2(1+s)^2}(5 + 6s9s^2 + 4s^3),$$

$$G_g'' = 2(3s - 1)A^{-1} - 2(3s - 2)s\alpha A^{-2} - 2s^2(1-s)\alpha A^{-3}$$
$$\underset{\text{GL}}{=} \frac{2}{(1+s)^3}(s^3 + 3s^2 + 3s - 1).$$

Notes and References

[1] J. Bardeen, L.N. Cooper, and J.R. Schrieffer, Phys. Rev. **108**, 1175 (1957).

[2] See the little textbook from Bogoliubov's school:
N.N. Bogoliubov, E.A. Tolkachev, and D.V. Shirkov, *A New Method in the Theory of Superconductivity*, Consultants Bureau, New York, 1959.

[3] A. Bohr, B.R. Mottelson, and D. Pines, Phys. Rev. **110**, 936 (1958).
B. Mottelson, *The Many-Body Problem*. Les Houches Lectures 1958, Dunod, Paris, 1959.

[4] L.N. Cooper, R.L. Mills, and A.M. Sessler, Phys. Rev. **114**, 1377 (1959).

[5] S.T. Belyaev, Mat. Fys. Medd. **31**, 11 (1959).
See also the author's 1964 M.S. thesis klnrt.de/0/0.pdf and recent papers;
E. Flowers, M. Ruderman, and P. Sutherland, Astrophys. J. **205**, 541 (1976);

M. Baldo, O. Elgaroey, L. Engvik, M. Hjorth-Jensen, and H.J. Schulze, Phys. Rev. C **58**, 1921 (1998);
D.J. Dean and M. Hjorth-Jensen, Rev. Mod. Phys. **75**, 607 (2003);
G. Baym, Lecture notes, www.conferences.uiuc.edu/bcs50/PDF/Baym.pdf.

[6] L.N. Pitaevski, Zh. Exp. Teor. Fiz. **37**, 1794 (1959). (Sov. Phys. JETP **10**, 1267 (1960).
K.A. Brueckner, T. Soda, P.W. Anderson, and P. Morel, Phys. Rev. **118**, 1442 (1960).
V.J. Emery and A.M. Sessler, Phys. Rev. **119**, 43 (1960).

[7] D.D. Osheroff, W.J. Gully, R.C. Richardson, and D.M. Lee, Phys. Rev. Lett. **29**, 920 (1972).
See also:
Z.M. Galasiewicz, Phys. Cond. Matter **18**, 141 (1974);
A.J. Leggett, Rev. Mod. Phys. **47**, 331 (1975).

[8] P.W. Anderson and P. Morel, Phys. Rev. **123**, 1911 (1960).

[9] R. Balian and N.R. Werthamer, Phys. Rev. **131**, 1553 (1963).

[10] For a review see:
P.W. Anderson and W.F. Brinkmann, Lecture at the Fifteenth Scottish Universities Summer School, 1944, in *The Helium Liquids*, Edited by J.G.J. Armitage and I.W. Farquhar, Academic Press.

[11] N.F. Berk and J.R. Schrieffer, Phys. Rev. Lett. **17**, 433 (1966);
A.J. Layzer and D. Fay, Proc. Int. Low Temp. Conf. (LT-11), 1968.

[12] D.S. Greywall, Phys. Rev. B **33**, 7520 (1986).

[13] J.W. Wheatley, Rev. Mod. Phys. **47**, 415 (1975).

[14] A.A. Abrikosov, L.P. Gorkov, and I.E. Dzyaloshinski, *Methods of Quantum Field Theory in Statistical Physics*, Dover, New York (1975);
L.P. Kadanoff and G. Baym, *Quantum Statistical Mechanics*, Benjamin, New York (1962);
A.L. Fetter and J.D. Walecka, *Quantum Theory of Many-Particle Systems*, McGraw-Hill, New York (1971).

[15] V. Ambegaokar, P.G. de Gennes, and D. Rainer, Phys. Rev. A **9**, 2676 (1974);
A **12**, 245 (1975).
V. Ambegaokar, Lectures presented at the 1974 Canadian Summer School.

[16] P.T. Mathews and A. Salam, Nuovo Cimento **12**, 563 (1954); **2**, 120 (1955).

[17] H. Kleinert, Fortschr. Phys. **26**, 397 (1978).

[18] H.E. Stanley, *Phase Transitions and Critical Phenomena*, Clarendon Press, Oxford (1971).
See also:
L.P. Kadanoff, Rev. Mod. Phys. **49**, 267 (1977);
F.J. Wegner, in *Phase Transitions and Critical Phenomena*, ed. by C. Domb and M.S. Green, Academic Press, New York 1976, p. 7.
E. Brézin, J.C. Le Guillou, and J. Zinn-Justin, ibid, p. 125;
H. Kleinert and V. Schulte-Frohlinde, *Critical Phenomena in ϕ^4-Theory*, World Scientific, 2001, pp. 1–487 (klnrt.de/b8).

[19] A.J. Leggett, Rev. Mod. Phys. **47**, 331 (1975).

[20] H. Kleinert, Phys. Lett. B **69**, 9 (1977); Lett. Nuovo Cimento **31**, 521, (1981); Phys. Lett. A **84**, 199 (1981); Phys. Lett. A **84**, 259, (1981).

[21] G. Barton and M.A. Moore, Journal of Physics C **7**, 4220 (1974).

[22] P.G. de Gennes, *The Physics of Liquid Crystals*, Clarendon Press, Oxford (1974).

[23] J.D. Jackson, *Classical Electrodynamics*, J. Wiley & Sons, New York (1967), p. 152.

[24] N.D. Mermin, lectures presented at the Erice Summer School on Low Temperature Physics, June 1977.

[25] N.D. Mermin, remarks prepared for the Sanibel Symposium on Quantum Fluids and Solids, 1977 and Physica **90 B**, 1 (1977).

[26] K. Maki and P. Kumar, Phys. Rev. Lett. **38**, 557 (1977);
K. Maki, Nucl. Phys. B **90**, 84 (1977).

[27] C.M. Gould and D.M. Lee, Phys. Rev. Lett. **37**, 1223 (1976).

[28] S. Coleman, Phys. Rev. D **11**, 2088 (1975);

[29] N.D. Mermin and T.L. Ho, Phys. Rev. Lett. **36**, 594 (1976).

[30] G. 't Hooft, Nucl. Phys. **B 79**, 276 (1974).
A.M. Polyakov, JETP Letters **20**, 194 (1974).

[31] K. Maki and T. Tsuneto, J. Low Temp. Phys. **27**, 635 (1976).
L.J. Buchholtz and A.L. Fetter, Phys. Rev. B **15**, 5225 (1977).

[32] G.E. Volovik and V.P. Mineev, Pis'ma Zh. Exsp. Teor. Fiz. **24**, 605 (1976).

[33] N. Steenrod, *The Topology of Fiber Bundles*, Princeton University Press, Princeton, 1951.

[34] V.P. Mineev and G.E. Volovik, Phys. Rev. B **18**, 3197 (1978).

[35] G.E. Volovik and V.P. Mineev, Zh. Eksp. Teor. Fiz. **73**, 767 (1977).

[36] P. Bhattacharyya, T.L. Ho, and N.D. Mermin, Phys. Rev. Lett. **39**, 1290 (1977).

[37] M.C. Cross, J. Low Temp. Phys. **21**, 525 (1975).

[38] M.C. Cross and M. Liu, J. Phys. C, Solid State Phys. **11**, 1795 (1978). G. Barton and M.A. Moore, Journal of Physics C **7**, 4220 (1974).

[39] H. Kleinert, Lectures presented at the Erice Summer School on Low Temperature Physics, June 1977;
See also:
H. Kleinert, *What Can a Particle Physicist Learn from Superliquid ^{3}He?*, lectures presented at the International School of Subnuclear Physics, Erice, 1978, in *The New Aspects of Subnuclear Physics, Plenum Press 1980*, A. Zichichi, ed. (klnrt.de/60/60.pdf).

[40] H. Kleinert, Y.R. Lin-Liu, and K. Maki, Phys. Lett. A **70**, 27 (1979); See also: A.L. Fetter, *ibid.* and Phys. Rev. Lett. **40**, 1656 (1978).

[41] Y.R. Lin-Liu, K. Maki, and D. Vollhardt, J. Phys. (Paris) Lett. 39, 381 (1978); D. Vollhardt, Y.R. Lin-Liu, and K. Maki, J. Low Temp. Phys. 37, 627 (1979).
See also:
A.L. Fetter, Phys. Rev. Lett. 40, 1656 (1978);
A.L. Fetter and M.R. Williams, Phys. Rev. B 23, 2186 (1981);
See also the textbook
D. Vollhardt and P. Wölfle, *The Superfluid Phases of Helium 3*, Taylor & Francis, London, 1990.

[42] R.L. Kleinberg, Phys. Rev. Lett. **42**, 182 (1979); J. Low Temp. Phys. **35**, 489 (1979).

[43] P. Wölfle, *Progress in Low Temperature Physics*, Vol. VII, ed. D.F. Brewer (North Holland, Amsterdam, 1978).

[44] H. Kleinert, *Multivalued Fields in Condensed Matter, Electromagnetism, and Gravitation*, World Scientific, 2009, pp. 1–497 (http://klnrt.de/b11).

[45] H. Kleinert, *Particles and Quantum Fields*, World Scientific, 2014, (http://klnrt.de/b6).

The supreme misfortune is when theory outstrips performance.
LEONARDO DA VINCI (1452–1519)

5

Liquid Crystals

In 1888, Friedrich Reinitzer investigated the thermodynamic properties of crystals of cholesteryl benzoate [1]. He observed that they melt at 145.5°C to form a cloudy liquid. This was stable up to 178.5°C, before it melted again to form a clear liquid. The cloudy liquid is a new phase of matter intermediate between a crystal and a liquid which is now referred to as a *liquid crystal*. A liquid crystal is a system of rod-like or disk-like molecules which behave under translations in the same way as the molecules in an ordinary liquid, while their molecular orientations can undergo phase transitions into states of long-range order, a typical property of crystals.

In this part of the book we shall focus our attention on molecules whose shape strongly deviates from spherical symmetry but which mechanically have no dipole properties, i.e., a reversal of the direction of the principal axis remains energetically negligible. Examples for such molecules are p, p'-azoxyanisole shortly called PAA, or p-methoxybenzylidene-p-n-butylaniline, usually abbreviated as MBBA. The chemical structure of the latter is shown in Fig. 5.1.

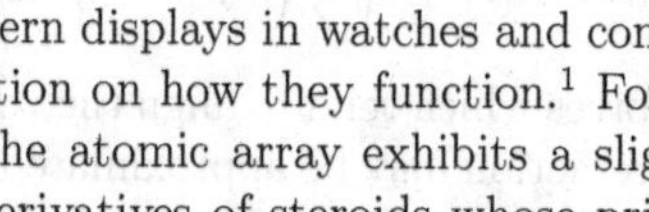

FIGURE 5.1 Molecular structure of PAA.

The long molecules in this and similar materials opened the way for the construction of all modern displays in watches and computers [2]. See the internet page for a movie illustration on how they function.[1] For some of these long molecules it may happen that the atomic array exhibits a slight screw-like structure. This is the case in many derivatives of steroids whose prime example is cholesterol. Such molecules violate mirror reflection symmetry.

A satisfactory description of the long-range correlations in such systems can again be given by means of a collective field theory. It is constructed by using the lowest non-vanishing multipole moment of the molecules as a local field characterizing the

[1] http://plc.cwru.edu/tutorial/enhanced/files/lcd/tn/tn.HTM.

orientation of the molecules and expanding the free energy in a power series in this field and its derivatives. The thermodynamic properties are then obtained by calculating the partition function for all fluctuating field configurations:

$$Z = \sum_{\substack{\text{field} \\ \text{configurations}}} e^{-\text{Energy}/k_B T}. \tag{5.1}$$

If the system is not extremely close to a critical point, where fluctuations become important, the partition function can be approximated by the field configuration which extremizes the energy (saddle point method). This is equivalent to considering the collective field as a mean-field variable of the Landau type [3, 4, 5, 6].

For a theoretical description of the system we derive first a mean-field theory in terms of a non-fluctuating order parameter [3], [4]. This is extended by derivative terms to find a Ginzburg-Landau type of field theory [5] which describes the physical properties resulting from the long-range fluctuations or the order field.

5.1 Maier-Saupe Model and Generalizations

The simplest microscopic model for the description of phase transitions in liquid crystals was constructed by Maier and Saupe. It is based on the standard molecular field approximation invented long time ago by Pierre Weiss [7] to explain ferromagnetism. By construction, the model is confined to nematic systems.

5.1.1 General Properties

The molecules are assumed to be non-polar, rod-like objects. If the direction of the body axis is denoted by the unit vector $\mathbf{n}(x)$, the instantaneous orientation may be characterized by the traceless tensor field

$$Q_{\alpha\beta}^{\text{mol}}(x) = \epsilon_{\alpha\beta}^{(0)}\left(\mathbf{n}(x)\right) = \sqrt{\frac{3}{2}}\left(n_\alpha(x)n_\beta(x) - \frac{1}{3}\delta_{\alpha\beta}\right). \tag{5.2}$$

In the normal phase, this field fluctuates around zero. Below the phase transition, however, there is a non-vanishing average order

$$Q_{\alpha\beta} = \langle Q_{\alpha\beta}^{\text{mol}}\rangle = S\epsilon^{(0)}(\bar{\mathbf{n}}) = S\sqrt{\frac{3}{2}}\left(\bar{n}_\alpha\bar{n}_\beta - \frac{1}{3}\delta_{\alpha\beta}\right). \tag{5.3}$$

This is due to the intermolecular forces which tend to align the vector field $n_\alpha(x)$ to a common average value $\bar{n}_\alpha$. The interaction may be approximated by an orientational energy

$$H_{\text{or}} = -A_0 Q_{\alpha\beta}^{\text{mol}} Q_{\alpha\beta} \tag{5.4}$$

with some coupling strength A_0. Inserting this into Boltzmann's distribution law, one finds the self-consistency relation

$$Q_{\alpha\beta} = \langle Q_{\alpha\beta}\rangle = Z^{-1}\int \frac{d^2\mathbf{n}}{4\pi} Q_{\alpha\beta}^{\text{mol}} e^{-H_{\text{or}}/k_B T}, \tag{5.5}$$

where

$$Z = \int \frac{d^2\mathbf{n}}{4\pi} e^{-H_{or}/k_BT}.$$ (5.6)

Orienting $\bar{\mathbf{n}}$ along the z-direction and setting $\mathbf{n} \cdot \bar{\mathbf{n}} \equiv z$, we see that

$$S = \frac{1}{Z} \frac{3}{2} \int_0^1 dz \left(z^2 - \frac{1}{3}\right) e^{\frac{3}{2}A^0 S\left(z^2 - \frac{1}{3}\right)/k_BT},$$ (5.7)

$$Z = \int_0^1 dz\, e^{\frac{3}{2}A_0 S\left(z^2 - \frac{1}{3}\right)/k_BT}.$$ (5.8)

Introducing $\sigma \equiv S/\kappa$ and $\kappa \equiv k_BT/\frac{3}{2}A^0$, this takes the form

$$\kappa\sigma = -\frac{1}{2} + \frac{3}{2} \frac{1}{J(\sigma)} \frac{d}{d\sigma} J(\sigma),$$ (5.9)

where

$$J(\sigma) = \int_0^1 dz\, e^{\sigma z^2}$$ (5.10)

is related to Dawson's integral

$$D(x) = \int_0^x e^{y^2} dy = x + \frac{x^3}{3 \cdot 1!} + \frac{x^5}{5 \cdot 2!} + \frac{x^7}{7 \cdot 3!} + \dots$$ (5.11)

by

$$J(\sigma) = \frac{1}{\sqrt{\sigma}} D(\sqrt{\sigma}).$$ (5.12)

After a partial integration we see that

$$J(\sigma) = e^\sigma - 2\sigma \frac{d}{d\sigma} J(\sigma),$$ (5.13)

so that Eq. (5.9) can be written as

$$\begin{aligned}
\kappa\sigma &= S(\sigma) \equiv -\frac{1}{2} + \frac{2}{3}\frac{1}{2}\left(\frac{e^\sigma}{\sqrt{\sigma}D(\sqrt{\sigma})} - \frac{1}{\sigma}\right) \\
&= -\frac{1}{2} + \frac{3}{2}\left(\frac{1}{3} + \frac{4}{45}\sigma - \frac{2^3}{3^3 \cdot 5 \cdot 7}\sigma^2 + \frac{2^4}{3^4 \cdot 5^2 \cdot 7}\sigma^3 + \dots\right).
\end{aligned}$$ (5.14)

This implicit equation is the extremum of the free energy density

$$f = \frac{1}{2}A_0 S^2 - k_BT \log Z = \frac{k_BT}{3}\left(\kappa\sigma^2 - 3\log Z\right),$$ (5.15)

where Z is the partition function of (5.7)

$$Z = \int_0^1 dz\, e^{\sigma\left(z^2 - \frac{1}{3}\right)} = e^{-\frac{1}{3}\sigma} \frac{1}{\sqrt{\sigma}} D(\sqrt{\sigma}).$$ (5.16)

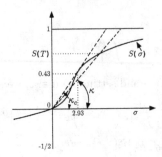

FIGURE 5.2 Graphical solution of the gap equation (5.14).

The solution of Eq. (5.14) is most easily found by choosing a parameter κ and determining the temperature T for various values of σ. Geometrically, the solution is given by the intersection of the straight lines $\kappa\sigma$ with the curves $S(\sigma)$ (see Fig. 5.2). For $\kappa > \kappa_c = 0.147$, the only solution is $S = \sigma = 0$ (normal phase).

At $\kappa = \kappa_0$, the order parameter jumps, in a first-order phase transition, to some finite value

$$S_c = 0.43, \quad \sigma_c = S_c/\kappa_c = 2.93. \tag{5.17}$$

In the limit $\kappa \to 0$ (i.e., $T \to 0$), σ approaches unity corresponding to a perfect order of the system.

5.1.2 Landau Expansion

For small values of σ we may use (5.16) and (5.11) to expand the free energy (5.15) in powers of σ:

$$\frac{3f}{k_B T} = \left(\kappa - \frac{2}{15}\right)\sigma^2 - \frac{8}{3^3 \cdot 5 \cdot 7}\sigma^3 + \frac{2^2}{3^3 \cdot 5^2 \cdot 7}\sigma^4 + \cdots . \tag{5.18}$$

Conventionally one denotes the temperature at which the quadratic term changes sign by T^*, i.e.,

$$\kappa - \frac{2}{15} \equiv \frac{2}{3}\frac{k_B T^*}{A_0}\left(\frac{T}{T^*} - 1\right) \equiv \kappa^*\left(\frac{T}{T^*} - 1\right), \tag{5.19}$$

so that

$$\kappa^* \equiv \frac{2k_B T^*}{3A_0} \equiv \frac{2}{15}. \tag{5.20}$$

With this notation, the expression (5.18) amounts to the Landau-de Gennes free energy expansion for the nematic liquid crystal. If terms beyond the fourth powers are neglected, the first-order nature of the transition H is seen to arise from the cubic term at a transition temperature T_1 determined by the equation

$$\frac{(\text{cubic term})^2}{4 \cdot \text{quadratic} \cdot \text{quartic term}} = \frac{\text{cubic term}}{2 \cdot \text{quartic term}}, \tag{5.21}$$

from which we find

$$\frac{T_1}{T^*} - 1 = \frac{(8/3^3 \cdot 5 \cdot 7)^2}{4(2/15)(4/3^3 \cdot 5^2 \cdot 7)} = \frac{10}{63}, \tag{5.22}$$

or $T_1/T^* = 73/63 \approx 1.159$. This lies quite a bit higher than the exact ratio $T_1/T^* = \kappa_c 15/2 \approx 1.1$ determined from the full gap equation (5.14). Experimentally, T_1 lies much closer to T^* ($T_1/T^* \approx 1.0025$) which shows that the cubic coefficient of the theory is somewhat too large with respect to fourth-order and quadratic coefficients in order to justify the Landau expansion, a well-known weakness of the model.

The κ-value at the transition point is

$$\kappa_c = \kappa^*(T_1/T^*) = 2 \cdot 73/15 \cdot 63. \tag{5.23}$$

The order parameter σ jumps from zero to

$$\sigma_c = 5, \tag{5.24}$$

so that S jumps from zero to

$$S_c = \kappa_c \sigma_c = \sigma_c = \frac{2}{3}\frac{73}{63} \approx 1.35. \tag{5.25}$$

5.1.3   Tensor Form of Landau-de Gennes Expansion

Let us rewrite the free energy density (5.15) in another form using the following auxiliary field quantity:

$$\sqrt{\frac{15}{8\pi}} n_\alpha n_\beta Q_{\alpha\beta} \equiv Q(\mathbf{n}). \tag{5.26}$$

Then the free energy density (5.15) can be written as

$$\frac{f}{k_B T} = \frac{4\pi}{3\kappa} \int \frac{d^2\mathbf{n}}{4\pi} Q^2 - \log Z, \tag{5.27}$$

with a partition function

$$Z = \int \frac{d^2\mathbf{n}}{4\pi} e^{\sqrt{\frac{4\pi}{5}} \frac{2}{3} Q(\mathbf{n})/\kappa}. \tag{5.28}$$

Expanding (5.27) in powers of $Q(\mathbf{n})$ we obtain

$$\begin{aligned}
\frac{f}{k_B T} &= \frac{1}{3\kappa}\left(\kappa - \frac{2}{15}\right) 4\pi \int \frac{d^2\mathbf{n}}{4\pi} Q^2 - \left(\frac{2}{3\kappa}\right)^3 \sqrt{\frac{4\pi}{5}}^3 \int \frac{d^2\mathbf{n}}{4\pi} Q^3 \\
&\quad + \left(\frac{2}{3\kappa}\right)^4 \sqrt{\frac{4\pi}{5}}^4 \left[\frac{1}{24} \int \frac{d^2\mathbf{n}}{4\pi} Q^4 - \frac{1}{8}\left(\int \frac{d^2\mathbf{n}}{4\pi} Q^2\right)^2\right] + \dots . \tag{5.29}
\end{aligned}$$

The angular integrals yield

$$\int \frac{d^2\mathbf{n}}{4\pi} Q^2 = \frac{15}{8\pi} \frac{2}{15} Q_{\alpha\beta} Q_{\alpha\beta},$$

$$\int \frac{d^2\mathbf{n}}{4\pi} Q^3 = \left(\frac{15}{8\pi}\right)^{3/2} \frac{8}{105} Q_{\alpha\beta} Q_{\beta\gamma} Q_{\gamma\alpha},$$

$$\int \frac{d^2\mathbf{n}}{4\pi} Q^4 = \left(\frac{15}{8\pi}\right)^2 \frac{36}{945} \left(Q_{\alpha\beta}{}^2\right)^2, \tag{5.30}$$

so that (5.29) takes the tensor form

$$\frac{f}{k_B T} = \frac{1}{3\kappa^2} \left(\kappa - \frac{2}{15}\right) \operatorname{tr} Q^2 - \frac{\sqrt{6}}{3 \cdot 5 \cdot 7} \left(\frac{2}{3\kappa}\right)^3 \operatorname{tr} Q^3$$

$$+ \frac{1}{700} \left(\frac{2}{3\kappa}\right)^4 (\operatorname{tr} Q^2)^2 + \dots \; . \tag{5.31}$$

Inserting here $Q_{\alpha\beta} = \kappa\sigma\epsilon^{(0)}(\mathbf{m})$ with $\operatorname{tr}\epsilon^{(0)2} = 1$, $\operatorname{tr}\epsilon^{(0)3} = 1/\sqrt{6}$, we recover (5.18).

In the sequel, we shall abbreviate the dimensionless reduced energy $f/k_B T$ by $\tilde{f}$.

5.2   Landau-de Gennes Description of Nematic Phase

The lowest non-vanishing multipole moment of the elongated molecules is of the quadruple type. Thus a traceless symmetric tensor $Q_{\alpha\beta}$ is the appropriate order parameter for a Landau expansion [3, 4]. To lowest approximation, any other physical property described by the same type of tensor must be a multiple of this order parameter $Q_{\alpha\beta}$. Examples are the deviations of the dielectric tensor $\epsilon_{\alpha\beta}$ or the magnetic permeability $\mu_{\alpha\beta}$ from the isotropic value

$$\delta\epsilon = \epsilon_{\alpha\beta} - \epsilon_0 \delta_{\alpha\beta},$$

$$\delta\mu = \mu_{\alpha\beta} - \mu_0 \delta_{\alpha\beta}. \tag{5.32}$$

If $Q_{\alpha\beta}$ vanishes, there can be no orientational preference. Thus $\delta^{\alpha\beta}\epsilon = 0$ and $\delta\mu = 0$. For small $Q_{\alpha\beta}$, one can expand

$$\delta\epsilon_{\alpha\beta} = M^\epsilon_{\alpha\beta\gamma\delta} Q_{\gamma\delta} + \dots, \tag{5.33}$$

where, from symmetry arguments, $M_{\alpha\beta\gamma\delta}$ can only have the general form

$$M_{\alpha\beta\gamma\delta} = a\,\delta_{\alpha\beta}\delta_{\gamma\delta} + \frac{b}{2}\left(\delta_{\alpha\gamma}\delta_{\beta\delta} + \delta_{\alpha\delta}\delta_{\beta\gamma}\right). \tag{5.34}$$

But applied to a symmetric traceless tensor $Q_{\gamma\delta}$, the a-term vanishes while the b-term gives simply $(b/2)\,Q_{\alpha\beta}$. Hence, the deviations of electric and magnetic permeability are proportional to $Q_{\alpha\beta}$. This makes all properties of the order parameter observable via an interaction Hamiltonian

$$H_{\text{int}} = \frac{1}{2} \int d^3x \left(\xi_E Q_{\alpha\beta} E_\alpha E_\beta + \xi_M Q_{\alpha\beta} H_\alpha H_\beta\right). \tag{5.35}$$

It will be convenient to choose the normalization of $Q_{\alpha\beta}$ such that

$$Q_{\alpha\beta} \equiv \delta\epsilon_{\alpha\beta} \quad ,\text{i.e.} \quad \xi_E \equiv 1. \tag{5.36}$$

Locally, the symmetric order parameter may be diagonalized by a rotation and has the form

$$Q_{\alpha\beta} = \begin{pmatrix} -Q_1 & & \\ & -Q_2 & \\ & & Q_1 + Q_2 \end{pmatrix}. \tag{5.37}$$

If $Q_1 \neq Q_2$, the order is called biaxial, if $Q_1 = Q_2$, it is called uniaxial. Suppose now that Q_1 and Q_2 are of similar magnitude and both are of equal sign, either positive or negative. In the first case, the dielectric tensor has two small and one larger component. This corresponds to an ellipsoid of rod-like shape. If they are of opposite sign $Q_1 \approx -Q_2$, the order corresponds to a disc. For the molecular systems discussed before we expect the rod-like option to have the lower energy. This will, in fact, emerge on very general grounds, except for small regions of temperature and pressure (close to the critical point in the phase diagram).

Let us now expand the free energy in powers of $Q_{\alpha\beta}$. On invariance grounds, we can have the folllowing terms

$$
\begin{aligned}
I_2 &= \operatorname{tr} Q^2, & (5.38) \\
I_3 &= \operatorname{tr} Q^3, & (5.39) \\
I_4 &= \operatorname{tr} Q^4, \quad I_2{}^2, & (5.40) \\
I_5 &= \operatorname{tr} Q^5, \quad I_2 I_3, & (5.41) \\
I_6 &= \operatorname{tr} Q^6, \quad I_3{}^2, I_2{}^3. & (5.42)
\end{aligned}
$$

$$\vdots$$

For traceless symmetric tensors, there is only one independent invariant of fourth and one of fifth order:

$$I_4 = \frac{1}{2} I_2{}^2, \quad I_5 = \frac{5}{6} I_2 I_3. \tag{5.43}$$

At sixth-order there are two invariants, which may be taken as $I_3{}^2$ and $I_2{}^3$. Then, for space- and time-independent order parameters, the free energy density may be expanded as [11]

$$f = \frac{1}{2}\left(a_2 I_2 + a_3 I_3 + \frac{a_4}{2} I_2{}^2 + a_5 I_2 I_3 + \frac{a_6}{2} I_3{}^2 + \frac{a_6'}{3} I_2{}^3 \right) + \mathcal{O}(Q^7). \tag{5.44}$$

Typical phase transitions take place roughly at room temperature. They are caused by the fact that the coefficient of the quadratic invariant vanishes at some temperature T^*, and can be expanded in a small neighborhood of T^* as

$$a_2 \approx a_2^0 \left(\frac{T}{T^*} - 1 \right). \tag{5.45}$$

The temperature T^* may be called *would-be critical temperature*. If the expansion (5.44) has only coefficients a_2, a_4 it is a a so-called *Landau expansion*. This has a second-order phase transition at the temperature T^*, which would then be a *critical temperature* T_c. The actual values of T^* and a_2^0 usually depend on pressure. In model calculations one typically finds that the other coefficients are of the same order as a_2^0. The only exception is a_3 which sometimes happens to be small. Then we define the dimensionless parameter

$$s_0 \equiv \frac{a_3^2}{12 a_2^0 a_4} \ll 1. \tag{5.46}$$

It can be used to characterize the strength of the first-order transition. By increasing the pressure to several hundred atmospheres, the parameter s_0 can be decreased so much that the point $a_3 = 0$ can be approached quite closely [12]. In the following we shall assume the existence of a point (P^*, T^*) in the (P, T)-diagram, where both a_2 and a_3 vanish. The neighborhood of this point will be particularly accessible to theoretical investigations. Within the (P, T)-diagram, the lines of constant a_2 and a_3 can be used to define a local coordinate frame whose axes cross at (P^*, T^*) at a non-zero angle (see Fig. 5.3). In some models, the coefficient a_3 is negative at low

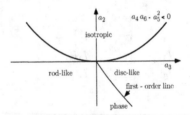

FIGURE 5.3 Phase diagram of general Landau expansion (5.44) of free energy in the (a_3, a_2)-plane.

pressure such that the a_3-axis points roughly in the direction of increasing P. With this mapping in mind we may picture all results directly in the (a_3, a_2)-plane with the a_3-axis pointing to the right, and only a slight distortion has to be imagined in order to transfer the phase diagrams to the (P, T)-plane.

Before starting it is useful to realize that the expansion (5.44), although it is a complicated sixth-order polynomial in the eigenvalues Q_1, Q_2 of the diagonalized order parameter, is a simple third order polynomial if treated as a function of the variables I_2, I_3. It is therefore convenient to treat it directly as such. One only has to keep in mind the allowed range of I_2, I_3: First of all, I_2 is positive definite. Second, I_3 is bounded by

$$I_3{}^2 \le \frac{1}{6} I_2{}^3. \tag{5.47}$$

The boundaries are reached for the uniaxial phase. This follows from the property

$$\mathrm{tr}\begin{pmatrix} -Q_1 & & \\ & -Q_1 & \\ & & 2Q_1 \end{pmatrix} = \pm\sqrt{6}\left[\mathrm{tr}\begin{pmatrix} -Q_1 & & \\ & -Q_1 & \\ & & 2Q_1 \end{pmatrix}^2\right]^{3/2} \quad \text{for} \quad Q_1 \lessgtr 0. \quad (5.48)$$

There is one boundary with $I_3 = I_2^{3/2}/\sqrt{6}$ where the order is positive or rod-like, the other has $I_3 = -I_2^{3/2}/\sqrt{6}$ where the order is negative or disc-like. Only between these boundaries are I_2, I_3 independent corresponding to a biaxial phase. The domain is shown in Fig. 5.4.

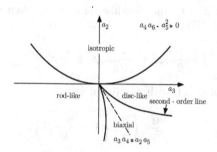

FIGURE 5.4 Biaxial regime in the phase diagram of the general Landau expansion (5.44) of free energy in the (a_3, a_2)-plane.

In this simplified view of the expansion (5.44) let us, for a moment, consider the expansion only up to the fourth power and look for the minimum in I_2 and I_3. Since $\partial f/\partial I_3 = a_3$, there is no extremum in the allowed domain of Fig. 5.4, except for $a_3 = 0$. There the transition is of second order: For $a_2 > 0$, $T > T^*$ one has only $I_2 = 0$ and hence $Q_{\alpha\beta} = 0$, which is the isotropic phase. For $a_2 < 0$, $T < T^*$ one finds $I_2 = -a_2/a_4$, and the system is ordered. Since I_3 is not specified, the order can be anywhere on the biaxial line in Fig. 5.4 between the rod-like and disc-like end points. The energy is

$$f = -\frac{a_2^2}{4a_4} = -\frac{a_2^{0^2}}{4a_4}\left(\frac{T}{T^*} - 1\right)^2. \quad (5.49)$$

The specific heat has the usual jump

$$\Delta c = -T\frac{\partial^2 f}{\partial T^2} = -\frac{1}{T^*}\frac{1}{2}\frac{a_2^{0^2}}{a_4}, \quad (5.50)$$

when passing from $T > T^*$ to $T < T^*$.

The situation is quite different in the presence of the cubic term $a_3 \neq 0$. Since there cannot be any minimum for independent I_2 and I_3, it must necessarily lie at the uniaxial boundaries (there must exist a minimum since F is continuous in

Q_1, Q_2 and eventually $F \to \infty$ for $Q_1, Q_2 \to \infty$). Let us insert the particular uniaxial parametrization

$$Q_{\alpha\beta} = \varphi\, \epsilon_{\alpha\beta}^{(0)}(\mathbf{n}) \equiv \varphi \sqrt{\frac{3}{2}} \left(n_\alpha n_\beta - \frac{1}{3} \delta_{\alpha\beta} \right), \tag{5.51}$$

where $\mathbf{n}$ is an arbitrary unit vector, $\varphi > 0$ the order parameter, and $\epsilon^{(0)}(\mathbf{n})$ the traceless polarization tensor

$$\epsilon^{(0)}(\mathbf{n}) \equiv \sqrt{\frac{3}{2}} \left(n_\alpha n_\beta - \frac{1}{3} \delta_{\alpha\beta} \right). \tag{5.52}$$

The order is rod-like for $\varphi > 0$, and disc-like for $\varphi < 0$. Then we find, using the traces $\mathrm{tr}\,(\epsilon^{(0)2}) = 1$, $\mathrm{tr}\,(\epsilon^{(0)3}) = 1/\sqrt{6}$,

$$f = \frac{1}{2} a_2 \varphi^2 + \frac{1}{2\sqrt{6}} a_3 \varphi^3 + \frac{a_4}{4} \varphi^4. \tag{5.53}$$

This energy is minimal at $\varphi = 0$ with $f = 0$ and at

$$\varphi_{\substack{> \\ <}} = -\frac{3a_3}{4\sqrt{6}a_4} \left(1 \pm \sqrt{1 - \frac{96 a_2 a_4}{9 a_3^2}} \right), \tag{5.54}$$

which are the solutions of

$$f' = \left(a_2 + \frac{3}{2\sqrt{6}} a_3 \varphi + a_4 \varphi^2 \right) \varphi = 0. \tag{5.55}$$

Combining (5.55) with (5.53) we see that the energy at $\varphi_{\substack{> \\ <}}$ is

$$f = -\frac{1}{4} \varphi^3 \left(\frac{a_3}{\sqrt{6}} + a_4 \varphi \right). \tag{5.56}$$

The energy vanishes at a point $\varphi \neq 0$, if $\varphi_{\substack{> \\ <}}$ satisfies

$$\varphi = -\frac{a_3}{\sqrt{6}a_4}. \tag{5.57}$$

From (5.54) we see that this happens at a temperature T_1 at which

$$a_2 = a_2^0 \left(\frac{T_1}{T^*} - 1 \right) = \frac{a_3^2}{12 a_4}, \tag{5.58}$$

i.e., at which

$$\frac{T_1}{T^*} - 1 = s_0. \tag{5.59}$$

At this point the potential has the usual symmetric double-well form entered around $\varphi_>/2$ (see Fig. 5.5).

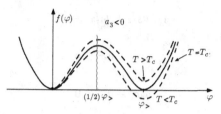

FIGURE 5.5 Jump of the order parameter φ from zero to a nonzero value $\varphi_>$ in a first-order phase transition at $T = T_1$.

The quantity (5.59) tells us how much earlier the first-order transition takes place with respect to the would-be critical temperature. It will be referred to as the *precocity* of the first-order transition due to the cubic term $a_3 \neq 0$ in the energy expansion (5.44).

Once a_0 is nonzero then, as T passes the temperature T_1 which lies *above* T^*, the order jumps discontinuously from the old minimum at $\varphi = 0$ to the new one at $\varphi = \varphi_>$. At that point, the entropy changes by

$$\Delta s = -T\left(\frac{\partial f}{\partial T}\bigg|_{T=T_1+\epsilon} - \frac{\partial f}{\partial T}\bigg|_{T=T_1-\epsilon}\right) = \frac{1}{2}\frac{T}{T^*}a_2^0\varphi_>^2 = -\frac{1}{T^*}\frac{a_2^0 a_2}{a_4}, \quad (5.60)$$

giving a *latent heat*,

$$\Delta q = \frac{T_1}{T^*}\frac{a_2^{02}}{a_4}\left(\frac{T_1}{T^*}-1\right) \approx \frac{a_2^{02}}{a_4}s_0. \quad (5.61)$$

It is proportional to the precocity s_0 which may therefore also be viewed as the *strength* of the first-order transition. From Eq. (5.54) it follows that, for $a_3 < 0$, the order is positive uniaxial, for $a_3 > 0$ negative uniaxial.

Energetically, the higher powers of the free energy are negligible as long as $\varphi_>$ is sufficiently small. From Eqs. (5.57) and (5.59) we see that, at the transition, the order parameter has the value

$$\varphi_> = \sqrt{\frac{2a_2^0}{a_4}}\sqrt{\frac{T_1}{T^*}-1}. \quad (5.62)$$

Since a_2^0 and a_4 are of the same order of magnitude, the corrections in the energy are of order $\mathcal{O}\left(\sqrt{T_1/T^*-1}\right)$. Experimentally, the temperature precocity of the first order transition $T_1/T^* - 1 = s_0$ is extremely small, typically $\approx 1/400$. First-order transitions with this property is usually referred to as being *weakly first-order*. For $a_2 > 0$ and close to the critical point (P^*, T^*), the higher orders are rather insignificant. They do become relevant for $a_2 < 0$, in particular in some neighborhood of the $a_3 = 0$ line where the different phases are unspecified. In order to get a qualitative picture, let us neglect the a_6'-term which could give only slight quantitative changes but which would make the following discussion much more clumsy. Varying f independently with respect to I_2 and I_3 we find the extremality conditions

$$a_2 + a_4 I_2 + a_5 I_3 = 0,$$

$$a_3 + a_5 I_2 + a_6 I_3 \;=\; 0. \tag{5.63}$$

For

$$a_4 a_6 - a_5{}^2 > 0, \tag{5.64}$$

or

$$a_4 a_6 - a_5{}^2 < 0, \tag{5.65}$$

this can be solved by

$$\begin{pmatrix} I_2 \\ I_3 \end{pmatrix} = \frac{1}{a_4 a_6 - a_5{}^2} \begin{pmatrix} -a_6 & a_5 \\ a_5 & -a_4 \end{pmatrix} \begin{pmatrix} a_2 \\ a_3 \end{pmatrix}. \tag{5.66}$$

We shall exclude the accidental equality sign since a_4, a_5, a_6 are rather invariable material constants. The extremum is a minimum only under the condition (5.64). We then have to see whether I_2 and I_3 remain inside the allowed domain $I_2{}^3 \geq 6 I_3{}^2$. For this we simply map the position of the boundaries into the (a_3, a_2)-plane. On the rod-like and disc-like boundaries, we have

$$a_2 \;=\; -a_4 I_2 \mp a_5 \frac{1}{\sqrt{6}} I_2^{3/2},$$

$$a_3 \;=\; -a_5 I_2 \mp a_6 \frac{1}{\sqrt{6}} I_2^{3/2}. \tag{5.67}$$

We may form two combinations

$$a_2 a_5 - a_3 a_4 \;=\; \pm (a_4 a_6 - a_5{}^2) \frac{1}{\sqrt{6}} I_2^{3/2}, \tag{5.68}$$

$$a_2 a_6 - a_3 a_5 \;=\; -(a_4 a_6 - a_5{}^2) I_2. \tag{5.69}$$

Eliminating I_2 from these equations gives

$$a_2 a_5 - a_3 a_4 = \pm \left(a_4 a_6 - a_5{}^2 \right) \frac{1}{\sqrt{6}} \left[\frac{a_2 a_6 - a_3 a_5}{-(a_4 a_6 - a_5{}^2)} \right]^{3/2}. \tag{5.70}$$

If the right-hand side is absent, this yields a straight line

$$a_2 = \frac{a_4}{a_5} a_3 \tag{5.71}$$

in the (a_3, a_2)-plane. It is easy to see that the right-hand side of (5.70) gives only a correction of order $a_2^{3/2}$ to this result. Indeed, inserting the lowest-order approximation (5.71), the right-hand side of (5.70) becomes

$$\pm (a_4 a_6 - a_5{}^2) \frac{1}{\sqrt{6}} \left(-\frac{a_2}{a_4} \right)^{3/2}, \tag{5.72}$$

so that, up to order $a_2^{3/2}$,

$$a_2 a_5 - a_3 a_4 = \pm (a_4 a_6 - a_5{}^2) \frac{1}{\sqrt{6}} \left(-\frac{a_2}{a_4} \right)^{3/2} + \dots \;. \tag{5.73}$$

The two boundary curves are displayed in Fig. 5.4. Between these branches the order is biaxial with a well determined ratio Q_1/Q_2. One may envisage the effect of higher powers in the free energy expansion as having slightly rotated the vertical degenerate line in Fig. 5.3, and opened it up into the two branches of Fig. 5.4, thereby generating an entire domain for the biaxial phase. Since the order parameter moves continuously towards the uniaxial boundary, the transition uniaxial to biaxial is of second order.

If the determinant $a_4a_6 - a_5{}^2$ becomes smaller, the biaxial region shrinks. For negative sign, it disappears and the two uniaxial regions overlap. Since only one of them can have the lower energy, there must be a line at which the transition takes place. This is found most easily by considering the uniaxial energy in the parametrization (5.51) which reads

$$2f = a_2\varphi^2 + \frac{a_3}{\sqrt{6}}\varphi^3 + \frac{a_4}{2}\varphi^4 + \frac{a_5}{\sqrt{6}}\varphi^5 + \frac{a_6}{12}\varphi^6 + \dots . \tag{5.74}$$

Here $a_6/12$ can be thought of as containing also a parameter $a_6'/3$ coming from the last term in (5.44). In the Landau approximnation, only the terms a_2 and a_4 are present.

The minimum lies at a nonzero field φ which satisfies the equation

$$\left(a_2 + \frac{3}{2\sqrt{6}}a_3\varphi + a_4\varphi^2 + \frac{5}{2\sqrt{6}}\varphi^3 + a_6\varphi^4\right)\varphi = 0. \tag{5.75}$$

Keeping the coefficients up to a_4, this is solved by

$$\varphi_{\gtrless} = -\frac{3}{4\sqrt{6}}\frac{a_3}{a_4}\left(1 \pm \sqrt{1 - \frac{96a_2a_4}{9a_3^2}}\right). \tag{5.76}$$

Only $\varphi_>$ gives a minimum, the other a maximum. As a_5, a_6 are turned on, the maximum may become a minimum. In order to see where this happens let us assume a_3 to be very small, as compared with a_2a_4. Then φ is given by

$$\varphi = \pm\sqrt{\frac{-a_2}{a_4}}\left(1 \pm \frac{3}{4\sqrt{6}}\sqrt{\frac{a_3^2}{-a_2a_4}} + \dots\right). \tag{5.77}$$

Inserting this back into the energy we find that the two energies become equal at

$$a_2 = a_3\frac{a_4}{a_5} + \frac{1}{\sqrt{6}}a_3{}^{3/2}a_5{}^{-1/2} + \mathcal{O}(a_3^2). \tag{5.78}$$

For small a_5 this reduces back to the line $a_3 = 0$. The latent heat is now

$$\Delta q = -\frac{a_3}{2a_4}T\frac{\partial a_2}{\partial T} - T\frac{\partial a_3}{\partial T}\frac{1}{2a_5}(a_2a_5 - a_3a_4). \tag{5.79}$$

Let us finally calculate the correction to the isotropic-uniaxial curve in Figs. 5.4 and 5.3. For small a_3 we find

$$a_2 = \frac{1}{12}\frac{a_3^2}{a_4} + \frac{1}{36}\frac{a_3^3 a_5}{a_4^3} + \mathcal{O}(a_3{}^4), \tag{5.80}$$

which may be used to calculate a small correction to the latent heat (5.61).

All ordered phases described here are referred to as *nematic*. In the Landau approximation, the minimum of the energy lies at $\varphi_L^2 = -a_2/a_4$ and has the value

$$f_{c,L} = -\frac{1}{4}\frac{a_2^2}{a_4} \tag{5.81}$$

called the Landau *condensation energy*.

5.3 Bending Energy

The order parameters discussed in the last section were independent of space and time. In the laboratory, such configurations are difficult to realize. External boundaries usually do not permit a uniform order but enforce spatial variations. The system tries, however, to keep the variations as smooth as possible. It exerts resistance to local deformations. In order to parametrize the restoring forces one expands the free energy in powers of the derivatives of the collective field $Q_{\alpha\beta}$. If the fields bend sufficiently smooth, the expansion may be terminated after the lowest derivative.

Due to rotational invariance, there can only be the following *bending energies*

$$f_{\text{bend}} = \frac{b}{2}\nabla_\gamma Q_{\alpha\beta}\nabla_\gamma Q_{\alpha\beta} + \frac{c_1}{2}\nabla_\alpha Q_{\alpha\gamma}\nabla_\beta Q_{\beta\gamma} + \frac{c_2}{2}\nabla_\alpha Q_{\beta\gamma}\nabla_\beta Q_{\alpha\gamma}. \tag{5.82}$$

As far as the total energy $F = \int d^3x f$ is concerned, the latter two terms may be collected by a partial integration into one, say the first, by substituting $c_1 \rightarrow c_1+c_2 \equiv c$.

In the ordered phase which is usually of the rod-like type we may use the parametrization (5.51) and split the gradient of $Q_{\alpha\beta}$ into variation of the size φ and the direction $\mathbf{n}$. In the bulk liquid the size of the order parameter φ is caught in the potential minimum at $\varphi_>$ (see Fig. 5.5) and only the direction $\mathbf{n}$ will vary from point to point. Then we can find from (5.82) the purely directional bending energy:

$$\begin{aligned} f_{\text{bend,dir}} = \ &\frac{3}{4}\varphi^2\,[b\nabla_\gamma(n_\alpha n_\beta)\nabla_\gamma(n_\alpha n_\beta) + c_1\,\nabla_\alpha(n_\alpha n_\gamma)\nabla_\beta(n_\beta n_\gamma) \\ &+ c_2\,\nabla_\alpha(n_\beta n_\gamma)\nabla_\beta(n_\alpha n_\gamma)]\,. \end{aligned} \tag{5.83}$$

Since $n_\alpha{}^2 = 1$, we may use the orthogonality property $n_\alpha\nabla_\gamma n_\alpha = 0$, and write

$$\begin{aligned} f_{\text{bend,dir}} = \ &\frac{3}{4}\varphi^2\left\{2b\,n_{\alpha,\beta}^2 + c_1\left[(\nabla\cdot\mathbf{n})^2 + (\mathbf{n}\cdot\nabla n_\gamma)^2\right]\right. \\ &\left.+ c_2\left[(\nabla_\alpha n_\beta)(\nabla_\beta n_\alpha) + (\mathbf{n}\cdot\nabla n_\gamma)^2\right]\right\}. \end{aligned} \tag{5.84}$$

We now rewrite

$$n_{\alpha,\beta}^2 = (\nabla\cdot\mathbf{n})^2 + [\mathbf{n}\cdot(\nabla\times\mathbf{n})]^2 + [\mathbf{n}\times(\nabla\times\mathbf{n})]^2, \tag{5.85}$$

and

$$n_{\alpha,\gamma} n_{\gamma,\alpha} = (\mathbf{\nabla})^2 + \nabla_\alpha(n_\beta \nabla_\beta n_\alpha) - \nabla_\beta(n_\beta \nabla_\alpha n_\alpha),$$ (5.86)

so that

$$f_{\text{bend,dir}} = \frac{3}{2}\varphi^2 \left\{ \left(b + \frac{c}{2}\right)(\mathbf{\nabla} \cdot \mathbf{n})^2 + b\left[\mathbf{n} \cdot (\mathbf{\nabla} \times \mathbf{n})\right]^2 + \left(b + \frac{c}{2}\right)\left[\mathbf{n} \times (\mathbf{\nabla} \times \mathbf{n})\right]^2 \right.$$
$$\left. + c_2 \left[\nabla_\alpha(n_\beta \nabla_\beta n_\alpha) - \nabla_\beta(n_\beta \nabla_\alpha n_\alpha)\right] \right\}.$$ (5.87)

The last term is a pure surface term. The coefficients

$$K_1 \equiv K_s = 3\left(b + \frac{c}{2}\right)\varphi^2,$$
$$K_2 \equiv K_t = 3\ b\ \varphi^2,$$
$$K_3 \equiv K_b = 3\left(b + \frac{c}{2}\right)\varphi^2$$ (5.88)

are known as *Frank constants* of textural bending. The subscripts s, t, b stand for *splay, twist, and bend* and indicate that each term dominates a certain class of distortions of the directional field. They are shown in Fig. 5.6. The experimental

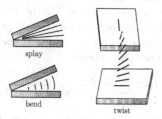

splay

bend

twist

FIGURE 5.6 Different configurations of textures in liquid crystals.

values of $K_{1,2,3}$ are of the order of 5 to 10×10^{-7} dynes, for example [6]:

MBBA : $T \approx 22^0 C$ $K_{1,2,3} = (5.3 \pm 0.5\ ,\ 2.2 \pm 0.7\ ,\ 7.45 \pm 1.1) \times 10^{-7}$dynes,
PAA : $T \approx 125^0 C$ $K_{1,2,3} = (4.5\ ,\ 2.9\ ,\ 9.5) \times 10^{-7}$dynes. (5.89)

For topological reasons, the field configurations may have singularities called *defects*. In their neighborhood, also the size φ has spatial variations. This is also true near boundaries or at the interface between two phases. The derivative terms for these variations are found from (5.82) by calculating

$$\nabla_\gamma Q_{\alpha\beta} \nabla_\gamma Q_{\alpha\beta} = (\mathbf{\nabla}\varphi)^2 + \mathcal{O}(\nabla\mathbf{n})^2$$ (5.90)
$$\nabla_\alpha Q_{\alpha\gamma} \nabla_\beta Q_{\beta\gamma} = \nabla_\alpha\varphi \nabla_\beta\varphi \frac{1}{2}\left(n_\alpha n_\beta + \frac{1}{3}\delta_{\alpha\beta}\right)$$ (5.91)
$$+ 3\left[(\nabla_\alpha n_\alpha)n_\gamma + n_\alpha(\nabla_\alpha n_\gamma)\right]\left(n_\beta n_\gamma - \frac{1}{3}\delta_{\beta\gamma}\right)\nabla_\beta\varphi + \mathcal{O}(\nabla\mathbf{n})^2$$

$$= \frac{1}{2}(\mathbf{n} \cdot \boldsymbol{\nabla}\varphi)^2 + \frac{1}{6}(\boldsymbol{\nabla}\varphi)^2 + 2(\mathbf{n} \cdot \boldsymbol{\nabla}\varphi)\boldsymbol{\nabla}\mathbf{n} - (\mathbf{n} \cdot \boldsymbol{\nabla}n_\gamma)\boldsymbol{\nabla}_\gamma\varphi + \mathcal{O}(\boldsymbol{\nabla}\mathbf{n})^2,$$

$$\boldsymbol{\nabla}_\alpha Q_{\beta\gamma}\boldsymbol{\nabla}_\beta Q_{\alpha\gamma} = \boldsymbol{\nabla}_\alpha\varphi\boldsymbol{\nabla}_\beta\varphi \frac{1}{2}\left(n_\alpha n_\beta + \frac{1}{3}\delta_{\alpha\beta}\right) \tag{5.92}$$

$$+ 3[(\boldsymbol{\nabla}_\alpha n_\beta)n_\gamma + n_\beta(\boldsymbol{\nabla}_\alpha n_\gamma)]\left(n_\alpha n_\gamma - \frac{1}{3}\delta_{\alpha\gamma}\right)\boldsymbol{\nabla}_\beta\boldsymbol{\nabla}_\beta\varphi + \mathcal{O}(\boldsymbol{\nabla}\mathbf{n})^2$$

$$= \frac{1}{2}(\mathbf{n} \cdot \boldsymbol{\nabla}\varphi)^2 + \frac{1}{6}(\boldsymbol{\nabla}\varphi)^2 + 2(\mathbf{n} \cdot \boldsymbol{\nabla}n_\gamma)\boldsymbol{\nabla}_\gamma\varphi - (\mathbf{n} \cdot \boldsymbol{\nabla}\varphi)\boldsymbol{\nabla}\mathbf{n} + \mathcal{O}(\boldsymbol{\nabla}\mathbf{n})^2,$$

so that the combined bending energies are

$$f_{\text{bend}} = f_{\text{bend,dir}} + \frac{1}{2}\left(b + \frac{c}{6}\right)(\boldsymbol{\nabla}\varphi)^2 + \frac{c}{2}(\mathbf{n} \cdot \boldsymbol{\nabla}\varphi)^2$$

$$+ \frac{2c_1 - c_2}{2}[\varphi(\mathbf{n} \cdot \boldsymbol{\nabla}\varphi)(\boldsymbol{\nabla} \cdot \mathbf{n}) + \varphi(\mathbf{n} \cdot \boldsymbol{\nabla}n_\alpha)\boldsymbol{\nabla}_\alpha\varphi]. \tag{5.93}$$

5.4   Light Scattering

The bending energies determine the length scale at which local field fluctuations take place. These in turn are directly observable in light scattering experiments.

Consider at first the region $T > T_1$. There the order parameter vanishes such that the field $Q_{\alpha\beta}$ fluctuates around zero. If the temperature is sufficiently far above T_1 (precisely how far will soon be seen), the quadratic term in the energy strongly confines such fluctuations and we can study their properties by considering only the quadratic term in the free energy

$$f_{\text{bend}} = \frac{a_2}{2}Q_{\alpha\beta}^2 + \frac{b}{2}(\boldsymbol{\nabla}_\gamma Q_{\alpha\beta}) + \frac{c}{2}\boldsymbol{\nabla}_\alpha Q_{\alpha\gamma}\boldsymbol{\nabla}_\beta Q_{\beta\gamma} + \text{surface terms}. \tag{5.94}$$

Obviously, b/a_2 and c/a_2 have the dimension of a length square and it is useful to define the squares of the coherence lengths for $T > T^*$:

$$\begin{aligned} \xi_1^2(T) &\equiv \frac{b}{a_2} = \frac{b}{a_2^0}\left(\frac{T}{T^*} - 1\right)^{-1} \equiv \xi_1^{02}\left(\frac{T}{T^*} - 1\right)^{-1} \\ \xi_2^2(T) &= \frac{c}{a_2} = \frac{c}{a_2^0}\left(\frac{T}{T^*} - 1\right)^{-1} \equiv \xi_2^{02}\left(\frac{T}{T^*} - 1\right)^{-1} \end{aligned} \tag{5.95}$$

which increase as the temperature approaches T^* from above. These length scales will turn out to control the range of local fluctuations.

Let us expand $Q_{\alpha\beta}$ in plane waves

$$Q_{\alpha\beta}(\mathbf{x}) = \frac{1}{\sqrt{V}}\sum_{\mathbf{q}} e^{i\mathbf{q}\cdot\mathbf{x}}Q_{\alpha\beta}(\mathbf{q}) \tag{5.96}$$

where $Q_{\alpha\beta}{}^*(\mathbf{q}) = Q_{\alpha\beta}(-\mathbf{q})$. Then the total energy becomes

$$F_{\text{bend}} = \frac{1}{2}\sum_{\mathbf{q}} Q_{\alpha\beta}(-\mathbf{q})\left[(a_2 + b\mathbf{q}^2)\delta_{\alpha\alpha'} + cq_\alpha q_{\alpha'}\right]Q_{\alpha'\beta}(\mathbf{q}). \tag{5.97}$$

The spin orbit coupling term c can be diagonalized most easily on states of fixed helicity. The spin matrix for the tensor field is

$$(S_\gamma Q)_{\alpha\beta} = -i \left[\epsilon_{\gamma\alpha\alpha'} Q_{\alpha'\beta} + (\alpha \leftrightarrow \beta) \right]. \tag{5.98}$$

The helicity is defined as the projection of $\mathbf{S}$ along $\mathbf{q}$

$$H \equiv \mathbf{S} \cdot \hat{\mathbf{q}}. \tag{5.99}$$

We now calculate

$$
\begin{aligned}
{[HQ(-\mathbf{q})]}_{\alpha\beta} \, [HQ(\mathbf{q})]_{\alpha\beta} &= \left[\hat{q}_\gamma \epsilon_{\gamma\alpha\alpha'} Q_{\alpha'\beta}(-\mathbf{q}) + (\alpha \leftrightarrow \beta) \right] \left[\hat{q}_\delta \epsilon_{\delta\alpha\alpha''} Q_{\alpha''\beta}(\mathbf{q}) + (\alpha \leftrightarrow \beta) \right] \\
&= 4 Q_{\alpha\beta}(-\mathbf{q}) Q_{\alpha\beta}(\mathbf{q}) - 6 Q_{\alpha\beta}(-\mathbf{q}) \hat{q}_\alpha \hat{q}_{\alpha'} Q_{\alpha'\beta}(\mathbf{q}), \tag{5.100}
\end{aligned}
$$

so that (5.97) can be rewritten as

$$F_{\text{bend}} = \frac{1}{2} \sum_{\mathbf{q}} \left\{ \left[a_2 + \left(b + \frac{2}{3}c \right) q^2 \right] |Q_{\alpha\beta}(\mathbf{q})|^2 - \frac{c}{6} |HQ(\mathbf{q})|^2 \right\}. \tag{5.101}$$

This is obviously diagonal on eigenstates of helicity. These are easily constructed. First those of unit angular momentum: For this one simply takes the spherical combinations of unit vectors

$$
\begin{aligned}
\boldsymbol{\epsilon}^{(\pm)} &= \mp \frac{1}{\sqrt{2}} \left(\hat{\mathbf{x}} \pm i\hat{\mathbf{y}} \right), \\
\boldsymbol{\epsilon}^{(0)} &= \hat{\mathbf{z}}, \tag{5.102}
\end{aligned}
$$

which are eigenstates of S_3 and $\mathbf{S}^2$:

$$
\begin{aligned}
S_3 \boldsymbol{\epsilon}^{\pm} &= \pm \boldsymbol{\epsilon}^{\pm}, & S_3 \boldsymbol{\epsilon}^0 &= 0, \tag{5.103} \\
\mathbf{S}^2 \boldsymbol{\epsilon}^{\pm} &= 2 \boldsymbol{\epsilon}^{\pm}, & \mathbf{S}^2 \boldsymbol{\epsilon}^0 &= 0. \tag{5.104}
\end{aligned}
$$

We rotate them into the direction of $\mathbf{q}$ by a matrix

$$R(\hat{\mathbf{q}}) = e^{-i\varphi L_3} e^{-i\theta L_2} = \begin{pmatrix} \cos\varphi & -\sin\varphi & 0 \\ \sin\varphi & \cos\varphi & 0 \\ 0 & 0 & 1 \end{pmatrix} \begin{pmatrix} \cos\theta & 0 & \sin\theta \\ 0 & 1 & 0 \\ -\sin\theta & 0 & \cos\theta \end{pmatrix}, \tag{5.105}$$

where φ and θ are the polar angles of $\hat{\mathbf{q}}$. This turns $\hat{\mathbf{x}}, \hat{\mathbf{y}}, \hat{\mathbf{z}}$ into a local triped $\mathbf{l}^{(1)}$, $\mathbf{l}^{(2)}$, $\hat{\mathbf{q}}$ of unit vectors,

$$\mathbf{l}^{(1)} = R(\hat{\mathbf{q}})\hat{\mathbf{x}}, \ \mathbf{l}^{(2)} = R(\hat{\mathbf{q}})\hat{\mathbf{y}}, \ \mathbf{q}^{(1)} = R(\hat{\mathbf{q}})\hat{\mathbf{z}}. \tag{5.106}$$

From these we create the rotated polarization vectors $\boldsymbol{\epsilon}^{(\pm,0)}(\hat{\mathbf{q}}) \equiv R(\hat{\mathbf{q}})\boldsymbol{\epsilon}^{(\pm,0)}$ which diagonalize H with eigenvalues $\pm 1, 0$:

$$
\begin{aligned}
(H\boldsymbol{\epsilon}^{(1,-1)})_\alpha &= -i\hat{q}_\gamma \epsilon_{\gamma\alpha\beta} \left(l_\beta^{(1)} \pm i l_\beta^{(2)} \right) = i\hat{\mathbf{q}} \times \left(\mathbf{l}^{(1)} \pm i\mathbf{l}^{(2)} \right)_\alpha = \pm i \left(\mathbf{l}^{(1)} \pm i\mathbf{l}^{(2)} \right)_\alpha, \\
(H\boldsymbol{\epsilon}^{(0)})_\alpha &= -i\hat{q}_\gamma \epsilon_{\gamma\alpha\beta} \hat{q}_\beta = 0. \tag{5.107}
\end{aligned}
$$

Now we couple pairs of these vectors symmetrically to tensors and obtain the angular momentum helicity tensors of spin 2 and helicities $h = (-2, -1, 0, 1, 2)$:

$$\epsilon^{(2)}_{\alpha\beta}(\hat{\mathbf{q}}) = \epsilon^{(-2)*}_{\alpha\beta}(\hat{\mathbf{q}}) = \hat{l}_\alpha \hat{l}_\beta = \epsilon^{(2)}(-\hat{\mathbf{q}}),$$

$$\epsilon^{(1)}_{\alpha\beta}(\hat{\mathbf{q}}) = -\epsilon^{(-1)*}_{\alpha\beta}(\hat{\mathbf{q}}) = \frac{1}{\sqrt{2}}\left(l_\alpha \hat{q}_\beta + l_\beta \hat{q}_\alpha\right) = -\epsilon^{(1)}(-\hat{\mathbf{q}}),\qquad (5.108)$$

$$\epsilon^{(0)}_{\alpha\beta}(\hat{\mathbf{q}}) = \sqrt{\frac{3}{2}}\left(\hat{q}_\alpha \hat{q}_\beta - \frac{1}{3}\delta_{\alpha\beta}\right) \equiv \epsilon^{(0)}(-\hat{\mathbf{q}}),$$

where we have introduced the unit vector $\mathbf{l} \equiv \frac{1}{\sqrt{2}}\left(\mathbf{1}^{(1)} + i\mathbf{1}^{(2)}\right)$. Using its properties $\mathbf{l}^2 = 0$, $\mathbf{l} \cdot \mathbf{l}^* = 1$, we verify directly the orthogonality

$$\mathrm{tr}\left(\epsilon^{(h)}(\hat{\mathbf{q}})\epsilon^{(h')*}(\hat{\mathbf{q}})\right) = \delta_{hh'}.\qquad (5.109)$$

The completeness relation is found to be

$$\sum_{h=-2}^{2} \epsilon^{(h)}_{\alpha\beta}(\hat{\mathbf{q}})\epsilon^{(h)}_{\gamma\delta}(\hat{\mathbf{q}}) = I_{\alpha\beta,\gamma\delta},\qquad (5.110)$$

where

$$I_{\alpha\beta,\gamma\delta} \equiv \frac{1}{2}\left(\delta_{\alpha\gamma}\delta_{\beta\delta} + \delta_{\alpha\delta}\delta_{\beta\gamma}\right) - \frac{1}{3}\delta_{\alpha\beta}\delta_{\gamma\delta}\qquad (5.111)$$

is the projection into the space of symmetric traceless tensors of spin 2, as it should.[2]
The energy can now be diagonalized by expanding $Q_{\alpha\beta}(\mathbf{x})$ in terms of the $\epsilon^{(h)}_{\alpha\beta}(q)$-eigenmodes:

$$Q_{\alpha\beta}(\mathbf{x}) = \sum_{\mathbf{q},h}\left[e^{i\mathbf{q}x}\epsilon^{(h)}_{\alpha\beta}(\hat{\mathbf{q}})\,\varphi^{(h)}(\mathbf{q}) + \text{c.c.}\right].\qquad (5.112)$$

This yields

$$F = \int d^3x\, f = \sum_{\mathbf{q},h} f^{(h)}(\mathbf{q}) = \sum_{\mathbf{q},h}\tau^{(h)}(\mathbf{q})|\varphi^{(h)}(\mathbf{q})|^2,\qquad (5.113)$$

with

$$\begin{aligned}
\tau^{(h)}(\mathbf{q}) &= a_2 + \left[b + \left(\frac{3}{2} - \frac{h^2}{6}\right)c\right]q^2\\
&= a_2\left\{1 + \left[\xi_1{}^2 + \left(\frac{2}{3} - \frac{h^2}{6}\right)\xi_2{}^2\right]q^2\right\}.
\end{aligned}\qquad (5.114)$$

We can now calculate the correlation functions of the field. If we express the partition function

$$Z = \sum_Q e^{-F/k_BT}\qquad (5.115)$$

---

[2] Compare with the gravitational polarization tensors in Subsection 4.10.6 of the textbook [5].

in terms of the diagonalized modes, we have

$$
\begin{aligned}
Z &= \prod_{\mathbf{q}} \int d\varphi^{(h)}(\mathbf{q}) \exp\left\{-\frac{1}{k_B T} \sum_{\mathbf{q},h} f^{(h)}(\mathbf{q})\right\} \\
&= \prod_{\mathbf{q}} \int d\varphi^{(h)}(\mathbf{q}) \exp\left\{-\frac{1}{k_B T} \sum_{\mathbf{q},h} \tau^{(h)}(\mathbf{q}) |\varphi^{(h)}(\mathbf{q})|^2\right\}. \quad (5.116)
\end{aligned}
$$

From the "*equipartition theorem*"[3], we deduce that the thermal expectation values of the correlation functions are[4]

$$
\langle \varphi^{(h)}(\mathbf{q}) \varphi^{(h)*}(\mathbf{q}') \rangle = \delta_{\mathbf{q},\mathbf{q}'} \frac{k_B T/2}{\tau^{(h)}(\mathbf{q})}. \quad (5.117)
$$

For the amplitude $Q_{\alpha\beta}$, this implies the expectation values in $\mathbf{q}$-space

$$
\langle Q_{\alpha\beta}(\mathbf{q})[Q_{\gamma\delta}(\mathbf{q}')]^* \rangle = \delta_{\mathbf{q},\mathbf{q}'} \sum_h \frac{k_B T}{\tau^{(h)}(\mathbf{q})} \epsilon^{(h)}_{\alpha\beta}(\hat{\mathbf{q}}') \epsilon^{(h)*}_{\gamma\delta}(\hat{\mathbf{q}}), \quad (5.118)
$$

so that in $\mathbf{x}$-space,

$$
\langle Q_{\alpha\beta}(\mathbf{x}) Q_{\gamma\delta}(\mathbf{x}') \rangle = k_B T \sum_{\mathbf{q}} \frac{e^{i\mathbf{q}(\mathbf{x}-\mathbf{x})}}{\tau^{(h)}(\mathbf{q})} \epsilon^{(h)}_{\alpha\beta}(\hat{\mathbf{q}}) \epsilon^{(h)}_{\gamma\delta}(\hat{\mathbf{q}})^*. \quad (5.119)
$$

The correlation function reads

$$
G^{\gamma\delta}_{\alpha\beta}(\mathbf{q}) = k_B T \sum_h \frac{\epsilon^{(h)}_{\alpha\beta}(\mathbf{q}) \epsilon^{(h)}_{\gamma\delta}(\mathbf{q})^*}{\tau^{(h)}(\mathbf{q})} \quad (5.120)
$$

These can be rewritten as

$$
\tau^{(\pm 1)} = a_2 \left[\left(1 - \frac{1}{4(\xi_1^2 + \frac{1}{2}\xi_2^2)} \frac{\xi_2^2}{\xi_h^2} \right) + \left(\xi_1^2 + \frac{1}{2}\xi_2^2 \right) \left(q \mp \frac{d}{2b+c} \right)^2 \right], \quad (5.121)
$$

$$
\tau^{(\pm 2)} = a_2 \left[1 - \frac{\xi_1^2}{\xi_h^2} + \xi_1^2 \left(q \mp \frac{d}{b} \right)^2 \right]. \quad (5.122)
$$

The correlations are observable in inelastic scattering of visible light. Recall that we assumed in (5.33) that the deviations $\delta\epsilon_{\alpha\beta}$ of the dielectric tensor from the average isotropic value are equal to the order parameter fluctuations $Q_{\alpha\beta}$, so that the interaction energy of the liquid crystal with the vector potential is, under the assumption (5.36), equal to

$$
\sigma H_{\text{int}} = \frac{1}{2} \int d^3 x E_\alpha Q_{\alpha\beta} E_\beta. \quad (5.123)
$$

---

[3]See page 327 in the textbook [10].
[4]The factor $\frac{1}{2}$ is due to the dependence of φ and φ^*, $\varphi^{(h)}(-\mathbf{q}) = \varphi^{(h)*}(\mathbf{q})$.

Let E_{in} be the field of incoming light with momentum k_{in} and frequency ω. For a given fixed dielectric configuration $\epsilon(\mathbf{x})$, the polarization of the medium is given by

$$P(\mathbf{x})e^{-i\omega t} = \frac{1}{4\pi}[\epsilon(\mathbf{x}) - 1]E_{\text{in}}e^{-i(\omega t - \mathbf{k}_{\text{in}}\mathbf{x})}. \tag{5.124}$$

Since $P(\mathbf{x})$ may be considered as a density of radiating dipoles, these will emit light in a spherical wave which, at a large distance R away from the sample, has an electric field strength

$$E_{\text{out}}(\mathbf{x}') = \frac{\omega^2}{c^2}\frac{1}{R}e^{ikR}P_\perp(\mathbf{x}), \tag{5.125}$$

where $k = \omega/c$, $R = |\mathbf{x}' - \mathbf{x}|$. We have accounted for the dipole nature of the radiation by putting on the right-hand side $P_\perp$, which is the component of the polarization transverse to the direction of the outgoing wave. We now expand R around $x = 0$, so that $kR \approx kR_0 - \mathbf{k}_{\text{out}}\mathbf{x}$, integrate over the entire sample, and obtain the scattering amplitude A for incoming and outgoing polarization directions ϵ_{in}, ϵ_{out}:

$$A = \frac{\omega^2}{4\pi c^2}\epsilon_{\text{out}}^* \left\{ \int d^3x e^{-i\mathbf{q}\mathbf{x}} \left[\epsilon(\mathbf{x}) - 1\right] \right\} \epsilon_{\text{in}}, \tag{5.126}$$

where $\mathbf{q} \equiv \mathbf{k}_{\text{out}} - \mathbf{k}_{\text{in}}$ is the momentum transfer. See Fig. 5.7 for the experimental setup. The outgoing electric field is given by

$$\epsilon_{\text{out}} \cdot \mathbf{E}_{\text{out}}(\mathbf{x}') = \frac{E_{\text{in}}}{R_0}e^{ikR_0}A. \tag{5.127}$$

The square of A gives the differential cross section per unit solid-angle:

$$\frac{d\sigma}{d\Omega} = |A|^2. \tag{5.128}$$

Eliminating the direct beam associated with the spatially constant part of the $\epsilon_{\alpha\beta}(\mathbf{x})$, we may write

$$\frac{d\sigma}{d\Omega} = \frac{\omega^4}{(4\pi c^2)^2} \left[\epsilon_{\text{out}}^\dagger \delta\epsilon(\mathbf{q})\epsilon_{\text{in}}\right] \left[\epsilon_{\text{in}}^\dagger \delta\epsilon^*(\mathbf{q})\epsilon_{\text{out}}\right]. \tag{5.129}$$

In the present case, the dielectric tensor has thermodynamic fluctuations and we have to replace $\delta\epsilon(\mathbf{q})\delta\epsilon^*(\mathbf{q})$ by the correlation function (5.123). This gives

$$\frac{d\sigma}{d\Omega} = \frac{\omega^4}{(4\pi c^2)^2}\frac{k_B T}{2}\sum_h \frac{1}{\tau^{(h)}(\mathbf{q})}|\epsilon_{\text{out}}^\dagger \epsilon^{(h)}(\mathbf{q})\epsilon_{\text{in}}|^2. \tag{5.130}$$

Let the incoming beam run in the z-direction with the outgoing beam being rotated by an angle θ towards the y-axis (see Fig. 5.7). Then

$$k_{\text{in}} = k(0,0,1), \quad k_{\text{out}} = k(0,\sin\theta,\cos\theta), \tag{5.131}$$

and the momentum transfer is

$$\mathbf{q} = q\left(0, \cos\frac{\theta}{2}, \sin\frac{\theta}{2}\right), \tag{5.132}$$

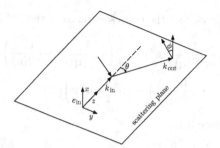

FIGURE 5.7 Experimental setup of the light-scattering experiment.

with

$$\mathbf{q}^2 = 2\mathbf{k}^2(1 - \cos\theta). \tag{5.133}$$

For an incoming polarization vertical to the scattering plane, i.e., along the x-axis, we have

$$\boldsymbol{\epsilon}_{\text{in}} = \boldsymbol{\epsilon}_V = (1, 0, 0). \tag{5.134}$$

Let the final polarization be inclined by an angle φ against the vertical direction, then

$$\mathbf{E}_{\text{out}} = E_{\text{out}}(\cos\varphi, -\sin\varphi\cos\theta, \sin\varphi\sin\theta). \tag{5.135}$$

The tensors $\epsilon^{(h)}(\hat{\mathbf{q}})$ are all given in terms of $\hat{\mathbf{q}}$ and $\mathbf{l}$, which may be taken as

$$\mathbf{l} = \frac{1}{\sqrt{2}}\left(1, -i\sin\frac{\theta}{2}, -i\cos\frac{\theta}{2}\right) = (\mathbf{l}^*)^*. \tag{5.136}$$

In this way we find

$$\begin{aligned}
\boldsymbol{\epsilon}_{\text{out}}^{\dagger}\epsilon^{(\pm 2)}(\hat{\mathbf{q}})\boldsymbol{\epsilon}_V &= \frac{1}{2}\left(\cos\varphi \mp \varphi\sin\varphi\sin\frac{\theta}{2}\right), \\
\boldsymbol{\epsilon}_{\text{out}}^{\dagger}\epsilon^{(\pm 1)}(\hat{\mathbf{q}})\boldsymbol{\epsilon}_V &= -\frac{1}{2}\sin\varphi\cos\frac{\theta}{2}, \\
\boldsymbol{\epsilon}_{\text{out}}^{\dagger}\epsilon^{(0)}(\hat{\mathbf{q}})\boldsymbol{\epsilon}_V &= -\frac{1}{\sqrt{6}}\cos\varphi.
\end{aligned} \tag{5.137}$$

If the initial polarization is horizontal

$$\boldsymbol{\epsilon}_{\text{in}} = \boldsymbol{\epsilon}_H = (0, 1, 0), \tag{5.138}$$

then the scalar products (5.137) read

$$\begin{aligned}
\boldsymbol{\epsilon}_{\text{out}}^{\dagger}\epsilon^{(\pm 2)}(\hat{\mathbf{q}})\boldsymbol{\epsilon}_H &= \frac{1}{2}\sin\frac{\theta}{2}\left(\mp i\cos\varphi - \sin\varphi\sin\frac{\theta}{2}\right), \\
\boldsymbol{\epsilon}_{\text{out}}^{\dagger}\epsilon^{(\pm 1)}(\hat{\mathbf{q}})\boldsymbol{\epsilon}_H &= \frac{1}{2}\cos\frac{\theta}{2}\cos\varphi, \\
\boldsymbol{\epsilon}_{\text{out}}^{\dagger}\epsilon^{(0)}(\hat{\mathbf{q}})\boldsymbol{\epsilon}_H &= -\frac{1}{\sqrt{6}}\sin\varphi\left(1 + \cos^2\frac{\theta}{2}\right).
\end{aligned} \tag{5.139}$$

Inserting this into (5.130) we find the cross section for a vertical incidence:

$$\frac{d\sigma_V}{d\Omega} = \frac{\omega^4}{(4\pi c^2)^2}\frac{k_B T}{2}\left[\frac{1}{6\tau^{(0)}(\mathbf{q})}\cos^2\varphi + \frac{1}{4}\left(\frac{1}{\tau^{(1)}(\mathbf{q})} + \frac{1}{\tau^{(-1)}(a)}\right)\cos^2\frac{\theta}{2}\sin^2\varphi\right.$$
$$\left.+\frac{1}{4}\left(\frac{1}{\tau^{(2)}(\mathbf{q})} + \frac{1}{\tau^{(-2)}(\mathbf{q})}\right)\left(-\sin^2\frac{\theta}{2}\sin^2\varphi\right)\right]. \tag{5.140}$$

For the horizontal incidence it is

$$\frac{d\sigma_H}{d\Omega} = \frac{\omega^4}{(4\pi c^2)^2}\frac{k_B T}{2}\left[\frac{1}{6\tau^{(0)}(\mathbf{q})}\sin^2\varphi\left(1 + \cos^2\frac{\theta}{2}\right)\right.$$
$$+ \frac{1}{4}\left(\frac{1}{\tau^{(1)}(\mathbf{q})} + \frac{1}{\tau^{(-1)}(a)}\right)\cos^2\frac{\theta}{2}\cos^2\varphi$$
$$\left.+ \frac{1}{4}\left(\frac{1}{\tau^{(2)}(\mathbf{q})} + \frac{1}{\tau^{(-2)}(\mathbf{q})}\right)\left(1 - \cos^2\frac{\theta}{2}\sin^2\varphi\right)\right]. \tag{5.141}$$

The experimental results show very little $\mathbf{q}$-dependence. In fact, for visible light of long wavelength with

$$\xi_1 q \ll 1, \quad \xi_2 q \ll 1, \tag{5.142}$$

i.e., for which the wavelength is much larger than both coherence lengths, we may neglect ξ_1, ξ_2 for a moment and see that

$$\tau^{(0)} \approx \tau^{(\pm 1)} \approx \tau^{(\pm 2)}. \tag{5.143}$$

Therefore the intensity of the scattered light goes like

$$I_V \sim \frac{1}{a_2}\left[\frac{1}{6}\cos^2\varphi + \frac{1}{2}\cos^2\frac{\theta}{2}\sin^2\varphi + \frac{1}{2}\left(1 - \sin^2\frac{\theta}{2}\sin^2\varphi\right)\right]. \tag{5.144}$$

For final polarizations vertical or horizontal to the scattering plane at a scattering angle $= 90^0$, the result implies that

$$I_{VV} \sim \frac{2}{3a_2},$$
$$I_{HV} \sim \frac{1}{2a_2}, \tag{5.145}$$

so that

$$\frac{I_{VV}}{I_{HV}} \sim \frac{4}{3}. \tag{5.146}$$

This ratio is approximately observed experimentally for T sufficiently above T^* [13]. As T approaches T^*, the coherence length grows larger, and the $\mathbf{q}$-dependence has a chance of becoming observable. Expanding $1/\tau^{(h)}(\mathbf{q})$ to lowest order in $\xi^2 q^2$ we find

$$\frac{1}{\tau^{(h)}(\mathbf{q})} = \frac{1}{a_2}\left\{1 - \left[\xi_1^2 + \left(\frac{2}{3} - \frac{h^2}{6}\right)\xi_2^2\right]q^2 + \ldots\right\}, \tag{5.147}$$

such that the intensities I_{VV}, I_{HV} behave like

$$I_{VV} \sim \frac{1}{6}\left[1 - \left(\xi_1^2 + \frac{2}{3}\xi_2^2\right)q^2\right] + \frac{1}{2}\left(1 - \xi_1^2 q^2\right) + \ldots,$$

$$I_{HV} \sim \frac{1}{4}\left[1 - \left(\xi_1^2 + \frac{2}{3}\xi_2^2\right)q^2\right] + \frac{1}{4}\left(1 - \xi_1^2 q^2\right) + \ldots, \tag{5.148}$$

with their ratio being

$$\frac{I_{VV}}{I_{HV}} \sim \frac{4}{3}\left(1 + \frac{1}{12}\xi_2^2 q^2 + \ldots\right). \tag{5.149}$$

For a comparison with the data it is most convenient to plot the inverse intensity against temperature which must behave for large enough T (a few ^0C above T^*) like [recall (5.45)].

$$I^{-1} \propto a_2^0 \left(\frac{T}{T^*} - 1\right)\left(1 + \xi^2 q^2 + \ldots\right), \tag{5.150}$$

i.e., it grows like a straight line, where ξ^2 is a combination of ξ_1^2 and ξ_2^2 depending on the polarizations (see Fig. 5.8). As the temperature drops towards T^*, the intensity of scattered light increases like a_2^{-1}, which is a manifestation of increasing fluctuations. This result is in agreement with experiment [13], with $\xi^2 > 0$.

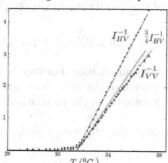

FIGURE 5.8 Inverse light intensities as a function of temperature. Note the small difference between $3I_{HV}/4$ and I_{VV}^{-1}, which will be commented on after Eq. (5.219).

Comparing such lines at different q values it is possible to deduce the size of the coherence lengths, for example in MBBA:

$$\xi(T) \approx 5.5 \times \left(\frac{T}{T^*} - 1\right)^{-1/2} \text{Å}. \tag{5.151}$$

As the temperature hits T_1 which usually lies one half to one ^0C above T^*, the inverse square of the coherence length jumps down to very small values, as shown in Fig. 5.9. This is where the intensity grows large in Fig. 5.8. The sample looks milky all of a sudden (critical opalescence).

It is easy to understand this behavior. At T_1, the size of the order parameter jumps from $\varphi = 0$ to $\varphi = \varphi_1 \neq 0$. Due to rotational invariance of the energy, there

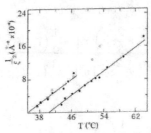

FIGURE 5.9 Behavior of inverse square of coherence length as a function of temperature near the weakly first-order phase transition in MBBA (see Ref. [13]).

is an infinite number of points in the $Q_{\alpha\beta}$ parameter space with the same energy, namely all those which differ only by a rotation of the direction vector $\mathbf{n}$. The associated continuous degeneracy causes strong directional fluctuations, and these result in strong fluctuations of the dielectric tensor that is proportional to $Q_{\alpha\beta}$. The latter can be observed by scattering light on the material.

Let us calculate the cross section of a small deviation $\delta Q_{\alpha\beta}(\mathbf{x})$ of the order parameter from the homogenous field configuration of the ground state. Expressing this in terms of the deviations of the director $n_\alpha(\mathbf{x})$ via Eq. (5.51), the cross section is

$$\frac{d\sigma}{d\Omega} = \frac{\omega^4}{(4\pi^2 c^2)^2} k_B T \frac{3}{4} \varphi^2 \left|\epsilon_{\text{out}\,\beta}^{\dagger}(\mathbf{k}_{\text{out}})\delta[(n_\alpha n_\beta)(\mathbf{q})]\epsilon_{\text{in}\,\alpha}(\mathbf{k}_{\text{in}})\right|^2, \tag{5.152}$$

where $\mathbf{q} \equiv \mathbf{k}_{\text{out}} - \mathbf{k}_{\text{in}}$ is the momentum transfer. For unpolarized incoming light, or if we do not measure the polarization of the outcoming light, the thermal average of the right-hand side requires knowledge of the correlation function

$$\langle\delta(n_\alpha n_\beta)\delta(n_\alpha n_\beta)\rangle = \langle\delta n_\alpha^* \delta n_\gamma\rangle n_\beta n_\delta + \langle\delta n_\alpha \delta n_\delta\rangle n_\beta n_\gamma + \langle\delta n_\beta \delta n_\gamma\rangle n_\alpha n_\delta + \langle\delta n_\beta \delta n_\delta\rangle n_\alpha n_\gamma. \tag{5.153}$$

To find $\langle\delta n_\alpha \delta n_\beta\rangle$, we consider the bending energy (5.87) for the Fourier transformed field

$$\delta\mathbf{n}(x) = \frac{1}{\sqrt{V}} \sum_{\mathbf{q}} e^{i\mathbf{q}\mathbf{x}} \delta\mathbf{n}(\mathbf{q}). \tag{5.154}$$

It has the form

$$F_{\text{bend}} = \frac{1}{2} \sum_{\mathbf{q}} \left[K_1 q_\alpha q_\beta + K_2 (\mathbf{n} \times \mathbf{q})_\alpha (\mathbf{n} \times \mathbf{q})_\beta + K_3 (\mathbf{n} \cdot \mathbf{q})^2 \delta_{\alpha\beta}\right] \delta n_\alpha(-\mathbf{q})\delta n_\alpha(\mathbf{q}). \tag{5.155}$$

In order to simplify the discussion suppose the system has an average orientation $\mathbf{n}\|\hat{\mathbf{z}}$. Then

$$F_{\text{bend}} = \frac{1}{2} \sum_{\mathbf{q}} \left[K_1 q_\alpha q_\beta + K_2 q_{\perp\alpha} q_{\perp\beta} + K_3 q_z^2 \delta_{\alpha\beta}\right] \delta n_\alpha(-\mathbf{q})\delta n_\alpha(\mathbf{q}), \tag{5.156}$$

where $q_\perp \equiv (-q_2, q_1, 0)$. The fluctuations can have only x- and y-components. This follows from the trivial equation $\frac{1}{2}\delta(n_\alpha n_\alpha)^2 = \delta n_\alpha n_\alpha = 0$. We can diagonalize this expression by introducing two orthogonal unit vectors $\mathbf{e}_1(\mathbf{q}) = (\hat{q}_1, \hat{q}_2, 0)$ and $\mathbf{e}_2(\mathbf{q}) = \hat{q}_\perp$. If we decompose

$$\delta\mathbf{n}(\mathbf{q}) = \mathbf{e}_1(\mathbf{q})\delta n_1(\mathbf{q}) + \mathbf{e}_2(\mathbf{q})\delta n_2(\mathbf{q}), \tag{5.157}$$

as illustrated in Fig. 5.10, we find the diagonal form

$$F = \frac{1}{2}\sum_{a=1,2}\left(K_a q_\perp^2 + K_3 q_z^2\right)|\delta n_a(\mathbf{q})|^2. \tag{5.158}$$

Thus the fluctuations of δn_1 and δn_2 diverge for $q \to 0$. The liquid crystal becomes opaque.

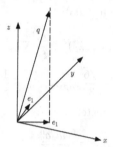

FIGURE 5.10 Relevant vectors of the director fluctuation (5.157).

In this fashion, the bending constants K_1, K_2, K_3 can be measured with values for which examples were quoted before.

5.5 Interfacial Tension between Nematic and Isotropic Phases

At the different lines of first-order phase transition, the order parameter moves from one value to another. Due to the derivative terms in the free energy, this change cannot take place abruptly but must be distributed over a length scale of the order of ξ in order to save gradient energies. It is a simple application of mean-field theory to calculate the energy stored in the interface.

Experimentally this quantity can be measured in the form of a surface tension [16]. This may be deduced to light scattering experiments [17] or, more directly, by looking at the curvature radius of a droplet of one phase embedded inside the other [18]. In this way, the surface tension was found for MBBA to be

$$\sigma \approx 2.3 \times 10^{-2} \ \text{erg/cm}^2, \quad [17] \tag{5.159}$$

$$\sigma \approx 1.6 \times 10^{-2} \ \text{erg/cm}^2. \quad [18] \tag{5.160}$$

For the calculation it is convenient to go to natural dimensionless quantities and introduce a renormalized field

$$\varphi_{\alpha\beta} = -4\sqrt{\frac{2}{3}\frac{a_4}{a_3}}Q_{\alpha\beta}. \tag{5.161}$$

We further measure the energy density in units of

$$f_1 \equiv \frac{9}{2^9}\frac{a_3^4}{a_4^3}, \tag{5.162}$$

and find

$$f = f_1\tilde{f}. \tag{5.163}$$

Then the energy density consisting of the sum of the Landau expansion f of Eq. (5.44) and the bending energy (5.94), $f_{tot} = f_{bend} + f$, corresponds to a dimensionless energy density [recall (5.95)]

$$
\begin{aligned}
\tilde{f}_{tot} &= 2\tau_0\xi_1^2\left[(\nabla_\gamma\varphi_{\alpha\beta})^2 + \frac{\xi_2^2}{\xi_1^2}\nabla_\alpha\varphi_{\alpha\gamma}\nabla_\beta\varphi_{\beta\gamma}\right] \\
&+ \tau\varphi_{\alpha\beta}^2 - \frac{\sqrt{6}}{3}\varphi_{\alpha\beta}\varphi_{\beta\gamma}\varphi_{\gamma\alpha} + \frac{1}{8}\left(\varphi_{\alpha\beta}^2\right)^2 + \dots ,
\end{aligned}
\tag{5.164}
$$

where τ_0 is a dimensionless parameter

$$\tau_0 \equiv \frac{4a_4a_2^0}{3a_3^2} \tag{5.165}$$

whose inverse $s \equiv \tau_0^{-1}$ measures how strongly the phase transition is of first order. It is useful to introduce the temperature-dependent dimensionless parameter

$$\tau = \frac{4a_4a_2}{3a_3^2} \equiv \frac{4a_4a_2^0}{3a_3^2}\frac{a_2}{a_2^0} = \tau_0\left(\frac{T}{T^*}-1\right). \tag{5.166}$$

Then we may write

$$f_1 = \frac{1}{8\tau^2}f_{c,L}. \tag{5.167}$$

where $f_{c,L}$ is the the Landau *condensation energy* (5.81)

$$f_{c,L} = -\frac{1}{4}\frac{a_2^2}{a_4}. \tag{5.168}$$

The nematic phase with the order parameter

$$\varphi_{\alpha\beta} = \varphi\epsilon_{\alpha\beta}^{(0)}(\mathbf{n}) \tag{5.169}$$

has a dimensionless potential energy

$$\tilde{f} = \tau\varphi^2 - \frac{1}{3}\varphi^3 + \frac{1}{8}\varphi^4, \tag{5.170}$$

if it is measured in units of f_1 of Eq. (5.162). This has a first-order transition at

$$\tau_1 = \frac{(1/3)^2}{4(1/8)} = \frac{2}{9}, \tag{5.171}$$

where φ jumps from $\varphi = 0$ to

$$\varphi = \varphi_1 = \frac{1/3}{2(1/8)} = \frac{4}{3}. \tag{5.172}$$

Note that at that temperature

$$\tau_1 = \tau_0 \left(\frac{T_1}{T^*} - 1 \right) = \frac{2}{9}. \tag{5.173}$$

The energy density (5.164) can be used to study a planar interface between the nematic and the disordered phase in the xy-plane. Let the region $z \gg 0$ be nematic and $z \ll 0$ be disordered. For symmetry reasons, we assume all gradients to point along the z-axis, leading to a bending energy

$$\tilde{f}_{\text{bend}} = 2\tau_0 \xi_1^2 \left[(\nabla_z \varphi_{\alpha\beta})^2 + \frac{\xi_2^2}{\xi_1^2} \nabla_z \varphi_{z\gamma} \nabla_z \varphi_{z\gamma} \right]. \tag{5.174}$$

With the order parameter (5.169) and $\varphi \neq 0$, this is minimized by letting **n** point orthogonal to the z-axis. Then (5.174) becomes

$$2\tau_0 \xi_1^2 \left(1 + \frac{1}{6} \frac{\xi_2^2}{\xi_1^2} \right) (\nabla_z \varphi)^2. \tag{5.175}$$

Therefore, the total energy density across the interface reads at $T = T_1$:

$$\tilde{f} = \xi_{\text{tr}}^2 (\nabla_z \varphi)^2 + \tau_c \varphi^2 - \frac{1}{3} \varphi^3 + \frac{1}{4} \varphi^4, \tag{5.176}$$

where we have introduced the transverse coherence length at $T = T_1$:

$$\xi_{\text{tr}}^2 = 2\tau_0 \xi_1^2 \left(1 + \frac{1}{6} \frac{\xi_2^2}{\xi_1^2} \right) \bigg|_{T=T_1}. \tag{5.177}$$

If we adopt ξ_{tr} as our transverse length scale, we may rewrite

$$\begin{aligned} \tilde{f} &= (\nabla_z \varphi)^2 + V(\varphi) \\ &= (\nabla_z \varphi)^2 + V_0 \varphi^2 (\varphi - \varphi_1)^2, \end{aligned} \tag{5.178}$$

where $V_0 = 1/8$, $\varphi_1 = 4/3$. The potential term has the standard form of a symmetric double well, with minima at $\varphi = 0$ and $\varphi_1 = 4/3$ (see Fig. 5.5). Inside the interface, the order parameter moves from one value to the other while keeping the total interface energy minimal, i.e., satisfying the Euler-Lagrange differential equation [10] which amounts here to

$$\nabla_z{}^2\varphi = V'(\varphi).\tag{5.179}$$

This is precisely the same as the equation of motion of a point particle at position φ as a function of pseudotime z, but in the reversed potential: The solution corresponds to a mass point rolling "down" the hill from $\varphi = 0$ through the "valley" at $\varphi = \varphi_1/2$ up to the other hill at $\varphi = \varphi_1$. The total pseudoenergy of this motion is conserved, i.e.,

$$(\nabla_z\varphi)^2 - V(\varphi) = \text{const.}\tag{5.180}$$

Far away from the interface, the field tends against $\varphi = 0$ or $\varphi = \varphi_1$. Having there the value $V = 0$, the constant is equal to zero, and we may integrate

$$z = \int_0^\varphi \frac{d\varphi'}{\sqrt{V(\varphi')}}.\tag{5.181}$$

to

$$\varphi(z) = \frac{1}{2}\varphi_1\left(1 + \tanh\frac{z\sqrt{V_0}\varphi_1}{2}\right).\tag{5.182}$$

This is the same as

$$\varphi(z) = \frac{2}{3}\left(1 + \tanh\frac{z}{3\sqrt{2}}\right).\tag{5.183}$$

The total free energy for this situation is found from the integral

$$\begin{aligned}
\tilde{f} &= \int_{-\infty}^\infty dz\left[(\nabla_z\varphi)^2 + V(\varphi)\right] = 2\int_0^\infty dz V(\varphi) = 2\int_0^{\varphi_1} d\varphi\sqrt{V(\varphi)}\\
&= 2\sqrt{V_0}\int_0^{\varphi_1} d\varphi\,\varphi(\varphi - \varphi_1) = \frac{\sqrt{V_0}}{3}\varphi_1^3 = \frac{16}{81}\sqrt{2}.
\end{aligned}\tag{5.184}$$

This is the surface tension which, back in physical units, reads

$$\sigma = \frac{16}{81}\sqrt{2}\,\xi_{\text{tr}}f_1 \, .\tag{5.185}$$

The value of f_1 involves a_3 and a_4, which are both somewhat hard to determine experimentally. But there is a simple experimental quantity which contains f_1 rather directly: the latent heat of the transition. In MBBA, for example, one measures [19]:

$$\Delta q = 0.3\,\frac{\text{kJ}}{\text{mol}} \approx 1.2\,\frac{\text{J}}{\text{g}} = 1.2 \cdot 10^7\frac{\text{erg}}{\text{g}}.\tag{5.186}$$

Within the present natural units, the latent heat is found from (5.170) as

$$\begin{aligned}
\Delta q &= f_1\frac{\partial\tau}{\partial T}T\left(\left.\frac{\partial\tilde{f}}{\partial\tau}\right|_{\varphi=\varphi_1} - \left.\frac{\partial\tilde{f}}{\partial\tau}\right|_{\varphi=0}\right)\\
&= f_1\tau_0\frac{T_1}{T^*}\varphi_1^2 = \frac{16}{9}\tau_0\frac{T_1}{T^*}f_1 = \frac{32}{81}\frac{T_1}{T^*}\left(\frac{T_1}{T^*} - 1\right)^{-1}f_1.
\end{aligned}\tag{5.187}$$

Comparing this with (5.185) we find the simple relation

$$\sigma = \sqrt{2}\Delta q\, \xi_{\text{tr}}\frac{T^*}{T_1}\left(\frac{T_1}{T^*}-1\right). \tag{5.188}$$

For MBBA we may insert Δq of (5.186) on the right-hand side and estimate

$$\xi_{\text{tr}} \approx 150\text{Å}, \quad \frac{T_1}{T^*}-1 \approx \frac{1}{400}, \tag{5.189}$$

so that $\sigma \approx 1.5 \times 10^{-2}$ erg/cm^2, in reasonable agreement with the experimental values (5.159), (5.160).

5.6 Cholesteric Liquid Crystals

The collective field theory developed up to this point is able to describe an ensemble of rod-like, disc-like or biaxial order. In the introduction it was mentioned that, in cholesterol and similar compounds, the molecular build-up exhibits a slight screw-like distortion. This violates mirror reflection symmetry. In order to describe such systems we have to add a parity violating piece to the energy. To lowest order there exists the following quadratic term with this property:

$$f_{\text{pv}} = -d\,\epsilon_{\alpha\beta\gamma}Q_{\alpha\beta}\nabla_\gamma Q_{\beta\gamma}. \tag{5.190}$$

This may be written alternatively in terms of the spin matrix (5.97) as

$$f_{\text{pv}} = -i\,d\,Q_{\alpha\beta}\left(\mathbf{S}\boldsymbol{\nabla}Q\right)_{\alpha\beta}. \tag{5.191}$$

For the Fourier transformed field, we can write (with $q \equiv |\mathbf{q}|$):

$$f_{\text{pv}} = -d\sum_q Q_{\alpha\beta}(-\mathbf{q})q\left(HQ(\mathbf{q})\right)_{\alpha\beta}. \tag{5.192}$$

In a notation slightly different from that in Subsection 5.1.3, this can also be written as

$$f_{\text{pv}} = -id\int d^2\mathbf{n}\, Q\,\hat{\mathbf{S}}\cdot\boldsymbol{\nabla}Q, \tag{5.193}$$

where $\hat{\mathbf{S}}$ is the operator

$$\hat{\mathbf{S}} = -i\mathbf{n}\times\boldsymbol{\nabla}_{\mathbf{n}}. \tag{5.194}$$

The total free-energy density $f_{\text{tot}} = f + f_{\text{bend}} \equiv f + f_{\text{der}} + f_{\text{pv}}$ describes the cholesteric phase transition. Let us construct the cholesteric ground state. For this we consider small fluctuations, and expand Q into normal modes as

$$Q_{\alpha\beta} = \sum_{\mathbf{q}}\left(\sum_h \epsilon_{\alpha\beta}^{(h)}(\hat{\mathbf{q}})e^{i\mathbf{q}\cdot\mathbf{x}}S^{(h)}(\mathbf{q}) + \text{c.c.}\right), \tag{5.195}$$

where $\epsilon^{(\pm 2)}(\hat{\mathbf{q}}), \epsilon^{(\pm 1)}(\hat{\mathbf{q}}), \epsilon^{(0)}(\hat{\mathbf{q}})$ are the five polarization tensors of helicities $h = -2, \ldots, 2$ of Eq. (5.108). Inserting (5.195) into $f_{\text{bend}} = f_{\text{der}} + f_{\text{pv}}$ of (5.94) and (5.190), the bending energy becomes

$$\frac{f_{\text{bend}}}{k_B T} = \frac{1}{k_B T} \sum_{\mathbf{q},h} f^{(h)}(\mathbf{q}) \tag{5.196}$$

$$= \sum_{\mathbf{q}} \left\{ \frac{1}{2} \left[a_2 + \left(b + \frac{2}{3}c \right) q^2 \right] |S^{(0)}(\mathbf{q})|^2 \right.$$

$$+ \frac{1}{2} \left[\left(a_2 - \frac{d^2}{4 \left(b + \frac{c}{2} \right)} \right) + \left(b + \frac{c}{2} \right) \left(q \mp \frac{d}{2 \left(b + \frac{c}{2} \right)} \right)^2 \right] |S^{(\pm 1)}(\mathbf{q})|^2$$

$$\left. + \frac{1}{2} \left[\left(a_2 - \frac{d^2}{b} \right) + b \left(q \mp \frac{d}{b} \right)^2 \right] |S^{(\pm 2)}(\mathbf{q})|^2 \right\}. \tag{5.197}$$

where

$$A = \frac{2}{3\kappa^2} \left(\kappa - \frac{2}{15} \right) = \frac{4}{45} \left(\frac{T}{T^*} - 1 \right) \frac{1}{\kappa^2}. \tag{5.198}$$

From light scattering experiments we find $S^{(2)}$ and $S^{(0)}$ to be the modes of largest fluctuations. The first has zero momentum $\mathbf{q}^{(0)} = 0$, the second has a non-vanishing momentum $\mathbf{q}^{(0)}$ pointing in an arbitrary direction, whose size is given by the minimum of the bending energy at

$$q^{(2)} = \frac{d}{b} = \frac{1}{\xi_h}. \tag{5.199}$$

This gives rise to a normal reflection of circularly polarized light of wavelength $\lambda_R = 4\pi\xi_h$.

The cholesteric ground state may now be found from a superposition of the dominant $h = 2$ and $h = 0$ modes

$$Q_{\alpha\beta} = S^{(0)} \epsilon_{\alpha\beta}^{(0)}(\hat{\mathbf{q}}^{(2)}) + S^{(2)}(\hat{\mathbf{q}}^{(2)}) \left[\epsilon_{\alpha\beta}^{(2)} e^{i(\mathbf{q}^{(2)} \cdot \mathbf{x} + \delta)} + \text{c.c.} \right], \tag{5.200}$$

where δ is an arbitrary phase. If we set $\mathbf{n} \cdot \mathbf{q}^{(2)} \equiv z$, the quantity defined in Eq. (5.26) becomes

$$\sqrt{\frac{4\pi}{5}} \frac{2}{3} Q(\mathbf{n}) = \left(z^2 - \frac{1}{3} \right) S^{(0)} + \sqrt{\frac{2}{3}} \left(1 - z^2 \right) S^{(2)} \cos \left(\mathbf{q}^{(2)} \cdot \mathbf{x} + \delta \right). \tag{5.201}$$

Averaging over all directions and a period along $\hat{\mathbf{q}}^{(2)}$, and replacing $S^{(h)}$ by $\kappa\sigma^{(h)}$, this results in a free energy

$$\frac{f}{k_B T} = \frac{1}{3}\kappa\sigma^{(0)2} + \frac{2}{3}\kappa \left(1 - \frac{d^2}{\kappa b} \right) \sigma^{(2)}$$

$$- \frac{1}{2\pi} \int_0^{2\pi} d\delta \log \int_0^1 dz\, e^{\left(z^2 - \frac{1}{3} \right) \sigma^{(0)} + \sqrt{\frac{2}{3}} (1 - z^2) \sigma^{(2)} \cos \delta} \tag{5.202}$$

$$= \frac{\kappa}{3}\sigma^{(0)2} + \frac{1}{3}\sigma^{(0)} + \frac{2}{3}\kappa \left(1 - \frac{d^2}{\kappa b} \right) \sigma^{(2)2} - \frac{1}{2\pi} \int_0^{2\pi} d\delta \log J\left(\sigma^{(0)}, \sigma^{(2)}, \delta \right),$$

where

$$J\left(\sigma^{(0)},\, \sigma^{(2)},\, \delta\right) \equiv \int_0^1 dz\, e^{z^2\left(\sigma^{(0)} - \sqrt{\frac{2}{3}}\sigma^{(2)}\cos\delta\right)} \tag{5.203}$$

is the generalization of the previous integral (5.10). In equilibrium, we now have the equations

$$\kappa\sigma^{(0)} = -\frac{1}{2} + \frac{2}{3}\frac{1}{2\pi}\int_0^{2\pi} d\delta \frac{1}{J}\frac{\partial}{\partial\sigma^{(0)}} J, \tag{5.204}$$

$$2\kappa\left(1 - \frac{d^2}{\kappa b}\right)\sigma^{(2)} = \frac{3}{2}\frac{1}{2\pi}\int_0^{2\pi} d\delta\frac{1}{J}\frac{\partial}{\partial\sigma^{(2)}} J. \tag{5.205}$$

The first equation can again be expressed in the same fashion as before in (5.14), except that σ has to be replaced by $\sigma^{(0)} - \sqrt{\frac{2}{3}}\sigma^{(2)}\cos\delta$, and an average has to be taken over all δ:

$$\kappa\sigma^{(0)} = \frac{1}{2\pi}\int_0^{2\pi} d\delta\, S\left(\sigma^{(0)} - \sqrt{\frac{2}{3}}\sigma^{(2)}\cos\delta\right). \tag{5.206}$$

The other equation has an additional weight factor $\sqrt{\frac{2}{3}}\cos\delta$, to be averaged as

$$2\kappa\left(1 - \frac{d^2}{\kappa b}\right)\sigma^{(2)} = \sqrt{\frac{2}{3}}\frac{1}{2\pi}\int_0^{2\pi} d\delta\cos\delta\, S\left(\sigma^{(0)} - \sqrt{\frac{2}{3}}\sigma^{(2)}\right). \tag{5.207}$$

Remember (5.14) for the definition of the function $S(\mathbf{x})$.

In order to establish contact with the previous calculations of the cholesteric free energy, it is useful to go to the natural variables to find

$$\kappa - \frac{2}{15} \equiv \frac{2}{21}\left(\tau + 2\alpha\right), \quad 2\alpha \equiv \frac{21}{2}\frac{3}{2}\frac{d^2}{b}, \quad \frac{f}{k_B T} \equiv \frac{25}{56}\tilde{f}, \tag{5.208}$$

$$\sigma^{(0)} \equiv \frac{15}{4}x, \quad \sigma^{(2)} \equiv \frac{15}{4}\frac{y}{\sqrt{2}}. \tag{5.209}$$

Then $\tilde{f}$ has the simple expansion

$$\tilde{f} = \left(\tau + 2\alpha\right)x^2 + \tau y^2 - \frac{1}{3}x^3 + xy^2 + \frac{1}{8}\left(x^2 + y^2\right)^2 + \dots, \tag{5.210}$$

and the field equations (5.206), (5.207) read

$$\left[\left(\tau + 2\alpha\right) + \frac{7}{5}\right]x = \frac{21}{5}\frac{1}{2\pi}\int_0^{2\pi} d\delta\, S\left(\tfrac{15}{4}\left(x - \tfrac{1}{\sqrt{3}}y\cos\delta\right)\right),$$

$$2\left(\tau + \frac{7}{5}\right)y = \sqrt{3}\frac{21}{5}\frac{1}{2\pi}\int_0^{2\pi} d\delta\cos\delta\, S\left(\tfrac{15}{4}\left(x - \tfrac{1}{\sqrt{3}}y\cos\delta\right)\right), \tag{5.211}$$

which may be solved by iteration. The results for $\tilde{f}$, x, and y as functions of α and τ are shown in Fig. 5.11 as contour plots.

The main defect of the Maier-Saupe model is that the size of the cubic term is too large in comparison with the physical transition. In Appendix 5A we show how this aspect can be improved by a biaxial version of the Maier-Saupe model.

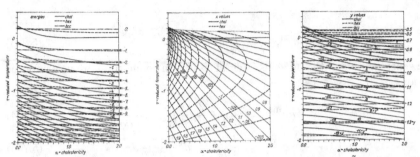

FigURE 5.11 Contour plots of constant reduced free energy density $\tilde{f}_{\rm ext}$ with order parameters $x = \varphi^{(0)}$ and $y = \varphi^{(2)}$.

5.6.1   Small Fluctuations above T_1

If the temperature lies far enough above T_1 (say a few ^0C) the fluctuations are dominated by the quadratic part of the free energy. The normal modes are still given by the different helicity tensors $\epsilon^{(2)}(\hat{\mathbf{q}})$, and energies behave on the average like [see (5.196)]

$$\tau^{(0)}(\mathbf{q}) = a_2\left(1 + \frac{2}{3}\xi_1^2 q^2\right), \tag{5.212}$$

$$\tau^{(\pm 1)}(\mathbf{q}) = a_2\left[1 + \left(\xi_1^2 + \frac{1}{2}\xi_2^2\right)\left(q^2 \pm \frac{d}{b + \frac{c}{2}}q\right)\right], \tag{5.213}$$

$$\tau^{(\pm 2)}(\mathbf{q}) = a_2\left[1 + \xi_1^2\left(q^2 \pm 2\frac{d}{b}q\right)\right]. \tag{5.214}$$

Another way of writing the last two equations is

$$\tau^{(\pm 1)} = a_2\left[\left(1 - \frac{1}{4(\xi_1^2 + \frac{1}{2}\xi_2^2)}\frac{\xi_1^2}{\xi_h^2}\right) + \left(\xi_1^2 + \frac{1}{2}\xi_2^2\right)\left(q \pm \frac{d}{2b + c}\right)^2\right], \tag{5.215}$$

$$\tau^{(\pm 2)} = a_2\left[1 - \frac{\xi_1^2}{\xi_h^2} + \xi_1^2\left(q \pm \frac{d}{b}\right)^2\right]. \tag{5.216}$$

The quantity d/b is equal to $1/\xi_h = q^{(2)}$ by Eq. (5.199). Similarly we set

$$d/(2b + c) \equiv q^{(1)} = (1 + \xi_2^2/2\xi_1^2)q^{(1)}. \tag{5.217}$$

The behavior of $\tau^{(h)}(\mathbf{q})$ for $h = 0, \pm 1, \pm 2$, is sketched in Fig. 5.12.

While $\tau^{(0)}(\mathbf{q})$ is unaffected by the parity-violating d-term, the helicity one and two fluctuations now are strongest for non-vanishing momenta (recall Fig. 5.8)

$$q^{(1)} \equiv \frac{1}{2\xi_h}\frac{1}{1 + \xi_2^2/2\xi_1^2},$$

$$q^{(2)} = \frac{1}{\xi_h}. \tag{5.218}$$

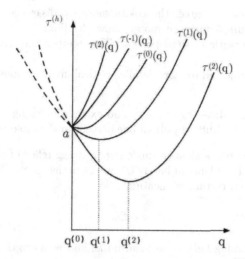

FIGURE 5.12 Momentum dependence of the gradient coefficients $\tau^{(h)}(\mathbf{q})$ of the modes of helicity h, as specified in Eqs. (5.212), (5.215), and (5.216).

This fact will be seen to give rise to a number of distinctive physical properties of cholesteric systems.

5.6.2 Some Experimental Facts

As far as Rayleigh scattering far above T_1 is concerned, the momentum transfers are so small that the result of (5.148),

$$\frac{I_{VH}^{-1}}{I_{VV}^{-1}} \approx \frac{4}{3}, \tag{5.219}$$

is still expected to be true. Experimentally a slight deviation $(1.448\pm.94)$ is observed which has not yet been explained (see Fig. 5.8).

The most striking difference with respect to the nematic case, however, consists in the following. The data points of I^{-1} no longer end at a precocious phase transition at $T_1 > T^*$. Instead, they turn off the straight line and can now be followed down to below T^* (see Fig. 5.8) by half a degree Celsius, where they suddenly jump down to small values as the ordered phase is reached. These values are, however, much (≈ 10 times) larger than those in the nematic ordered phase, i.e., the scattered light intensity is much smaller. This indicates a lower level of degeneracy of orientational degrees of freedom as compared to the nematic phase. There is another characteristic feature which was already observed by Reinitzer [1] in his first investigations of such systems. The liquid appears in a bright blue color. For this reason, this temperature regime is referred to as the *blue phase* [20].

When pressed into a thin layer between two glass plates, the liquid forms a great number of domains, called plaquelets, some of them blue [21]. As the temperature is lowered by one more degree, the colors suddenly disappear and the intensity of scattered light jumps up once more. Now the liquid shows the same degree of opaqueness as nematic ordered phases. This temperature regime is called the *cholesteric phase.*

If the liquid is subjected to more detailed optical investigations, it reveals several important phenomena:

1. The refractive indices for ordinary and extraordinary light rays are equal in the blue phase but differ by about one percent in the cholesteric phase [21, 22].

2. The cholesteric phase shows a single strong Bragg reflex of circularly polarized light at normal incidence at barely UV wavelengths. Thus, the liquid is capable of transferring a certain momentum[5],

$$q = 2k_0 = \frac{4\pi}{\lambda_R}, \qquad (5.220)$$

upon the incoming light of momentum k_0 and wavelength λ_R. The quantity $P = 4\pi/q$ is referred to as *optical pitch.*

3. For oblique incidence there are also reflexes of higher order $2q$, $3q$, at Bragg angles θ:

$$\lambda_R = \frac{P}{m} \sin\theta \qquad (5.221)$$

($\theta = 90^0$, normal incidence). But now the polarizations are elliptical.

4. Also the blue phase gives Bragg reflexes but with a larger pitch P_{blue} which is about two times larger than that in the cholesteric phase (this is why the color is blue rather than UV). Moreover, the plaquelets described above reflect light at wavelengths which are integer fractions of the above pitch P_{blue} and of $P_{\text{blue}} \cdot \sqrt{2}$. As a matter of fact, the directions of reflexes can be fitted by the same Bragg condition as those in a bcc lattice:

$$\left(\frac{\sin\theta}{\lambda_R/P_b}\right)^2 = \frac{m_1{}^2 + m_2{}^2 + m_3{}^2}{2}, \qquad (5.222)$$

where the Miller indices can take integer values with even numbers. The presence of lattice planes $(1, 1, 0)$, $(2, 0, 0)$, $(2, 0, 0)$ has apparently been established [22].

5. There is one more important observation [22]. The wave length of reflected light remains constant for about half a degree Celsius. Then it has a jump to a higher value and increases even more for another half degree before it falls back to a low value as the cholesteric phase is reached. The jump is present only for samples of shorter pitch. We shall now try to understand these properties theoretically.

---

[5]If the light is observed outside the medium, λ_R has to be replaced by λ_R/n.

5.6.3 Mean-Field Description of Cholesteric Phase

In the presence of the parity violating term (5.191), the ground state is much harder to determine than in nematics, even at the mean-field level. The reason is that a constant field configuration can no longer give the lowest energy. For the following discussion let us truncate the free energy after the quartic term, for simplicity. In the natural energy units (5.162) introduced before we may write the free energy density as

$$
\begin{aligned}
\tilde{f} &= (\tau + 2\alpha)\varphi_{\alpha\beta}{}^2 - \frac{\sqrt{6}}{3}\varphi_{\alpha\beta}\varphi_{\beta\gamma}\varphi_{\gamma\alpha} + \frac{1}{4}\left(\varphi_{\alpha\beta}^2\right)^2 \\
&\quad + 2\alpha\xi_h^2\left[(\nabla_\gamma\varphi_{\alpha\beta})^2 + \frac{\xi_2^2}{\xi_1^2}\nabla_\alpha\varphi_{\alpha\gamma}\nabla_\beta\varphi_{\beta\gamma}\right] \\
&\quad - 4\alpha\xi_h\epsilon_{\alpha\beta\gamma}\varphi_{\alpha\beta}\nabla_\gamma\varphi_{\beta\delta}.
\end{aligned}
\tag{5.223}
$$

Here we have introduced the additional dimensionless parameter

$$
2\alpha \equiv \frac{4a_4 b}{3a_3^2}\frac{d^2}{b^2} = \frac{2}{9}\frac{\xi_1^2(T_1)}{\xi_h^2},
\tag{5.224}
$$

where $\xi_1(T_1) = \xi_1(T_1/T^* - 1) = \sqrt{6a_4 b}/a_3$ is the coherence length at the first-order phase transition [recall (5.95)]. The cholesteric phase condenses when $\tau + 2\alpha$ becomes negative. Recalling (5.170) and (5.171), we identify

$$
\tau + 2\alpha \equiv \frac{4a_4 b}{3a_3^2}\frac{a_2}{b} \equiv \frac{4a_4 b}{3a_3^2}\frac{a_2^0}{b}\left(\frac{T}{T^*} - 1\right) = \tau_0\left(\frac{T}{T^*} - 1\right) \equiv \frac{2}{9}\frac{\xi_1^2(T_1)}{\xi_1^2(T)}.
\tag{5.225}
$$

The parameter $\sqrt{\alpha}$ measures the coherence lenght $\xi_1 = \sqrt{b/a_2}$ at T_1 in units of the cholesteric length scale ξ_h, apart from a trivial factor $1/3$. For this reason we call α the *cholestericity* of the liquid crystal. Obviously, the limit $\alpha \to 0$ which is reached for $d^2 \to 0$, restores the nematic case [see (5.224)], in which case (5.225) coincides with the previous definition (5.166), and $2\alpha\xi_h^2$ becomes $2\tau_0\xi_1^2$.

We have seen in the last chapter that, at the level of small fluctuations, the last term in (5.223) gives a preference to the helicity-two ($q \approx q^{(2)}$) mode with $q \approx q^{(2)}$ (see Fig. 5.12). Thus we may expect a lower energy for an ansatz:

$$
\varphi_{\alpha\beta} = \frac{1}{\sqrt{2V}}\left(\epsilon^{(2)}(\hat{\mathbf{q}})e^{i\mathbf{q}\mathbf{x}}\varphi^{(2)} + \text{c.c.}\right).
\tag{5.226}
$$

Inserting this into (5.223) we find

$$
\tilde{f} = \tau\varphi^{(2)2} + \frac{1}{8}\varphi^{(2)4} + 2\alpha\left(\frac{q}{q^{(2)}} - 1\right)^2\varphi^{(2)2}.
\tag{5.227}
$$

There is no cubic term since the product of three $\epsilon^{(2)}(\mathbf{q})$, $\epsilon^{(2)*}(\mathbf{q})$ tensors vanishes. The energy is minimized by setting $q = q^{(2)}$, where it becomes

$$
\tilde{f} = \tau\varphi^{(2)} + \frac{1}{8}\varphi^{(2)4}.
\tag{5.228}
$$

This is to be compared with the helicity-zero expression:

$$\tilde{f} = (\tau + 2\alpha)\varphi^{(0)2} - \frac{1}{3}\varphi^{(0)3} + \frac{1}{8}\varphi^{(0)4}$$
$$+ 2d\xi_h{}^2\left(1 + \frac{2}{3}\frac{\xi_1{}^2}{\xi_2{}^2}\right)q^2\varphi^{(0)2} \qquad (5.229)$$

which is minimal at $q = 0$.

We now realize that for large enough α the energy (5.228) is always lower than (5.229). For if $2\alpha > 2/g$, the energy (5.229) vanishes for $\tau > (2/g) - 2\alpha$ while (5.227) has a second-order phase transition at $\tau = 0$ and starts being negative for $\tau < 0$. But this is by far not the lowest possible energy. In order to see this let us combine both helicities linearly and take

$$\varphi_{\alpha\beta} = \frac{1}{\sqrt{V}}\left[\epsilon^{(0)}(\mathbf{n})\varphi^{(0)} + \frac{1}{\sqrt{2}}\left(\epsilon^{(2)}(\hat{\mathbf{q}})e^{i\mathbf{q}\mathbf{x}}\varphi^{(2)} + \text{c.c.}\right)\right], \qquad (5.230)$$

where $\hat{\mathbf{q}}$ points in an arbitrary direction and the direction vector $\mathbf{n}$ may be parametrized as

$$\mathbf{n} = (n_x, n_y, n_z) = \sin\theta(\cos\phi\,\hat{\mathbf{x}} + \sin\phi\,\hat{\mathbf{y}}) + \cos\theta\hat{\mathbf{z}}. \qquad (5.231)$$

Now the energy has the form

$$\tilde{f} = (\tau + 2\alpha)\varphi^{(0)2} + \tau|\varphi^{(2)}|^2 - \frac{\varphi^{(0)3}}{3} - \varphi^{(0)}|\varphi^{(2)}|^2\left(3|\hat{\mathbf{l}}\cdot\hat{\mathbf{n}}|^2 - 1\right)$$
$$+ \frac{1}{8}\left[\left(\varphi^{(0)2} + |\varphi^{(2)}|^2\right) + 6\varphi^{(0)}|\varphi^{(2)}|^2\right] + \mathcal{O}(|\varphi^{(0)}|^4). \qquad (5.232)$$

The two modes are coupled at the cubic level. This gives rise to a linear asymmetry for the $\varphi^{(0)}$-amplitude such that it is pulled out of the equilibrium position to a new minimum thereby reducing the remaining quartic potential for $\varphi^{(2)}$. This effect is strongest if the cubic term is maximal and the quartic term minimal, which happens for

$$\hat{\mathbf{n}} \cdot \hat{\mathbf{l}} = 0. \qquad (5.233)$$

The associate energy density is

$$\tilde{f} = (\tau + 2\alpha)x^2 + \tau y^2 - \frac{x^3}{3} + xy^2 + \frac{1}{8}\left(x^2 + y^2\right)^2. \qquad (5.234)$$

Here we have changed variables from $\varphi^{(0)}$ and $\varphi^{(2)}$ to x and y, for convenience. We now minimize $\tilde{f}$ with respect to x and y and find

$$(\tau + 2\alpha)\,x - \frac{1}{2}x^2 + \frac{1}{2}y^2 + x(x^2 + y^2) = 0, \qquad (5.235)$$
$$\tau y + xy + y(x^2 + y^2) = 0. \qquad (5.236)$$

From these two equations we obtain

$$y^2 = 3x^2 - 4\alpha x, \tag{5.237}$$

which, after inserting it back into (5.233), gives

$$x^2 + (1 - \alpha)x + \tau = 0, \tag{5.238}$$

which has the two solutions

$$x_{1,2} = -\left[\frac{1-\alpha}{2} \pm \sqrt{\frac{(1-\alpha)^2}{4} - \tau}\right]. \tag{5.239}$$

At the extrema, the energy is

$$
\begin{aligned}
\tilde{f}_{\text{ext}} &= 2\left(x^2 + \frac{x^3}{3} - \alpha\tau x\right) \\
&= 2\left[-\tau^2 + \frac{1}{3}(1-\alpha)\tau - \frac{4}{3}\left(\frac{(1-\alpha)^2}{4} - \tau\right)\left(\frac{1-\alpha}{2} \pm \sqrt{\frac{(1-\alpha)^2}{4} - \tau}\right)\right],
\end{aligned} \tag{5.240}
$$

and we see that the $+$ sign corresponds to the lower value.

The phase transition takes place at $\tau_c = \tau_c(\alpha)$, where $\tilde{f}_{\text{ext}}$ vanishes. Instead of solving $\tilde{f}_{\text{ext}} = 0$ from (5.240) it is more convenient to combine $\tilde{f}_{\text{ext}} = 0$ with (5.238) to get two linear equations:

$$x = -\frac{\tau + \alpha - \alpha^2}{\alpha + \frac{1}{3}}, \tag{5.241}$$

and

$$x = \frac{\left(\alpha + \frac{1}{3}\right)\tau}{\tau + (\alpha - 1)/3}. \tag{5.242}$$

Using these we eliminate once the lowest and once the highest power of x in $\tilde{f}_{\text{ext}} = 0$. Combining the resulting equations we obtain

$$g\tau^2 + 2(g\alpha - 1)\tau - 3\alpha(1 - \alpha)^2 = 0, \tag{5.243}$$

which determines the curve in the (α, τ)-plane, where $\tilde{f}$ vanishes. For $\alpha < 0$, this happens first at a value $\tau_c > 0$, which for $\alpha = 0$ takes the nematic value $2/9$, and which decreases down to zero at $\alpha = 1$. Above $\alpha = 1$, the curve (5.241) does not correspond to a minimum. In that region, the phase transition takes place at $\tau = 0$ and is of second order, as can be seen directly from (5.240). Above $\alpha = 1$, the energy becomes for small $\tau \leq 0$:

$$\tilde{f}_{\text{ext}} = -2\tau^2\left(1 + \frac{4}{3(\alpha - 1)}\right) + O(\tau^3). \tag{5.244}$$

The full behaviour of $\tilde{f}_{\text{ext}}$ as a function of temperature τ and cholestericity α is shown in the form of contour plots in Fig. 5.11.

The order parameters x and y are also displayed in the contour plots of Fig. 5.11. The lines of constant x are straight: $\epsilon = x\alpha - (x^2 + x)$. For $\alpha > 0$ one winds up in the *cholesteric phase*. Notice that for $\alpha \to \infty$, the helicity-zero component becomes more and more suppressed, and only $\varphi^{(2)} = y$ survives.

What happens if also the helicity-one component is admitted? In order to study this let us assume all fields to vary only along the z-axis. For symmetry reasons, we may take $\hat{\mathbf{l}} = \frac{1}{\sqrt{2}} (\hat{\mathbf{x}} + i\hat{\mathbf{y}})$. Then we have $\mathbf{n} = \hat{\mathbf{z}}$ from (5.233), and we may expand

$$\varphi_{\alpha\beta}(z) = \frac{1}{\sqrt{V}} \left[\epsilon_{\alpha\beta}^{(0)}(\hat{\mathbf{z}})\varphi^{(0)}(z) + \frac{1}{\sqrt{2}} \left(\epsilon_{\alpha\beta}^{(1)}(\hat{\mathbf{z}})\varphi^{(1)}(z) + \epsilon_{\alpha\beta}^{(2)}(\hat{\mathbf{z}})\varphi^{(2)}(z) + \text{c.c.} \right) \right], \quad (5.245)$$

with a real field $\varphi^{(0)}(z)$ and two complex fields $\varphi^{(1)}(z)$, $\varphi^{(2)}(z)$. The energy density becomes using (5.232)

$$
\begin{aligned}
\tilde{f} &= (\tau + 2\alpha) \left(\varphi^{(0)2} + |\varphi^{(1)}|^2 + |\varphi^{(2)}|^2 \right) \\
&\quad - \frac{1}{3}\varphi^{(0)3} - \frac{1}{2}\varphi^{(0)} \left(|\varphi^{(1)}|^2 - 2|\varphi^{(2)}|^2 \right) - \frac{\sqrt{3}}{4} \left(\varphi^{(2)*}\varphi^{(1)2} + \text{c.c.} \right) \\
&\quad + \frac{1}{8} \left(\varphi^{(0)2} + |\varphi^{(1)}|^2 + |\varphi^{(2)}|^2 \right)^2 \\
&\quad + 2\alpha\xi_h^2 \left[r_0 \left(\nabla_z \varphi^{(0)} \right)^2 + r_1 \left(|\nabla_z \varphi^{(1)}|^2 + |\nabla_z \varphi^{(2)}|^2 \right) \right. \\
&\quad \left. - 2\alpha\xi_h (\varphi^{(1)*} \frac{\nabla_z}{2} \varphi^{(1)} + 2\varphi^{(2)*} \frac{\nabla_z}{2} \varphi^{(2)}) \right].
\end{aligned}
\quad (5.246)
$$

Here we have introduced the convenient abbreviations

$$
\begin{aligned}
r_0 &\equiv 1 + \frac{2}{3} \frac{c_1 + c_2}{b} = \frac{4r_1 - 1}{3}, \\
r_1 &\equiv 1 + \frac{c_1 + c_2}{2b} = 1 + \frac{c}{2b}.
\end{aligned}
\quad (5.247)
$$

Both are experimentally accessible in the ordered phase by measuring the ratio of Frank constants

$$r_1 \equiv \frac{K_1 + K_3}{2K_2} = \frac{K_3 + K - b}{2K_t}. \quad (5.248)$$

In momentum space, the quadratic terms can be rewritten after a quadratic completion as

$$
\begin{aligned}
\tilde{f} = \sum_{\mathbf{q}} &\left[\left(\tau + 2\alpha + 2\alpha r_0 \xi_h^2 q^2 \right) |\varphi^{(0)}(\mathbf{q})|^2 \right. \\
&+ \left(\tau + 2\alpha \left(1 - \frac{1}{4r_1} \right) + 2\alpha r_1 \left(q\xi_h \mp \frac{1}{2r_1} \right)^2 \right) |\varphi^{(\pm 1)}(\mathbf{q})|^2 \\
&\left. + \left(\tau + 2\alpha \left(q\xi_h \mp 1 \right)^2 \right) |\varphi^{(\pm 2)}(\mathbf{q})|^2 \right],
\end{aligned}
\quad (5.249)
$$

where $q \equiv q_z$. For very large α, this is certainly minimal at the former solution with $q = 1/\xi_h$, and no $\varphi^{(0)}, \varphi^{(2)}$ components can be present. Experimentally, however, α

is not so large to justify ignoring $\varphi^{(0)}, \varphi^{(2)}$: A typical cholesteric system has $\xi_1^0 \approx 11\text{Å}$ and $\xi_h \approx 2000/4\pi\text{Å}$, so that $\alpha \approx 0.21$. Therefore, $\varphi^{(1)}$ will be present. From the energy we see that the amplitude $\varphi^{(1)}$ enters only in higher orders. Thus there can be a second-order phase transition with $\varphi^{(1)} \neq 0$ developing from the previous solution with $\varphi^{(0)}, \varphi^{(2)} \neq 0$ along a line in the $\alpha - \tau$ plane where the coefficient of the quadratic term becomes negative:

$$D \equiv \tau + 2\alpha \left(1 - \frac{1}{4r_1}\right) + 2\alpha r_1 \left(1 - \frac{1}{2r_1}\right)^2 - \frac{1}{2}x - \frac{\sqrt{3}}{2}y + \frac{1}{4}\left(x^2 + y^2\right) \leq 0. \quad (5.250)$$

Inserting the solutions (5.237) and (5.239) we find that this cannot happen. At $\alpha = 0$ one has $x = -\frac{1}{2} + \sqrt{\frac{1}{2} - \tau}$, $y = -\sqrt{3}x$, and $x^2 + x + \tau = 0$, implying that $D = 0$. But for all allowed $\alpha > 0$, and τ in the cholesteric phase, we can verify that $\tau + \alpha - \frac{x}{2} - \frac{\sqrt{3}}{2}y + \frac{1}{4}\left(x^2 + y^2\right)$ starts out with $\mathcal{O}(\alpha^2)$ and is always > 0. This ensures also $D > 0$ since the first line in (5.250) is $\tau + \alpha\left(1 + \frac{r_1}{2}\right)$ and $r_1 > 0$.

Let us take a look at the cholesteric order parameter with $\varphi^{(0)}, \varphi^{(2)} \neq 0$. It may be written in a matrix form as

$$\varphi_{\alpha\beta} = \varphi^{(0)}\frac{1}{\sqrt{6}}\begin{pmatrix} -1 & & \\ & -1 & \\ & & 2 \end{pmatrix}_{\alpha\beta} + \frac{1}{\sqrt{2}}\varphi^{(2)}\frac{1}{2}\begin{pmatrix} 1 & i & 0 \\ \epsilon & -1 & 0 \\ 0 & 0 & 0 \end{pmatrix}_{\alpha\beta} e^{iqz} + \text{c.c.}$$

$$= \begin{pmatrix} -\frac{1}{\sqrt{6}} + \frac{1}{\sqrt{2}}\cos qz, & -\frac{1}{\sqrt{2}}\varphi^{(2)}\sin qz & 0 \\ -\frac{1}{\sqrt{2}}\varphi^{(2)}\sin qz & -\frac{1}{\sqrt{6}}\varphi^{(0)} - \frac{1}{\sqrt{2}}\varphi^{(2)}\cos qz & 0 \\ 0 & 0 & \frac{2}{\sqrt{6}}\varphi^{(0)} \end{pmatrix}_{\alpha\beta}. \quad (5.251)$$

This has to be added to $\epsilon_0\delta_{\alpha\beta}$ in order to obtain the dielectric tensor which is usually parametrized as

$$\epsilon_{\alpha\beta} = \begin{pmatrix} \bar{\epsilon} + \delta\cos 2kz & -\delta\sin 2kz & 0 \\ -\delta\sin 2kz & \bar{\epsilon} - \delta\cos 2kz & 0 \\ 0 & 0 & \epsilon_3 \end{pmatrix}_{\alpha\beta}. \quad (5.252)$$

Note that the mixing between $\varphi^{(0)}$ and $\varphi^{(2)}$ induces, in general, biaxiality. The local eigenvalues are now all three different $\bar{\epsilon} + \delta, \bar{\epsilon} - \delta, \epsilon_3$.

In order to interpret the order parameter (5.254) physically it is useful to realize the following: Suppose the helicity zero rod-like form $\epsilon_{\alpha\beta}^{(0)}(\mathbf{n}) = \sqrt{\frac{3}{2}}\left(n_\alpha n_\beta - \frac{1}{3}\delta_{\alpha\beta}\right)$ is taken in the direction

$$\mathbf{n}(z) = (\cos kz, -\sin kz, 0), \quad (5.253)$$

that is rotated away from the z-direction into the xy-plane. Then $\epsilon_{\alpha\beta}^{(0)}(\mathbf{n}(z))$ becomes

$$\epsilon_{\alpha\beta}^{(0)}(\mathbf{n}(z)) = \sqrt{\frac{3}{2}}\begin{pmatrix} \cos^2 kz - \frac{1}{3} & -\sin kz\cos kz & 0 \\ \sin kz\cos kz & \sin^2 kz - \frac{1}{3} & 0 \\ 0 & 0 & -\frac{1}{3} \end{pmatrix}_{\alpha\beta}$$

$$
= \sqrt{\frac{3}{2}} \begin{pmatrix} \frac{1}{6} + \frac{1}{2}\cos 2kz & -\sin 2kz & 0 \\ \sin 2kz & \frac{1}{6} - \frac{1}{2}\cos 2kz & 0 \\ 0 & 0 & -\frac{2}{6} \end{pmatrix}_{\alpha\beta}
$$

$$
= \frac{1}{2}\left(-\epsilon^{(0)}(z) + \sqrt{3}\frac{1}{\sqrt{2}}\left(\epsilon^{(2)}(z)e^{iqz} + \text{c.c.}\right)\right)_{\alpha\beta}. \tag{5.254}
$$

Thus it has precisely the form (5.230), with the particular ratio

$$
\frac{\varphi^{(2)}}{\varphi^{(0)}} = \frac{y}{x} = \sqrt{3}. \tag{5.255}
$$

In this case we may interpret the solution (5.248) as a purely transverse helical configuration of rod-like molecules. These parameters can be measured in optical experiments. They show that the biaxiality remains usually small: the eigenvalue $\bar{\epsilon} - \delta$ is usually equal to ϵ_3 (typical example: $\bar{\epsilon} = 2.745$, $\delta = 0.315$, $\epsilon_3 = 2.430$ [23]).

Thus experimentally, the ratio (5.255) is observed. Looking back at (5.237) we notice that, for $\alpha = 0$, this is automatically true (as it should be since $\alpha = 0$ corresponds to the absence of the parity violating term). The uniaxiality remains approximately true for the typical experimental value $\alpha \approx 0.21$. Thus we find for the ratio of the dielectric eigenvalues of (5.252):

$$
\frac{\bar{\epsilon} - \delta}{\epsilon_3} = 1 + \frac{3}{8}\left(\frac{1}{\sqrt{\frac{1}{4} - \tau}} - \frac{2}{3}\right). \tag{5.256}
$$

5.7   Other Phases

There are several other possible configurations of momenta where we can expect a low total energy.

One is the *hexatic phase*, in which the order parameter contains one component $\varphi^{(0)}(\mathbf{n})$ and three components $\varphi^{(2)}(\hat{\mathbf{q}}_i)$ where $\mathbf{q}_1, \mathbf{q}_2, \mathbf{q}_3$ form an equilateral triangle, for whose directions we may choose

$$
\hat{\mathbf{q}}_1 = \hat{\mathbf{x}}, \quad \hat{\mathbf{q}}_{2,3} = -\frac{1}{2}\hat{\mathbf{x}} \pm \frac{\sqrt{3}}{2}\hat{\mathbf{y}}. \tag{5.257}
$$

The polarization vectors associated with these momenta may be taken as

$$
\mathbf{l}_1 = \frac{1}{\sqrt{2}}(\hat{\mathbf{y}} + i\hat{\mathbf{z}})e^{i\gamma_1/2} \tag{5.258}
$$

$$
\mathbf{l}_{2,3} = \frac{1}{\sqrt{2}}\left(\mp\frac{\sqrt{3}}{2}\hat{\mathbf{x}} - \frac{1}{2}\hat{\mathbf{y}} + i\hat{\mathbf{z}}\right)e^{i\gamma_{2,3}/2}. \tag{5.259}
$$

For symmetry reasons, the chirality of all polarization vectors has to be the same, i.e., they must form a positively oriented triped.

Another possible phase is a body-centered cubic (bcc) phase in which the momenta and polarization vectors are oriented, as shown in Fig. 5.13.

The regimes where the four possible phases, cholesteric, hexatic, or bcc, are the lowest are shown in the phase diagram in Fig. 5.14. A discussion of all possible phases was given in 1981 by Kleinert and Maki [24] (the development reviewed in Ref. [25]). The most interesting phase is the icosahedral phase in which the momenta and polarizations are arranged as shown in Fig. 5.15. Such a phase is not periodic in space but it displays a fivefold symmetry under rotation. It is called *quasicrystalline*. For the liquid crystal with a Landau-de Gennes expansion of the order field, this phase has so far not been shown to be a stable configuration. Higher powers in the field seem to be necessary to achieve stabilization.

Such a phase would have an interesting density profile displayed in Fig. 5.16.

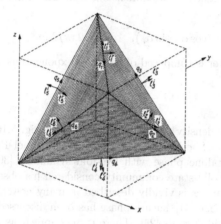

FIGURE 5.13 Momenta and polarization vectors for a body-centered cubic (bcc) phase of a cholesteric liquid crystal.

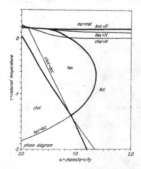

FIGURE 5.14 Regimes in the plane of α, τ, where the phases cholesteric, hexatic, or bcc have the lowest energy.

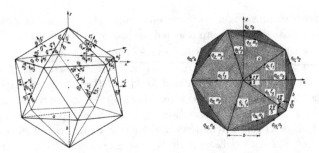

FIGURE 5.15 Momenta and polarization vectors for an icosahedral phase of a cholesteric liquid crystal.

This profile can be obtained in two dimensions from the simplest set of momenta with fivefold symmetry

$$\mathbf{p}_k = (\cos\alpha_k, \sin\alpha_k), \quad \alpha_k = 2\pi k/N. \tag{5.260}$$

A possible order parameter is composed of a sum of exponentials

$$\phi(\mathbf{x}) = \sum_{k=0}^{N-1} e^{i\mathbf{x}\mathbf{p}_k}. \tag{5.261}$$

This leads directly to the density $\rho(\mathbf{x}) = |\phi(\mathbf{x})|^2$ shown in Fig. 5.16.

In the year 2011, Dan Shechtman was awarded the Nobel Prize for his 1984 discovery of a quasicrystalline phase with five-fold symmetry [30] in a sputtered Al-Mn alloy. While the solid-state community considered his observation for some time with great skepticism, as is vividly described in many newspaper articles and Shechtman's Wikipedia page [31], such a phase has been discussed three years earlier in the context of liquid crystals [24]. There it may appear as the so called *blue phase*, as calculated by Seidemann [25] and by Rokhsar and Sethna [26, 27]. Thermodynamics of various phases in cholesteric liquid crystals is shown in Fig. 5.18. For a comprehensive analysis of the blue phases see Ref. [28]. Recent experiments are discussed in [29].

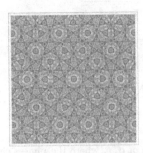

FIGURE 5.16 Density profile $\rho(\mathbf{x}) = |\phi(\mathbf{x})|^2$ with five-fold symmetry.

In two dimensions, we may also take $N = 7$ and expect to find a heptagonal density distribution shown in Fig. 5.17 [33].

For more details on quasicrystals see the textbooks [34, 35, 36].

Appendix 5A   Biaxial Maier-Saupe Model

In order to improve the Maier-Saupe model with respect to the large-a_3 coefficient, let us try a modified version in which the basic molecules are biaxial. In the general discussion of the free energy in Section 5.2 we have seen that the cubic term, a_3 produces a region of biaxial order. Thus the large size of a_3 in the model seems to be connected with the basic assumption of uniaxial molecules at the microscopic level. Let us see whether this is, in fact, true. Consider again the nematic free energy (5.29). The integral over $\mathbf{n}$ corresponds to averaging over all microscopic

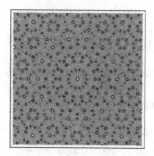

FIGURE 5.17 Density profile $\rho(\mathbf{x}) = |\phi(\mathbf{x})|^2$ with seven-fold symmetry.

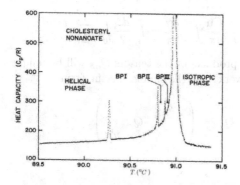

FIGURE 5.18 Blue phases in a cholesteric liquid crystal.

orientations of the rod-like uniaxial molecules such that the orientational energy (5.4) is proportional to

$$Q(\mathbf{n}) = \sqrt{\frac{15}{8\pi}}\sqrt{\frac{2}{3}}Q_{\alpha\beta}^{\text{mol}}Q_{\alpha\beta} = \sqrt{\frac{15}{8\pi}}\left(n_\alpha n_\beta - \frac{1}{3}\delta_{\alpha\beta}\right)Q_{\alpha\beta}. \tag{5A.1}$$

Suppose now the microscopic order parameter is biaxial. Then it contains an extra term in addition to the axial order parameter (5.2):

$$Q_{\alpha\beta}^{\text{mol}} = \sqrt{\frac{3}{2}}\left[\left(n_\alpha n_\beta - \frac{1}{3}\delta_{\alpha\beta}\right) + \epsilon\left(m_\alpha m_\beta - \frac{1}{3}\delta_{\alpha\beta}\right)\right], \tag{5A.2}$$

where $\mathbf{m}$ is another unit vector orthogonal to $\mathbf{n}$. If $\mathbf{n}$, $\mathbf{m}$ point in z- and x-directions, respectively, we have the explicit matrices:

$$\sqrt{\frac{2}{3}}Q_{\alpha\beta}^{\text{mol}} = \begin{pmatrix} -\frac{1}{3} & & \\ & -\frac{1}{3} & \\ & & \frac{2}{3} \end{pmatrix} + \epsilon\begin{pmatrix} -\frac{2}{3} & & \\ & -\frac{1}{3} & \\ & & -\frac{1}{3} \end{pmatrix}$$

$$= \frac{1}{3}\begin{pmatrix} 2\epsilon - 1 & & \\ & -(1+\epsilon) & \\ & & -(2-\epsilon) \end{pmatrix}. \tag{5A.3}$$

By an appropriate choice of ϵ we can now simulate any desired ratio for the three principal axes of the molecules. The spatial averages are a little more involved. Let us parametrize n and m in terms of angles as

$$\mathbf{n} = (\sin\theta\cos\varphi,\ \sin\theta\sin\varphi,\ \cos\theta) \tag{5A.4}$$

$$\mathbf{m} = (\cos\theta\cos\varphi\cos\gamma - \sin\varphi\sin\gamma,\ \cos\theta\sin\varphi\cos\gamma + \cos\varphi\sin\gamma,\ -\sin\theta\cos\gamma).$$

Then the directional average must be performed as a product of integrals

$$\langle\ldots\rangle = \langle\ldots\rangle_z\langle\ldots\rangle_\gamma$$

$$= \int\frac{d^2\mathbf{n}}{4\pi}\int_0^{2\pi}\frac{d\gamma}{2\pi} = \int_1^1\frac{dz}{2}\int_0^{d\pi}\frac{2\varphi}{2\pi}\int_0^{2\pi}\frac{d\gamma}{2\pi}. \tag{5A.5}$$

The resulting invariants of products of the tensors $Q_{\alpha\beta}$ will be unique up to fourth power. We may therefore work with the simple specific form

$$Q_{\alpha\beta} = \begin{pmatrix} -Q & & \\ & -Q & \\ & & 2Q \end{pmatrix}, \tag{5A.6}$$

and substitute, at the end:

$$Q^2 \to \frac{1}{6}\text{tr}\,Q^2,\quad Q^3 \to \frac{1}{6}\text{tr}\,Q^3,\quad Q^4 \to \frac{1}{36}\left(\text{tr}\,Q^2\right)^2. \tag{5A.7}$$

With (5A.6) we see that the source term $Q^{\text{mol}}_{\alpha\beta}Q_{\alpha\beta}$ satisfies

$$
\begin{aligned}
\sqrt{\frac{2}{3}}Q^{\text{mol}}_{\alpha\beta}Q_{\alpha\beta} &= Q\left[\left(-n_x{}^2 - n_y{}^2 + 2n_z\right) + \epsilon\left(-mx^2 - m_y{}^2 + 2m_z{}^2\right)\right] \\
&= Q\left[3n_z{}^2 - 1 + \epsilon\left(3m_z{}^2 - 1\right)\right] \\
&= Q\left\{3z^2 - 1 + \epsilon\left[3\cos^2\alpha\left(1 - z^2\right) - 1\right]\right\} \\
&= Q\left(3z^2 a - b\right),
\end{aligned}
\tag{5A.8}
$$

where we have set

$$
a \equiv 1 - \epsilon\cos^2\gamma\,, \quad b \equiv 1 - \epsilon\left(3\cos^2\gamma - 1\right).
\tag{5A.9}
$$

The averages over z are now easily performed, yielding:

$$
\begin{aligned}
\left\langle Q^2(n)\right\rangle &= \frac{15}{8\pi}\,\frac{\text{tr}\,Q^2}{6}\left\langle\frac{9}{5}a^2 - 2ab + b^2\right\rangle_\gamma \\
\left\langle Q^3(n)\right\rangle &= \left(\frac{15}{8\pi}\right)^{3/2}\frac{\text{tr}\,Q^3}{6}\left\langle\frac{27}{7}a^3 - \frac{27}{5}a^2 b + 3ab^2 - b^3\right\rangle_\gamma \\
\left\langle Q^4(n)\right\rangle &= \left(\frac{15}{8\pi}\right)^2\frac{(\text{tr}\,Q^2)^2}{36}\left\langle\frac{81}{9}a^4 - \frac{108}{7}a^3 b + \frac{54}{5}a^2 b^2 - \frac{12}{3}ab^3 + b^4\right\rangle_\gamma.
\end{aligned}
\tag{5A.10}
$$

For $\epsilon = 0$ one has $a = b = 1$, leading back to the previous results (5.30) for nematic liquid crystals. For cholesteric liquid crystals, we must perform a remaining nontrivial average over γ. This is easily done using the basic average formula

$$
\left\langle\cos^{2n}\gamma\right\rangle_\gamma = \frac{(2n-1)!!}{2n!!},
\tag{5A.11}
$$

which yields explicitly the averages $1/2$, $3/8$, $5/16$, $35/(8\cdot 16)$ for $n = 1, 2, 3, 4$. If we write $a = 1 - \epsilon\alpha$, $b = 1 - \epsilon\beta$ with $\alpha = \cos^2\gamma$, $\beta = 3\cos^2\gamma - 1$, we calculate

$$
\begin{aligned}
&\langle\alpha\rangle = \langle\beta\rangle = \frac{1}{2}, \quad \langle\alpha^2\rangle = \frac{3}{8}, \quad \langle\alpha\beta\rangle = \frac{5}{8}, \quad \langle\beta^2\rangle = \frac{11}{8}, \\
&\langle\alpha^3\rangle = \frac{5}{16}, \quad \langle\alpha^2\beta\rangle = \frac{9}{16}, \quad \langle\alpha\beta^2\rangle = \frac{17}{16}, \quad \langle\beta^3\rangle = \frac{29}{16}, \\
&\langle\alpha^4\rangle = \frac{35}{128}, \quad \langle\alpha^3\beta\rangle = \frac{65}{128}, \quad \langle\alpha^2\beta^2\rangle = \frac{123}{128}, \quad \langle\alpha\beta^3\rangle = \frac{233}{128}, \quad \langle\beta^4\rangle = \frac{467}{128}.
\end{aligned}
\tag{5A.12}
$$

Hence we find

$$
\begin{aligned}
\left\langle a^2\right\rangle &= 1 - \epsilon + \frac{3}{8}\epsilon^2 \langle ab\rangle = 1 - \epsilon + \frac{5}{8}\epsilon^2 \langle b^2\rangle = 1 - \epsilon + \frac{11}{8}\epsilon^2, \\
\left\langle a^3\right\rangle &= 1 - \frac{3}{2}\epsilon + \frac{9}{8}\epsilon^2 - \frac{5}{16}\epsilon^3, \\
\left\langle a^2 b\right\rangle &= 1 - \frac{3}{2}\epsilon + \frac{13}{8}\epsilon^2 - \frac{9}{16}\epsilon^3,
\end{aligned}
$$

$$\langle ab^2 \rangle = 1 - \frac{3}{2}\epsilon + \frac{21}{8}\epsilon^2 - \frac{17}{16}\epsilon^3,$$

$$\langle b^3 \rangle = 1 - \frac{3}{2}\epsilon + \frac{33}{8}\epsilon^2 - \frac{29}{16}\epsilon^3,$$

$$\langle a^4 \rangle = 1 - 2\epsilon + \frac{18}{8}\epsilon^2 - \frac{5}{4}\epsilon^3 + \frac{35}{128}\epsilon^4,$$

$$\langle a^3 b \rangle = 1 - 2\epsilon + 3\epsilon^2 - 2\epsilon^3 + \frac{65}{128}\epsilon^4,$$

$$\langle a^2 b^2 \rangle = 1 - 2\epsilon + \frac{34}{8}\epsilon^2 - \frac{52}{16}\epsilon^3 + \frac{123}{8 \cdot 16}\epsilon^4,$$

$$\langle a b^3 \rangle = 1 - 2\epsilon + 6\epsilon^2 - \frac{80}{16}\epsilon^3 + \frac{233}{8 \cdot 16}\epsilon^4,$$

$$\langle b^4 \rangle = 1 - 2\epsilon + \frac{66}{8}\epsilon^2 - \frac{29}{4}\epsilon^3 + \frac{467}{8 \cdot 16}\epsilon^4. \tag{5A.13}$$

Combining these we obtain the following correction factors to the $\epsilon = 0$ -terms of (5A.10)

$$\left(1 - \epsilon + \epsilon^2\right),$$

$$\left(1 - \frac{3}{2}\epsilon - \frac{3}{2}\epsilon^3 + \epsilon^3\right), \tag{5A.14}$$

$$\left(1 - 2\epsilon + 3\epsilon^2 - 2\epsilon^3 + \epsilon^4\right) = \left(1 - \epsilon + \epsilon^2\right)^2.$$

Going back to (5.29) we see that the first two coefficients multiply directly the coefficients in the Landau expansion (5.31) with $\epsilon = 0$, while the quartic term receives a combined correction factor produced by the third row of (5A.14):

$$a_4 \rightarrow a_4 \left[\frac{7}{2}\left(1 - \epsilon + \epsilon^2\right)^2 - \frac{5}{2}\left(1 - \epsilon + \epsilon^2\right)^2\right] = a_4 \left(1 - \epsilon + \epsilon^2\right)^2. \tag{5A.15}$$

Since the cubic factor may be written as $\left(\epsilon - \frac{1}{2}\right)(\epsilon + 1)(\epsilon - 2)$ we see that we can indeed make it arbitrarily small, for example by choosing $\epsilon \approx \frac{1}{2}$. Note that the values $\epsilon = \frac{1}{2}$, $\epsilon = -1$, $\epsilon = 2$ correspond to

$$\sqrt{\frac{3}{2}}Q^{\mathrm{mol}} = \frac{1}{2}\begin{pmatrix} 0 & & \\ & -1 & \\ & & 1 \end{pmatrix}, \begin{pmatrix} -1 & & \\ & 0 & \\ & & 1 \end{pmatrix}, \begin{pmatrix} 1 & & \\ & -1 & \\ & & 0 \end{pmatrix}. \tag{5A.16}$$

Notes and References

[1] F. Reinitzer, Monatsh. Chemie **9**, 421 (1888).

[2] These are the so-called twisted nematic (TN) displays of
     M. Schadt and W. Helfrich, App. Phys. Lett. **18**, 127 (1971).

[3] L. Landau, Phys. Z. Sovjetunion **11**, 26, 545 (1937); and *The Collected Papers of L. Landau*, ed. by D. ter Haar, Gordon and Breach, Pergamon, N.Y., 1965.

[4] P.G. de Gennes, Mol. Cryst. Liq. Cryst. **12**, 193 (1971).

[5] H. Kleinert, *Particles and Quantum Fields*, World Scientific, 2016 (klnrt.de/b6).

[6] P.G. de Gennes, *The Physics of Liquid Crystals*, Clarendon Press, Oxford, 1974;
M.J. Stephen and J.P. Straley, Rev. Mod. **46**, 617 (1974);
Hans Kelker and Volker Hatz, *Handbook of Liquid Crystals*, Verlag Chemie, Weinheim, 1980.

[7] P. Weiss, J. Phys. Theor. Appl. **6**, 661 (1907).

[8] H.E. Stanley, *Introduction to Phase Transitions and Critical Phenomena*, Oxford University Press, Oxford, 2007.

[9] W. Maier and A. Saupe, Z. Naturforsch. **A 13**, 564 (1958), **A 14**, 882 (1959), **A15**, 287 (1960). See also the reference on p. 623 of the second paper of these.

[10] See the general discussion in the textbook:
H. Kleinert, *Path Integrals in Quantum Mechanics, Statistics, Polymer Physics, and Financial Markets*, 5th ed., World Scientific, 2009 (klnrt.de/b8).

[11] P.B. Vigman, A.I. Larkin, and V.M. Filev, JETP **41**, 944 (1975).

[12] V.Y. Paskakov, V.K. Semenchonko, and V.M. Byankin, JETP **39**, 3a83 (1975).

[13] T.W. Stinson and J.D. Litster, Phys. Rev. Lett. **25**, 503 (1970); **30**, 688 (1973);
B. Chu, C.S. Bak, and F.L. Lin, Phys. Rev. Lett. **28**, 1111 (1972).

[14] J.D. Litster, *Critical Phenomena*, McGraw-Hill, N.Y. (1971), p. 393.

[15] E. Gulari and B. Chu, J. Chem. Phys. **62**, 798 (1975).

[16] K. Kahlweit and W. Ostner, Chem. Phys. Lett. **18**, 589 (1973).

[17] D. Langevin and M.A. Bouchiat, Mol. Cryst. Liq. Cryst. **22**, 317 (1973).

[18] R. Williams, *ibid.*, **35**, 349 (1976).

[19] T. Shinoda, Y. Maeda, and H. Enokido, J. Chem. Thermod. **6**, 92 (1974).

[20] T. Harada and P.O. Crooker, Phys. Rev. Lett. **34**, 1259 (1975).

[21] F. Porsch and H. Stegemeyer, Chem. Phys. Lett. **155**, 620 (1989).

[22] D.L. Johnson, H.H. Flack, and P.O. Crooker, Phys. Rev. Lett. **45**, 641 (1980). See also
S. Meiboom and M. Sammon, Phys. Rev. Lett. **44**, 882 (1980).

[23] D.W. Berremann and T.J. Scheffer, Phys. Rev. Lett. **25**, 577 (1970).

[24] H. Kleinert and K. Maki, Fortschr. Phys. **29**, 219 (1981) (klnrt.de/75).

[25] T. Seideman, Rep. Prog. Phys. **59**, 659 (1990).

[26] D.S. Rokhsar and J.P. Sethna, Phys. Rev. Lett. **56**, 1727 (1986).

[27] For more details see http://klnrt.de/icosahedral.

[28] D.C. Wright and N.D. Mermin, Rev. Mod. Phys. **61**, 385 (1989).

[29] O. Henrich, D. Marenduzzo, K. Stratford, and M. E. Cates, Phys. Rev. E **81**, 031706 (2010).

[30] D. Shechtman, I. Blech, D. Gratias, and J. W. Cahn, Phys. Rev. Lett. **53**, 1951 (1984).

[31] Here a short collection of comments by his colleagues:

- The head of the NBS group of researchers told him: "Dr. Shechtman, you are embarrassing the group and I have to ask you to transfer to another research group."

- The editors of the Journal of Applied Physics wrote: "The paper will not interest physicists. We recommend that you send it to a metallurgical journal."

- "Shechtman is talking nonsense", at a scientific conference where Shechtman was sitting in the audience. "There are no such things as quasicrystals — there are only quasi-scientists."

See http://www1.technion.ac.il/en/technion/scitech/111006-nobel.

[32] http://en.wikipedia.org/wiki/Dan_Shechtman.
On this page he recounts his experience of "several years of hostility".

[33] H. Kleinert, EJTP **8**, 169 (2012) (klnrt.de/394).

[34] C. Junot, *Quasicrystals*, Oxford University Press, London, 1995.

[35] Z.M. Stadnik, *Physical Properties of Quasicrystals*, Springer Series in Solid-State Sciences, Berlin, 1999.

[36] T. Fujuwara and Y. Ishii, *Quasicrystals*, Elsevier, Amsterdam, 2008.

6

Exactly Solvable Field-Theoretic Models

The techniques developed in the previous chapters can be understood better by observing how they work in some simple exactly solvable models whose physics is well known on the basis of conventional techniques. This will be illustrated in this chapter in several typical cases.

6.1 Pet Model in Zero Plus One Time Dimensions

Consider the extremely simple case of a fundamental theory with a Hamiltonian

$$H = (a^\dagger a)^2/2, \tag{6.1}$$

where $a^\dagger$ and a denote the creation and annihilation operator of either¡ a boson or a fermion. In the first case, the eigenstates are

$$|n\rangle = \frac{1}{\sqrt{n!}}(a^\dagger)^n|0\rangle, \quad n = 0, 1, 2, \dots , \tag{6.2}$$

with the energies

$$E_n = \frac{n^2}{2}. \tag{6.3}$$

In the fermionic case, there are only two eigenstates

$$|0\rangle, \quad |1\rangle = a^\dagger|0\rangle, \tag{6.4}$$

with the energy eigenvalues

$$E_0 = 0, \quad E_1 = \frac{1}{2}. \tag{6.5}$$

The Lagrangian corresponding to H is

$$\mathcal{L}(t) = a^\dagger(t)i\partial_t a(t) - \tfrac{1}{2}\big[a^\dagger(t)a(t)\big]^2, \tag{6.6}$$

371

and the generating functional of all Green functions reads

$$
\begin{aligned}
Z[\eta^\dagger, \eta] &= \langle 0|T\exp\left[i\int dt(\eta^\dagger a + a^\dagger \eta)\right]|0\rangle \\
&= N\int \mathcal{D}a^\dagger \mathcal{D}a \exp\left[i\int dt\left(\mathcal{L} + \eta^\dagger a + a^\dagger \eta\right)\right].
\end{aligned} \tag{6.7}
$$

A collective field may be introduced via the formula

$$
\exp\left\{-\frac{i}{2}\int dt[a^\dagger a(t)]^2\right\} = \int \mathcal{D}\rho(t)\exp\left\{\frac{i}{2}\int dt\left[\rho^2(t) - 2\rho(t)a^\dagger a(t)\right]\right\}. \tag{6.8}
$$

Equivalently we may add to the exponent of (6.7) a term

$$
\frac{i}{2}\int dt\left[\rho(t) - a^\dagger(t)a(t)\right]^2,
$$

and integrate the generating functional Z over the ρ-field [compare the Hubbard-Stratonovich transformation in Eq. (1.79)].

The resulting Z can be rewritten as

$$
\begin{aligned}
Z[\eta^\dagger, \eta] = N\int \mathcal{D}a^\dagger \mathcal{D}a\, \mathcal{D}\rho \\
\times \exp\left\{dt\left[a^\dagger(t)i\partial_t a(t) - \rho(t)a^\dagger(t)a(t) + \frac{\rho^2(t)}{2} + \eta^\dagger(t)a(t) + a^\dagger(t)\eta(t)\right]\right\}.
\end{aligned} \tag{6.9}
$$

The collective field $\rho(t)$ describes the particle density. Indeed, a functional derivative of the Lagrangian density in the exponent of (6.9) displays the dependence

$$
\rho(t) = a^\dagger(t)a(t) \tag{6.10}
$$

which holds exactly at the classical level.

Integrating out the $a^\dagger, a$ fields gives

$$
Z[\eta^\dagger, \eta] = N\int \mathcal{D}\rho \exp\left\{i\mathcal{A}[\rho] - \int dtdt'\eta^\dagger(t)G_\rho(t,t')\eta(t')\right\}, \tag{6.11}
$$

with the collective field action [see once more (1.79)]

$$
\mathcal{A}[\rho] = \pm i\text{Tr}\log\left(iG_\rho^{-1}\right) + \int dt\frac{\rho^2(t)}{2}, \tag{6.12}
$$

where G_ρ denotes the propagator of the fundamental particles in a classical $\rho(t)$ field satisfying

$$
[i\partial_t - \rho(t)]\,G_\rho(t,t') = i\delta(t - t'). \tag{6.13}
$$

The solution can be found by introducing an auxiliary field

$$
\varphi(t) = \int^t \rho(t')dt', \tag{6.14}
$$

in terms of which

$$G_\rho(t, t') = e^{-i\varphi(t)} e^{i\varphi(t')} G_0(t - t'), \tag{6.15}$$

with G_0 being the free-field propagator of the fundamental particles. At this point one has to specify the boundary condition on $G_0(t - t')$. They have to be adapted to the physical properties of the system. The generating functional is supposed to describe the amplitude for vacuum to vacuum transitions in the presence of the source fields $\eta^\dagger, \eta$. The propagation of the free particles must take place in the same vacuum. If $a_0^\dagger, a_0$ describes a free particle, their time ordered product in the free vacuum is

$$G_0(t - t') = \langle 0 | T\left(a_0(t) a_0^\dagger(t')\right) | 0 \rangle = \Theta(t - t'). \tag{6.16}$$

Using (6.15), we find

$$G_\rho(t, t') = e^{-i\varphi(t)} e^{i\varphi(t')} \Theta(t - t'). \tag{6.17}$$

Equipped with this knowledge we can readily calculate the Tr log term in (6.12). The functional derivative is certainly

$$\frac{\delta}{\delta \rho(t)} \left\{ \pm i \mathrm{Tr} \log(i G_\rho^{-1}) \right\} = \mp G_\rho(t, t') \Big|_{t'=t+\epsilon} = 0, \tag{6.18}$$

where the $t' \to t$ limit is specified in such a way that the field $\rho(t)$ in (6.9) couples to

$$a^\dagger(t) a(t) = \pm T\left(a(t) a^\dagger(t')\right) \Big|_{t'=t+\epsilon} \stackrel{.}{=} \pm G^\rho(t, t') \Big|_{t'=t+\epsilon}. \tag{6.19}$$

Hence, the Θ-function in (6.17) makes the functional derivative vanish and the Tr log becomes an irrelevant constant. The generating functional reduces to the simple expression

$$Z[\eta^\dagger, \eta] = N \int \mathcal{D}\varphi(t) \exp\left\{ \frac{i}{2} \int dt \dot\varphi(t)^2 - \int dt dt' \eta^\dagger(t) \eta(t') e^{-i\varphi(t)} e^{-i\varphi(t')} \Theta(t - t') \right\}, \tag{6.20}$$

where we have used the relation

$$\mathcal{D}\rho = \mathcal{D}\varphi \det\left(\dot\delta(t - t')\right) = \mathrm{const} \cdot \mathcal{D}\varphi. \tag{6.21}$$

Observe that it is the field $\varphi(t)$ which becomes a convenient dynamical plasmon variable, not $\rho(t)$ itself.

The original theory has been transformed into a new one involving plasmons of zero mass. At this point we take advantage of the equivalence between functional and quantized operator formulation by considering the plasmon action in the exponent of (6.20) directly as a quantum field theory. The first term may be associated with a Lagrangian

$$\mathcal{L}_0(t) = \frac{1}{2}\dot{\varphi}(t)^2, \tag{6.22}$$

describing free plasmons.

The Hilbert space of the corresponding Hamiltonian $H = p^2/2$ consists of plane waves which are eigenstates of the functional momentum operator $p = -i\partial/\partial\varphi$:

$$\{\varphi|p\} = \frac{1}{\sqrt{2\pi}}e^{ip\varphi}, \tag{6.23}$$

normalized according to

$$\int_{-\infty}^{\infty} d\varphi\,\{p|\varphi\}\,\{\varphi|p'\} = \delta(p - p'). \tag{6.24}$$

In the operator version, the generating functional reads

$$Z[\eta^\dagger, \eta] = \frac{1}{\{0|0\}}\left\{0\left|T\exp\left[-\int dt dt' \eta^\dagger(t)\eta(t')e^{-i\varphi(t)}e^{i\varphi(t')}\Theta(t - t')\right]\right|0\right\}, \tag{6.25}$$

where $\varphi(t)$ are free field operators. Note that it is the zero functional momentum states between which the operator is evaluated. Due to the norm (6.24) there is an infinite normalization factor which has formally been taken out.

We can now verify the generation of all Green functions of fundamental particles from the functional derivatives with respect to $\eta^\dagger, \eta$. First

$$\langle 0|Ta(t)a^\dagger(t')|0\rangle = -\frac{\delta^{(2)}Z}{\delta\eta^\dagger(t)\delta\eta(t')}\bigg|_{\eta^\dagger,\eta=0}$$

$$= \frac{1}{\{0|0\}}\left\{0\left|e^{-i\varphi(t)}e^{i\varphi(t')}\right|0\right\}\Theta(t - t'). \tag{6.26}$$

Inserting the time translation operator

$$e^{iHt} = e^{i\frac{p^2}{2}t}, \tag{6.27}$$

the matrix element (6.26) becomes

$$\frac{1}{\{0|0\}}\left\{0\left|e^{-ip^2}2e^{-i\varphi(0)}e^{-i\frac{p^2}{2}(t-t')}e^{i\varphi(0)}e^{-i\frac{p^2}{2}t'}\right|0\right\} = \frac{1}{\{0|0\}}\left\{0\left|e^{-i\varphi(0)}e^{-i\frac{p^2}{2}(t-t')}e^{i\varphi(0)}\right|0\right\}. \tag{6.28}$$

But the state $e^{i\varphi(0)}|0\}$ is an eigenstate of the functional momentum p with $p = 1$, so that (6.28) equals

$$\frac{1}{\{0|0\}}\{1|1\}\,e^{-i(t-t')/2} = e^{-i(t-t')/2}, \tag{6.29}$$

and the Green function (6.26) becomes

$$\langle 0|Ta(t)a^\dagger(t')|0\rangle = e^{-i(t-t')/2}\Theta(t-t'). \tag{6.30}$$

This coincides exactly with the result of a calculation within the fundamental field operators $a^\dagger(t), a(t)$:

$$
\begin{aligned}
\langle 0|Ta(t)a^\dagger(t')|0\rangle &= \Theta(t-t')\langle 0|e^{i(a^\dagger a)^2 t/2}a(0)e^{-\frac{i}{2}(a^\dagger a)^2(t-t')}a^\dagger(0)e^{-i(a^\dagger a)^2 t'/2}|0\rangle \\
&= \Theta(t-t')e^{-i(t-t')/2}.
\end{aligned} \tag{6.31}
$$

Observe that nowhere in the calculation have Fermi- or Bose- statistics been used. This becomes relevant only for higher Green functions. Expanding the exponent (6.25) to nth order gives

$$
\begin{aligned}
Z^{[n]}\left[\eta^\dagger,\eta\right] &= \frac{1}{\{0|0\}}\frac{(-)^n}{n!}\int dt_1 dt_1' \cdots dt_n dt_n' \eta^\dagger(t_1)\eta(t_1')\cdots\eta^\dagger(t_n)\eta(t_n') \\
&\times \left\{0\left|Te^{-i\varphi(t_1)}e^{i\varphi(t_1')}\cdots e^{-i\varphi(t_n)}e^{i\varphi(t_n')}\right|0\right\}\Theta(t_1-t_1')\cdots\Theta(t_n-t_n').
\end{aligned} \tag{6.32}
$$

The Green function

$$\langle 0|Ta(t_1)\cdot\cdots\cdot a(t_n)a^\dagger(t_n')\cdot\cdots\cdot a^\dagger(t_1')|0\rangle \tag{6.33}$$

is obtained by forming the derivative

$$(-i)^{2n}\delta^{(2n)}Z[\eta^\dagger\eta]/\delta\eta^\dagger(t_1)\cdot\cdots\cdot\delta\eta^\dagger(t_n)\delta\eta(t_n')\cdot\cdots\cdot\delta\eta(t).$$

There are $(n!)^2$ contributions due to the product rule of differentiation, $n!$ of them being identical, thereby canceling the factor $1/n!$ in (6.32). The others correspond, from the point of view of combinatorics, to all Wick contractions of (6.33), each being associated with a factor $e^{-i\varphi(t)}e^{i\varphi(t')}$. In addition, the Grassmann nature of source fields η causes a minus sign to appear in all contractions which deviate from the natural order $11', 22', 33', \ldots$ by an odd permutation. For example

$$
\begin{aligned}
&\langle 0|Ta(t_1)a(t_2')a^\dagger(t_2')a^\dagger(t_1')|0\rangle \\
&= \langle 0|T(\overbrace{a(t_1)a(t_2)}\,\overbrace{a(t_2')}a^\dagger(t_1'))\,|0\rangle \pm \langle 0|T(\overbrace{a(t_1)a(t_2)}\,a(t_2')a^\dagger(t_1'))\,|0\rangle \tag{6.34} \\
&= \frac{1}{\{0|0\}}\left\{0\left|Te^{-i\varphi(t_1)}e^{-i\varphi(t_2)}e^{i\varphi(t_2')}e^{i\varphi(t_1')}\right|0\right\} \\
&= [\Theta(t_1-t_1')\Theta(t_2-t_2') \pm \Theta(t_1-t_2')\Theta(t_2-t_1')], \tag{6.35}
\end{aligned}
$$

where the upper sign holds for bosons, the lower for fermions. The lower sign enforces the Pauli exclusion principle: If $t_1 > t_2 > t_2' > t_1'$, the two contributions cancel, reflecting the fact that no two fermions $a^\dagger(t_2')a^\dagger(t_1')$ can be created successively on the particle vacuum. For bosons one may insert again the time translation operator (6.27) and complete sets of states $\int dp|p\rangle\langle p| = 1$, with the result:

$$
\begin{aligned}
&\frac{1}{\{0|0\}}\int dp\,dp'\{0|e^{-i\varphi(0)}e^{-i\frac{p^2}{2}(t_1-t_2)}e^{-i\varphi(0)}e^{-i\frac{p^2}{2}(t_2-t_2')}e^{i\varphi(0)}e^{-i\frac{p^2}{2}(t_2'-t_1')}e^{i\varphi(0)}|0\rangle \\
&= e^{-i(t_1-t_2)/2}e^{-i2(t_2-t_2')}e^{-i(t_2'-t_1')/2}. \tag{6.36}
\end{aligned}
$$

Here the expectation values $\{0|e^{-i\varphi(0)}|p\} = \delta(1-p)$, $\{p|e^{-i\varphi(0)}|p'\} = \delta(p+1-p')$ have been used. The result agrees again with an operator calculation of the type (6.31).

We now understand how the collective quantum field theory works in this model. Its Hilbert space is very large consisting of states of *all* functional momenta $|p\rangle$. When it comes to calculating the Green functions of the fundamental fields, however, only a small portion of this Hilbert space is used. A fermion can make plasmon transitions back and forth between the ground state $|0\}$ and the unit momentum state $|1\}$ only, due to the anticommutativity of the fermion source fields $\eta^\dagger, \eta$. Bosons, on the other hand, can connect states of any integer momentum $|n\}$. No other states can be reached. The collective basis is over-complete as far as the description of the underlying system is concerned. Strong selection rules, $p \to p \pm 1$, together with the source statistics make sure that only a small subspace becomes involved in the dynamics of the fundamental system. The compatibility of such a projection with unitarity is ensured by the conservation law $a^\dagger a = \text{const}$. In higher dimensions, there have to be infinitely many conservation laws (one for every space point).

Actually, in the boson case, the overcompleteness can be removed by defining the collective Lagrangian in (6.20) on a cyclic variable, i.e., one takes (6.22) on $\varphi \in [0, 2\pi)$ and extends it periodically. The path integral (6.20) is then integrated accordingly. In this case, the Hilbert space would be grated containing only integer momenta $p = 0, \pm 1, \pm 2, \ldots$ coinciding with the multi-boson states.

The following observations may be helpful in understanding the structure of the collective theory: It may sometimes be convenient to build all Green functions not on the vacuum state $|0\rangle$ but on some other reference state $|R\rangle$ for which we may choose the excited state $|n\rangle$. In the operator language this amounts to a generating functional

$$^n Z[\eta^\dagger, \eta] = \langle n|T \exp\left\{i \int dt \left[\eta^\dagger(t)a(t) + a^\dagger(t)\eta(t)\right]\right\}|n\rangle. \tag{6.37}$$

This would reflect itself in the boundary condition of G_0 for bosons

$$\begin{aligned} ^n G_0(t-t') &= \langle n|T\left(a_0(t)a_0^\dagger(t')\right)|n\rangle \\ &= (n+1)\Theta(t-t') + n\Theta(t'-t). \end{aligned} \tag{6.38}$$

For fermions, only $n = 1$ would be an alternative, with

$$^1 G_0(t-t') = \langle 1|T\left(a_0(t)a_0^\dagger(t')\right)|1\rangle = -\Theta(t'-t). \tag{6.39}$$

As a consequence of (6.38) or (6.39), formula (6.18) would become

$$\frac{\delta}{\delta\rho(t)}\left\{\pm i \text{Tr}\log\left(iG_\rho^{-1}\right)\right\} = -\left\{\begin{array}{c} n \\ 1 \end{array}\right\}. \tag{6.40}$$

Integrating this functionally gives

$$\pm i \text{Tr}\log\left(iG_\rho^{-1}\right) = -\left\{\begin{array}{c} n \\ 1 \end{array}\right\}\int_{-\infty}^{\infty}\rho(t)dt, \tag{6.41}$$

so that the functional form of (6.37) reads, according to (6.12):

$$\begin{Bmatrix} n \\ 1 \end{Bmatrix} Z[\eta^\dagger, \eta] = \int \mathcal{D}\varphi \exp\left[i \int dt \left(\frac{\dot{\varphi}^2}{2} - \begin{Bmatrix} n \\ 1 \end{Bmatrix} \dot{\varphi}\right) dt\right]$$

$$\times \exp\left\{-\int dt dt' \eta^\dagger(t)\eta(t') e^{-i\varphi(t)} e^{i\varphi/t')} \left[\begin{Bmatrix} n+1 \\ 0 \end{Bmatrix} \Theta(t-t') + \begin{Bmatrix} n \\ -1 \end{Bmatrix} \Theta(t'-t)\right]\right\}. \quad (6.42)$$

The collective Lagrangian of this model is

$$\mathcal{L}(t) = \frac{\dot{\varphi}^2}{2} - \begin{Bmatrix} n \\ 1 \end{Bmatrix} \dot{\varphi}$$

$$= \frac{1}{2}\left(\dot{\varphi} - \begin{Bmatrix} n \\ 1 \end{Bmatrix}\right)^2 - \frac{1}{2}\begin{Bmatrix} n^2 \\ 1 \end{Bmatrix}. \quad (6.43)$$

With the help of the functional canonical field momentum

$$p = \dot{\varphi} - \begin{Bmatrix} n \\ 1 \end{Bmatrix}$$

we find the Hamiltonian

$$H = \left(\dot{\varphi} - \begin{Bmatrix} n \\ 1 \end{Bmatrix}\right)\dot{\varphi} - \mathcal{L}$$

$$= \frac{\dot{\varphi}^2}{2} = \frac{\left(p + \begin{Bmatrix} n \\ 1 \end{Bmatrix}\right)^2}{2}. \quad (6.44)$$

Thus the spectrum is the same as before, but the momenta are shifted by n (or 1) units accounting for the fundamental particles contained in the reference state $|R\rangle$ of (6.37). In the collective quantum field theory, this reference state has a functional momentum zero:

$$\begin{Bmatrix} n \\ 1 \end{Bmatrix} Z[\eta^\dagger, \eta] = \frac{1}{\{0|0\}}\{0|T \exp\left[-\int dt dt' \eta^\dagger(t)\eta(t') e^{-i\varphi(t)} e^{i\varphi(t')}\right.$$

$$\times \left[\begin{Bmatrix} n \\ -1 \end{Bmatrix} \Theta(t-t') + \begin{Bmatrix} n \\ -1 \end{Bmatrix} \Theta(t'-t)\right]\right] |0\}. \quad (6.45)$$

In fact, the one-particle Green function becomes

$$\begin{Bmatrix} n \\ 1 \end{Bmatrix} G(t,t') = -\frac{\delta^{(2)}}{\delta\eta^\dagger(t)\delta\eta(t')}\begin{Bmatrix} n \\ 1 \end{Bmatrix} Z[\eta^\dagger, \eta]$$

$$= \frac{1}{\{0|0\}}\{0|T e^{-i\varphi(t)} e^{i\varphi(t')}|0\}$$

$$\times \left[\begin{Bmatrix} n+1 \\ 0 \end{Bmatrix} \Theta(t-t') + \begin{Bmatrix} n \\ -1 \end{Bmatrix} \Theta(t'-t)\right]. \quad (6.46)$$

Inserting the times translation operator corresponding to (6.44) this yields, for $t > t'$,

$$
\{{}^n_1\}G(t,t') = \exp\left[-i\left\{\begin{array}{c} n+1/2 \\ 3/2 \end{array}\right\}(t-t')\right]\left\{\begin{array}{c} n+1 \\ 0 \end{array}\right\}
$$

$$
= \left\{\begin{array}{c} (n+1)\exp[-i(n+1/2)(t-t')] \\ 0 \end{array}\right\}, \tag{6.47}
$$

and for $t < t'$

$$
\{{}^n_1\}G(t,t') = \exp\left[-i\left\{\begin{array}{c} n-1/2 \\ 1/2 \end{array}\right\}(t-t')\right]\left\{\begin{array}{c} n \\ -1 \end{array}\right\}
$$

$$
= \left\{\begin{array}{c} n\exp[-i(n-1/2)(t-t')] \\ -e^{-i(t-t')/2} \end{array}\right\}, \tag{6.48}
$$

in agreement with a direct operator calculation.

The appearance of the additional derivative term $\dot\varphi$ in the Lagrangian (6.43) can be understood in an alternative fashion. The reference state $|n\rangle$ of ${}^nZ$ in (6.37) can be generated in the original generating functional by applying successively derivatives $-\delta^{(2)}/\delta\eta^\dagger(t)\delta\eta(t')$, letting $t' \to -\infty$, $t \to \infty$ and absorbing the infinite phase $\exp[-i\Delta E \times (\infty - (-\infty))]$ into the normalization constant where ΔE is the energy difference between $|n\rangle$ and $|0\rangle$:

$$
{}^nZ[\eta^\dagger,\eta]\big|_{\eta^\dagger=\eta=0} \propto \frac{\delta^{(n)}}{(\delta\eta^\dagger(+\infty))^n}\frac{\delta^{(n)}}{(\delta\eta(-\infty))^n}{}^0Z[\eta^\dagger,\eta]\Big|_{\eta^\dagger=\eta=0}. \tag{6.49}
$$

Each such pair of derivatives brings down a Green function

$$
e^{-i\varphi(t)}e^{i\varphi(t')}\Theta(t-t') = \exp\left[-i\int_{t'}^{t}\dot\varphi(t'')dt''\right]\Theta(t-t'). \tag{6.50}
$$

In the limits $t' \to -\infty$, $t \to \infty$ we obtain, for n such factors,

$$
\exp\left[-in\int_{-\infty}^{\infty}\dot\varphi(t)dt\right], \tag{6.51}
$$

in agreement with the derivative term in (6.42).

While the functional Schrödinger picture is useful in understanding what happens in the Hilbert space of the collective field theory, it is quite awkward to apply to more than one dimension, in particular to the relativistic situation where the time does not play a special role. A more direct and easily generalizable method for the evaluation of fermion propagators in the collective theory consists in the following procedure: One brings the products of exponentials in (6.32) to normal order by using Wick's contraction formula in the functional form. Let the "charges" of the incoming and outgoing fermions be $q_i = +1$ and $q_i - 1$, respectively. Then the matrix element to be calculated in (6.32) are

$$
\{0|T\exp\left[i\sum_i q_i\varphi(t_i)\right]|0\} = \{0|T\exp\left[i\int dt\varphi(t)\partial_i q_i(t-t_i)\right]|0\}, \tag{6.52}
$$

where we have re-numbered the times in the exponents as $t_1, t_2, t_3, t_4, \ldots$ rather than $t_1, t'_1, t_2, t'_2, \ldots$, etc. Now, from Wick's contraction rule one has

$$\{0|Te^{i\sum_i q_i\varphi(t_i)}|0\} = \exp\left[-\frac{1}{2}\int dtdt' \sum_i q_i\delta(t - t_i) \overline{\dot\varphi(t)\dot\varphi}(t') \sum_j q_j\delta(t - t_j)\right]$$

$$\times\{0|T : \exp\left[i\int dt\varphi(t)\sum_i q_i\delta(t' - t_i)\right] : |0\}$$

$$= \exp\left[-\frac{1}{2}\sum_{ij} q_iq_j \overline{\dot\varphi(t_i)\dot\varphi}(t_j)\right], \qquad (6.53)$$

where a contraction denotes again the propagator of a φ-field. This is well defined after introducing a small regulator mass κ:

$$\dot\varphi(t)\dot\varphi(t') = \int \frac{dE}{2\pi} \frac{i}{E^2 - \kappa^2 + i\epsilon} e^{-iE(t-t')}$$

$$= \frac{1}{2\kappa} e^{-\kappa|t-t'|} = \frac{1}{2\kappa} - \frac{i}{2}|t - t'| + \mathcal{O}(\kappa). \qquad (6.54)$$

As $\kappa \to 0$ this expression vanishes unless the sum of all charges is zero: $\sum_i q_i = 0$. Thus one finds the general result for (6.32):

$$\{0|T\exp\left[i\sum_{q_i}\varphi(t_i)\right]|0\} = \delta_{\Sigma q_i, 0}\exp\left[\frac{i}{2}\sum_{i>j} q_iq_j|t_i - t_j|\right]. \qquad (6.55)$$

In particular, the two-point function (6.26) agrees with the Schrödinger calculation (6.30).

6.1.1 The Generalized BCS Model in a Degenerate Shell

A less trivial but completely transparent example is provided by the BCS degenerate-shell model used in nuclear physics to describe the energy levels of some nuclei in which pairing forces are dominant (for example Sn and Pb isotopes [31]). For understanding the structure of the collective theory it is useful to consider at first both bosons and fermions as well as a more general interaction, and impose the restriction to fermions and to the particular BCS pairing force at a later stage. This more general Hamiltonian reads

$$H = H_0 + H_{\text{int}} = \epsilon \sum_{i=1}^{\Omega}(a_i{}^\dagger a_i + b_i{}^\dagger b_i) - \frac{V}{2}\sum_{i,j} a_i{}^\dagger b_i{}^\dagger b_j a_j$$

$$\pm\frac{V}{4}g\left[\sum_i(a_i{}^\dagger a_i + b_i{}^\dagger b_i) \pm \Omega\right], \qquad (6.56)$$

where $g = 0$ reduces to the actual BCS model in the case of fermions. The model can be completely solved by introducing quasi-spin operators

$$L^\dagger = \sum_{i=1}^{\Omega} a_i{}^\dagger b_i{}^\dagger, \qquad L^- = \sum_{i=1}^{\Omega} b_i a_i = (L^\dagger)^\dagger, \qquad (6.57)$$

$$L_3 = \frac{1}{2}\left\{ \sum_i (a_i{}^\dagger a_i + b_i{}^\dagger b_i) \pm \Omega \right\} = \frac{1}{2}\sum_i a_i{}^\dagger a_i \pm b_i b_i{}^\dagger = \frac{1}{2}[N \pm \Omega],$$

where N counts the total number of particles. These operators generate the group $SU(1,1)$ or $SU(2)$ for bosons or fermions, respectively:

$$\begin{aligned}
[L_3, L^\pm] &= \pm L^\pm, \\
[L^+, L^-] &= \mp 2L_3.
\end{aligned} \tag{6.58}$$

Using

$$L^+ L^- = L^2 \mp L_3 \pm L_3{}^2 \tag{6.59}$$

we can write

$$\begin{aligned}
H &= 2\varepsilon L_3 \mp \epsilon\Omega - V(L^2 \pm L_3{}^2 \mp gL_3{}^2) \\
&= 2\varepsilon L_3 - V[L^2 \pm (1-g)L_3{}^2] \mp \varepsilon\Omega.
\end{aligned} \tag{6.60}$$

Note that for $g = 1$, the interaction term is $SU(1,1)$- or $SU(2)$-symmetric. The irreducible representation of the algebra (6.58) consists of the states

$$|n[\Omega, \nu]\rangle = N_n (L^+)^n |0[\Omega, \nu]\rangle, \tag{6.61}$$

where the seniority label ν denotes the presence of ν unpaired particles $a_i{}^\dagger$ or $b_j{}^\dagger$, i.e. those which are orthogonal to the configurations $(L^\dagger)^n|0\rangle$. For $\nu = 0$ the spectrum of L_3 in an irreducible representation is

$$\pm\frac{\Omega}{2}, \pm\frac{\Omega}{2}+1, \pm\frac{\Omega}{2}, +2, \dots \; . \tag{6.62}$$

This continues ad infinitum for bosons due to the non-compact topology of $SU(1,1)$ while it terminates for fermions at $\Omega/2$ corresponding to a finite spin $\Omega/2$. The invariant Casimir operator

$$L^2 \equiv L_1{}^2 + L_2{}^2 \mp L_3{}^2 \tag{6.63}$$

characterizing the representation has the eigenvalue $\Omega/2(1 \mp \Omega/2)$ showing, in the fermion case, again the quasi-spin $\Omega/2$. If ν unpaired particles are added to a vacuum, the eigenvalues start at $\pm(\Omega + \nu)/2$. Thus the quasi-spin is reduced to $(\Omega - \nu/2)$. If $\nu = \Omega$ unpaired fermions are present, the state is quasi-spin symmetric, for example:

$$|0[\Omega, \Omega]\rangle = b_1{}^\dagger b_2{}^\dagger \cdots \cdot b_\Omega{}^\dagger |0\rangle. \tag{6.64}$$

Due to the many choices of unpaired particles with the same total number, the levels show considerable degeneracies and one actually needs another label for their distinction. This has been dropped for brevity.

On the states $|n[\Omega\nu]\rangle$, the energies are taken from (6.60) and become, after inserting $N = 2n + \nu$:

$$E = e(N \pm \Omega) - V\left[\frac{\Omega \pm \nu}{2}\left(1 \mp \frac{\Omega \pm \nu}{2}\right) \pm \frac{(1-g)}{4}(N \pm \Omega)^2\right] \mp \varepsilon\Omega. \tag{6.65}$$

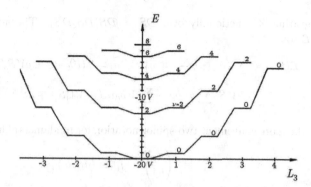

FIGURE 6.1 Level scheme of the BCS model in a single degenerate shell of multiplicity $\Omega = 8$. The abscissa denotes the third component of quasi-spin. The index ν at each level stands for the number of unpaired particles ("seniority").

In Fig. 6.1, a typical level scheme is displayed for fermions of $\Omega = 8$ with $\varepsilon = 0$. If the single particle energy ε is non-vanishing, the scheme is distorted via a linear dependence on L_2 lifting the right- and depressing the left-hand side. For an attractive potential and given total particle number N, the state with $\nu = 0$ is the ground state, with the higher seniorities having higher energies:

$$E_{N\Omega\nu} - E_{N\Omega 0} = V \left(\Omega \mp 1 \pm \frac{\nu}{2} \right) \nu. \tag{6.66}$$

The Lagrangian of the model is from (6.56)

$$\begin{aligned}
\mathcal{L}(t) &= \sum_i (a_i^\dagger(t)(i\partial_t - \varepsilon)a_i(t) + b_i^\dagger(t)(i\partial_t - \varepsilon)b_i(t)) \\
&\quad + \frac{V}{2} \left\{ \sum_{i,j} a_i^\dagger b_i^\dagger b_j a_i \right\} \mp \frac{V}{4} g \left\{ \sum_i (a_i^\dagger a_i \pm b_i b_i^\dagger) \right\}^3,
\end{aligned} \tag{6.67}$$

implying the generating functional

$$\begin{aligned}
Z[\eta^\dagger, \eta, \lambda] &= \int \prod_i \mathcal{D}a_i^\dagger \mathcal{D}a_i \mathcal{D}b_i^\dagger \mathcal{D}b_i \\
&\quad \times \exp\left[i \int dt \left\{ \mathcal{L} + \sum_i \eta_i^\dagger a_i + a_i^\dagger \eta_i + \lambda_i^\dagger b_i + b_i^\dagger \lambda_i \right\} \right].
\end{aligned} \tag{6.68}$$

The fourth-order terms in the exponential can be removed by introducing a complex field $S = S_1 + iS_2$, $S^\dagger = S_1 - iS_2$ and a real field S_3', adding

$$-V \left\{ \left| S(t) - \sum_i a_i^\dagger b_i^\dagger \right|^2 \mp g \left[S_3'(t) - \frac{1}{2} \sum_i (a_i^\dagger a_i \pm b_i b_i^\dagger) \right]^2 \right\} \tag{6.69}$$

and integrating Z functionally over $\mathcal{D}S = \mathcal{D}S_1\mathcal{D}S_2\mathcal{D}S_3$. The addition of (6.69) changes $\mathcal{L}$ to:

$$\mathcal{L}(t) = \sum_i \{a_i^\dagger(i\partial_t - \varepsilon \mp gVS_3')a_i \mp b_i(i\partial_t + \varepsilon \pm gVS_3')b_i^\dagger\}$$
$$+ VS^\dagger \sum_i a_i^\dagger b_i^\dagger + \sum_i Vb_i a_i S - V(|S|^2 \mp gS_3'^2) \pm \varepsilon\Omega. \qquad (6.70)$$

By using the more convenient two-spinor notation for fundamental fields and sources

$$f_i \equiv \begin{pmatrix} a_i \\ b_i^\dagger \end{pmatrix}; \quad f_i^\dagger \equiv (a_i^\dagger, b_i)$$

$$j_i \equiv \begin{pmatrix} n_i \\ \lambda_i^\dagger \end{pmatrix}; \quad j_i^\dagger \equiv (\eta_i^\dagger, \lambda_i) \qquad (6.71)$$

the generating functional can be rewritten as

$$Z[j^\dagger, j] = \int \prod_i \mathcal{D}f_i^\dagger \mathcal{D}f_i \mathcal{D}S \exp\left[i \int dt \left\{\mathcal{L}(\mathbf{S}) + \sum_i (j_i^\dagger f_i + f_i^\dagger j_i)\right\}\right], \qquad (6.72)$$

with

$$\mathcal{L}(\mathbf{S}) = \sum_{i=1}^{\Omega} f_i^\dagger(t) \begin{pmatrix} i\partial_t - \varepsilon \mp gVS_3' & VS^\dagger \\ VS & \mp(i\partial_t + \varepsilon \pm gVS_3') \end{pmatrix} f_i(t)$$
$$- V(|S|^2 \mp gS_3'^2) \pm \varepsilon\Omega. \qquad (6.73)$$

Now the fundamental fields $f_i^\dagger, f_i$ can be integrated out in (6.72) yielding the collective action [32]

$$\mathcal{A}[\mathbf{S}] = \pm i\,\mathrm{Tr}\log(iG_\mathbf{S}^{-1}) - V(S_1^2 + S_2^2 \mp g - S_3'^2) \pm \varepsilon\Omega, \qquad (6.74)$$

where $G_\mathbf{S}$ is the matrix collecting the Green functions of the particles in the external field $\mathbf{S} = (S_1, S_2, S_3) = (S, S^\dagger, S_3)$:

$$G_\mathbf{S}(t, t')_{ij} = \begin{pmatrix} \overline{a_i(t)a_j^\dagger(t')} & \overline{a_i(t)b_j(t')} \\ \overline{b_i^\dagger(t)a_j^\dagger(t')} & \overline{b_i^\dagger(t)b_j(t')} \end{pmatrix}. \qquad (6.75)$$

The associated equation of motion of $G_\mathbf{S}(t, t')$ reads

$$\begin{pmatrix} i\partial_t - \varepsilon \mp gVS_3' & VS^\dagger \\ \mp VS & i\partial_t + \epsilon \pm gVS_3' \end{pmatrix} G_\mathbf{S}(t, t') = i\left\{\begin{matrix} \sigma^3 \\ 1 \end{matrix}\right\}\delta(t - t'). \qquad (6.76)$$

It may be solved by an ansatz

$$G_\mathbf{S}(t, t') = U^\dagger(t)G_0(t, t')U(t'), \qquad (6.77)$$

where G_0 is a solution of (6.76) for $S = 0$, $S_3' = 0, \varepsilon = 0$.

Before we proceed it is useful to absorb ε and g into S_3', by defining the more symmetric variable

$$\mp S_3 = \mp g S_3' - \frac{\varepsilon}{V}. \tag{6.78}$$

Then Eq. (6.76) reads

$$\left(i\partial_t + V \left\{ \begin{matrix} -iS_2 \\ S_1 \end{matrix} \right\} \sigma^1 + V \left\{ \begin{matrix} iS_1 \\ -S_2 \end{matrix} \right\} \sigma^2 \mp V S_3 \sigma^3 \right) U^\dagger(t) G_0 U(t') = i \left\{ \begin{matrix} \sigma^3 \\ 1 \end{matrix} \right\} \delta(t - t'). \tag{6.79}$$

This can be solved be parametrizing the matrix $U(t)$ in terms of Euler angles as

$$U(t) = e^{i\alpha\frac{\sigma_3}{2}} e^{\left\{ \begin{matrix} -\tilde{\beta} \\ i\tilde{\beta} \end{matrix} \right\} \frac{\sigma^2}{2}} e^{i\gamma\frac{\sigma_3}{2}}. \tag{6.80}$$

Then they satisfy the identity

$$U^\dagger(t) \left\{ \begin{matrix} \sigma^3 \\ 1 \end{matrix} \right\} U(t) = \left\{ \begin{matrix} \sigma^3 \\ 1 \end{matrix} \right\}. \tag{6.81}$$

Thus they form a subgroup of the rotation group $SU(2)$ in the fermion case, or of the Lorentz group $SU(1,1)$ in the Bose case. The differential equation (6.79) can be rewritten as

$$i\dot{U}^\dagger(t) U^\dagger(t)^{-1} = -V \left(\left\{ \begin{matrix} -iS_2 \\ S_1 \end{matrix} \right\} \sigma^1 + \left\{ \begin{matrix} iS_1 \\ S_2 \end{matrix} \right\} \sigma^2 \mp V S_3 \sigma^3 \right). \tag{6.82}$$

In the Bose case, the left-hand side can be expressed as an exponential $e^{i\tilde{\boldsymbol{\omega}}\cdot\boldsymbol{\sigma}}$ involving the angular velocities $\tilde{\boldsymbol{\omega}} = (\tilde{\omega}_1, \tilde{\omega}_2, \tilde{\omega}_3)$ of $SU(1,1)$ matrices. They depend on the Lorentz version of the Euler angles as

$$\begin{aligned} \tilde{\omega}_1 &\equiv \tilde{\beta}\sin\gamma + \dot{\alpha}\sinh\tilde{\beta}\cos\gamma = 2V S_1, \\ \tilde{\omega}_2 &\equiv \tilde{\beta}\cos\gamma - \dot{\alpha}\sinh\tilde{\beta}\sin\gamma = 2V S_2, \\ \tilde{\omega}_3 &\equiv \dot{\alpha}\cosh\tilde{\beta} + \dot{\gamma} = 2V S_3. \end{aligned} \tag{6.83}$$

In the Fermi-case, where the matrices (6.80) are of the rotation group $SU(2)$, the time-derivatives $\dot{U}^\dagger(t) U(t)$ can be expressed as exponentials $e^{i\boldsymbol{\omega}\cdot\boldsymbol{\sigma}}$ involving the ordinary angular velocities $\boldsymbol{\omega} = (\omega_1, \omega_2, \omega_3)$ depending on the standard Euler angles as

$$\begin{aligned} \omega_1 &\equiv -\dot{\beta}\sin\gamma + \dot{\alpha}\sin\beta\cos\gamma = -2V S_2, \\ \omega_2 &\equiv \dot{\beta}\cos\gamma + \dot{\alpha}\sin\beta\sin\gamma = -2V S_2, \\ \omega_3 &\equiv \dot{\alpha}\cos\beta + \dot{\gamma} = -2V S_3. \end{aligned} \tag{6.84}$$

The upper equations in (6.82) follow from the lower ones by replacing in (6.80) $\beta \to -i\tilde{\beta}$, and in (6.82) $S_1 \to -iS_2$, $S_2 \to iS_1$, $S_3 \to -S_3$. Since this transition can be done at any later stage it is convenient to avoid the clumsy distinction of different cases and focus attention upon the Fermi case only. Then Eq. (6.80) is unitary and coincides with the well-known representation matrix $D^{1/2}_{m'm}(\alpha\beta\gamma)$ of the rotation group[1]. The formal solution of Eq. (6.82) is

$$U(t) = e^{i\boldsymbol{\omega}\cdot\boldsymbol{\sigma}} = T\exp\left[-i\int_{\infty}^{t} 2V\,\mathbf{S}\cdot\boldsymbol{\sigma}dt'\right]. \tag{6.85}$$

Given this $U(t)$-matrix we can now proceed to evaluate the Trlog-term in (6.74). By differentiation with respect to S we find:

$$\frac{\delta}{\delta S_k(t)}[-i\mathrm{Tr}\log(iG_\mathbf{S}^{-1})] = V\sum_i \mathrm{Tr}(\sigma^k G_\mathbf{S}^{ii}(t,t')|_{t'=t+\varepsilon}. \tag{6.86}$$

The right-hand side can be calculated in terms of Euler angles by inserting (6.80). In addition one has to choose the reference state for $Z[\eta^\dagger, \eta]$ by specifying the boundary condition on G_0. Since G_0 represents the same matrix of Green functions as (6.75), except with free oscillators $a_0^\dagger$, $b_0^\dagger$ of zero energy, this is easily done.

Let us choose as our reference state $|R\rangle$ one of the quasi-spin symmetric states of seniority $\nu = \Omega$, say (6.64). Then G_0 has to have the form

$$G_0{}^{ij}(t,t') = \begin{pmatrix} \Theta(t-t') & 0 \\ 0 & \Theta(t-t') \end{pmatrix}\delta^{ij}. \tag{6.87}$$

As a consequence $G_0^{ij}(t,t')|_{t'=t+\varepsilon} = 0$ such that also (6.86) vanishes and $-i\mathrm{Tr}\log(iG_\mathbf{S}^{-1})$ becomes an irrelevant constant.

Hence the generating functional in the quasi-spin symmetric reference state (6.64) is

$$^R Z[j^\dagger, j] = \int \mathcal{D}\mathbf{S}\exp\left[\left[\int dt\ V\mathbf{S}(t)^2 - \int dt dt'\Theta(t-t')\sum_i j_i^\dagger(t)U^\dagger(t)U(t')j_i(t')\right]\right]. \tag{6.88}$$

As in the case of the trivial model it is now convenient to change variables and integrate directly over the Euler angles α, β, γ rather than the vectors $\mathbf{S} = (S_1, S_2, S_3)$. Using the derivatives

$$-\frac{1}{2V}\frac{\delta S_i(t)}{\delta q_j(t')} \equiv A(t)_{ij}\delta(t-t') + B(t)_{ij}\dot{\delta}(t-t')$$

$$= \begin{pmatrix} 0 & \dot{\alpha}\cos\beta\cos\gamma & -\dot{\beta}\cos\gamma - \dot{\alpha}\sin\beta\sin\gamma \\ 0 & \dot{\alpha}\cos\beta\sin\gamma & -\dot{\beta}\sin\gamma + \dot{\alpha}\sin\beta\cos\gamma \\ 0 & -\dot{\alpha}\sin\beta & 0 \end{pmatrix}_{ij} \tag{6.89}$$

$$\times\delta(t-t') + \begin{pmatrix} \sin\beta\cos\beta\gamma & -\sin\gamma & 0 \\ \sin\beta\sin\gamma & \cos\gamma & 0 \\ \cos\beta & 0 & 1 \end{pmatrix}_{ij}\delta(t-t'), \tag{6.90}$$

---

[1]For conventions see: A.R. Edmonds, *Angular Momentum in Quantum Mechanics*, Princeton University Press, 1960.

one calculates the functional determinant as the determinant of the second matrix B. This can be seen most easily by multiplication with the constant (functional) matrix $\int dt'\Theta(t'-t'')$ which diagonalizes the $\dot{\delta}(t-t')$ and brings the $\delta(t-t')$-term completely to the right of the functional diagonal: $\delta\Theta' = \Theta$. The determinant of such a matrix equals the determinant of the diagonal part only. Thus, up to an irrelevant factor, one has

$$\mathcal{D}\mathbf{S} = \text{const.} \times \mathcal{D}\alpha\mathcal{D}\beta\mathcal{D}\gamma\sin\beta \qquad (6.91)$$

corresponding to the standard measure of the rotation group. Inserting now (6.84) into (6.88) we find

$$Z[j^\dagger, j] = \int \mathcal{D}\alpha\mathcal{D}\cos\beta\mathcal{D}\gamma \exp\left[i\int dt\left\{-\frac{1}{4V}\left[\omega_1{}^2 + \omega_2{}^2 + \frac{1}{g}(\omega_3 - 2\varepsilon)^2\right] - \varepsilon\Omega\right\}\right.$$
$$\left. \times \exp\left[i\int dtdt'\Theta(t-t')\sum_i j_i{}^\dagger(t)U^\dagger(t)U(t')j_i(t')\right]. \right. \qquad (6.92)$$

The collective Lagrangian becomes

$$\mathcal{L} = -\frac{1}{4V}\left\{(\dot{\beta}^2 + \dot{\alpha}^2\sin^2\beta) + \frac{1}{g}(\dot{\gamma} + \dot{\alpha}\cos\beta)^2\right\} + \frac{\varepsilon}{Vg}(\dot{\gamma} + \dot{\alpha}\cos\beta) - \frac{\varepsilon^2}{Vg} - \varepsilon\Omega. \quad (6.93)$$

This has the standard form

$$\mathcal{L} = \frac{1}{2}\dot{q}^i g_{ij}(q)\dot{q}^j + a_i(q)\dot{q}^i - v(q), \qquad (6.94)$$

with the metric

$$g_{ij}(q) = -\frac{1}{2V}\begin{bmatrix} \sin^2\beta + \frac{1}{g}\cos^2\beta & 0 & \frac{1}{g}\cos\beta \\ 0 & 1 & 0 \\ \frac{1}{g}\cos\beta & 0 & \frac{1}{g} \end{bmatrix}, \qquad (6.95)$$

$$g^{ij}(q) \equiv (g^{-1}(q))^{ij} = -2V\frac{g}{\sin^2\beta}\begin{bmatrix} \frac{1}{g} & 0 & -\frac{1}{g}\cos\beta \\ 0 & 1 & 0 \\ -\frac{1}{g}\cos\beta & 0 & \sin^2\beta + \frac{1}{g}\cos^2\beta \end{bmatrix} \qquad (6.96)$$

of determinant

$$g \equiv \det(g_{ij}) = -\frac{1}{8V^3}\frac{1}{g}\sin^2\beta, \qquad (6.97)$$

in the space labelled again by $q^i \equiv (\alpha, \beta, \gamma)$.

In this curved space, the Hamiltonian is given by [33]

$$H = H_1 + H_2 + H_3 + v(q) + \frac{1}{2}a^i a_i(q), \qquad (6.98)$$

with the three terms

$$H_1 = -\frac{1}{2} g^{-1/2} \frac{\partial}{\partial q^i} \left(g^{1/2} g^{ij} \frac{\partial}{\partial q^j} \right),$$ (6.99)

$$H_2 = \frac{i}{2} g^{-1/2} \left(\frac{\partial}{\partial q^i} g^{1/2} g^{ij} a_j(q) \right),$$ (6.100)

$$H_3 = i a_i(q) g^{ij} g^{ij} \frac{\partial}{\partial q^i}.$$ (6.101)

Here we find H_1 as the standard asymmetric-top Hamiltonian,

$$H_1 = V \left[\frac{\partial^2}{\partial \beta^2} + \cot \beta \frac{\partial}{\partial \beta} + (g + \cot \beta) \frac{\partial^2}{\partial \gamma^2} + \frac{1}{\sin^2 \beta} \frac{\partial^2}{\partial \alpha^2} - 2 \frac{\cos \beta}{\sin^2 \beta} \frac{\partial^2}{\partial \alpha \partial \gamma} \right].$$ (6.102)

Since

$$a_i = \frac{\epsilon}{Vg} (\cos \beta, 0, 1),$$ (6.103)

the second part, H_2, vanishes and the third part becomes

$$H_3 = -2\varepsilon i \partial_\gamma.$$ (6.104)

The resulting Hamiltonian is exactly the Schrödinger version of the quasi-spin form (6.60) with

$$L^\pm = e^{\pm i\gamma} \left[\pm \partial_\beta + \cot \beta i \partial_a - i \frac{i}{\sin \beta} \partial_\gamma \right],$$

$$L_3 = -i \partial_\gamma.$$ (6.105)

The eigenfunctions of H coincide with the rotation matrices

$$D^j_{m'm}(\alpha, \beta, \gamma) = e^{i\alpha m' + \gamma m)} d^j_{m'm}(\beta).$$ (6.106)

The energy eigenvalues of H_1 are well-known

$$E^1_{jm} = -V[j(j+1) - m^2(1-g)],$$ (6.107)

such that the full energies are

$$E_{jm} = 2\varepsilon m - V[j(j+1) - (1-g)m^2] + \varepsilon \Omega.$$ (6.108)

This coincides with the fermion part of the spectrum (6.65) if m, j are set equal to

$$m = (N - \Omega)/2, \quad j = \frac{\Omega - \nu}{2}.$$ (6.109)

For $g = 1, \varepsilon = 0$ the spectrum is degenerate as the Lagrangian (6.93) is rotationally invariant. It may be worth mentioning that in this case the Lagrangian can also be written as a standard σ-model in the time dimension. In order to see this, we use the identity $i\dot{U}^\dagger(t)U(t) = -iU^\dagger(t)\dot{U}(t) = \omega_i(t)\sigma_i/2$ to bring (6.85) to the form

$$\mathcal{L} = -\frac{1}{4V}(\omega_i{}^2 + \omega_2{}^2 + \omega_3{}^2) = -\frac{1}{2V}\text{tr}(\dot{U}^\dagger U U^\dagger \dot{U}). \tag{6.110}$$

If we now define field $\sigma(t)$ and $\pi(t)$ by decomposing

$$U(t) = \sigma(t) + i\pi(t) \cdot \sigma, \tag{6.111}$$

where $\sigma^2(t) + \pi^2(t) = 1$ due to unitary of $U(t)$, the Lagrangian takes the familiar expression

$$\mathcal{L} = -\frac{1}{V}(\dot{\sigma}^2 + \pi^2). \tag{6.112}$$

It is instructive to exhibit the original quasi-spin operators and their algebra within the collective Lagrangian. For this we add to the Hamiltonian (6.56) a coupling to external currents:

$$\Delta H = -2V \int L_i(t)l_i(t)dt, \tag{6.113}$$

where L_i are the operators (6.57). In the Lagrangian (6.70), this amounts to

$$\Delta\mathcal{L}(t) = 2V L_i(t)l_i(t)dt, \tag{6.114}$$

which modifies (6.73) by adding the matrix

$$V f^\dagger(t) \begin{pmatrix} l_3 & l^\dagger \\ l & l_3 \end{pmatrix} f(t). \tag{6.115}$$

This has the effect of replacing

$$S_i \to \tilde{S}_i \equiv S_i + l_i \tag{6.116}$$

in the Tr log term in (6.74).

Performing a shift in the integration $\mathcal{D}\mathbf{S} \to \mathcal{D}\tilde{\mathbf{S}} = \mathcal{D}(\mathbf{S} + \mathbf{l})$ we can also write

$$\mathcal{A}[\mathbf{S},\mathbf{l}] = +i\text{Tr}\log(iG_{\mathbf{S}}^{-1}) - V\left[(S_1 - l_1)^2 + (S_2 - l_2)^2 - \frac{1}{g}\left(S_3 + \frac{\varepsilon}{V} - l_3\right)^2\right]. \tag{6.117}$$

The Green function involving angular momentum operators can now be generated by differentiating

$$Z[l_i] = \int \mathcal{D}\mathbf{S} \exp\{i\mathcal{A}[\mathbf{S},\mathbf{l}]\}$$

with respect to δl_i:

$$L_i \hat{=} -\frac{i}{2V}\frac{\delta}{\delta l_i}. \tag{6.118}$$

In the reference state $|R\rangle$ in which the $\mathrm{Tr}\log$-term vanishes, the derivatives $(-i/2V)\delta/\delta l_1, -(i/2V)\delta/\delta l_3$ generate the fields $S_{1,2} - l_{1,2}, (S_3 + \varepsilon/V - l_3)/g$ via the source terms (6.114) in the functional integral.

In the fermion case, this implies for $l = 0$, using Eq. (6.80),

$$L^{\pm} = -\frac{1}{2V}(\omega_1 \pm i\omega_2) = -\frac{1}{2V}(\pm i\dot\beta + \dot\alpha \sin\beta)e^{\pm i\gamma}, \tag{6.119}$$

$$L_3 = -\frac{1}{2Vg}(\omega_3 - 2\varepsilon) = -\frac{1}{2Vg}(\dot\alpha \cos\beta + \dot\gamma - 2\varepsilon), \tag{6.120}$$

which are exactly the angular momenta of the Lagrangian (6.93) with moments of inertia

$$I_{1,2} = -\frac{1}{2V}, \quad I_3 = -\frac{1}{2Vg}. \tag{6.121}$$

Inserting the canonical momenta of (6.93)

$$P_\alpha = -\frac{1}{2V}\left[\dot\alpha \sin^2\beta + \frac{1}{g}(\dot\gamma + \dot\alpha \cos\beta - 2\varepsilon)\cos\beta\right]$$

$$= -\frac{1}{2V}\dot\alpha \sin^2\beta + \cos\beta p_\gamma = -i\partial_\alpha, \tag{6.122}$$

$$P_\beta = -\frac{1}{2V}\dot\beta = -i\sin^{-1/2}\beta\partial_\beta \sin^{1/2}\beta = -i\partial_\beta - \frac{i}{2}\cot\beta, \tag{6.123}$$

$$P_\gamma = -\frac{1}{2V_g}(\dot\gamma + \dot\alpha \cos\beta - 2\varepsilon) = -i\partial_\gamma,$$

we recover the differential operators (6.105).

The quasi-spin algebra can now be verified by applying the derivatives:

$$-\frac{1}{4V^2}\left(\frac{\delta}{\delta l_j(t + \varepsilon)}\frac{\delta}{\delta l_i(t)} - \frac{\delta}{\delta l_i(t)}\frac{\delta}{\delta l_i(t + \varepsilon)}\frac{\delta}{\delta l_j(t)}\right)Z\Big|_{t=0} = \frac{1}{2V}\varepsilon_{ijk}\frac{\delta}{\delta l_k}Z\Big|_{l=0}. \tag{6.124}$$

What would have happened in this model if we had not chosen the symmetric reference state $|R\rangle$ to specify the boundary condition on G_0? Consider for example the vacuum state $|0\rangle$. Then the Green function becomes, for $\mathbf{S} = 0$,

$$G_0^{ij}(t, t') - \begin{pmatrix} \Theta(t - t') & 0 \\ 0 & -\Theta(t' - t) \end{pmatrix}\delta^{ij} \tag{6.125}$$

rather than (6.87). In this case *there is* a contribution of $-i\mathrm{Tr}\log(iG_s^{-1})$ since from (6.86) and (6.77):

$$\frac{\delta}{\delta S_i}[-i\mathrm{tr}\log(iG_{\mathbf{s}}^{-1})] = -V\Omega\mathrm{tr}\left(\sigma^i U^\dagger(l)\frac{-1 + \sigma^3}{2}U(t')\right)\Big|_{t'=t}. \tag{6.126}$$

Now (6.80) implies

$$U^\dagger(t)\sigma^3 U(t) = \cos\beta\sigma_3 + \sin\beta(\cos\gamma\sigma_1 + \sin\gamma\sigma_2)$$

yielding for the right-hand side of (6.126) the expression

$$-V\Omega \left\{ \begin{array}{c} n_1 \\ n_2 \\ n_3 \end{array} \right\} \equiv -V\Omega \left\{ \begin{array}{c} \sin\beta\cos\gamma \\ \sin\beta\sin\gamma \\ \cos\beta \end{array} \right\}. \tag{6.127}$$

Observe that due to the differential equations (6.84), the unit vector n_i can be found to satisfy the equation of motion

$$\dot{\mathbf{n}} = 2V\mathbf{n} \times \mathbf{S}. \tag{6.128}$$

We can now proceed and find $-i\mathrm{Trlog}iG_{\mathbf{S}}^{-1}$ by functionally integrating (6.126). We shall do so in terms of the Euler variables $\alpha\beta\gamma$. Using (6.126), (6.127), (6.90), and the chain rule of differentiation

$$
\begin{aligned}
\frac{\delta}{\delta q_j(t')}[-i\mathrm{Trlog}G_{\mathbf{S}}^{-1}] &= \sum_i \int dt \frac{\delta S_i(t)}{\delta q_j(t')} \frac{\delta}{\delta S_i(t)}[-i\mathrm{Trlog}iG_{\mathbf{S}}^{-1}] \\
&= -V\Omega \sum_i \int dt n_i(t) \frac{\delta S_i(t)}{\delta q_j(t')},
\end{aligned} \tag{6.129}
$$

we find

$$
\begin{aligned}
\frac{\delta}{\delta q_i(t)}[-i\mathrm{Trlog}iG_{\mathbf{S}}^{-1}] &= \frac{\Omega}{2} \sum_i \int dt\, (n_i(t)A_{ij}(t)\delta(t-t') + n_i(t)B_{ij}(t)\delta(t-t')) \\
&= \frac{\Omega}{2}[(0,0,-\dot{\beta}\sin\beta(t'))_j + \int dt(1,0,\cos\beta(t))_j\delta(t-t')].
\end{aligned} \tag{6.130}
$$

The second part in brackets yields upon a partial integration

$$(1,0,\cos\beta(t))\delta(t-t')|_{t=-\infty}^{t=\infty} + (0,0,\dot{\beta}\sin\beta(t')). \tag{6.131}$$

With the boundary condition $\cos\beta(\pm\infty) = 1$, one has therefore

$$\frac{\delta}{\delta(\alpha,\beta,\gamma)(t)}[-i\mathrm{Trlog}iG_{\mathbf{S}}^{-1}] = \frac{\Omega}{2}(1,0,1)[\delta(\infty-t) - \delta(-\infty-t)]. \tag{6.132}$$

This pure boundary contribution can immediately be functionally integrated with the result

$$-i\mathrm{Trlog}iG_{\mathbf{S}}^{-1} = \frac{\Omega}{2} \int_{\infty}^{\infty} [\dot{\alpha}(t) + \dot{\gamma}(t)]dt. \tag{6.133}$$

Hence the exponent of the generating functional $Z[j^\dagger, j]$ on the reference state $|0\rangle$ becomes

$$
\begin{aligned}
&i\int dt \left\{ -\frac{1}{4V}\left[\omega_1{}^2 + \omega_2{}^2 + \frac{1}{g}(\omega_3 - 2\varepsilon)^2\right] + \frac{\Omega}{2}[\dot{\alpha} + \dot{\gamma}] - \varepsilon\Omega \right\} \\
&- \int dt dt' \sum_i j_i^\dagger(t) \left\{ U^\dagger(t)\frac{1+\sigma^3}{2}U(t')\Theta(t-t')U^\dagger(t)\frac{1-\sigma^3}{2}U(t')\Theta(t'-t) \right\} j_i(t'),
\end{aligned} \tag{6.134}
$$

rather than (6.92). As in the case of the Pet model in the last section, the Hamiltonian changes rather trivially. The canonical momenta P_α, P_γ become

$$
\begin{aligned}
P_\alpha &= -\frac{1}{2V}\left[\dot{\alpha}\sin^2\beta + \frac{\cos\beta}{g}(\dot{\gamma} + \dot{\alpha}\cos\beta - 2\varepsilon)\right] + \frac{\Omega}{2} \\
&= -\frac{1}{2V}\dot{\alpha}\sin^2\beta + \cos\beta\, p_\gamma - \frac{\Omega}{2}(\cos\beta - 1) = -i\partial_\alpha, \qquad (6.135) \\
P_\gamma &= -\frac{1}{2Vg}(\dot{\gamma} + \dot{\alpha}\cos\beta - 2\varepsilon) + \frac{\Omega}{2} = -i\partial_\gamma.
\end{aligned}
$$

The additional term can be removed by multiplying all eigenfunctions belonging to (6.136) by a phase $\exp[-i\Omega/2(\alpha+\gamma)]$ thereby reducing them to the previous case. In the present context it is really superfluous to discuss such trivial surface terms. We are doing this only because these terms become important at that moment where the transition to the true BCS model is made by going to the weak-coupling limit $g \to 0$. This will be discussed in the next section.

6.1.2   The Hilbert Space of the Generalized BCS Model

Let us now study in which fashion the Hilbert space of all rotational wave functions imbeds the fermion theory. For this consider the generation of Green functions by functional derivation of $^R Z[j^\dagger, j]$, with the reference state $|R\rangle$ being the quasi-spin symmetric one (6.62), for simplicity.

The resulting one-particle Green function will have to coincide with

$$
G^{ij}_{mm'}(t,t') = \langle 0|b_\Omega \cdots b_1 \begin{pmatrix} Ta_i(t)a_j^\dagger(t') & Ta_i(t)b_j(t') \\ Tb_i^\dagger(t)a_j^\dagger(t') & Tb_i^\dagger(t)b_j(t') \end{pmatrix}_{mm'} b_1^\dagger \cdots b_\Omega^\dagger|0\rangle. \quad (6.136)
$$

If we differentiate (6.92) accordingly, we find

$$
G^{ij}_{mm'}(t,t') = \int \mathcal{D}\alpha \mathcal{D}\cos\beta \mathcal{D}\gamma\, \delta^{ij}(U^\dagger(t)U(t')_{mm'}\Theta(t - t')\exp[i\int dt\mathcal{L}(t)]. \quad (6.137)
$$

This can be calculated most easily by going to the Schrödinger picture

$$
G^{ij}_{mm'}(t,t') = \sum_k \{R|D^{1/2}_{km}(\alpha\beta\gamma(t))D^{1/2}_{km'}(\alpha\beta\gamma(t'))|R\}\delta^{ij}\Theta(t - t'). \quad (6.138)
$$

Since the reference state is symmetric, it must be associated with the wave function $\{\alpha\beta\gamma(t)|R\} = D^0_{00}(\alpha\beta\gamma(t)) \equiv 1/\sqrt{8\pi^2}$

$$
E_R \equiv E_{0;0} = \varepsilon\Omega. \qquad (6.139)
$$

Inserting the time translation operator[2]

$$
D(\alpha\beta\gamma(t)) = e^{iHt}D(\alpha\beta\gamma(0))e^{iHt}, \qquad (6.140)
$$

---

[2]The Schrödinger angles α, β, γ coincide with the time dependent angles $\alpha(t), \beta(t), \gamma(t)$ at $t = 0$.

with H in the differential form (6.98) one finds a phase

$$e^{i\Delta E(t-t')}, \tag{6.141}$$

where ΔE is the energy difference between the state $|jm\rangle = |\frac{1}{2}\frac{1}{2}\rangle$ and the reference state $|R\rangle = |0,0\rangle$:

$$\Delta E = E_{\frac{1}{2}\frac{1}{2}} - E_{0,0} = \varepsilon - V\left(\frac{1}{2} + \frac{g}{4}\right). \tag{6.142}$$

The orthogonality relation is

$$\sum_k \int d\alpha d\cos\beta d\gamma \{R|\alpha\beta\gamma\} D_{km}^{1/2*}(\alpha\beta\gamma) D_{km'}^{1/2}(\alpha\beta\gamma)\{\alpha\beta\gamma|R\} = \delta_{mm'}. \tag{6.143}$$

This coincides exactly with the result one would obtain from (6.136) by using the original operator (6.56) and observing the energy spectrum (6.65).

Note that the orthogonality relation together with the Grassmann algebra ensure the validity of the anticommutation rules among the operators. For higher Green functions the functional derivatives amount again to the contractions as in (6.35), except that here the contractions are associated with

$$\begin{aligned}
\overline{f_{mi}(t)f^\dagger_{m'j}(t')} &= D_{mm'}^{1/2}(U^\dagger(t)U(t'))\Theta(t-t')\delta^{ij} \\
&= \sum_k D_{km}^{1/2*}(U^\dagger(t))D_{km'}^{1/2}(U(t))\Theta(t-t')\delta^{ij}, \tag{6.144}
\end{aligned}$$

where $f_{1/2i}, f_{-1/2i}$ stands for $(a_i, b_i^\dagger)$.

We can now proceed and construct the full Hilbert space by piling up operators $a_i^\dagger$ or b_j on the reference state $|R\rangle = b_1^\dagger \cdot \ldots \cdot b_\Omega^\dagger |0\rangle$. First we shall go to the true vacuum state $|0\rangle$ of $a^\dagger, b^\dagger$ which means that we calculate $^0Z[j^\dagger, j]$ in this state. For this we obviously have to bring down successively into the main integral line of the functional (6.92) the operators $b_1^\dagger(\infty) \cdot \ldots \cdot b_\Omega^\dagger(-\infty)b_\Omega(-\infty) \cdot \ldots \cdot b_1(-\infty)$. We do this by forming the functional derivatives:

$$Z^0[0,0] \propto \frac{\delta^{2\Omega}}{\delta j_{-1/2,1}(\infty) \cdot \ldots \cdot \delta j_{-1/2,1}^\dagger(-\infty)} {}^R Z[j^\dagger, j]\bigg|_{j=0}. \tag{6.145}$$

Of the resulting $n!$ contractions, only one combination survives, since all indices i, j are different and the Kronecker-δ^{ij} permits only one set of contractions. The result is

$$^0Z[0,0] = N \int \mathcal{D}s\alpha \mathcal{D}\cos\beta \mathcal{D}\gamma \exp\left[i\int dt\mathcal{L}(t)\right] [D_{-1/2-1/2}^{1/2}U^\dagger(\infty)U(-\infty))]^\Omega. \tag{6.146}$$

But from the coupling rules of angular momenta and the group property one has:

$$\begin{aligned}
[D_{-1/2-1/2}^{1/2}(U^\dagger(\infty)U(-\infty))]^\Omega &= D_{-\Omega/2-\Omega/2}^{\Omega/2}(U^\dagger(\infty)U(-\infty)) \\
&= \sum_k D_{k-\Omega/2}^{\Omega/2*}(U(\infty))D_{k-\Omega/2}(U(-\infty)). \tag{6.147}
\end{aligned}$$

Going to the Schrödinger picture and inserting the time translation operator (6.140) one finds an infinite phase $\exp[i(E_R - E_0)2\infty]$ which can be absorbed in the normalization factor N. Here $E_0 = E_{\Omega/2,-\Omega/2}$ is the energy of the ground state $|0\rangle$ which has $|jm\rangle = |\Omega/2 - \Omega/2\rangle$. The eigenfunctions $D(\alpha, \beta, \gamma)$ now appear both at $t = 0$. The functional (6.146) in the Schrödinger picture becomes,

$$^0Z[0,0] = \sum_{k=-\Omega/2}^{\Omega/2} \int d\alpha d\beta d\gamma \sin\gamma \{0k|\alpha\beta\gamma\}\{\alpha\beta\gamma|0k\}, \tag{6.148}$$

with the vacuum wave functions

$$\{\alpha\beta\gamma|0,k\} = D_{k,-\Omega/2}^{\Omega/2}(\alpha\beta\gamma) = e^{i(k\alpha-\Omega\gamma/2)}d_{k,-\Omega/2}^{\Omega/2}(\beta). \tag{6.149}$$

It is easy to verify, how an additional unpaired particle $a^\dagger$, added to the vacuum, decreases $\Omega/2 \to (\Omega - 1)/2$ and raises the third component of quasi-spin by $\frac{1}{2}$ unit. Differentiating (6.90) by $-\delta^2/\delta j_{\frac{1}{2}1}(\infty)\delta j_{\frac{1}{2}1}^\dagger(-\infty)$ in addition to (6.145) leads to a different set of contractions. Picturing them within the original fermion language, these are

$$\langle R|T(b_1{}^\dagger(+\infty)\cdot \ldots \cdot b_\Omega{}^\dagger(+\infty)a_1(+\infty)a_1{}^\dagger(-\infty)b_\Omega(-\infty)\cdot \ldots \cdot b_1(-\infty))|R\rangle$$

$$= \langle R|T(b_1{}^\dagger(\infty)\cdot \ldots \cdot b_\Omega{}^\dagger(\infty)\, a_1(\infty)a_1^\dagger(-\infty)\cdot \ldots \cdot b_\Omega(-\infty)\cdot \ldots \cdot b_1(-\infty))\,|R\rangle$$

$$\vdots$$

$$+ <\langle R|T(b_1{}^\dagger(\infty)\cdot \ldots \cdot b_\Omega{}^\dagger(\infty)\, a_1(\infty)a_1^\dagger(-\infty)\cdot \ldots \cdot b_\Omega(-\infty)\cdot \ldots \cdot b_1(-\infty))\,|R\rangle.$$

Under the functional integral (6.146), they lead to:

$$[D_{-\frac{1}{2}-\frac{1}{2}}^{\frac{1}{2}}(U^\dagger(\infty)U(-\infty))]^\Omega D_{\frac{1}{2}\frac{1}{2}}^{\frac{1}{2}}(U^\dagger(\infty)U(-\infty))$$

$$- [D_{-\frac{1}{2}-\frac{1}{2}}^{\frac{1}{2}}(U^\dagger(\infty)U(-\infty))]^{\Omega-1} D_{-\frac{1}{2}\frac{1}{2}}^{\frac{1}{2}}(U^\dagger(\infty)U(-\infty))D_{\frac{1}{2}-\frac{1}{2}}^{\frac{1}{2}}(U^\dagger(\infty)U(-\infty))$$

$$= [D_{-\Omega/2\,-\Omega/2}^{\Omega/2}(U^\dagger(\infty)U(-\infty))]D_{\frac{1}{2}\frac{1}{2}}^{\frac{1}{2}}(U^\dagger(\infty)U(-\infty)) \tag{6.150}$$

$$- D_{-(\Omega-1)/2\,-(\Omega-1)/2}^{(\Omega-1)/2}(U^\dagger(\infty)U(-\infty))D_{-\frac{1}{2}\frac{1}{2}}^{\frac{1}{2}}(U^\dagger(\infty)U(-\infty))D_{\frac{1}{2}-\frac{1}{2}}^{\frac{1}{2}}(U^\dagger(\infty)U(-\infty)).$$

Employing the explicit formulas

$$D_{-\Omega/2-\Omega/2}^{\Omega/2}(\alpha\beta\gamma) = e^{-\Omega(\alpha+\gamma)/2}\left(\cos\frac{\beta}{2}\right)^\Omega,$$

$$D_{\frac{1}{2}\frac{1}{2}}^{\frac{1}{2}}(\alpha\beta\gamma) = e^{(\alpha+\gamma)/2}\cos\frac{\beta}{2}, \tag{6.151}$$

$$D_{-\frac{1}{2}\frac{1}{2}}^{\frac{1}{2}}(\alpha\beta\gamma)D_{\frac{1}{2}-\frac{1}{2}}^{\frac{1}{2}}(\alpha\beta\gamma) = -\sin^2\frac{\beta}{2},$$

the r.h.s. of (6.150) becomes

$$e^{-\Omega(\alpha+\gamma)/2}\left(\cos\frac{\beta}{2}\right)^{\Omega}e^{(\alpha+\gamma)/2}\cos\frac{\beta}{2}+e^{-(\Omega-1)(\alpha+\gamma)/2}\left(\cos\frac{\beta}{2}\right)^{\Omega-1}\sin^2\frac{\beta}{2}$$

$$=e^{-(\Omega-1)(\alpha+\gamma)/2}\left(\cos\frac{\beta}{2}\right)^{\Omega-1}=D^{(\Omega-1)/2}_{-(\Omega-1)/2,-(\Omega-1)/2}(\alpha\beta\gamma),\qquad(6.152)$$

and therefore, by analogy with (6.146) and (6.148),

$$a_1^\dagger|0\rangle Z[j^\dagger,j]\Big|_{j=0}=N\int\mathcal{D}\alpha\mathcal{D}\cos\beta\mathcal{D}\gamma\,D^{(\Omega-1)/2}_{-(\Omega-1)/2,-(\Omega-1)/2}(\alpha\beta\gamma)\exp\left[\frac{i}{\hbar}\int dt\mathcal{L}\right]$$

$$=\sum_{k=-(\Omega-1)/2}^{(\Omega-1)/2}\int dt\,d\alpha\cos\beta d\gamma\{a_1k|\alpha\beta\gamma\}\{\alpha\beta\gamma|a_1^\dagger k\},\qquad(6.153)$$

with the Schrödinger wave functions

$$\{\alpha\beta\gamma|a_1{}^\dagger k\}\equiv D^{(\Omega-1)/2}_{k,-(\Omega-1)/2}(\alpha\beta\gamma).\qquad(6.154)$$

In a similar fashion we may work our way through the whole Hilbert space!

The method has been applied to field theories of nuclear excitations where they form the basis of a theory of supersymmetry in nuclei.[3]

6.2   Thirring Model in 1+1 Dimensions

Let us also study an example of a quantum field theory in two spacetime dimensions, the *Thirring model* [3, 4]. It is a model of a self-interacting spin-$\frac{1}{2}$ field with an action

$$\mathcal{A}=\int d^2x\left\{\bar{\psi}(x)i\hbar\gamma^\mu\partial_\mu\psi(x)-\frac{g}{2}[\bar{\psi}(x)\gamma^\mu\psi(x)]^2\right\}.\qquad(6.155)$$

In this model, the technique presented here leads to an exact translation from the Fermi fields ψ to collective Bose fields $\varphi(x),\lambda(x)$. Consider the partition function of the model

$$Z=\int\mathcal{D}\psi\mathcal{D}\bar{\psi}e^{i\mathcal{A}/\hbar},\qquad(6.156)$$

and let us perform a Hubbard-Stratonovich transformation à la Eq. (1.79), by adding to the action the complete square

$$\Delta\mathcal{A}=\frac{g}{2}\int d^2x\left[\bar{\psi}(x)\gamma^\mu\psi(x)-A^\mu\right]^2.\qquad(6.157)$$

---

[3]See the web pages http://klnrt.de/55/1978 and http://klnrt.de/55/1978/1978-4.gif where the theory is illustrated.

This removes the four-fermion interaction term and makes the action quadratic in the fermion fields. They can be integrated out to obtain the collective action as a functional of the vector field A_μ:

$$\mathcal{A}_{\text{coll}}[A_\mu] = -i\text{Tr}\log\left(i\hbar\slashed{\partial} - g\slashed{A}\right) + \frac{g}{2}\int d^4x A_\mu^2(x). \tag{6.158}$$

Now one can make use of the fact that in two dimensions, a vector field A^μ has only two components and can be expressed in terms of two scalar fields as

$$A^\mu(x) = \frac{1}{\sqrt{g}}\left(\partial^\mu\varphi(x) - \epsilon^{\mu\nu}\partial_\nu\lambda(x)\right), \tag{6.159}$$

so that

$$\frac{g}{2}A_\mu^2(x) = \frac{1}{2}[\partial_\mu\varphi(x)]^2 - \frac{1}{2}[\partial_\mu\lambda(x)]^2. \tag{6.160}$$

The trace log term can be expanded as in (2.15), with only the $n = 1$-term contributing. This is equal to

$$\frac{g^2}{2\pi}\left[\left(g^{\mu\nu} - \frac{\partial^\mu\partial^\nu}{\partial^2}\right)A_\nu(x)\right]^2. \tag{6.161}$$

Hence the collective action is simply

$$\mathcal{A}_{\text{coll}}[\varphi,\lambda] = \int d^4x\left[\frac{1}{2}(\partial\varphi(x))^2 - \frac{1}{2}\left(1 + \frac{g}{\pi}\right)(\partial\lambda)^2\right]. \tag{6.162}$$

Since this transformation from the ψ to the $\varphi-$ and λ-field description is exact, one can also calculate the Green functions of the original fermion fields ψ. For this, an external source term

$$\int d^4x\left[\bar{\psi}(x)\eta(x) + \text{c.c.}\right] \tag{6.163}$$

is added in the exponent of the generating functional. After a quadratic completion the source term leads to an additiaonal quadratic term in the collective action (6.162) (setting $\hbar = 1$ from here on, for brevity)

$$\mathcal{A}_{\text{ext curr}} = i\int d^4x d^4y\bar{\eta}(x)\left(\frac{i}{i\slashed{\partial} - g\slashed{A}}\right)(x,y)\eta(y). \tag{6.164}$$

Using the decomposition (6.159), the Green function in this expression can also be calculated exactly as follows:

$$\begin{aligned}
\frac{i}{i\slashed{\partial} - g\slashed{A}}(x,y) &= \frac{i}{i\slashed{\partial} - \sqrt{g}\slashed{\partial}(\varphi + \gamma_5\lambda)}(x,y) \\
&= e^{-i\sqrt{g}\varphi(x) + \gamma_5\lambda(x)}\frac{i}{i\slashed{\partial}}(x,y)e^{i\sqrt{g}(\varphi(y) + \gamma_5\lambda(y))} \\
&= e^{-i\sqrt{g}(\varphi(x) + \gamma_5\lambda(x))}\frac{1}{4\pi i}\frac{\slashed{x} - \slashed{y}}{(x-y)^2 + i\epsilon}e^{i\sqrt{g}(\varphi(y) + \gamma_5\lambda(y))}. \tag{6.165}
\end{aligned}$$

All Green functions of the theory can now be calculated by applying the functional derivatives $\frac{\delta}{\delta\bar{\eta}(x)}\frac{\delta}{\delta\eta(y)}$ and $\frac{\delta}{\delta\bar{\eta}(x)}\frac{\delta}{\delta\eta(y)}$ to the generating functional

$$Z[\bar{\eta}, \eta] = e^{i(\mathcal{A}+\mathcal{A}_{\text{ext curr}}/\hbar)}. \tag{6.166}$$

In particular, the original one-particle Green function is obtained from the derivative $\frac{\delta}{\delta\bar{\eta}(x)}\frac{\delta}{\delta\eta(y)}$ and reads

$$G(x,y) = \frac{1}{4\pi i}\frac{\not{x}}{x^2+i\epsilon}\langle 0|:e^{-i\sqrt{g}(\varphi(x)+\gamma_5\lambda(x))}::e^{i\sqrt{g}(\varphi(0)+\gamma_5\lambda(0))}:|0\rangle. \tag{6.167}$$

We now apply the standard rule for calculating the exponential of free fields

$$\frac{1}{4\pi i}\frac{\not{x}}{x^2+i\epsilon}\langle 0|:e^{-i\alpha\varphi(x)}::e^{i\alpha\varphi(y)}:|0\rangle = e^{\alpha^2\langle\varphi(x)\varphi(y)\rangle}, \tag{6.168}$$

that follows directly from Wick's theorem [see (1.253)], together with the use of the expectation values of the two-dimensional massless scalar fields [2]

$$\langle 0|\varphi(x)\varphi(0)|0\rangle = -\frac{1}{4\pi}\log(\mu^2 x^2), \tag{6.169}$$

$$\langle 0|\lambda(x)\lambda(0)|0\rangle = \frac{1}{1+g/\pi}\frac{1}{4\pi}\log(\mu^2 x^2). \tag{6.170}$$

In this way we find for the vacuum expectation value in the Green function (6.167)

$$\left(\frac{x^2}{\mu^2}\right)^{\left(-g+\frac{g}{1+g/\pi}\right)\frac{1}{4\pi}} = \left(\frac{x^2}{\mu^2}\right)^{-\frac{g^2}{1+g/\pi}\frac{1}{4\pi}}. \tag{6.171}$$

Hence we find the exact Green function of the Thirring model

$$G(x,0) = \frac{1}{4\pi i}\frac{\not{x}}{x^2+i\epsilon}\left(\frac{x^2}{\mu^2}\right)^{-\frac{g^2}{1+g/\pi}\frac{1}{4\pi}}. \tag{6.172}$$

The result is very interesting. It is scale-invariant and for this reason it contains an arbitrary mass parameter μ that can be chosen freely. The physical reason for this freedom is the absence of a mass term in the Thirring action (6.157). Such a mass term would destroy the exact solvability of the theory. It would make it calculable only approximately in perturbation theory, order by order in the coupling strength g. In this case the above exact solution would be the result of a strong-coupling limit [1]. This limit is of special interest in all quantum field theories. Take for instance the Heisenberg model of ferromagnetism. In the classical limit it is a theory of an N-component scalar field with $O(N)$ rotational symmetry. That model can be studied experimentally in the strong-coupling limit by going to a second-order phase transition. Then the model possesses scale-invariant correlation functions which have a pure power form and contain an arbitrary mass scale. The powers reflect a dynamically generated anomalous dimension of the field. This vanishes, of course, in the free-field limit.

Note that a mass term in the original action $m\bar{\psi}\psi$ could be obtained from the generating functional (6.166) by a derivative

$$m\frac{\delta}{\delta\mu(x)}\frac{\delta}{\delta\bar{\mu}(x')}\bigg|_{x=x'}. \tag{6.173}$$

Due to Eq. (6.165), this is equivalent to the replacement

$$m\bar{\psi}\psi = m\left(\psi_1^*\psi_2 + \psi_2^*\psi_1\right) \to m\left(e^{i2\sqrt{g}\lambda(x)}\psi_{01}^*\psi_{02} + \text{c.c.}\right), \tag{6.174}$$

where ψ_0 are free fields. In the two-dimensional model world all matrix elements of products of many Fermi fields $\psi_{01}^*\psi_{02}$, $\psi_{02}^*\psi_{01}$ can also be calculated with the help of exponentials of the massless Bose fields $\varphi(x)$ and $\lambda(x)$, for instance

$$\psi_{01}^*\psi_{02} \simeq e^{i\sqrt{4\pi}\varphi}. \tag{6.175}$$

Moreover, the matrix elements of

$$e^{i(2\sqrt{g}\lambda + \sqrt{4\pi}\varphi)} \tag{6.176}$$

are, again due to (6.168), (6.169), (6.170), the same as those of

$$e^{-i\sqrt{\frac{4\pi}{1+g/\pi}}\varphi}. \tag{6.177}$$

Thanks to this, the mass term of the Thirring model can be expressed with the help of the scalar field $\varphi(x)$ as

$$m\bar{\psi}(x)\psi(x) \simeq 2m\cos\left(\sqrt{\frac{4\pi}{1+(g/\pi)}}\varphi(x)\right). \tag{6.178}$$

In this way we arrive at the well-known sine-Gordon bosonic description of the massive Thirring model [4].

In a similar way, the *Schwinger model* can be treated exactly. It is a two-dimensional version of QED with the action

$$\mathcal{A} = \int d^2x\left\{\bar{\psi}(x)i\hbar\left(\partial_\mu - eA_\mu\right)\psi(x) - \frac{1}{4}F_{\mu\nu}F^{\mu\nu}\right\}, \tag{6.179}$$

where $F_{\mu\nu} \equiv \partial_\mu A_\nu - \partial_\nu A_\mu$ is the tensor collecting the electric and magnetic field strengths. The partition function

$$Z = \int \mathcal{D}\psi\mathcal{D}\bar{\psi}e^{i\mathcal{A}/\hbar} \tag{6.180}$$

can be calculated by integrating out the Fermi fields, which leads to the new collective action

$$\mathcal{A}_{\text{pl}} = -i\text{Tr}\log\left(i\hbar\slashed{\partial} - g\slashed{A}\right) - \frac{1}{4}\int d^2x F_{\mu\nu}F^{\mu\nu}. \tag{6.181}$$

The subscript emphasizes the similarity of the new vector potential to the collective plasmon field in Chapter 2. The trace log term can be evaluated exactly as before and the action becomes

$$\mathcal{A}_{\text{pl}} = \int d^2x \left(-\frac{1}{4} F_{\mu\nu} F^{\mu\nu} + \frac{m_{\text{pl}}^2}{2} A_\mu A^\mu \right), \qquad (6.182)$$

with

$$m_{\text{pl}} = \frac{e^2}{\pi}.$$

It describes a single free *plasmon* Bose field of mass m_{pl} (see also Section 14.12 in the textbook [5]).

6.3 Supersymmetry in Nuclear Physics

We may consider the algebra formed by the creation operators $a_i^\dagger$, $b_i^\dagger$ and their annihilation operators a_i, b_i as well as the quasi-operators (6.70). Then the eigenstates contain even and odd nuclei. They form a broken supersymmetry. The level scheme looks like a generalization of Fig. 6.1 which includes half-integer nuclei. For more details see Refs. [6, 7]. This model has been the answer to a question posed to Sergio Ferrara after his lecture on supersymmetry in elementary-particle physics by the student Yuan K. Ha at the Erice summer school.[4]

Notes and References

[1] H. Kleinert and V. Schulte-Frohlinde, *Critical Properties of ϕ^4-Theories*, World Scientific, 2001, pp. 1–489 (klnrt.de/b8).

[2] J.A. Swieca, Fortschr. Phys. **25**, 303 (1977).

[3] W. Thirring, Annals of Physics **3**, 91 (1958).

[4] B. Klaiber, *The Thirring Model*, published in his Boulder Lecture ed. by A.O. Barut and W. Brittin, *Lectures in Theoretical Physics*, Gordon and Breach, N.Y., 1968.

[5] H. Kleinert, *Particles and Quantum Fields*, World Scientific, 2016.

[6] H. Kleinert, *Collective Quantum Fields*, Lectures presented at the First Erice Summer School on Low-Temperature Physics, 1977, Fortschr. Physik **26**, 565-671 (1978).

[7] F. Iachello and A. Arima, *The Interacting Boson Model*, Cambridge University Press, Cambridge, 1987.

---

[4]See the web pages http://klnrt.de/55/1978 and http://klnrt.de/55/1978/1978-4.gif where the theory is illustrated.

Index

ABRIKOSOV, A.A. 40, 49, 138, 320
ADRIANOV, V.A. 139
AMBEGAOKAR, V. 320
ANDERSON, P.W. 320
ARIMA, A. 397
axial vector field 34

BABAEV, E. 140
BAK, C.S. 369
BALDO, M. 320
BALIAN, R. 320
BARDEEN, J. 138, 319
BARTON, G. 321
BARUT, A.O. 397
BASOV, D. 141
BAYM, G. 40, 49, 138, 320
BCS
    model, generalized 379
    theory 74–76, 145
BECHGAARD, K. 138
BEDNORZ, J.G. 138
BELYAEV, S.T. 319
bend configuration 337
BENDER, C.M. 143
bending energy 171
BEREZINSKII, V.L. 141
BERK, N.F. 320
BHATTACHARYYA, P. 322
Bianchi identity 31
bilocal 1
BLECH, I. 370
Bogoliubov
    theory 145
    transformation 79, 195
BOGOLIUBOV, N.N. 138, 319
BOHM, H.V. 139
BOHR, A. 319
boojum 179
Bose distribution 19

BOUCHIAT, M.A. 369
BRÉZIN, E. 139, 321
BREWER, D.F. 322
BRINKMANN, W.F. 320
BRITTIN, W. 397
BROUT, R. 142
BRUECKNER, K.A. 320
BUCHER, B. 141
BUCHHOLTZ, L.J. 321
BULGAC, A. 144
BUTLER, D. 139
BYANKIN, V.M. 369

CAHN, J. W. 370
CATES, M. E. 370
cholesteric
    liquid crystals 351
    phase 351, 355, 356
CHOMAZ, P. 142
CHU, B. 369
classical
    collective field 113
    Heisenberg model 169
Clausius-Clapeyron equation 147
coherence
    current 230
    length 87, 162
    velocity 230
COLEMAN, S. 40, 141, 142, 321
collective
    classical field 113
    pair field 160
    quantum field 112
components
    hard 151
    soft 151
    spherical field, higher l 158
Composite Bosons 122
compound, MBAA 323

compound, PAA 323
condensation energy 71, 88,
    89, 109, 110, 118, 121, 125, 202,
    241, 276, 283, 348
configuration
    bend 337
    splay 337
    twist 337
constant, dielectric 46, 47
contraction 12
Cooper pair 1, 78, 145
    p-wave 146
COOPER, F. 143
COOPER, L.N. 138, 319
correlation function 22
coupling, electromagnetic 69
CREUTZ, M. 143
critical
    point 324
    temperature 63, 330
    temperature, would-be 330
CROSS, M.C. 322
crossover parameter 94
crystal, liquid 323

DAGOTTO, E. 140
DE GENNES, P.G. 320, 321
DEAN, D.J. 320
Debye
    frequency 63
    temperature 63
defects 105, 337
density, superfluid 89
dielectric constant 46, 47
DING, H. 140
dipole-locked 226
Dirac
    equation 36
    field 36
    matrices 36
distribution
    Bose 19
    Fermi 19
DOMB, C. 321
double superfluid 249
DRECHSLER, M. 141
DRUT, J.E. 144

dual field tensor 32
DYSON, F. 13
DZYALOSHINSKI, I.E. .. 40, 49, 138, 320

EAGLES, D.M. 142
effective mass 2
EICHTEN, E. 143
electromagnetic coupling 69
ELGAROEY, O. 320
EMERY, V.J. 141, 320
energy, condensation 71, 88,
    89, 109, 110, 118, 121, 125, 202,
    241, 276, 283, 348
ENGELBRECHT, J.R. 140, 142, 144
ENGLERT, F. 142
ENGVIK, L. 320
ENOKIDO, H. 369
equal-time amplitude 78
equation
    Dirac 36
equation, gap 162, 167, 194–198
Euclidean
    four-momentum 20
    spacetime coordinate 20
expansion, hopping 122

FAY, D. 320
Fermi
    distribution 19
    momentum 63, 153
    temperature 63
    velocity 66
Feshbach resonance 117
FETTER, A.L. 40, 138, 320–322
FEYNMAN, R.P. 39
field
    collective classical 113
    collective pair 160
    collective quantum 112
    Dirac 36
    momentum 2
    operator 3
    pair 52
    tensor 31
FILEV, V.M. 369
FIOLHAIS, M. 142
FLETCHER, R.J. 142

flow enthalpy 219
FLOWERS, E. 319
Fock space 3
FORBES, M.M. 144
four-curl 31
four-momentum, Euclidean 20
Frank constants 337
FREDERICO, T. 142
FUJUWARA, T. 370
function
    correlation 22
    Riemann 65
    Yoshida 83, 86, 210
functional
    methods 2

GALASIEWICZ, Z.M. 320
GAMMAL, A. 142
GANDOLFI, S. 144
gap
    equation 73, 162, 167, 194–198
    pseudo 91
GEZERLIS, A. 144
ghost field 32
Goldstone theorem 78
GORKOV, L.P. 40, 49, 138, 320
GOULD, C.M. 321
GRADSHTEYN, I.S. 64
GRATIAS, D. 370
Green function
    thermal 18
GREEN, M.S. 321
GREYWALL, D.S. 320
GRIFFIN, A. 142
group
    Poincaré 33
    representation theory 35
GULARI, E. 369
GULLY, W.J. 320
GURALNIK, G.S. 142, 143
GUSYNIN, V.P. 141

HADZIBABIC, Z. 142
HAGEN, C.R. 142
HALDANE, F.D.M. 140
hard components 151
HARRIS, J.M. 140

HELFRICH, W. 368
HENRICH, O. 370
HIBBS, A.R. 39
HIGGS, P.W. 142
HJORTH-JENSEN, M. 320
HO, T.L. 321, 322
HOHENBERG, P.C. 141
HOMES, C.C. 141
hopping
    expansion 122
    parameter 122
HUBBARD, J. 40, 139
Hubbard-Stratonovich transformation
    15, 41, 90, 112, 116, 117
hydrodynamic limit 88

IACHELLO, F. 397
INGUSCIO, M. 143
integrability condition 31, 32
intrinsic angular momentum 24
ISHII, Y. 370
ITO, T. 140

JACKSON, J.D. 321
JÉRÔME, D. 138
JOSÉ, J.V. 140
JUNOT, C. 370

KADANOFF, L.P. . 40, 49, 138, 320, 321
KAHLWEIT, K. 369
KETTERLE, W. 143
KIBBLE, T.W.B. 142
KIVELSON, S.A. 141
KLAIBER, B. 397
KLEINBERG, R.L. 322
KLEINERT, H. 39, 49, 138–140, 142, 143,
    320–322, 369, 370, 397
KOSTERLITZ, J.M. 141
KUMAR, P. 321

Lagrangian density 28
LANGER, J. 40, 139
LANGEVIN, D. 369
largest subgroup 24
LARKIN, A.I. 369
LAYZER, A.J. 320
LE GUILLOU, J.C. 139, 321
LEBED, A.G. 138

LEE, D.M. 320, 321
LEGGETT, A.J. 140, 320, 321
Levi-Città tensor 31
LIN, F.L. 369
LIN-LIU, Y.R. 322
liquid crystal 323
LITSTER, J.D. 369
little group 24
LIU, M. 322
LOKTEV, V.M. 141
London limit 88
LOPES, R. 142
LORAM, J. 141
Lorentz group
    full 27
    proper 23

macroscopic superfluid velocity 171
MAEDA, Y. 369
MAGIERSKI, P. 144
Maier-Saupe Model 324
MAKI, K. 321, 322, 370
MAN, J. 142
MARENDUZZO, D. 370
MARINI, M. 141
MARSHALL, D.S. 142
mass
    effective 2, 150, 222
    matrix 113
MATHEWS, P.T. 40, 139, 320
matrices, Dirac 36
MAZAUD, A. 138
MBAA compound 323
MERMIN, N.D. ...49, 141, 321, 322, 370
meson 1
MILLS, R.L. 319
MINEEV, V.P. 321, 322
minimal coupling 69
Minkowski
    metric 24
mirror reflection 27
momentum
    Fermi 63, 153
    field 2
    rest 23
MOORE, M.A. 321
MOREL, P. 320

MORSE, R.W. 139
MOTTELSON, B.R. 319
MÜHLSCHLEGEL, B. 40, 139
MÜLLER, K.A. 138

NAKANISHI, N. 139
natural units 30
NAVON, N. 142
NAZARENKO, A. 140
nematic phase 336
NIKOLIĆ, P. 143
NISHIDA, Y. 143
Noether theorem 165
non-inert phase 243
nonrelativistic particles 2
normal product 10
NOZIÈRES, P. 141

order field 164
ORENSTEIN, J. 140
OSHEROFF, D.D. 320
OSTNER, W. 369

PAA compound 323
PÉCSELI, H.L. 49
pair field 52, 56
paramagnon 146
parameter
    crossover 94
    hopping 122
parity transformation 27
PASKAKOV, V.Y. 369
phase
    cholesteric 351
    nematic 336
    non-inert 243
PINES, D. 319
PISTOLESI, F. 141
PITAEVSKI, L.N. 320
plasma 1
Poincaré group 33
point, critical 324
POLYAKOV, A.M. 321
Pomeranchuk effect 147
POPOV, V.N. 139
positronium 2
PRESKILL, J. 143
product, normal 10

proper Lorentz transformations 23
pseudogap 91
PUCHKOV, A. 141

quanta 152
quantization, second 3
quantum, collective field 112
quasiparticle 151

RANDERIA, M. 139, 140, 142, 144
rapidity 23
RAY, P.J. 138
REBBI, C. 143
reference momentum 23
REINITZER, F. 323, 368
representation theory, group 35
resonance, Feshbach 117
rest momentum 24
RIBAULT, M. 138
RICE, T.M. 40, 139
RICHARDSON, R.C. 320
RICHARDSON, R.W. 139
Riemann, zeta function 65
ROATI, G. 143
ROKHSAR, D.S. 370
ROSKIES, P. 143
ROTTER, L.D. 140
RUDERMAN, M. 319
RYZHIK, I.M. 64
RZEWUSKI, J. 40

SÁ DE MELO, C.A.R. 141, 142, 144
SACHDEV, S. 143
SAINT-JAMES, D. 139
SALAM, A. 40, 139, 320
SARMA, G. 139
SCHADT, M. 368
SCHMITT-RINK, S. 141
SCHRIEFFER, J.R. 138, 319, 320
SCHULTE-FROHLINDE, V. . 140, 321, 397
SCHULZE, H.J. 320
SEMENCHONKO, V.K. 369
semiconductor 1
SESSLER, A.M. 319, 320
SETHNA, J.P. 370
SHARAPOV, S.G. 141
SHARP, D.H. 143
SHECHTMAN, D. 370

SHERRINGTON, D. 40, 139
SHINODA, T. 369
SHIRKOV, D.V. 138, 319
SMITH, R.P. 142
SNOKE, D.W. 142
soft components 151
SON, D.T. 143
space
    Fock 3
    inversion 27
spacetime coordinate
    Euclidean 20
special orthogonal group 25
spherical
    components, higher l 158
spin 24
splay configuration 337
STADNIK, Z.M. 370
STANLEY, H.E. 39, 139, 321, 369
STEENROD, N. 321
Stern-Gerlach experiment 34
STINSON, T.W. 369
STRATFORD, K. 370
STRATONOVICH, R.L. 40
STRINATI, G.C. 141
STRINGARI, S. 142, 143
substantial change 26
superconductor 1
    type I 72
    type II 72
superfluid
    density 89, 211, 212
        with Fermi liquid corrections . 223
    double 249
supersymmetry in nuclear physics .. 397
SUTHERLAND, P. 319
SVIDZINSKIJ, A.V. 40, 139
SWIECA, J.A. 397

TAYLOR, E. 140
temperature
    critical 63, 330
    critical, would-be 330
    Debye 63
    Fermi 63
temperature-dependent coherence length
    229

tensor, Levi-Cività 31
theorem, Wick 13
theory
    Bardeen-Cooper-Schrieffer(BCS) 74–76, 145
    Bogoliubov 145
Thirring model 393
THIRRING, W. 397
THOMAS, E.J. 139
'T HOOFT, G. 321
THOULESS, D.J. 140, 141
TIMUSK, T. 141
TINKHAM, M. 142
TOLKACHEV, E.A. 138, 319
TOMIO, L. 142
transformation
    Bogoliubov 79, 195
    Hubbard-Stratonovich 15, 41
TSUNETO, T. 321
twist configuration 337
TYTGAT, M. 143

UEMURA, Y.J. 141

VALATIN, J.G. 139
velocity, Fermi 66
VIGMAN, P.B. 369
VOLLHARDT, D. 322

VOLOVIK, G.E. 321, 322

WAGNER, H. 141
WALECKA, J.D. 40, 138, 320
weakly first order 333
WEGNER, F.J. 39, 139, 321
WEINBERG, E. 142
WEINBERG, S. 143
WEISS, P. 369
WERTHAMER, N.R. 320
WHEATLEY, J.W. 320
Wick theorem 13
WICK, G.C. 13
WIEGEL, F.W. 139
Wigner group 24
WILLIAMS, M.R. 322
would-be critical temperature 330
WÖLFLE, P. 322
WRIGHT, D.C. 370

XUE, S.-S. 142, 143

Yoshida function 83, 86, 210

zeta function, Riemann 65
ZICHICHI, A. 138
ZINN-JUSTIN, J. 139, 321
ZWERGER, W. 139, 141, 143
ZWIERLEIN, M.W. 139, 142

Printed in the United States
By Bookmasters